NUCLEAR PHYSICS AT STORAGE RINGS

Related Titles from AIP Conference Proceedings

521 Synchrotron Radiation Instrumentation: Eleventh US National Conference
Edited by Piero Pianetta, John Arthur, and Sean Brennan, May 2000, 1-56396-941-6

514 Theory and Computation for Synchrotron Radiation Spectroscopy
Edited by M. Benfatto, C. R. Natoli, and E. Pace, April 2000, 1-56396-936-X

513 Nuclear and Condensed Matter Physics: VI Regional CRRNSM Conference
Edited by Antonino Messina, April 2000, 1-56396-929-7

482 RHIC Physics and Beyond: Kay Kay Gee Day
Edited by Berndt Müller and Robert Pisarski, July 1999, 1-56396-878-9

473 Heavy Ion Accelerator Technology: Eighth International Conference
Edited by Kenneth W. Shepard, April 1999, 1-56396-806-1

472 Advanced Accelerator Concepts: Eighth Workshop
Edited by Wes Lawson, Carol Bellamy, and Dorothea F. Brosius, July 1999,
CD ROM included, 1-56396-889-4

459 Heavy Quarks at Fixed Target
Edited by Harry W. K. Cheung and Joel N. Butler, February 1999, 1-56396-864-9

451 Beam Instrumentation Workshop
Edited by Robert O. Hettel, Stephen R. Smith, and Jennifer D. Masek, December 1998,
1-56396-794-4

439 Production and Neutralization of Negative Ions and Beams: Eighth International Symposium/Production and Application of Light Negative Ions: Seventh European Workshop
Edited by Claude Jacquot, August 1998, 1-56396-737-5

421 Polarized Gas Targets and Polarized Beams: Seventh International Workshop
Edited by Roy J. Holt and Michael A. Miller, January 1998, 1-56396-700-6

417 Synchrotron Radiation Instrumentation: Tenth US National Conference
Edited by E. Fontes, December 1997, 1-56396-742-1

To learn more about these titles, or the AIP Conference Proceedings Series, please visit the
webpage **http://www.aip.org/catalog/aboutconf.html**

NUCLEAR PHYSICS AT STORAGE RINGS

Fourth International Conference: STORI99

Bloomington, Indiana   12–16 September 1999

EDITORS
Hans-Otto Meyer
Peter Schwandt
Indiana University

Melville, New York
AIP CONFERENCE PROCEEDINGS ■ 512

Editors:

Hans-Otto Meyer
Peter Schwandt

Indiana University Cyclotron Facility
2401 Milo B. Sampson Lane
Bloomington, IN 47405
USA

E-mail: meyer@iucf.indiana.edu
schwandt@iucf.indiana.edu

Authorization to photocopy items for internal or personal use, beyond the free copying permitted under the 1978 U.S. Copyright Law (see statement below), is granted by the American Institute of Physics for users registered with the Copyright Clearance Center (CCC) Transactional Reporting Service, provided that the base fee of $17.00 per copy is paid directly to CCC, 222 Rosewood Drive, Danvers, MA 01923. For those organizations that have been granted a photocopy license by CCC, a separate system of payment has been arranged. The fee code for users of the Transactional Reporting Service is: 1-56396-928-9/00/$17.00.

© 2000 American Institute of Physics

Individual readers of this volume and nonprofit libraries, acting for them, are permitted to make fair use of the material in it, such as copying an article for use in teaching or research. Permission is granted to quote from this volume in scientific work with the customary acknowledgment of the source. To reprint a figure, table, or other excerpt requires the consent of one of the original authors and notification to AIP. Republication or systematic or multiple reproduction of any material in this volume is permitted only under license from AIP. Address inquiries to Office of Rights and Permissions, Suite 1NO1, 2 Huntington Quadrangle, Melville, N.Y. 11747-4502; phone: 516-576-2268; fax: 516-576-2450; e-mail: rights@aip.org.

L.C. Catalog Card No. 00-102026
ISBN 1-56396-928-9
ISSN 0094-243X
Printed in the United States of America

Contents

Preface xi
Committees and Sponsors xv

THE NN INTERACTION: SCATTERING AND PION PRODUCTION

pp Elastic Scattering: New Results from EDDA (COSY) 3
    W. *Scobel* for the EDDA Collaboration
Pion Production with Polarized Beam and Polarized Target at IUCF 18
    J. T. Balewski
Spin Dependence of the Reaction $\vec{n}p \to pp\pi^-$ 34
    H. Lacker for the Collaboration
Spin Correlation Coefficients $\vec{p}\vec{p} \to pn\pi^+$ from 325 to 400 MeV 37
    S. K. Saha, W. W. Daehnick, R. W. Flammang, J. T. Balewski,
    H. O. Meyer, R. E. Pollock, B. v. Przewoski, T. Rinckel,
    P. Thörngren-Engblom, B. Lorentz, F. Rathmann, B. Schwartz,
    T. Wise, and P. V. Pancella
Pion Production in pN Collisions near Threshold 43
    J. Złomańczuk for the PROMICE/WASA Collaboration

MESON PRODUCTION FROM NUCLEI AND PRODUCTION OF HEAVY MESONS

Pion Production and Entropy Experiments at Storage Rings 49
    B. Jakobsson and the CHIC Collaboration
Meson Production in $p+d$ Reactions 60
    H. Machner for the GEM Collaboration
Heavy Meson Production at COSY-11 65
    P. Moskal, H.-H. Adam, J. T. Balewski, A. Budzanowski,
    C. Goodman, D. Grzonka, L. Jarczyk, M. Jochmann, A. Khoukaz,
    K. Kilian, P. Kowina, M. Köhler, T. Lister, W. Oelert,
    C. Quentmeier, R. Santo, G. Schepers, U. Seddik, T. Sefzick,
    S. Sewerin, J. Smyrski, A. Strzałkowski, M. Wolke, and P. Wüstner

THEORETICAL ISSUES IN MESON PRODUCTION

Meson Exchange Models for Meson Production 81
    C. Hanhart
Diagrammatic Approach to Meson Production in Proton-Proton Collisions near Threshold 96
    N. Kaiser

*Italicized name indicates author who presented the paper.

Off-Mass-Shell πN Scattering and $pp \to pp\pi^0$ 111
  M. T. Peña, S. A. Coon, J. Adam, Jr., and A. Stadler
Observables for the $pd \to {}^3H\pi^+$ and $pd \to {}^3He\pi^0$ Reactions
in a $pp \to d\pi^+$ Model 114
  W. R. Falk
Pion Production Mechanism in Nucleon-Nucleon Collisions 117
  K. Tamura, Y. Maeda, and N. Matsuoka
Near Threshold Λ and Σ^0 Production in pp Collisions 120
  A. Gasparian, J. Haidenbauer, C. Hanhart, L. Kondratyuk, and J. Speth

STRANGENESS PRODUCTION

Theoretical Issues in Strangeness Production 125
  J.-M. Laget
First Results on Strangeness Production from the ANKE Facility 138
  S. Barsov, U. Bechstedt, G. Borchert, W. Borgs, M. Büscher,
  M. Debowski, M. Drochner, W. Erven, R. Eßer, P. Fedorets,
  D. Gotta, M. Hartmann, H. Junghans, A. Kacharava, B. Kamys,
  F. Klehr, H. R. Koch, V. I. Komarov, V. Koptev, P. Kulessa,
  A. Kulikov, V. Kurbatov, G. Macharashvili, R. Maier, S. Mikirtytiants,
  S. Merzliakov, H. Müller, A. Mussgiller, M. Nioradze, H. Ohm,
  A. Petrus, D. Prasuhn, K. Pysz, F. Rathmann, B. Rimarzig, Z. Rudy,
  R. Schleichert, C. Schneider, H. Schneider, O. W. B. Schult,
  H. Seyfarth, K. Sistemich, H. J. Stein, *H. Ströher*,
  and P. Wüstner for the ANKE Collaboration
Hyperon and Charged Kaon Pair Production Close to Threshold 143
  M. Wolke, H.-H. Adam, J. T. Balewski, A. Budzanowski,
  C. Goodman, D. Grzonka, L. Jarczyk, M. Jochmann, A. Khoukaz,
  K. Kilian, M. Köhler, P. Kowina, T. Lister, P. Moskal, N. Lang,
  W. Oelert, C. Quentmeier, R. Santo, G. Schepers, U. Seddik,
  T. Sefzick, S. Sewerin, M. Siemaszko, J. Smyrski, A. Strzałkowski,
  P. Wüstner, and W. Zipper
Determination of the Lifetime of Heavy Λ-Hypernuclei at COSY Jülich 157
  H. Ohm, W. Borgs, W. Cassing, M. Hartmann, L. Jarczyk,
  B. Kamys, H. R. Koch, P. Kulessa, H. J. Maier, R. Maier,
  M. Matoba, D. Prasuhn, K. Pysz, Z. Rudy, O. Schult,
  H. Ströher, A. Strzalkowski, Y. Uozumi, and I. Zychor
Near Threshold Two Meson Production with the $pd \to {}^3He\pi^+\pi^-$
and $pd \to {}^3HeK^+K^-$ Reactions 168
  R. Jahn for the COSY-MOMO Collaboration
Associated Strangeness Production at COSY-TOF 171
  S. Wirth for the COSY-TOF Collaboration

*Italicized name indicates author who presented the paper.

POLARIZED BEAMS AND POLARIZED TARGETS

Polarization in Meson Production Reactions 177
    L. D. Knutson

Review of Polarized Internal Gas Targets 193
    F. Rathmann

Nuclear Polarization of H_2 Molecules Formed from Polarized Atoms 210
    T. Wise, J. T. Balewski, W. W. Daehnick, J. Doskow, D. Friesel,
    W. Haeberli, H. Kolster, B. Lorentz, H. O. Meyer, P. V. Pancella,
    R. E. Pollock, B. v. Przewoski, P. A. Quin, F. Rathmann,
    T. Rinckel, S. K. Saha, B. Schwartz, and A. Wellingshausen

Acceleration and Storage of Polarized Proton Beam at RHIC 213
    T. Roser

Preparing the COSY-Ring for a Test of Time-Reversal-Invariance 224
    P. D. Eversheim for the TRI Collaboration

NEW FACILITIES AND TECHNIQUES

WASA Detector: Towards Rare Pion and Eta Decays 229
    H. Calén for the CELSIUS/WASA Collaboration

Tagged Neutron Production with a Storage Ring 235
    T. Peterson for the TNT Collaboration

Stochastic Cooling and Extraction at COSY 243
    R. Maier, U. Bechstedt, J. Dietrich, K. Henn, A. Lehrach,
    B. Lorentz, D. Prasuhn, A. Schnase, H. Schneider, R. Stassen,
    H. Stockhorst, and R. Tölle

Status of the ESR and Prospects for Radioactive Ion Beams 246
    B. Franzke, K. Beckert, L. Groening, F. Nolden, and M. Steck

Calorimetric Low Temperature Detectors for High Resolution X-ray Spectroscopy on Stored Highly Stripped Heavy Ions 259
    A. Bleile, P. Egelhof, H.-J. Kluge, U. Liebisch, D. Mc Cammon,
    H. J. Meier, O. Sebastián, C. K. Stahle, T. Stöhlker, and M. Weber

Schottky Mass Spectrometry at the ESR with a New Data Acquisition System 263
    M. Falch, K. E. G. Löbner, T. Kerscher, F. Attallah, F. Bosh,
    B. Franzke, H. Geissel, M. Hausmann, O. Klepper, H. Kluge,
    C. Kozhuharov, G. Münzenberg, F. Nolden, Y. Novikov,
    Z. Patyk, T. Radon, C. Scheidenberger, M. Steck, M. Winkler,
    and H. Wollnik

Considerations for a Collector Ring of Nuclear Fragments at GSI 266
    F. Nolden, A. Dolinsky, and B. Franzke

Particle Detectors for Beam Diagnosis and for Experiments with Stable and Radioactive Ions in the Storage-Cooler Ring ESR 269
    O. Klepper and C. Kozhuharov

*Italicized name indicates author who presented the paper.

PHYSICS WITH STORED HEAVY IONS

Gross Properties of Exotic Nuclei Investigated at Storage Rings and Ion Traps 275
    C. *Scheidenberger*, G. Bollen, F. Bosch, A. Casares,
    H. Geissel, A. Kholomeev, G. Münzenberg, H. Weick,
    and H. Wollnik

Reaction Studies with Exotic Nuclei in Storage Rings 293
    G. *Münzenberg* and G. Schrieder

First Isochronous Time-of-Flight Mass Measurements of Short-Lived Projectile Fragments in the ESR 305
    J. *Stadlmann*, H. Geissel, M. Hausmann, F. Nolden, T. *Radon*,
    H. Schatz, C. Scheidenberger, F. Attallah, K. Beckert, F. Bosch,
    M. Falch, B. Franczak, B. Franzke, T. Kerscher, O. Klepper,
    H.-J. Kluge, C. Kozhuharov, K. E. G. Löbner, G. Münzenberg,
    Y. N. Novikov, M. Steck, Z. Sun, K. Sümmerer, H. Weick,
    and H. Wollnik

Studying Phase Transitions in Nuclear Collisions 308
    I. N. *Mishustin*

Experiments with RHIC 324
    G. D. *Westfall*

PHYSICS WITH STORED ELECTRONS

The HERMES Experiment 339
    R. G. *Milner*

Longitudinally Polarized Electrons in a Storage Ring Below 1 GeV 353
    I. *Passchier*, D. J. Boersma, M. Harvey, D. W. Higinbotham,
    H. R. Poolman, E. Six, R. Alarcon, P. W. van Amersfoort,
    T. S. Bauer, H. Boer Rookhuizen, J. F. J. van den Brand,
    L. D. van Buuren, H. J. Bulten, R. Ent, M. Ferro-Luzzi,
    D. Geurts, P. Heimberg, C. W. de Jager, P. Klimin, I. Koop,
    F. Kroes, J. van der Laan, G. Luijckx, A. Lysenko, B. Militsyn,
    I. Nesterenko, J. Noomen, B. E. Norum, M. J. J. van den Putte,
    Y. Shatunov, J. J. M. Steijger, D. Szczerba, and H. de Vries

Electron Scattering Experiments with Polarized Hydrogen/Deuterium Internal Targets 356
    L. D. *van Buuren* for the 97-01 Collaboration

Multi-Target Operation at the HERA-B Experiment 359
    Y. *Vassiliev*, V. Aushev, K. Ehret, M. Funcke, S. I. Sever,
    Y. Pavlenko, V. Pugatch, S. Spratte, M. Symalla, N. Tkatch,
    and D. Wegener

---

*Italicized name indicates author who presented the paper.

**Recent Results from the Internal Polarized Deuterium Target
Experiment at the Electron Storage Ring VEPP-3** 362
    I. A. Rachek, H. Arenhoevel, L. M. Barkov, V. F. Dmitriev,
    M. V. Dyug, S. L. Belostotsky, R. Gilman, R. J. Holt, L. G. Isaeva,
    C. W. de Jager, E. R. Kinney, R. S. Kowalczyk, B. A. Lazarenko,
    A. Y. Loginov, S. I. Mishnev, V. V. Nelyubin, D. M. Nikolenko,
    A. V. Osipov, D. H. Potterveld, Y. V. Shestakov, A. A. Sidorov,
    V. N. Stibunov, D. K. Toporkov, D. K. Vesnovsky, V. V. Vikhrov,
    H. de Vries, and S. A. Zevakov

**Photoneutron Cross Section Measurements on ^{9}Be Using
Laser-Induced Compton-Backscattered Photons** 365
    H. Utsunomiya, Y. Yonezawa, H. Akimune, T. Yamagata,
    M. Ohta, M. Fujishiro, H. Toyokawa, and H. Ohgaki

STORED, COOLED BEAMS

Storage Conundra 371
    R. E. Pollock

News in Electron Cooling: Highlights from ECOOL'99 380
    D. Reistad

Storage Rings: Past, Present, and Future 394
    A. D. Krisch

APPENDICES

A.   **Program** 419
B.   **List of Participants** 423
C.   **Author Index** 433

*Italicized name indicates author who presented the paper.

PREFACE

The 4th International Conference on Nuclear Physics at Storage Rings, STORI99, was held September 12–16, 1999, on the campus of Indiana University in Bloomington, Indiana, USA. This conference was the fourth in a series that started in 1991 in Lund and was continued in 1994 in St. Petersburg and in 1996 in Bernkastel-Kues, Germany.

The aim of the STORI99 conference was to bring together physicists from a diverse international research community connected by the common technology of storage rings and review the topics of current interest in nuclear physics research with stored, cooled ion beams and electron beams. Specifically, the scientific program of STORI99 focused on recent results from a wide variety of experimental programs at existing stored-beam facilities, on progress in associated theoretical issues, and on discussion of new facilities and experimental techniques. In addition to the traditional physics topics covered by the series of STORI conferences (nucleon physics & meson production, physics with stored heavy-ion beams, polarized beams & targets), new physics topics introduced at STORI99 included strangeness production with high-energy stored proton beams and physics with stored electron beams. The nine physics & technology topics covered in this conference can be found in the table of contents for these proceedings.

We are grateful to the members of the International Advisory Committee and the Local Program Committee who actively and expertly contributed to the selection of topics and invited speakers. Members of these committees are listed on page xv.

At this conference eight plenary sessions were held in the Indiana Memorial Union. The meeting was attended by 91 participants from 11 countries. Of these participants, 29 were invited speakers and 15 others presented short oral contributions. A listing of the conference participants is provided in Appendix B to these proceedings.

A poster session was arranged for Tuesday afternoon; 14 posters were presented and entries were judged by a "Best Posters"awards committee. The conference opened on Sunday after a Welcome Lunch with a "Welcome to Indiana University" address by the Indiana University Vice President for Research, George Walker (a nuclear theorist). STORI99 concluded with a fascinating, personal look at "Storage Rings: Past, Present and Future" by Alan Krisch, an active high-energy physicist and long-term accelerator user. Among other things, these closing remarks reminded us that some limited storage-ring experiments were done on otherwise conventional accelerators long before dedicated storage rings came into existence.

The conference was sponsored by the Indiana University Cooler Facility (IUCF), Indiana University, and the US National Science Foundation. The latter provided much-appreciated funds to underwrite the cost of travel and attendance at the conference for several foreign graduate students and Russian scientists without institutional support. We are also grateful for the generous financial support received from the Indiana University offices of the Vice President for Research, the University Vice President & Chancellor, and the International Programs office. Without these contributions this conference would not have been possible.

These conference proceedings contain the written versions of most of the presented talks and posters. The manuscripts are grouped according to subject matter as to reflect the concept and structure of the conference; their order in this volume does not always follow the program given in Appendix . We greatly appreciate the cooperation of the authors in preparing their manuscripts so well in both content and format that we, the editors, were given very little opportunity to wield our blue editorial pencils. An author index is provided in Appendix C.

Several social events were organized during this conference for the participants and their companions: an IUCF Open House and tour of the laboratory; an afternoon picnic outing to the shore of Lake Monroe with a pig roast; a conference banquet in downtown Bloomington, attended by nearly all conference participants who enjoyed live jazz during the reception and plenty of California red wine during dinner — and NO after-dinner speeches, save for the awarding of cash prizes to the winners of the "Best Posters" selection: 1. Pavel Golubev (Lund), 2. Otto Klepper (Darmstadt), 3. Igor Vassiliev (Dortmund). In parallel with the daytime physics sessions, there were activities for companions that included trips to nearby points of interest.

We would like to express our gratitude to the people willing (or unable to resist our coercion) to chair the plenary sessions. The names of the session chairpersons are included in the conference program in Appendix A. Thanks also to A. Wellinghausen and T. Peterson who organized the poster session. We also join the conference participants in expressing our appreciation for the efforts of the small local organizing committee whose members arranged all non-program aspects of STORI99: Edward Stephenson for the minimalist single-fold conference poster in eye-catching colors and for the computer and office machine set-up in the conference office; Barbara v. Przewoski for maintaining the conference web site and dealing with the publisher of these proceedings; Janet Meadows for her untiring and cheerful efforts to provide essential secretarial help before, during and after the meeting. One of us (Peter Schwandt) worked with the the Indiana University Conferences office in attending to the numerous details of local arrangements in general and the social events & companion program in particular, and we wish to acknowledge the considerable contributions made to this conference by Karin Reece of the IU Conferences office. We are also obliged to two IUCF staff members, Teresa Jones for solving participants' travel problems and Julia Mobley for handling financial matters.

Finally, the location of the first STORI meeting in the new millennium was announced at the close of the last meeting in this millennium: STORI02 will be held at Uppsala, Sweden in 2002; it will be organized under the sponsorship of CELSIUS by Curt Ekstrom. We wish him well in this endeavor and are confident he will fulfill our fond hopes for another memorable meeting.

Until we meet again in Uppsala,

Hans-Otto Meyer
Peter Schwandt
Conference Organizers & Editors

INTERNATIONAL ADVISORY COMMITTEE

E. Bosch	Darmstadt, Germany
P. Egelhof	Darmstadt, Germany
H. Ejiri	Osaka, Japan
D. Habs	Heidelberg, Germany
W. Haeberli	Wisconsin, USA
B. Jacobsson	Lund, Sweden
P. Kienle	München, Germany
K. Kilian	Jülich, Germany
H.-J. Kluge	Darmstadt, Germany
O. Lozhkin	St. Petersburg, Russia
H.O. Meyer	Indiana, USA
L. Nilsson	Uppsala, Sweden
W. Oelert	Jülich, Germany
N. Saito	Brookhaven, USA
I. Tanihata	Tokyo, Japan
L. Tecchio	INFN-LNL, Italy
A.W. Thomas	Adelaide, Australia
C. Wilkin	London, UK

PROGRAM COMMITTEE

L.C. Bland	Indiana
H.O. Meyer (Chair)	Indiana
R.E. Pollock	Indiana
P. Schwandt	Indiana
E.J. Stephenson	Indiana
A. Szczepaniak	Indiana
V.E. Viola	Indiana

SPONSORS

Indiana University Cyclotron Facility
US National Science Foundation
Vice President for Research, Indiana University
Vice President & Chancellor, Indiana University
Office of International Programs, Indiana University

THE NN INTERACTION: SCATTERING AND PION PRODUCTION

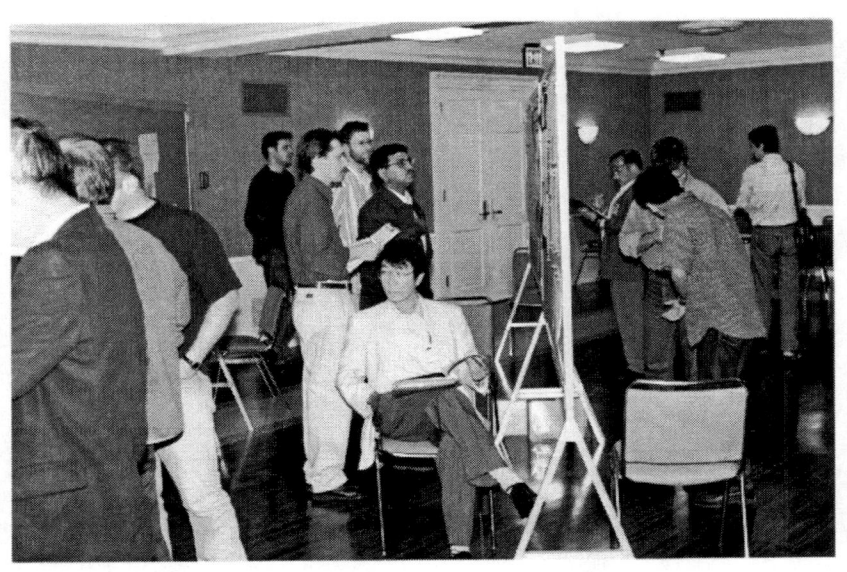

pp Elastic Scattering: New Results from EDDA (COSY) *

W. Scobel for the EDDA collaboration¶

I. Institut für Experimentalphysik, Universität Hamburg, Luruper Chaussee 149, 22761 Hamburg, Germany

Abstract. In the EDDA experiment excitation functions of proton–proton elastic scattering are studied with narrow steps in the projectile momentum range from 0.8 to 3.4 GeV/c and the angular range $35° \leq \Theta_{cm} \leq 90°$ with a detector providing $\Delta\Theta_{cm} \approx 1.4°$ resolution and 85% solid angle coverage. Measurements are performed continuously during projectile acceleration in the Cooler Synchrotron COSY. In phase 1 of the experiment spin–averaged differential cross sections $d\sigma/d\Omega$ have been measured with an internal CH_2 fiber target; background corrections were derived from measurements with a carbon fiber target and from Monte Carlo simulations of inelastic pp contributions. The results provide excitation functions and angular distributions of high precision and internal consistency. In phase 2 of the experiment excitation functions of the analyzing power A_N have been measured using a polarized ($P \geq 75\%$) atomic beam target, and those of the polarization correlation parameters A_{NN}, A_{SS} and A_{SL} will be measured lateron with the polarized COSY beam. The measured excitation functions are compared to recent phase shift analyses, and their impact on them is discussed. So far evidence for narrow structures was neither found in the spin averaged cross sections nor in the analyzing powers.

INTRODUCTION

Proton-proton elastic scattering [1] is fundamental to the understanding of the strong interaction. At present, the long- and medium range parts of the NN interaction seem to be well studied. The data are well represented by phase shift solutions

---

*) Work supported by the BMBF and the Forschungszentrum Jülich GmbH
¶) M. Altmeier[1], F. Bauer[2], J. Bisplinghoff[1], M. Busch[1], K. Büßer[2], T. Colberg[2], L. Demirörs[2], O. Diehl[1], F. Dohrmann[2], H.P. Engelhardt[1], P.D. Eversheim[1], O. Felden[3], R. Gebel[3], M. Glende[1], J. Greiff[2], R. Groß-Hardt[1], F. Hinterberger[1], R. Jahn[1], E. Jonas[2], H. Krause[2], R. Langkau[2], T. Lindemann[2], J. Lindlein[2], R. Maier[3], R. Maschuw[1], A. Meinerzhagen[1], O. Nähle[1], D. Prasuhn[3], H. Rohdjeß[1], D. Rosendaal[1], P. von Rossen[3], N. Schirm[2], V. Schwarz[1], W. Scobel[2], H.J. Trelle[1], E. Weise[1], A. Wellinghausen[2], K. Woller[2], R. Ziegler[1]
(1) Inst. f. Strahlen- und Kernphysik, Univ. Bonn
(2) I. Inst. f. Experimentalphysik, Univ. Hamburg
(3) Inst. f. Kernphysik, KFA Jülich

CP512, *Nuclear Physics at Storage Rings*, edited by H.-O. Meyer and P. Schwandt
© 2000 American Institute of Physics 1-56396-928-9/00/$17.00

[2,3] and meson exchange potential models [4–6] provide adequate descriptions of the data up to the pion production thresholds. At smaller distances (< 0.8 fm), however, genuinely new processes may occur involving the dynamics of the quark-gluon constituents. A related problem is the nature of the strong repulsive core of the NN interaction that prevents a close contact at low energies. Therefore, in order to reach such small distances essentially higher bombarding energies are needed. Accordingly, proton-proton scattering at 0.5 – 2.5 GeV is ideally suited to study the short range part of the NN interaction. For the meson based potential models it provides a sharp focus on the role of the ω meson [7,8]. Apart from true ω exchange additional quark-gluon processes might occur when two nucleons overlap. Interestingly, a four momentum transfer in the order of 1 – 1.5 GeV/c is reached which has the benefit of a rather high spatial resolution (0.13 – 0.2 fm).

Another important problem related to the quark-gluon dynamics is the question of existence or nonexistence of dibaryons [1]. Various QCD inspired models [9–13] predict dibaryonic resonances with c.m. resonance energies ranging between 2.1 and 2.9 GeV. Not any resonance has been identified so far. Excitation functions of spin correlation coefficients, i.e. cross sections depending on the relative spin orientation in the entrance channel, provide an especially sensitive test for dibaryonic resonances that may show up in the elastic channel due to the interference between resonant and nonresonant amplitudes.

When EDDA started, the experimental data base was rather poor [3] for energies $T_p > 1.2\ GeV$, see Figure 1, and there are inconsistencies and normalization errors in the angular distributions measured at different discrete energies. In addition, energy dependent structures have been observed in certain spin observables [14]. Therefore, in order to provide consistent data and to scan the proton-proton interaction at short distances excitation functions should be measured at intermediate energies in narrow energy steps with a high relative accuracy. Precise data are also needed for the improvement and extension of potential models [7] and phase shift analyses [3] to higher energies. For these purposes it is mandatory to measure not only spin-averaged cross sections, but also spin observables such as analyzing powers A_N and spin correlation coefficients A_{NN}, A_{SS} and A_{SL}.

DETECTOR AND MEASUREMENTS

The EDDA experiment [15] is designed to measure excitation functions of the unpolarized differential cross sections $d\sigma/d\Omega$, analyzing powers A_N and spin correlation parameters A_{NN}, A_{SS} and A_{SL} with a high relative accuracy over a wide momentum range. It was conceived as an internal target experiment using the recirculating COSY beam in order to obtain luminosities high enough for use of a polarized atomic beam target [16]. A smooth momentum variation is achieved by measuring during the beam acceleration ramp of the cooler synchrotron COSY [17]. Data collection proceeds during synchrotron acceleration in a multi-pass technique, so that a full excitation function is measured in each acceleration cycle. Statistical

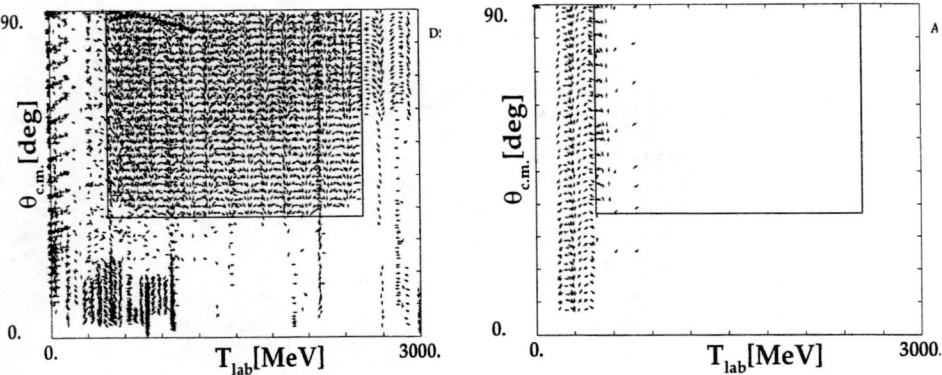

FIGURE 1. World data set of pp elastic scattering cross sections (left, 6264 entries) and spin correlation coefficients A_{SS} (right, 419 entries) in the Θ_{cm} vs. T_p plane entering into the global phase shift analysis (solution FX98 from [3]). The area EDDA contributes to is indicated.

accuracy is obtained by averaging over many cycles. For this technique it is very important that COSY has proven to operate very stable and reproducable.

In *phase 1* unpolarized differential cross sections have been measured using 4 μm × 5 μm CH$_2$ fiber targets and 5 μm diameter C fiber targets for background subtraction. Both targets were covered by a 20 μg/cm^2 aluminum film in order to make the CH$_2$ targets electrically conducting. In *phase 2* the analyzing power A_N and the polarization correlation parameters A_{NN}, A_{SS} and A_{SL} are measured as a function of momentum using a polarized atomic beam target and a polarized proton beam.

The EDDA detector has been designed for large solid angle coverage in a cylindrical geometry. It provides a fast trigger on low multiplicity events, that is based on a scintillator structure matching in its granularity the signature of elastic pp scattering. Elastic events are identified by

- coplanarity with the beam ($\varphi_1 - \varphi_2 = 180°$) and
- elastic scattering kinematics imposing $\tan\Theta_1 \cdot \tan\Theta_2 = 2m_p c^2/(2m_p c^2 + T_p)$.

These conditions are employed – with different degrees of stringency – both in triggering and in event identification during off-line analysis.

The EDDA detector shown in Figure 2 in a schematic fashion consists of two cylindrical double layers covering 30° to 150° in Θ_{cm} for the elastic pp channel and about 82 % of the full solid angle. The inner layers are composed of scintillating fibers (H) which are helically wound in opposing directions and connected to 16-channel multianode photomultipliers at backward angles. They are not required and were not installed for measuring spin-averaged cross sections with a CH$_2$ fiber target, but are essential for measuring spin observables with the polarized atomic beam target, as they faciliate the vertex reconstruction. The outer layers consist

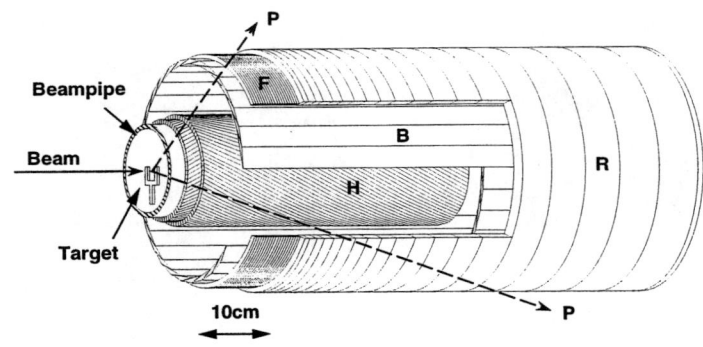

FIGURE 2. The EDDA detector (not to scale).

of 32 scintillator bars (B) which are running parallel to the beam axis and which are read out at both ends. They are surrounded by scintillator semi-rings (R; F). The scintillator cross sections were designed so that each particle traversing the outer layers produces a signal in two neighbouring bars and rings. Analysis of the fractional light output is used to determine the angles of incidence much more accurately (by a factor of 5) than would be possible on the basis of detector granularity alone. The resulting polar and azimuthal FWHM angle resolution is about 1° and 1.9°, respectively. Combined with the spatial resolution of the 2.5 mm scintillating fibers (H), this provides for vertex reconstruction to a resolution of about 2 mm in the x-, y- and z-direction. Details are given in [16,18].

The beam parameters are continuously measured during the acceleration ramp. The beam momentum is derived from the RF of the cavity and the circumference of the closed orbit with an uncertainty of 0.25 to 2.0 MeV/c for the lowest and highest momentum, respectively. Beam position and beam width are deduced from the proton-proton elastic scattering data. The horizontal and vertical beam width (FWHM) is about 3 mm and 5 mm, respectively.

Phase 1: A crucial point of the differential cross section measurements was the luminosity monitor. Relative normalization as a function of momentum was accomplished by concurrently measuring (i) the secondary electron current emanating from the fiber target and (ii) the δ-electrons from the elastic proton-electron scattering. The secondary electron current scales as the energy deposited into the target fiber. It is proportional to the restricted energy loss rate. The δ-electrons were measured at $\Theta = 40°$ using four silicon detectors located in a left-right, up-down symmetrical arrangement behind thin windows in the beam pipe. The δ-electron rate scales as the Rosenbluth cross section. The differential cross section excitation functions were taken with a constant momentum ramp of 1.15 (GeV/c)/s and an average luminosity of 5×10^{29} cm^{-2}s^{-1}. Absolute normalization is established at 1455 MeV/c with reference to the angular distribution measured by Simon et al.

[19] with 1% total uncertainty.

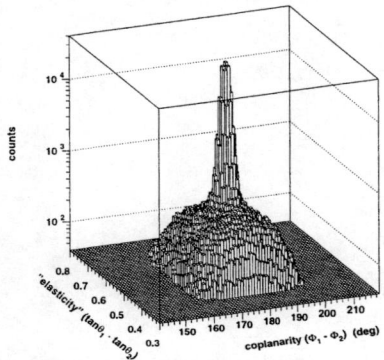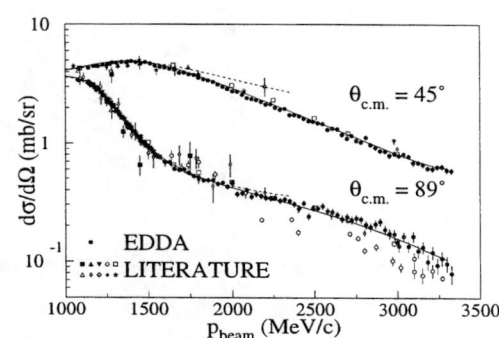

FIGURE 3. Left: Kinematic distribution of EDDA events as obtained with a CH$_2$ fiber target at T$_p$ = 1500 MeV. Right: Excitation functions for elastic pp scattering: Comparison of results from [15] (solid dots) at Θ_{cm} = 45°±1° and 89°±1° in Δp=28 MeV/c bins with all data compiled in [3], and with the phase shift solutions SM94 (dashed lines) and SM97 (solid lines) [3].

Figure 3 illustrates the signature of elastic proton-proton events by kinematic coincidence. The events in the peak originate from elastic pp-scattering. Nonelastic background results in a wide and structureless distribution, which is measured separately with the carbon fiber target and subtracted on a statistical basis.

Phase 2: The analyzing power excitation functions A$_N$(Θ_{cm},p$_p$) have been measured by bombarding a polarized atomic beam target with the unpolarized proton beam. The polarized target [20] is shown schematically in Figure 4. Hydrogen atoms with nuclear polarization are prepared in an atomic beam source with dissociator, cooled nozzle, permanent sexpole magnets and RF-transition units, where the former remove one of the two electron spin states and the latter induce a transition to a thus unpopulated hyperfine state, with one nuclear spin state remaining. This preparation provides an atomic beam of ~12 mm width (FWHM) at the intersection with the COSY beam, a peak polarization of 85 % (see Figure 5) and an effective target thickness of about $1.8 \cdot 10^{11}$ atoms/cm^2. The polarization is continuously monitored in the atomic beam dump with a Breit-Rabi polarimeter. The magnetic holding field (B$_x$, B$_y$ ~ 1mT, B$_z$ ~ 10mT) can be switched between $\pm x$, $\pm y$ and $\pm z$. Spin misalignment with respect to the detector coordinates is negligible across the fiducial vertex volume.

Measurements of the excitation functions A$_N$(Θ_{cm}, p$_p$) were performed in cycles of about 13 s duration with data acquisition extending over the acceleration, the

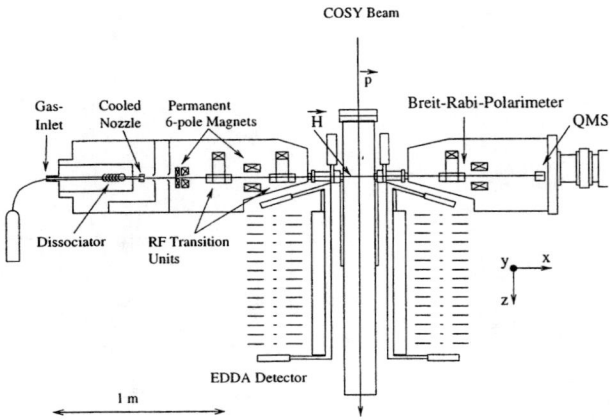

FIGURE 4. Scheme of the atomic beam target in combination with the EDDA detector.

flattop at 3300 MeV/c, and over the deceleration as well. With an average of $2.8 \cdot 10^{10}$ protons in the ring, luminosities of about $0.9 \cdot 10^{28} \text{cm}^{-2}\text{s}^{-1}$ were achieved, yielding so far an integrated luminosity of $\sim 10^{34} \text{cm}^{-2}$. The direction of the target polarization was changed between subsequent cycles between $\pm x$ and $\pm y$; the two sets of runs with opposite polarizations $P_{\pm x}$ and $P_{\pm y}$ were used to apply a proper spin–flip correction [21] to the measured left–right and bottom–top asymmetries.

The vertex reconstruction by means of the multilayer structure of the EDDA detector allowed to study polarization parameters in detail. The average polarization experienced by a COSY beam of fixed energy swept on purpose vertically (y) across the atomic beam shows a broad profile with typical maximum values of 75% for P_x and P_y (see Figure 5), and no statistically significant component perpendicular to the nominal holding field direction. An example for polarization profiles $P_{\pm y}(z)$ along the COSY beam is given in Figure 5, too. The indicated software cuts were applied at $z = \pm 12.5$mm; outside, the figure–of–merit $N \cdot P^2$ has dropped by more than 3 orders of magnitude. The target was operating very stable and with constant polarization during each cycle as well as over run periods of up to two weeks, and altogether $29 \cdot 10^6$ elastic scattering events were collected.

RESULTS AND DISCUSSION

Excitation Functions of Spin Averaged Cross Sections

Figure 3 shows excitation functions for two selected angular intervals after a third of all fiber target data taken by EDDA [15]; for the interval $\Theta_{cm}=88°$–$90°$ good agreement is obtained with the fixed angle ($\Theta_{lab}=40.4°$) experiment [22] performed

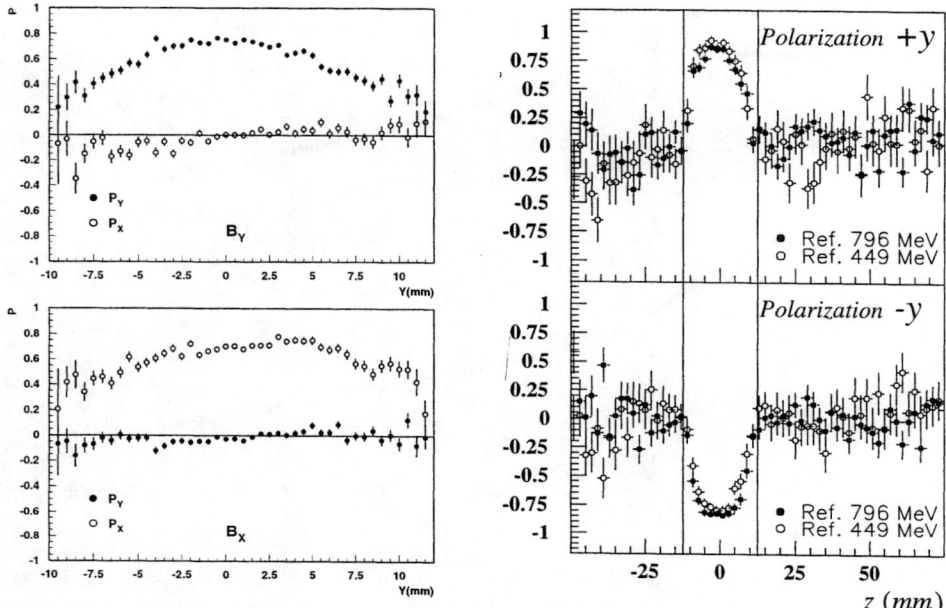

FIGURE 5. Left: Polarization profiles $P_x(y)$ and $P_y(y)$ of the atomic beam obtained from left–right and bottom–top asymmetries at $T_p = 796$ MeV with A_N taken from [23,24] for holding fields B_y and B_x. Right: Polarization profiles $P_y(z)$ and $P_{-y}(z)$ measured during acceleration at $T_p = 796$ MeV and $\Theta_{cm} = 58°$ with the A_N reference at 796 MeV or 449 MeV [25], resp.

at SATURNE with the multipass technique, too. In contrast the data of single energy runs taken from the SAID compilation [3] scatter considerably. For a more quantitative comparison Arndt et al. have generated the solution SM97 [3] that is based on an updated data set including these EDDA data and a modified formalism. The new fit showed an improvement in χ^2, with the EDDA data contributing as little as 1.07 per datum in spite of their small uncertainties. It is obvious from Figure 3 that the EDDA experiment has provided a solid basis for a substantial extension of the global solution as compared to the last published one, SM94, that was limited to $T_p \leq 1600$ MeV.

Meanwhile the whole set of $50 \cdot 10^6$ elastic events is analyzed; the comparison of Figure 6 with Figure 3 demonstrates the statistical improvement. It remains to be seen if this progress will already have visible impact on the NN potentials deduced, in particular on the short range part of meson exchange potentials. Based on comparisons of the angular distributions to calculations with one–boson–exchange potentials [7], Machleidt suggests a gradual phaseout of the OBE model with the

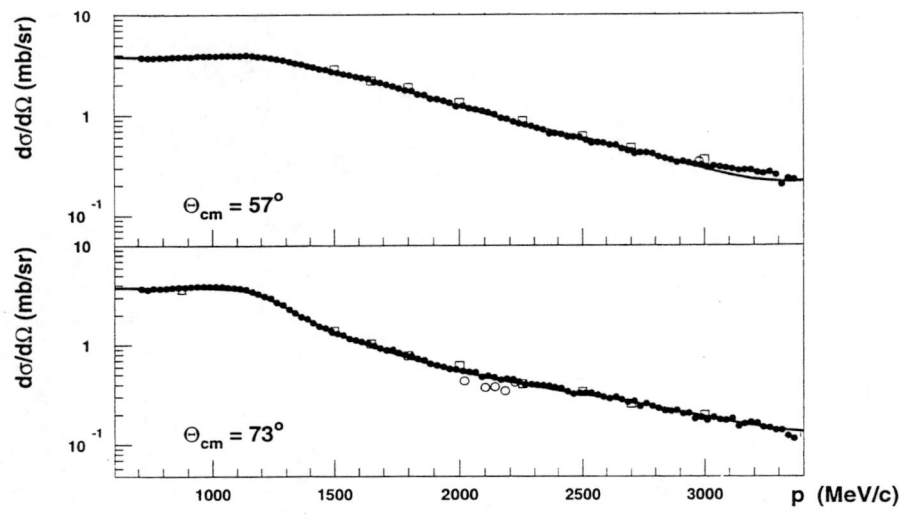

FIGURE 6. Experimental excitation function of the full data set for $\Theta_{cm} = 57° \pm 1°$ and $73° \pm 1°$ and $\Delta p_p = 25$ MeV/c bins. The solid curve is the SAID solution SP99 containing the same fractional EDDA data set as SM97.

projectile momentum increasing from 1 to 10 GeV/c; at 3 GeV/c a phaseout by 40% is needed [8] to adjust the model to the experiment. In the regime $p \geq 2$ GeV/c of our data, the ω meson is dominant in the theoretical description and the cross section scales approximately with the ω coupling constant, that must therefore be accordingly reduced. A shrinking of the short range repulsive spin–orbit interaction in this region with increasing momentum is discussed in [26], too.

Resonant Excursions

These excitation functions can be examined for evidence of resonant excursions by rebinning them with $\Delta\Theta_{cm} = 5°$ or $10°$ and $\Delta p_p = 10$ MeV. We assume, that a single, isolated resonance will modify the smooth momentum dependence of the excitation function as it is obtained from a global phase shift analysis. It must have quantum numbers that are allowed for the resonating pp system; among them are spin S and orbital angular momentum L coupling to J, and the resonance energy W_r. The decay into the elastic channel occurs with a partial width Γ_{el} of the total width Γ. Such a resonance can be introduced [27,28] without unitarity violation into the scattering formalism in phase shift parametrization by superimposing a Breit-Wigner amplitude onto the corresponding partial wave amplitude, viz.

$$\eta'_{LJ}e^{2i\delta'_{LJ}} = \eta_{LJ}e^{2i\delta_{LJ}}\left(1 - \frac{i\Gamma_{el}}{(W-W_r)+i\Gamma/2}\right) \quad (1)$$

Here, η_{LJ} and δ_{LJ} denote absorption parameter and real phase shift of the non-resonant background at the total cm energy W; η'_{LJ} and δ'_{LJ} are the corresponding modified parameters. For η_{LJ} and δ_{LJ} we take the results of the global phase shift solution SM97 [3].

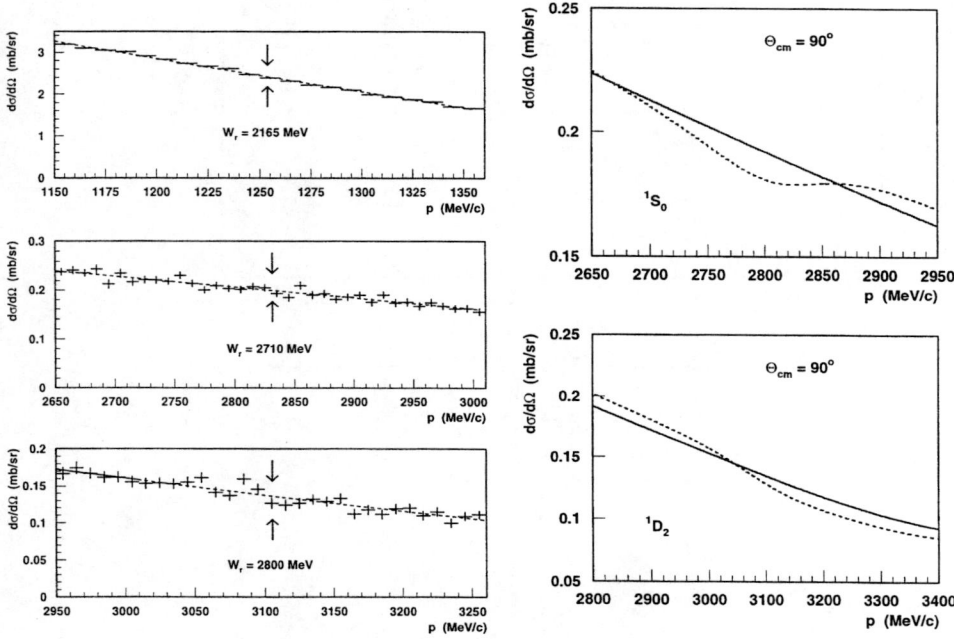

FIGURE 7. Left: Enlarged display of the elastic scattering excitation function for $\Theta_{cm} = 80°$ - $90°$ and $\Delta p_p = 10$ MeV/c in the neighbourhood of the predicted $^3F_3$ (top), $^1S_0$ (middle) and $^1D_2$ (bottom) resonance. The dashed curve represents the smooth reference (see text). Right: Excitation function calculated with the phases from the solution SM97 without (solid lines) and with (dashed) Breit-Wigner modification for the $^1S_0$ and $^1D_2$ resonance. Maximum sensitivities are 0.60 and 0.33, respectively.

As a quantitative measure of the response of an observable σ to such a modification we introduce the sensitivity S

$$S = \frac{\Delta\sigma/\sigma}{(2J+1)\Gamma_{el}/\Gamma} \quad (2)$$

with the relative enhancement of the observable in the numerator and the *elasticity* $\eta_{el} = \Gamma_{el}/\Gamma$ multiplied with the statistical weight (2J+1) of the resonance in the denominator. The sensitivity will depend on the resonance parameters Γ and η_{el}, but also on W_r, L and J. Calculations have been performed for the resonances $^1S_0$ at W_r = 2710 MeV (p = 2.82 MeV/c) and $^1D_2$ at W_r = 2800 MeV (p = 3.10 MeV/c) predicted in [11,29] with $\Gamma \approx$ 50 MeV, $\eta_{el} \approx$ 0.1 and $\Gamma \approx$ 100 MeV, $\eta_{el} \approx$ 0.05, respectively, and a narrow ($\Gamma \approx$ 15 MeV, $\eta_{el} \approx$ 0.1) $^3F_3$ resonance at W_r = 2170 MeV (p = 1.25 MeV/c), for that some evidence is claimed in [27].

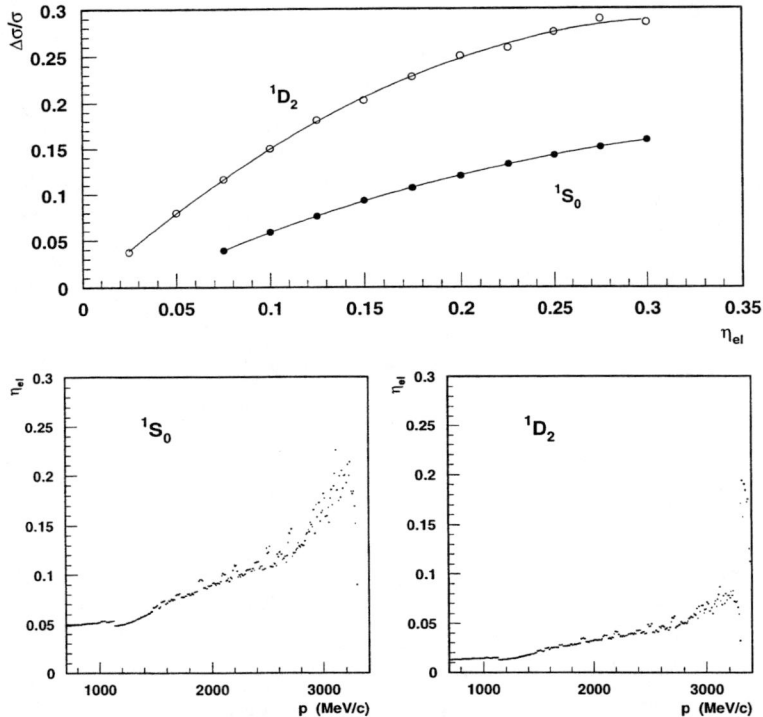

FIGURE 8. Top: Variation of relative enhancement $\frac{\Delta \sigma}{\sigma}$ with elastcity η_{el} of the predicted $^1S_0$ and $^1D_2$ resonances. Bottom: Upper limits of the elasticity $\eta_{el}(p_p)$ resulting from the total excitation function (Figure 7) for σ and its pointwise threefold statistical uncertainty for $\Delta\sigma$.

For the excitation functions $\sigma(\Theta_{cm}, p_p)$ the sensitivity reaches for all three resonances its maximum S_{max} with highest momentum transfer, i. e. for Θ_{cm} = 90°. Figure 7 shows our data in the regions of the predicted resonances as well as the calculations with scattering phases without and with resonant admixture, Eq.(1). It is obvious that the maximum sensitivities S_{max} are at the edge of being insufficient for the observation of a weak excursion; in particular the low associated sensitivity $S \leq 0.08$ of the $^3F_3$ resonance prevents a visibility in our data if $\eta_{el} < 0.1$.

A hypothetical $^1S_0$ ($^1D_2$) resonance in our data may deviate from the predicted parameters W_r, Γ and η_{el}. Calculations of $\Delta\sigma/\sigma$ with modification of one parameter at a time and corresponding phase shifts show that the maximum sensitivity S_{max} varies less than 15% (30%) over the whole momentum range of study (corresponding to 2000 MeV $\leq W_r \leq$ 2900 MeV). S_{max} does not depend on Γ as long as η_{el} is kept constant, although the width of the excursion does of course scale with Γ. It is, however, sensitive to variations of η_{el}; as a result $\Delta\sigma/\sigma$ increases monotonically with η_{el}, see Figure 8. This bijective relation can be reversed to deduce lower limits of η_{el} for a hypothetical resonance to be visible in our excitation function $\sigma(\Theta_{cm}, p_p)$ partly shown in Figure 7. For this purpose we take conservatively [30] for $\Delta\sigma$ the threefold statistical uncertainty. These lower limits of η_{el} for $^1S_0$ and $^1D_2$ resonances are shown in the lower part of Figure 8.

The search for resonant excursions was then extended to the whole angular range $\Theta_{cm} \geq 35°$; data were binned in $\Delta p_p = 10$ MeV/c and $\Delta\Theta_{cm} = 5°$ intervals. The product of the phase shift solution SM97 and a polynomial in p_p free of any narrow resonances was fitted to these altogether 11 excitation functions. The deviations of the data points from the smooth fits (see Figure 7), divided by their statistical errors, follow a normal distribution with a standard deviation of ≈ 1. Among the 2830 data points are 24 with three standard deviations or more away from the fit, but not with any significant concentration at a center of mass energy W.

Based on this evidence the existence of a $^1S_0(^1D_2)$ resonance with $\eta_{el} \geq 0.05$ can be excluded for $p_p < 1000$ MeV/c (2500 MeV/c), with $\eta_{el} \geq 0.1$ for $p_p < 2400$ MeV/c (3300 MeV/c) and $\eta_{el} \geq 0.15$ for $p_p < 3000$ MeV/c, cf. Figure 8.

The quoted limits for η_{el} refer to total widths Γ ranging from 10 MeV to 100 MeV. The lower limit results from the bin width $\Delta p_p = 10$ MeV/c corresponding to $\Delta W_r \approx 3$ MeV and the request for a minimum of two or three adjacent bins to depart by three standard deviations or more from the smooth trend. The upper limit is a more qualitative estimate based on the luminosity uncertainties; in addition such broad resonances may to some extent be absorbed in the momentum dependence of the global phase shift analysis taken as reference for deviations.

Analyzing Power $A_N(\Theta_{cm}, p_p)$

The absolute polarization values P_x and P_y are established by normalizing the observed asymmetries $\epsilon = P \cdot A_N$ for **one** momentum bin $\Delta p_p = 60$ MeV/c around the energy $T_p = 730$ MeV to the analyzing power values from McNaughton et al. [31]. The extent of variation introduced by changing the reference to e.g. $T_p = 449$ MeV [25] or 796 MeV [23,24] can be estimated from Figure 5.

The COSY beam changes its shape and position during acceleration and deceleration, though very reproducible in each cycle. Prior to merging all data in one final set it was necessary to perform consistency checks on subsets obtained under different conditions of polarization and acceleration cycle. Some of them are shown in Figure 9; they demonstrate that the holding field is properly aligned to the detector coordinates, and that vertex reconstruction plus proper flip elimination of false asymmetries work well. Therefore all data was considered compatible and combined in one set.

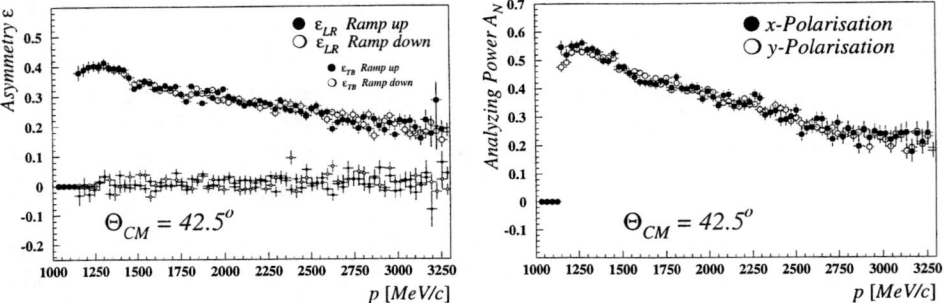

FIGURE 9. Left: Comparison of asymmetries measured for a target polarization defined by a holding field in y–direction during the acceleration and deceleration part of the cycle. Right: Comparison of analyzing powers A_N acquired in full cycles with holding fields in x– and y–direction, respectively.

Preliminary excitation functions $A_N(\Theta_{cm}, p_p)$ have been deduced from our experimental results by grouping them into $\Delta\Theta_{cm} = 6°$ and $\Delta p_p = 30$ MeV/c wide bins; four out of ten resulting excitation functions are shown in Figure 10. The comparison to the recent phase shift solution SP99 [3] shows agreement in the general size and momentum dependence, but also systematic deviations for momenta from 1800 – 2500 MeV/c. In this region data is particularly scarce in the SAID data base for A_N, that does not yet contain the results of this paper. Differences are also visible for the highest momenta ≥ 3000 MeV/c, but that region is sensitive to details of the treatment of background from inelastic pp interactions that is not finished yet.

For a first test of the impact the excitation functions A_N of this paper may eventually have on the scattering phases, we have added them to the present SAID data base and searched for a global solution. It turns out, that the χ^2 per datum is not changed and that the triplet phases $^3F_3$ and $^3F_2$ experience the largest relative modifications, which are located in the momentum range with the systematic deviations of A_N, see Figure 11. This is due to the fact, that the analyzing power is proportional to invariant amplitudes that include only triplet partial waves. It implies that the excitation functions for A_N are less sensitive to resonant excursions in siglet than to those in triplet partial waves. Indeed did an application of Eqs. (1) and (2) for A_N and its threefold statistical uncertainty taken as observable $\Delta\sigma/\sigma$ show, for the hypothetical resonances discussed before, the highest sensitivity in the $^3F_3$ case. Our elasticity limit $\eta_{el} \leq 0.16$, however, is still too high for a test of the predicted value ≈ 0.1.

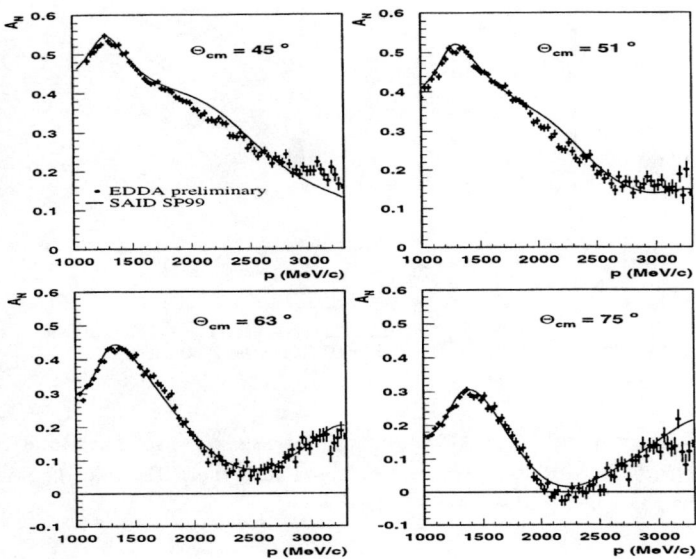

FIGURE 10. PRELIMINARY experimental excitation functions $A_N(p_p)$. Elastic pp scattering data is for $\Delta\Theta_{cm} = 6°$ and $\Delta p_p = 30$ MeV/c bins. The SAID solution SP99 is given as solid line.

SUMMARY AND OUTLOOK

The cyclic accumulation of excitation functions for pp elastic scattering with an internal target during acceleration, in the flattop and deceleration of the COSY synchrotron beam works well. The results for $d\sigma/d\Omega(\Theta_{cm}, p_p)$ and $A_N(\Theta_{cm}, p_p)$ have extended the data base for phase shift analyses, or will soon do so. The impact on the SAID solution SM97 was considerable, and we anticipate smaller modifications of the solution SP99 as well. The EDDA collaboration will next proceed to the measurement of spin correlation coefficients A_{NN}, A_{SS}, and A_{SL} with the polarized COSY beam [32]. The aim is to get more exclusive information on the short range part of the NN interaction, and more restrictive limits on the elasticity η_{el} for resonant excursions in the elastic channel by a combined analysis of all excitation functions.

ACKNOWLEDGMENTS

The EDDA collaboration gratefully acknowledges the great support received from the COSY accelerator group. Helpful discussions with and comments from R. A. Arndt, E. L. Lomon and R. Machleidt are very much appreciated.

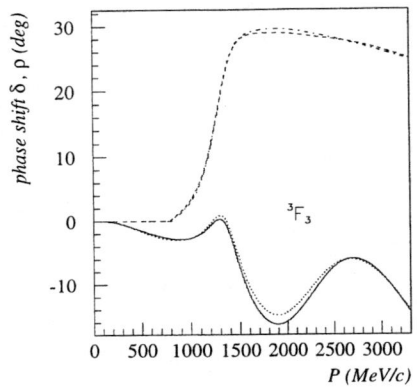
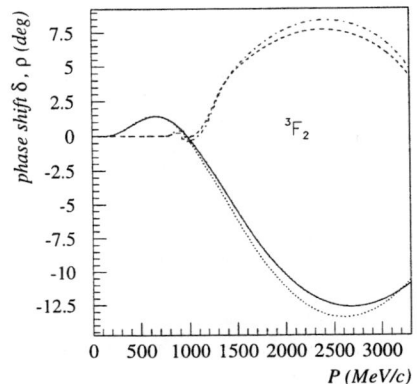

FIGURE 11. Triplet phase shifts $^3F_3$ and $^3F_2$. SAID solution SP99 is shown as solid line (real part) and dashed (imaginary part); the adjacent lines are obtained with the PRELIMINARY excitation functions A_N of this paper being added to the data base.

REFERENCES

1. Lechanoine-Leluc, C., and Lehar, F., *Rev. Mod. Phys.* **65**, 47 (1993).
2. Stoks, V. G. J. et al., *Phys. Rev. C* **48**, 792 (1993).
3. R.A. Arndt, R. A. et al., *Phys. Rev. C* **56**, 3005 (1997), SAID SM94, SM97, SP98 and SP99.
4. Lacombe, M. et al., *Phys. Rev. C* **21**, 861 (1980).
5. Machleidt, R., Holinde, K., and Elster, Ch., *Phys. Rep.* **149**, 1 (1987).
6. Stoks, V. G. J. et al., *Phys. Rev. C* **49**, 2950 (1994).
7. Machleidt, R., *Adv. in Nucl. Phys.* **19**, 189 (1989).
8. Machleidt, R., *Progress Report of the Workshop on Intermediate Spin Physics, Jülich, 1998*, 169.
9. Aerts, A. Th. M. et al., *Phys. Rev. D* **17**, 260 (1978); Mulders, P. J. et al., *Phys. Rev. D* **21**, 2653 (1980).
10. Wong, C. W., *Prog. Part. Nucl. Phys.* **8**, 223 (1982).
11. LaFrance, P., and Lomon, E. L., *Phys. Rev. D* **34**, 1341 (1986); González, P., and Lomon, E. L., *ibid.* **34**, 1351 (1986); González, P., LaFrance, P., and Lomon, E. L., *ibid.* **35**, 2142 (1987).
12. Konno, N. et al., *Phys. Rev. D* **35**, 239 (1987).
13. Goldman, T. et al., *Phys. Rev. C* **39**, 1889 (1989).
14. Ball, J. et al., *Phys. Lett.* **B320**, 206 (1994).
15. Albers, D. et al., *Phys. Rev. Lett.* **78**, 1652 (1997) and references therein.
16. Bisplinghoff, J., and Hinterberger, F., *Particle Production Near Threshold, AIP Conf. Proc.* **221**, 312 (1991); Scobel, W., *Phys. Scr.* **48**, 92 (1993); Rohdjeß, H., *Proc. Int. Conf. on Physics with GeV-Particle Beams, Jülich, 1994*,

World Scientific, 334 (1995).
17. Maier, R., *Nucl. Instr. and Meth. in Phys. Res. A* **390**, 1 (1997).
18. Altmeier, M. et al., *Nucl. Instr. and Meth. in Phys. Res. A* **431**, 428 (1999).
19. Simon, A. J. et al., *Phys. Rev. C* **48**, 662 (1993); **53**, 30 (1996).
20. Eversheim, P. D. et al., *Nucl. Phys.* **A626**, 117c (1996).
21. Ohlsen, G. G., and Keaton, P. W., *Nucl. Instr. and Meth.* **109**, 41 (1973).
22. Garçon, M. et al., *Nucl. Phys.* **A445**, 669 (1985).
23. Bevington, P. R. et al., *Phys. Rev. Lett.* **41**, 384 (1978).
24. McNaughton, M. W. et al., *Phys. Rev. C* **23**, 1128 (1981).
25. Przewoski, B.v. et al., *Phys. Rev. C* **58**, 1897 (1998).
26. Nagata, J. et al., in: *KoK, L. P. et al (Eds.), Handbook XVth Int. Conf. on Few Body Problems in Physics, Groningen*, 70 (1997); Matsuda, M., *ibid.* p. 68.
27. Kobayashi, Y. et al., *Nucl. Phys.* **A569**, 791 (1994).
28. Nagata, J. et al., Phys. Rev. C **45**, 1432 (1992).
29. Lomon, E. L., *preprint MIT-CTP-2680* (1997).
30. Garçon, M. et al, *Phys. Lett.* **B183**, 273 (1987).
31. McNaughton, M. W. et al., *Phys. Rev. C* **41**, 2809 (1990).
32. Lehrach, A. et al., in: de Jager, C. W. et al. (Eds.), *Proceedings 12th Int. Symposium on High-Energy Spin Physics, Amsterdam*, Singapore: World Scientific, 1997, p. 416.

Pion Production with Polarized Beam and Polarized Target at IUCF

J.T.Balewski[1,2]

[1] *Indiana University Cyclotron Facility, Bloomington, Indiana 47408, USA*
[2] *Institute of Nuclear Physics, Cracow 31-342, Poland*

Abstract. The PINTEX experimental apparatus used for detection of $\vec{p}\vec{p} \longrightarrow pp\pi^0$ and $\vec{p}\vec{p} \longrightarrow pn\pi^+$ at IUCF is described. Measured spin observables, integrated as well as angle- or energy-dependent, are compared with the Jülich model. Significant discrepancies are found for the angular distributions.

INTRODUCTION

A precise measurement of the $pp \longrightarrow pp\pi^o$ reaction in the early '90's triggered the development of various theoretical models of pion production. Since then a variety of theoretical approaches have been proposed, as summarized e.g. in [1], but only the recent Jülich model [2] is able to calculate polarization observables. This model incorporates standard pion production mechanisms (direct production and rescattering) as well as heavy meson exchange and Δ excitation. The known NN and πN amplitudes are used as an input and final-state partial waves up to $l_\pi, L_{NN} \leq 2$, are considered. Here, l_π is the pion angular momentum in the three-body center-of-mass system (CMS), while L_{NN} refers to the angular momentum in the NN-subsystem.

The experimental energy dependence of the unpolarized cross sections for four different reaction channels $pp \longrightarrow pn\pi^+$, $pp \longrightarrow pp\pi^0$, $pp \longrightarrow d\pi^+$, and $pn \longrightarrow pp\pi^-$ are well described by the Jülich model for energies $\eta \leq 1$, where η is the maximal momentum of the pion in the center-of-mass system. The predictions of the analyzing power $A(\theta_\pi)$ are less accurate, but still within limits given by the existing experimental data [3]- [6].

This talk [7] focuses on recent measurements of pion production at IUCF using a polarized beam and a polarized target as well as on a comparison with the Jülich model.

The new generation of experiments utilizing polarized beams and targets is now yielding precise polarization data. Even though the measurement is kinematically complete, in order to present the data one integrates usually over one of two center-of-mass directions: (i) the direction of the relative momentum between the two nucleons Ω_{NN}, or (ii) the direction of the pion Ω_π. This leads to two independent sets of the two-body type analyzing powers A_{0y}, A_{y0} and spin correlation coefficients A_{ij} describing a three-body reaction.

Lets integrate over $\int d\Omega_{NN}$ to illustrate how one set of A_y and A_{ij} couples to the polarized differential cross section $d^3\sigma_o(\vec{P},\vec{Q})/d\Omega_\pi dp_\pi$

$$\frac{d^3\sigma_o}{d\cos\theta_\pi dp_\pi} \cdot \left[1 + (\mathbf{U}\vec{P})^\dagger \begin{pmatrix} 0 \\ A_{y0} \\ 0 \end{pmatrix} + (\mathbf{U}\vec{Q})^\dagger \begin{pmatrix} 0 \\ A_{0y} \\ 0 \end{pmatrix} + (\mathbf{U}\vec{P})^\dagger \begin{pmatrix} A_{xx} & 0 & A_{xz} \\ 0 & A_{yy} & 0 \\ -A_{xz} & 0 & A_{zz} \end{pmatrix} (\mathbf{U}\vec{Q}) \right]$$

This "polarized" cross section may be decomposed [8] into the unpolarized cross section, σ_o, terms involving the beam or target analyzing powers A_{0y}, A_{y0}, and terms involving the spin correlation coefficients. $\vec{P}$ and $\vec{Q}$ are the beam and the target polarizations, respectively. The U-matrix

$$\mathbf{U} = \begin{pmatrix} \cos\phi_\pi & \sin\phi_\pi & 0 \\ -\sin\phi_\pi & \cos\phi_\pi & 0 \\ 0 & 0 & 1 \end{pmatrix},$$

rotating the coordinate system by $-\phi_\pi$ around the Z-axis, carries the whole ϕ_π dependence of the differential cross section.

One of the objectives of the experiments performed at IUCF was to determine A_y and A_{ij}. For the partial waves with a final state where the pion as well as the NN angular momentum is either 0 or 1, it is sufficient to use the associated Legendre polynomials P_2^0, P_2^1, and $P_2^2(\cos\theta)$ to describe the θ dependence of A_y and A_{ij}. The expansion coefficients, if extracted from the experiment, contain information on the strength of individual partial waves and may be compared with the corresponding theoretical values, thus selectively testing certain ingredients of the model.

PINTEX EXPERIMENT

The PINTEX facility [9] at IUCF consists of an internal polarized target mounted at the Indiana Cooler storage ring and a ϕ-symmetric detector, shown in fig. 1.

Polarized Beam

The Cooler is capable to deliver a polarized proton beam with P=0.7 and with energies from 100 to 450 MeV. Normally, the stable spin direction in a storage ring is vertical. To get longitudinal beam polarization at the target in the A region, (see fig. 1) spin-precessing solenoids are added to the Cooler lattice. We use three solenoids that are already present in the electron-cooling system in the C region (fig. 1), and a newly installed superconducting solenoid in the T region. The fully longitudinal polarization ($P_z = P$) at the A-region may be obtained up to 200 MeV, but the maximum available strength of the solenoids typically limits P_z to $0.6P$ with $P_y = 0.8P$ and a small sideways component $P_x = 0.1P$ (the values are for 400 MeV stored protons).

Polarized Atomic Beam Source

Polarized hydrogen atoms from an atomic beam source (ABS) are injected into a target cell. To produce the polarized atoms, first a slow atomic beam is generated using a dissociator with a cold nozzle. This beam then passes through the inhomogeneous field of a sextupole magnet which defocuses two of the four possible hyperfine states. Subsequently, RF fields are used to induce a transition between spin states, followed by another sextupole magnet, which focuses a pure atomic spin state. The atomic beam source [10] employed in this experiment delivers 6×10^{16} atom/s at the entrance to the storage cell with polarization Q=0.7.

The orientation of polarized atoms from the source follows adiabatically the direction of the ambient magnetic field, ending up in the direction of the guide field established at the target cell. This guide field, generated by one of three orthogonal Helmholtz-type coils, was alternated during experiment between $\{\pm X, \pm Y, \pm Z\}$ directions.

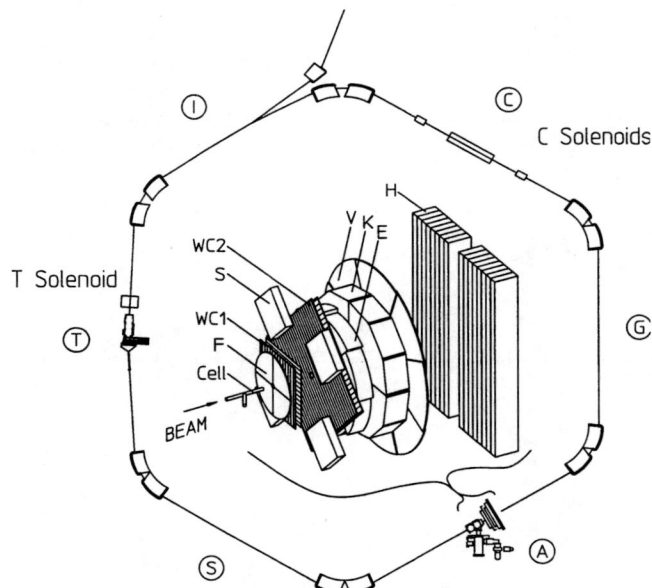

FIGURE 1. Overview of the Cooler, a storage ring and synchrotron for polarized protons. The PINTEX experimental area is located in the straight section (A). Detailed insert shows the Forward Detector Stack (items F through K), which is described in details in ref. [9].

Target Cell

When the atomic beam is injected into a storage cell, as shown in fig. 2, the target thickness increases by a factor of several hundred, compared to a gas jet target crossing the stored beam. The target cell consists of an open-ended tube through which the stored beam passes, with a fill tube attached to the side through which the incident atomic beam is directed (see fig. 2) The polarized atoms accumulate in the cell because of the flow impedance along the tubes. The density of the target along the storage cell is triangular reaching zero at the ends of the cell. The thickness of the entire target is a few times 10^{13} atoms/cm^2. The typical luminosity during the run is a few times 10^{28} cm^{-2}s^{-1}.

For pion production it was desirable to have a low-mass cell so that low-energy reaction particles emerging from any point along the cell axis into the detector acceptance are unaffected. We wanted also to have uniform azimuthal acceptance over the full 360 degrees. This rules out any massive supporting elements.

The present target cell is a 12 mm diameter, 250 mm long tube constructed from 25 μm aluminum foil. Depolarization during the roughly 300 wall collisions is minimized by coating the cell with Teflon.

To minimize background from beam halo, the beam should be centered in the cell. Therefore, a motorized cell adjustment mechanism was constructed, that maintains alignment of the fill tube to the atomic beam. Motion in the vertical (y) direction

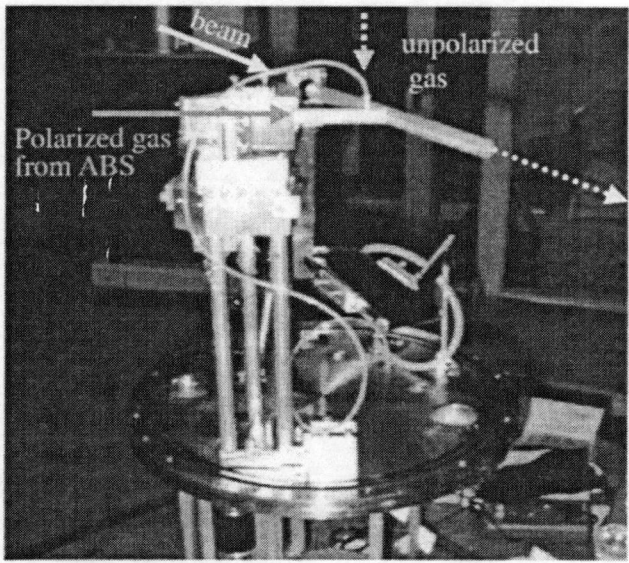

FIGURE 2. Storage cell for pion production is made of 12 mm diameter 250 mm long aluminum tube with 25 μm walls.

is achieved by pivoting the cell about the fill tube entrance. Horizontal (x) motion is achieved by translating the target along the fill tube axis.

Detector Setup

The detector setup consisting of scintillators and wire chambers is shown in fig. 1. The purpose of the main scintillators in the detector stack (marked as F, E, K and V in fig. 1) is to generate a fast trigger for pion production, to measure the kinetic energy of stopped, charged particles in (E+K), and to determine the time of flight of these particles (F) in order to distinguish between charged pions, protons, and deuterons. The typical particle-ID spectrum is shown in fig. 3. The veto detector (V) is used to confirm that a given charged particle has stopped in the preceding E and K detectors.

The purpose of the wire chambers (WC1, WC2 in fig. 1) is to establish the directions of multiple charged-particle tracks. There are two units, each containing two wire planes with wires running orthogonal to each other, to resolve ambiguities with multiple tracks. The wire chambers are of an unique design featuring a hole in the center in order to admit the beam pipe.

Event Reconstruction

Combining energy of protons stopped in scintillators with their directions, obtained from wire chambers, the four-momenta of both charged particles produced in the reaction were reconstructed. Since the four-momentum on the initial state is known, we can calculate the four-momentum of the unobserved pion for each event, using energy and momentum conservation. Its missing mass square (MM^2)

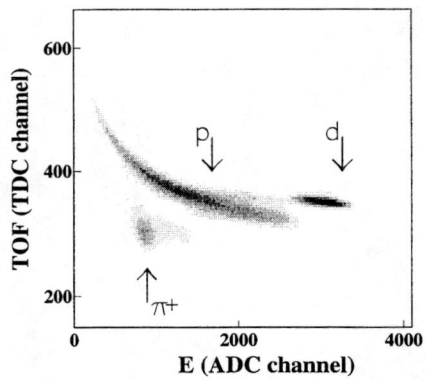

FIGURE 3. Particle identification. Shown is the time of flight between scintillators F and E versus the particle energy deduced from the amplitude from stopping detector E.

was required to be equal to the squared mass of a π^0 within 3 standard deviations (SDEV), otherwise the event was discarded. Typically, the $\sqrt{MM^2}$ peak has SDEV of 3-5 MeV and the underlying background of 5-10%.

To reduce the background beneath the MM peak the following procedure was applied: The topology of any two-track event (the neutral pion is not observed directly) is described by 7 independent parameters, e.g. 3-D vertex position (3 par.) and two 3-D directions (2x2 par.). Since the wire chambers used in this experiment delivered 8 independent hit coordinates, we were able to reconstruct the 3-D vertex position on an event-by-event basis. Both panels of fig. 4 show the density distribution of reconstructed vertices in the plane perpendicular to the storage cell and for some limited range of its length (few cm). The cell wall is marked by a circle. The $pp \to pp\pi^0$ events (identified via MM-cut) are plotted on the left panel of fig. 4 and peak in the middle of the storage cell, where the beam was located.

The background events (those rejected by MM-cut) are shown on the right panel and a large fraction of them originates from the wall of the storage cell. A most probable source of such background events is the $p + Al \to ppX$ reaction.

In the final analysis a cylindrical cut along the beam axis was applied to reduce this type of background.

Polarization Monitor

Four scintillators (S in fig. 1) were used to trigger on pp elastic scattering events near $90°_{c.m.}$, observed concurrently with pion production, in order to determine the

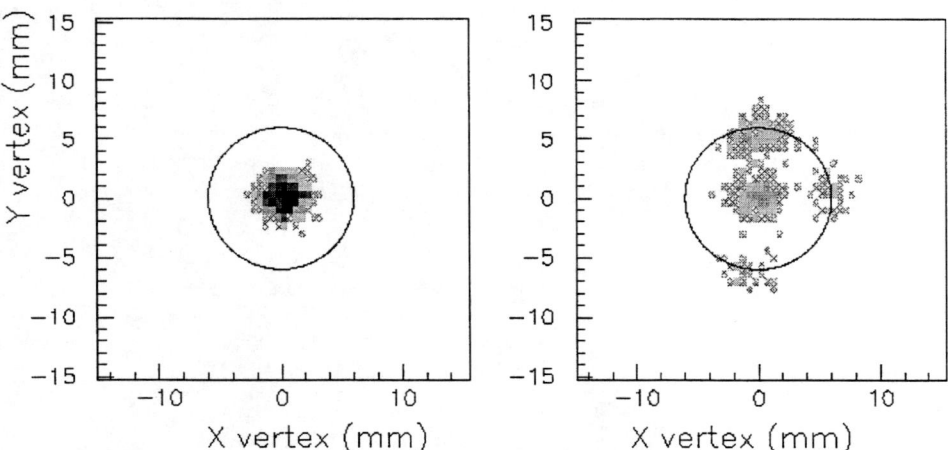

FIGURE 4. Reconstructed vertex in a plane perpendicular to the beam direction. The wall of the storage cell is marked as circle. $\vec{p}\vec{p} \to pp\pi^0$ events selected by means of the MM cut, are in the center (left). The background events originate frequently from the cell wall (right).

product of beam and target polarization (P·Q) and the integrated luminosity.

For vertically polarized beam the measured, spin-dependent asymmetry is proportional to the (large) difference $A_\Delta \equiv A_{xx} - A_{yy}$ of two spin correlation parameters, (At 350 MeV e.g. $A_\Delta = -1.35$). For longitudinally polarized beam the measured, spin-dependent asymmetry is proportional to A_{zz} (at 350 MeV, $A_{zz} = 0.35$). The method of diagonal scaling [8] was used to analyze the data. This method minimizes systematic uncertainties due to spin-dependent differences in luminosity and due to variations in detector efficiency. The values used for the $\vec{p}\vec{p} \to pp$ spin correlation parameters for each beam energy were obtained from the SM97 partial-wave analysis of the Virginia group [11].

RESULTS

ϕ-dependent distributions

Spin correlation observables are deduced by sorting data as a function of ϕ_π for specific combination of beam and target spins. Fig. 5, taken from ref. [12],

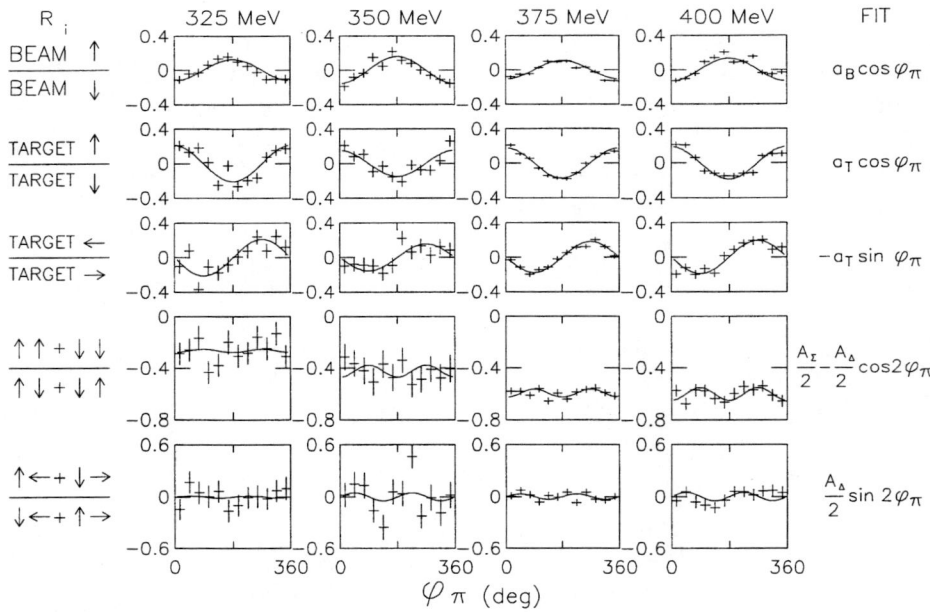

FIGURE 5. Asymmetries for different spin combinations R_i as a function of the azimuthal angle of the π^+. One expects only selected trigonometric functions of $\phi_{\pi+}$, as shown on the right hand side. Each column corresponds to different beam energy.

proves that the performance of the detector is well understood and the expected ϕ_π dependences are observed.

Angle integrated observables

It is useful to define the following quantities: $A_\Sigma(\theta_\pi) \equiv A_{xx}(\theta_\pi) + A_{yy}(\theta_\pi)$ and $A_\Delta(\theta_\pi) \equiv A_{xx}(\theta_\pi) - A_{yy}(\theta_\pi)$. Fig. 6, taken from Saha et al. [12], shows integrated over the θ_π angle values of A_Σ, A_{zz}, and A_y for the $\vec{p}\vec{p} \longrightarrow pn\pi^+$ reaction, compared with the recent calculation of the Jülich model [13]. The agreement between data and model is very good. A similar agreement was found for the $A_y(\theta_\pi)$ distributions measured at different beam energies.

The geometric acceptance for the $p\pi^+$ coincidences ranges from 21% at 325 MeV down to 15% at 400 MeV. The main reason for the low acceptance was that the kinematics does not constrain the opening angle of the π^+ at this excess energies.

For the $pp \longrightarrow pp\pi^0$ measurement the situation is much better: protons are heavier than π^+ and the transverse component of momentum vector in LAB is always

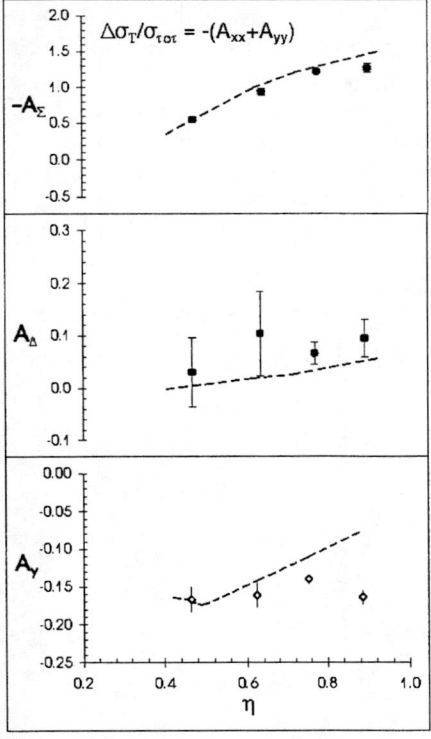

FIGURE 6. The integrated $\vec{p}\vec{p} \longrightarrow pn\pi^+$ observables. Points are PINTEX data, dashed line represents the Jülich model corrected for the detector acceptance.

smaller then the longitudinal one. Thus, the geometric acceptance for detecting both protons was between 90% and 70%.

The analogous experimental results for $\vec{p}\vec{p} \longrightarrow pp\pi^0$ were obtained by Meyer et al. [14], [15]. Fig. 7 shows integrated over θ_π functions $A_\Sigma(\theta_\pi)$ and $A_{zz}(\theta_\pi)$, measured with the same experimental setup. Here the model calculation [13] follows the data only approximately.

Moreover, as demonstrated in [15], those polarized cross sections combined with the unpolarized world data are sufficient to extract the Ps ($L_{NN} = 1, l_\pi = 0$) and Pp partial wave cross sections. Data shown in fig. 8, are in perfect agreement with η^6 and η^8 dependences expected from the expansion of spherical Bessel functions for small arguments and integrated with the three-body phase space. The solid lines, resulting from Jülich calculations, miss significantly the data and even the slope of the σ_{Pp} seems to differ from η^8. A deviation from η^8 could be an artifact of the incorrect treatment of FSI.

Angular distributions

In order to extract $A_\Sigma(\theta_\pi)$ for $\vec{p}\vec{p} \longrightarrow pp\pi^0$ we used four measurements with the *transversally* polarized target $\vec{Q} = \pm\{0, Q_y, 0\}$ and alternated beam polarization $\vec{P} = \pm\{P_x, P_y, P_z\}$. In practice P_x was close to zero and its presence was ignored in the analysis. Let index $k = \{++, +-, -+, --\}$ denote those four combinations of orientations of $\{\vec{P}, \vec{Q}\}$.

After kinematics for each event was reconstructed, only events with $MM \sim m_{\pi^0}$ were accepted. Let $Y_k^H(\theta_\pi)$ denotes distribution of those events as a function the pion angle θ_π in CMS.

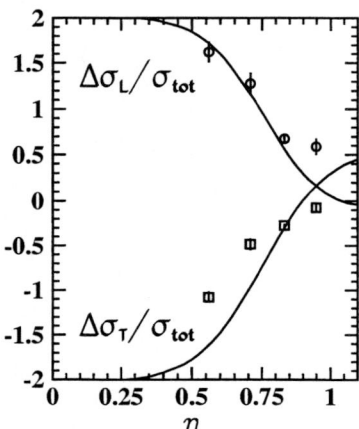

FIGURE 7. The integrated $\vec{p}\vec{p} \longrightarrow pp\pi^0$ observables. Solid line results from the Jülich model. Notation: $\Delta\sigma_T/\sigma_{tot} \equiv -\int \sigma_0 A_\Sigma(\theta)/\int \sigma_0(\theta);\quad \Delta\sigma_L/\sigma_{tot} \equiv -2\int \sigma_0 A_{zz}(\theta)/\int \sigma_0(\theta)$

In order to subtract background below the π^0 peak for each θ_π bin, an additional measurement with a N_2 target was used. We assumed the distribution of recorded $p + N \to ppX$ events is very close to the background measured during regular H-target run, where the background originates from $p + Al$ or $p + O$ reactions on the wall of the storage cell or in the rest-gas. Therefore, by applying an identical analysis scheme as for H-target data, we got the background distribution $Y^{N_2}(\theta_\pi)$.

To normalize the integrated luminosities (L) of H- and N_2-target runs we used the ratio of the number of background events ($N(bg)$) with $MM \in [0, 100]$ MeV, i.e. significantly below the pion mass, recorded in both runs.

$$\alpha_k \equiv \frac{L_k^H}{L^{N_2}} = \frac{N_k^H(bg)}{N^{N_2}(bg)}$$

The background-corrected $\vec{pp} \to pp\pi^0$ yields $Y_k(\theta_\pi)$ were calculated for each θ_π bin

$$Y_k(\theta_\pi) = Y_k^H(\theta_\pi) - \alpha_k \cdot Y^{N_2}(\theta_\pi)$$

Using the definition of A_Σ and assuming the ratio of luminosities with target polarized + and − was the same for both directions of the beam polarization, it may be shown, that

$$A_\Sigma(\theta_\pi) = \frac{-2Z_y(\theta_\pi)}{P_y Q_y} \; ; \quad Z_y(\theta_\pi) = \frac{1 - \sqrt{R_y(\theta_\pi)}}{1 + \sqrt{R_y(\theta_\pi)}} \; ; \quad R_y(\theta_\pi) = \frac{Y_{++}(\theta_\pi) \cdot Y_{--}(\theta_\pi)}{Y_{-+}(\theta_\pi) \cdot Y_{+-}(\theta_\pi)}$$

The procedure described above yields the $A_\Sigma(\theta_\pi)$ data shown in fig. 9a.

To extract from the data the $A_{zz}(\theta_\pi)$ shown in fig. 9b, we used another four measurements with the *longitudinally* polarized target $\vec{Q} = \pm\{0, 0, Q_z\}$ and alternated

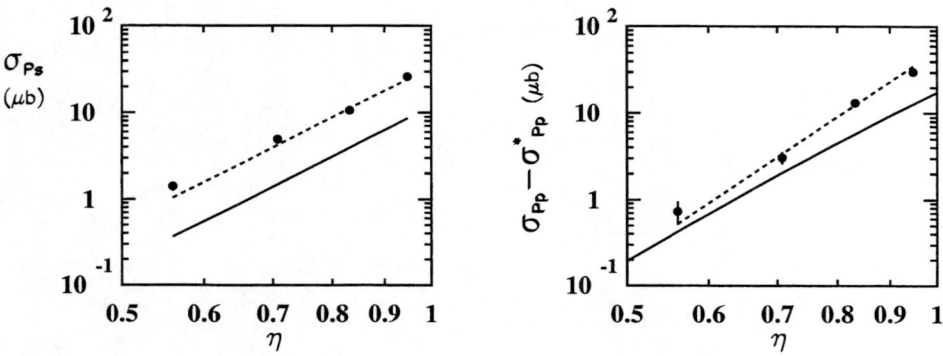

FIGURE 8. Decomposition of $\vec{pp} \to pp\pi^0$ cross section into partial waves. Dashed lines are fitted η^6 and η^8 functions to σ_{Ps} and σ_{Pp}, respectively. Solid lines result from the Jülich model [13].

beam polarization $\pm\vec{P}$. The whole analysis was the the same, except the factor 2 in the final formula (1).

$$A_{zz}(\theta_\pi) = \frac{-Z_z(\theta_\pi)}{P_z Q_z} \qquad (1)$$

Values of $A_\Sigma(\theta_{NN})$ and $A_{zz}(\theta_{NN})$ shown in fig. 9c,d were extracted in the analogous way, using the difference between CMS momenta of two protons as a definition for θ_{NN}. Since the two detected particles in the exit channel are identical, the θ_{NN} distribution is symmetric with respect to $\theta_{cm} = 90°$ and the distribution $0 \leq cos(\theta_{NN}) \leq 1$ contains all information.

In fact, from basic assumptions about limited number of partial waves involved in the reaction, one expects the θ-dependence to be of the form $\alpha + \beta \cdot P_2(cos\theta)$, for all 4 distributions in fig. 9, where $P_2(X)$ is the Legendre polynomial. The dashed lines in fig. 9 represent fits of this formula to the data.

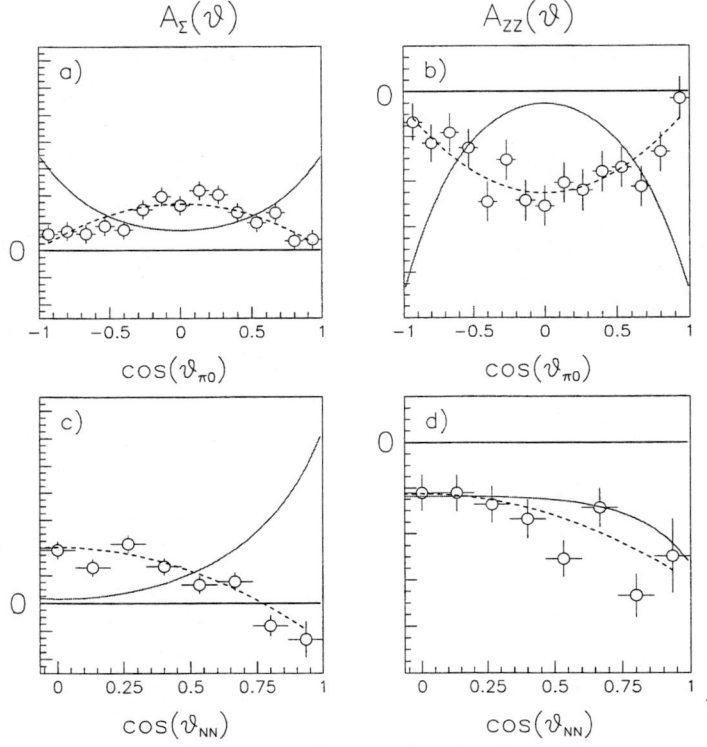

FIGURE 9. Angular distributions of $\vec{p}\vec{p} \longrightarrow pp\pi^0$ spin observables measured at 375 MeV. Dashed line is a fit of $\alpha + \beta \cdot P_2(cos\theta)$ to data, solid line - Jülich model. The data are preliminary: for this reason the vertical axes have been left unlabeled.

Note: (i) the expected shape is really present in the data. This proves that the
ϕ-symmetry of the detector efficiency was achieved,
(ii) the values of α's for $A_\Sigma(\theta_\pi)$ and $A_\Sigma(\theta_{NN})$ are the same and equal to $-\Delta\sigma_T/\sigma_{tot}$
at 375 MeV (compare with fig. 7),
(iii) also α's for $A_{zz}(\theta_\pi)$ and $A_{zz}(\theta_{NN})$ are equal to $-2\Delta\sigma_L/\sigma_{tot}$,
(iv) contrary, the four β-coefficients may have different values for each observable,
since they correspond to a different bilinear combinations of partial wave amplitudes.

The results of this detailed analysis of the $\vec{p}\vec{p} \longrightarrow pp\pi^0$ data were compared with
the Jülich model [16], plotted by means of the solid line in fig. 9. This model has
problems with the description of the $A_\Sigma(\theta_\pi)$, $A_{zz}(\theta_\pi)$, and $A_\Sigma(\theta_{NN})$. It looks like
the sign of β's for the Jülich model is opposite to the data (except for $A_{zz}(\theta_{NN})$
where the agreement is reasonable).

Dependence on energy sharing in the final state

The complementary approach of extracting the strength of individual partial
waves involved in the $\vec{p}\vec{p} \longrightarrow pp\pi^0$ reaction uses their energy dependence (the

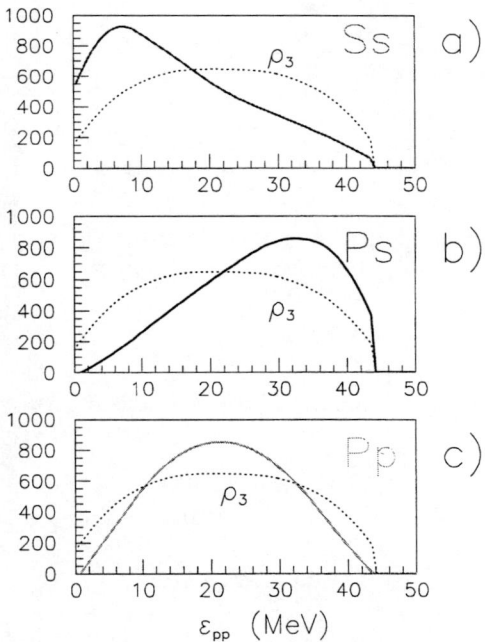

FIGURE 10. $pp \longrightarrow pp\pi^0$ energy distributions derived from the "power-law" model for the
beam energy T=375 MeV. The dashed line is a pure three-body phase space distribution.

power-law model) induced by expansion of the spherical Bessel functions. The similar method was earlier applied by e.g.: [18] or [19]. As the energy-sharing parameter we have chosen the energy in the NN-subsystem, $\epsilon_{pp} = \sqrt{s_{pp}} - 2m_p$, and integrated three-body kinematics over all angle observables.

In the case of a pure $L_{NN} = 0$, $l_\pi = 0$ (Ss) interaction one expects the Ss-matrix element to be constant. Only the $FSI_{pp}(\epsilon_{pp})$, depending explicitly on ϵ_{pp}, influences the final distribution of the outgoing particles. Numerical integration of the $FSI_{pp}(\epsilon_{pp})$ with the three-body phase space ρ_3 enabled the prediction of the ϵ_{pp}-dependence (but not a normalization) of the Ss cross section, as shown in fig. 10a:

$$\frac{d\sigma_{Ss}(T)}{d\epsilon_{pp}} \propto \int d\rho_3(T) \cdot FSI_{pp}(\epsilon_{pp}) \cdot A(p_1^{LAB}, p_2^{LAB}) \quad ,$$

where A is a geometrical acceptance depending on the LAB-momenta of the detected protons and T is the incident beam energy.

The matrix element for the $L_{NN} = 1$, $l_\pi = 0$ (Ps) interaction is proportional to p, ($M_{Ps} \propto p \propto \sqrt{\epsilon_{pp}}$). The resulting energy dependence of the Ps cross section is shown in fig. 10b.

$$\frac{d\sigma_{Ps}(T)}{d\epsilon_{pp}} \propto \int d\rho_3(T) \cdot p^2 \cdot A(p_1^{LAB}, p_2^{LAB})$$

Finally, the $L_{NN} = 1$, $l_\pi = 1$ (Pp) cross section, corresponding to matrix element $M_{Pp} \propto p \cdot q$, is plotted in fig. 10c.

$$\frac{d\sigma_{Pp}(T)}{d\epsilon_{pp}} \propto \int d\rho_3(T) \cdot p^2 \cdot q^2 \cdot A(p_1^{LAB}, p_2^{LAB})$$

Here q is the pion momentum in the CMS frame.

The power-law model described above describes the energy dependence, but not the strength of individual partial waves $\hat{\sigma}_i(T)$, where $i = \{Ss, Ps, Pp\}$.

$$\frac{d\sigma_i(T)}{d\epsilon_{pp}} = \hat{\sigma}_i(T) \cdot f_i(\epsilon_{pp}, T) \quad ; \quad \int f_i(\epsilon_{pp}, T) d\epsilon_{pp} \equiv 1$$

The partial cross sections and the cross sections measured in the experiment are related [15]:

$$\frac{d\sigma_0}{d\epsilon_{pp}} = \frac{d\sigma_{Ss}}{d\epsilon_{pp}} + \frac{d\sigma_{Ps}}{d\epsilon_{pp}} + \frac{d\sigma_{Pp}}{d\epsilon_{pp}} \quad ,$$

$$\frac{d\sigma_0 A_\Sigma}{d\epsilon_{pp}} = 2\left(\frac{d\sigma_{Ss}}{d\epsilon_{pp}} - \frac{d\sigma_{Ps}}{d\epsilon_{pp}} - \frac{d\sigma_{Pp}^*}{d\epsilon_{pp}}\right) \quad ,$$

$$\frac{d\sigma_0 A_{zz}}{d\epsilon_{pp}} = \left(-\frac{d\sigma_{Ss}}{d\epsilon_{pp}} - \frac{d\sigma_{Ps}}{d\epsilon_{pp}} + \frac{d\sigma_{Pp}}{d\epsilon_{pp}} - 2\frac{d\sigma_{Pp}^*}{d\epsilon_{pp}}\right) \quad ,$$

where σ_{Pp} and σ^*_{Pp} denotes two different, positive-definite bilinear forms of partial wave amplitudes (see also [17] for details). The beam energy T was omitted for clarity.

Inserting "power-law" distributions to the above relations yields

$$A_\Sigma(\epsilon_{pp}) \equiv \frac{\frac{d\sigma_0 A_\Sigma}{d\epsilon_{pp}}}{\frac{d\sigma_0}{d\epsilon_{pp}}} = 2\frac{\hat{\sigma}_{Ss} \cdot f_{Ss}(\epsilon_{pp}) - \hat{\sigma}_{Ps} \cdot f_{Ps}(\epsilon_{pp}) - \hat{\sigma}^*_{Pp} \cdot f_{Pp}(\epsilon_{pp})}{\hat{\sigma}_{Ss} \cdot f_{Ss}(\epsilon_{pp}) + \hat{\sigma}_{Ps} \cdot f_{Ps}(\epsilon_{pp}) + \hat{\sigma}_{Pp} \cdot f_{Pp}(\epsilon_{pp})}$$

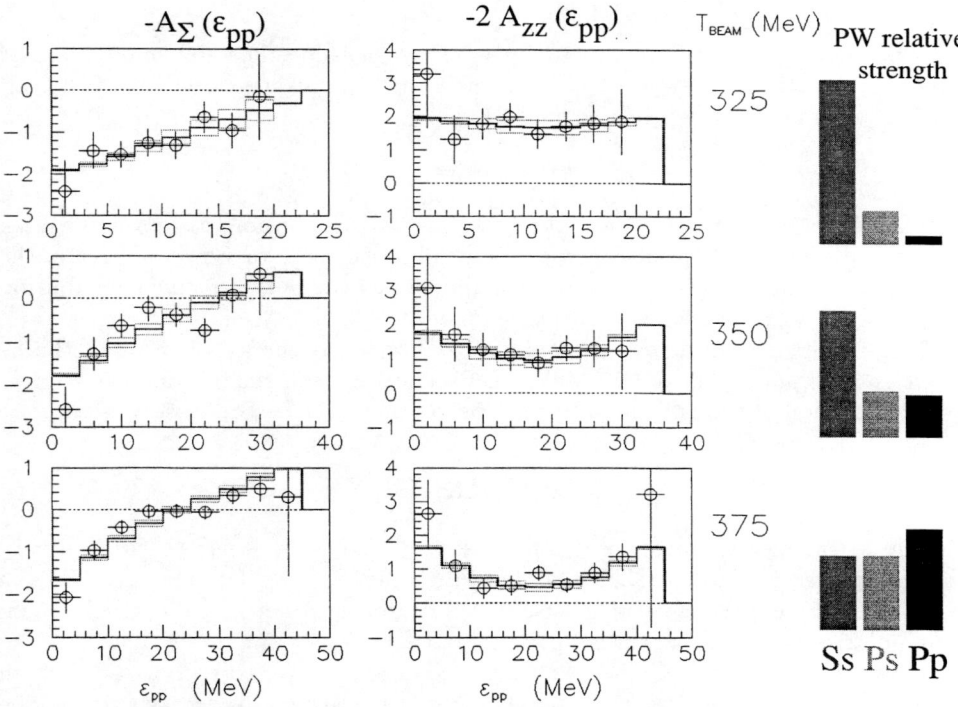

FIGURE 11. Dependence of the $\vec{p}\vec{p} \longrightarrow pp\pi^0$ spin observables on energy sharing parameter ϵ_{pp} in the final state. $\epsilon_{pp} = \sqrt{s_{pp}} - 2m_p$ is the energy in the NN-subsystem. The solid lines result from the "power law" model described in the text, with fitted strength of partial wave amplitudes P_1, P_2, P_3. The dotted lines represent the standard deviation of the fitted curves. The vertical bars on the right hand side are proportional to the relative strength of Ss, Ps, and Pp partial cross sections.

$$A_{zz}(\epsilon_{pp}) \equiv \frac{\frac{d\sigma_0 A_{zz}}{d\epsilon_{pp}}}{\frac{d\sigma_0}{d\epsilon_{pp}}} = \frac{-\hat{\sigma}_{Ss} \cdot f_{Ss}(\epsilon_{pp}) - \hat{\sigma}_{Ps} \cdot f_{Ps}(\epsilon_{pp}) + \left(\hat{\sigma}_{Pp} - 2\hat{\sigma}_{Pp}^*\right) \cdot f_{Pp}(\epsilon_{pp})}{\hat{\sigma}_{Ss} \cdot f_{Ss}(\epsilon_{pp}) + \hat{\sigma}_{Ps} \cdot f_{Ps}(\epsilon_{pp}) + \hat{\sigma}_{Pp} \cdot f_{Pp}(\epsilon_{pp})}$$

The $A_\Sigma(\epsilon_{pp})$ and $A_{zz}(\epsilon_{pp})$ on the left hand side were extracted from the experimental $\vec{p}\vec{p} \longrightarrow pp\pi^0$ data (at fixed beam energy T) in the same way as the $A_\Sigma(\theta_{NN})$ and $A_{zz}(\theta_{NN})$ in the previous chapter. The only difference is that the H- and N_2-target yields $Y_k^H(\epsilon_{pp})$ and $Y_k^{N_2}(\epsilon_{pp})$ were accumulated as a function of the energy in pp-subsystem ϵ_{pp} and integrated over all angle observables.

The right hand side enables determination on only 3 independent parameters

$$P_1 = \hat{\sigma}_{Ss}/(\hat{\sigma}_{Ss} + \hat{\sigma}_{Ps} + \hat{\sigma}_{Pp}) \;,$$
$$P_2 = \hat{\sigma}_{Pp}/(\hat{\sigma}_{Ss} + \hat{\sigma}_{Ps} + \hat{\sigma}_{Pp}) \;,$$
$$P_3 = \hat{\sigma}_{Pp}^*/(\hat{\sigma}_{Ss} + \hat{\sigma}_{Ps} + \hat{\sigma}_{Pp}) \;,$$

i.e. the relative strength of individual partial waves. Note, the value of unpolarized cross section $\sigma_o \equiv \hat{\sigma}_{Ss} + \hat{\sigma}_{Ps} + \hat{\sigma}_{Pp}$ is unconstrained by this measurement of $A_\Sigma(\epsilon_{pp})$ and $A_{zz}(\epsilon_{pp})$ and $\hat{\sigma}_{Ps}$ depends on P_1 and P_2

$$\hat{\sigma}_{Ps}/(\hat{\sigma}_{Ss} + \hat{\sigma}_{Ps} + \hat{\sigma}_{Pp}) = 1 - P_1 - P_2 \;.$$

Those parameters were fitted to the experimental distributions as shown in fig. 11. The middle solid line, resulting from the fit, follows the data very well. Note, the minimum in the -$A_{zz}(\epsilon_{pp})$ distribution builds up as the only positive contribution from the Pp wave increases with the beam energy T.

This results are intuitively reasonable: at low beam energy (325 MeV) the Ss wave is dominant, while at 375 MeV the Pp one start to be significant.

A prediction of the Jülich model are not yet available for this observables.

CONCLUSIONS

The resent measurements of the $\vec{p}\vec{p} \longrightarrow pp\pi^0$ and $\vec{p}\vec{p} \longrightarrow pn\pi^+$ delivered very precise polarized data, challenging the Jülich model in a detailed way. In the nearest future experiments with a polarized deuteron target and the development of a polarized deuteron beam are planed at IUCF.

A workshop held at IUCF in 1998 suggested possible future directions of research at the Indiana Cooler. A focal point of interest during this workshop was the three-nucleon force and its role in the p+d elastic scattering as well as in the p+d breakup. In particular, the spin dependence of the three-body force was questioned. The experiments proposed at IUCF hope to gain information about the three-body force and its spin dependence by measuring spin correlation effects in a kinematic region where Faddeev calculations fail to reproduce the measured cross section and where effects due to the three-body force are thought to be large.

In parallel, the $\vec{p}\vec{d} \longrightarrow pd\pi^0$ reaction will be used as a test of the spectator model of π^0 production, assuming the proton in the deuteron to be a spectator. In this model, close to threshold ($\eta < 0.5$) only two amplitudes are non-negligible, namely $a_1(^3P_1 \to {}^3S_1, l_\pi = 0)$ and $a_2(^1D_2 \to {}^3S_1, l_\pi = 1)$, where the a_2 amplitude is driven by the formation of an intermediate Δ. Certain relations between σ_{tot}, A_Σ and A_{zz} are predicted within this model. The purpose of this new experiment is to make use of the spin observables to study the question "to what extent can pion production in the three-nucleon system be understood in terms of *quasi-free* production in NN collisions, where one of the three nucleons acts as a spectator?".

To provide answers to those questions the development of the polarized deuteron target at IUCF is underway, including both vector and tensor polarization. This effort is accompanied by the study of the suitable polarimeter.

REFERENCES

1. H.O. Meyer, *Annu. Rev. Nucl. Part. Sci.* **47**, 235 (1997).
2. C.Hanhart et al., *Phys. Lett.* **B 545**, 176 (1999).
3. D.A.Hutcheon et al., *Phys. Rev. Lett.* **64**, 176 (1990).
4. E.L.Mathie et al., *Nucl. Phys.* **A 397**, 469 (1983).
5. G.Rappenecker et al., *Nucl. Phys* **A 590**, 763 (1995).
6. J.G.Hardie et al., *Phys. Rev.* **C 56**, 20 (1995).
7. transparences are available at
   http://www.iucf.indiana.edu/~balewski/stori99/index.htm
8. H.O.Meyer , *Phys. Rev.* **C 56** , 2074 (1997).
9. T.Rinckel et al., *Nucl. Inst. Meth.* (1999), accepted for publication.
10. T. Wise et al.,Nucl. Instr. and Meth. **A336** 410 (1993).
11. R.A. Arndt et al., Phys. Rev. **C56**, 3005 (1997).
12. S. Saha et al., *Phys. Lett.* **B 461**, 175 (1999).
13. C.Hanhart et al., *Phys. Lett.* **B 444**, 25 (1998).
14. H.O.Meyer et al., *Phys. Rev. Lett.* **81**, 3096 (1998).
15. H.O.Meyer et al., submitted to *Phys. Rev. Lett.*, available as nucl-ex/9907017.
16. C.Hanhart, private communication 1999, see also STORI'99 contribution.
17. L.D.Knutson, STORI'99 contribution.
18. J.Balewski et al., *Phys.Lett.* **B 420**, 211 (1998).
19. J.Złomanczuk et al., *Phys. Lett.* **B 436**, 251 (1998).

Spin dependence of the reaction $\vec{n}p \to pp\pi^-$

H. Lacker[1] for the collaboration

(1) Fakultät für Physik der Universität Freiburg, D-79104 Freiburg, Germany,
(2) Paul Scherrer Institut(PSI), (3) Charles University, Prague,
(4) Joint Institute for Nuclear Research, Dubna, (5) DAPNIA, CEN-Saclay

Abstract. The reaction $\vec{n}p \to pp\pi^-$ was investigated with a large acceptance time-of-flight spectrometer from threshold up to 570 MeV. The experiment was performed with the polarised neutron beam at PSI using a liquid hydrogen target. Preliminary results of invariant mass spectra, angular distributions and asymmetries are presented.

The cross sections of the various single pion production channels can be decomposed into cross sections $\sigma_{I_iI_f}$ according to the isospin of the two nucleons in the initial (I_i) and final state (I_f). The σ_{01} cross section appears only in np collisions. From a comparison between (i) $\sigma_{np \to pp\pi^-} = \frac{1}{2}(\sigma_{01} + \sigma_{11})$ and (ii) $\sigma_{pp \to pp\pi^0} = \sigma_{11}$ it was found that σ_{01} is small or compatible with zero. The angular dependence, historically parameterised by $\frac{d\sigma}{d\Omega} \propto \frac{1}{3} + b \cdot \cos^2\theta_\pi^{cms}$ gives different values of the anisotropy parameter b for reaction (i) [1–5] and (ii) [6–8] which indicates the presence of σ_{01}. Moreover, while the distributions of (ii) are forward-backward(FB) symmetric a FB-asymmetry for process (i) at low energies was observed [1]. This is a signal for an interference between the $I_i = 0$ and $I_i = 1$ initial state. Measurements of $\vec{p}n \to pp(^1S_0)\pi^-$, with the two protons in a relative S state, revealed the Sp(I_i =0) and the Ss(I_i =1) partial wave contributions [9,10] where s and p classify the angular momentum of the pion with respect to the two protons.

For the present experiment, we used the polarised neutron beam at PSI which has a continuous energy spectrum [11]. The setup consisted of a liquid hydrogen target, two drift chambers, a trigger hodoscope and a scintillator wall and allowed to cover almost the whole phase space. By measuring tracks and time-of-flights for each event the reaction was identified by a kinematical fit. Only 10 % of the total statistics has been analysed so far and preliminary results are presented in four neutron energy bins (T_n) between threshold and 570 MeV averaging over the two polarisation states of the beam.

The invariant pp mass (M_{pp}) distributions (Fig. 1, I) show deviations from phase space at higher neutron energies due to the contribution of higher partial waves and the influence of the Δ-resonance. A comparison to data of reaction (ii) at 500 MeV

[7] shows only small differences in shape. At low neutron energies the enhancement at small M_{pp}-values is caused by the $pp(^1S_0)$ final state interaction (FSI).

The θ_π^{cms} distributions have been fitted with $\frac{dN}{d\Omega} \propto \frac{1}{3} + a \cdot \cos\theta_\pi^{cms} + b \cdot \cos^2\theta_\pi^{cms}$. The anisotropy parameter b (Fig. 1, II) agrees with ref. [1,5] but disagrees with [2–4], which can be understood from the restricted phase space of those experiments. Being weakly energy dependent, b differs strongly from the values obtained for reaction (ii) [6,7], indicating the presence of σ_{01}. Above 500 MeV this difference is smaller if the recent data of ref. [8] are considered. The FB-parameter a shows a strong energy dependence (Fig. 1, II) and confirms the FB-asymmetry found at low energies and the zero results at high energies.

At low neutron energies ($T_n = 384\text{MeV}$) a cut of $M_{pp} \leq 2m_p + 6\text{MeV}$ selects events with a $pp(^1S_0)$ final state due to the FSI while the P-wave contamination should be small. The isotropy in $\theta_{\pi p}^{Rpp}$, the angle between the proton and pion direction in the pp rest system, confirms these assumptions. The θ_π^{cms} distribution (Fig. 1, III) shows a strong FB-asymmetry due to the interference between Ss and Sp waves as already found in [9,10] and agrees with their results in shape.

The energy dependence of b and a can be understood as follows. The parameter b is determined by the p-waves (Sp) which become more important with increasing energy as compared to Ss. In fact, the partial wave analysis in [10] shows this behaviour. The models of the Osaka [12] and the Jülich group [13] expect a suppression of the Ss wave at high energies due to a cancelation between the direct production and short range effects like heavy-meson exchanges. We remark that the M_{pp} distributions at high energies for reaction (i) and (ii) differ only slightly while b is large for (i) as compared to (ii) except for the data of ref. [8] at higher energies. If only Sp-waves determined the b parameter they would result in a σ_{01} contribution of about 15 % on $\sigma_{np\to pp\pi^-}$ at high energies.

So far, the results were obtained by averaging over the two polarisation states of the beam. First results for the analysing power A_{N0} are given in Fig. 1, IV. They show large negative values and a pronounced energy dependence. With expected $6.5 \cdot 10^6$ reconstructed events a finer binning in the neutron energy, a detailed study of the phase space dependence of the observables and the extraction of cross sections and σ_{01} will be performed. Besides A_{N0} the parameters A_{S0} and A_{L0} will be determined in the out-of-plane geometry.

This work was supported by the German BMBF. We thank K. Tamura, Y. Maeda, J. Zlomanczuk, N. Kaiser and C. Hanhart for the helpful discussions. The financial support given by the organizers of the STORI99 conference is appreciated.

REFERENCES

1. Handler, R., *Phys. Rev.* **138**, 1230 (1965).
2. Dzhelepov V. P. et al., *Sov. Phys. JETP* **23**, 993 (1966).
3. Kleinschmidt M. et al., *Z. Phys.* **A298**, 253 (1980).
4. Bannwarth A. et al., *Nucl. Phys.* **A567**, 761 (1993).

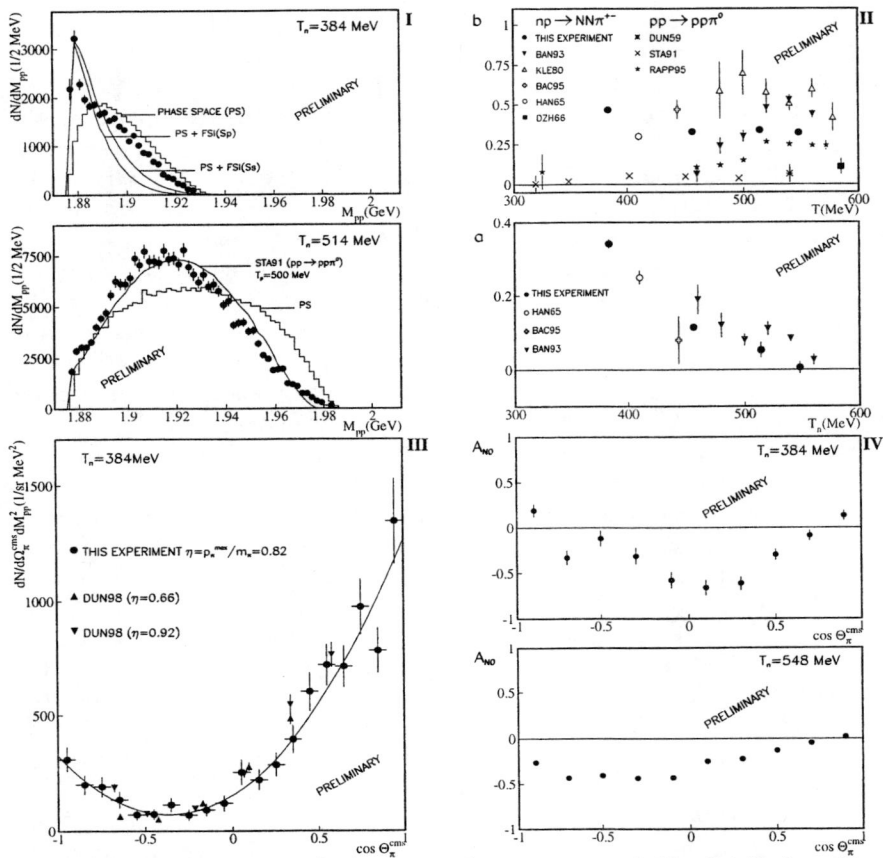

FIGURE 1. Preliminary results for the reaction np → ppπ^-. I: M_{pp} distributions; II: anisotropy b and FB-asymmetry a vs. beam energy; III: $^1S_0$-selection; IV: analysing power A_{N0}.

5. Bachman M. G. et al., *Phys. Rev. C* **52**, 495 (1995).
6. Dunaitsev A. F. et al., *Sov. Phys. JETP* **9**, 1179 (1959).
7. Stanislaus S. et al., *Phys. Rev. C* **44**, 2287 (1991).
8. Rappenecker G. et al., *Nucl. Phys. A* **590**, 763 (1995).
9. Duncan F. et al., *Phys. Rev. Lett.* **80**, 4390 (1998).
10. Hahn H. et al., *Phys. Rev. Lett.* **82**, 2256 (1999).
11. Arnold J. et al., *Nucl. Instr. Meth.* **A386**, 211 (1997).
12. Tamura K. et al., *Pion production in nucleon-nucleon collisions*, Proc. on the STORI99 conference, Bloomington, IN, Sept. 1999, and private communications.
13. Hanhart C., *Meson exchange models for Meson production*, Proc. on the STORI99 conference, Bloomington, IN, Sept. 1999, and private communications.

Spin Correlation Coefficients in $\vec{p}\vec{p} \to pn\pi^+$ from 325 to 400 MeV

Swapan K. Saha, W.W. Daehnick, R.W. Flammang,

Dept. of Physics and Astronomy, University of Pittsburgh, Pittsburgh, PA 15260

J.T. Balewski, H.O. Meyer, R.E. Pollock, B. v. Przewoski,
T. Rinckel, P. Thörngren-Engblom

Dept. of Physics and Cyclotron Facility, Indiana University, Bloomington, IN 47405

B. Lorentz, F. Rathmann, B. Schwartz, T. Wise

University of Wisconsin-Madison, Madison, WI, 53706

P.V. Pancella

Western Michigan University, Kalamazoo, MI, 49008

Abstract. The spin correlation coefficient combinations $S \equiv A_{xx} + A_{yy}$, $D \equiv A_{xx} - A_{yy}$ and the analyzing powers $A_y(\theta)$ were measured for $\vec{p}\vec{p} \to pn\pi^+$ at beam energies of 325, 350, 375 and 400 MeV. A polarized internal atomic hydrogen target and a stored, polarized proton beam were used. These polarization observables are sensitive to contributions of higher partial waves. A comparison with recent theoretical calculations is provided.

I INTRODUCTION

The $\vec{p}p \to pn\pi^+$ reaction involves a large momentum transfer, but near threshold only a few partial waves can contribute. The study of the short-range features of the NN interaction is greatly facilitated because it is possible to experimentally separate the contributions from different angular momentum states. In recent studies it was found that within 10-20 MeV of threshold, i.e. below 300 MeV, only one (Ss for $pp \to pp\pi^0$, [1]) or at most two partial waves (Ss and Sp for $\vec{p}p \to d\pi^+$ and $\vec{p}p \to pn\pi^+$ [2-5]) were important, and direct comparisons with simple theoretical approaches can be made. The present study in the near threshold region provides information on polarization observables in a region still subject to relatively parameter-free microscopic calculations. Microscopic calculations in the meson exchange picture, involving angular momentum states as large as $L_\pi = 2$ are now being published by the Jülich group [6].

In the present experiment we used a polarized beam on a polarized internal target to measure the $\vec{p}\vec{p} \to pn\pi^+$ reaction at 325, 350, 375 and 400 MeV. Bombarding energies were chosen to cover the range over which some higher partial waves gradually become significant. This is the first investigation of spin correlation coefficients in this reaction near threshold. We have measured A_y, $S \equiv A_{xx} + A_{yy}$, $D \equiv A_{xx} - A_{yy}$, integrated over all kinematic variables and the observed range of the of the polar angles. We also determined the differential analyzing powers $A_y(\theta)$.

II EXPERIMENT

The experiment was conducted at the Cooler ring of the Indiana University Cyclotron Facility (IUCF). The target consisted of 74±4% polarized atomic hydrogen in a storage cell, where it was maintained by an atomic beam [7]. The open-ended cell, a movable cylinder of 12 mm diameter with a 25 μm thick aluminum wall, minimized the background induced by the beam halo in the cell wall. Depolarization due to wall collisions was minimized by coating the cell with Teflon. The target gas density distribution along the storage cell was of triangular shape with a spread of ±12.5 cm about the center. The thickness of the target was $\approx 10^{13}$ atoms/cm^2. The target polarization was flipped every 2 sec, pointing in sequence up, down, left, right. Each change of direction took less than 10 ms.

Elastic proton-proton scattering events were observed in the PINTEX detector concurrently with pion production in order to determine the luminosity and beam and target polarization. Four plastic scintillators at azimuthal angles of ±45° and ±135° detected coincident protons elastically scattered near $\theta_{\rm lab} = 45°$. The product PQ of the beam (P) and target (Q) polarization was determined from the spin correlation $A_{xx} - A_{yy}$, which is large and well known for pp scattering. Typically, we saw PQ ≈ 0.5. The time-integrated luminosity was deduced from the known pp elastic cross section.

We estimate the angular uncertainty for the trajectories for protons and pions of about $\sigma = 0.5°$ and $\sigma = 1.0°$, respectively. The energy of the charged particles was measured by a stack of segmented plastic scintillators (E,K) of total thickness of 25.4 cm. The energy resolution of the stack was about $\sigma = 4\%$. Events are rejected if a charged particle reaches the veto detector behind the K-detector. At 325 MeV the resolution of the neutron missing mass peak was $\sigma = 1.6$ MeV. To assess the background shape we took data with pure N$_2$ gas in the target cell. A Monte Carlo simulation of the experiment was used to determine various limiting effects of the apparatus and to derive corresponding corrections. The simulation showed that the limited θ range of the detector caused a loss in count rate which is significant;

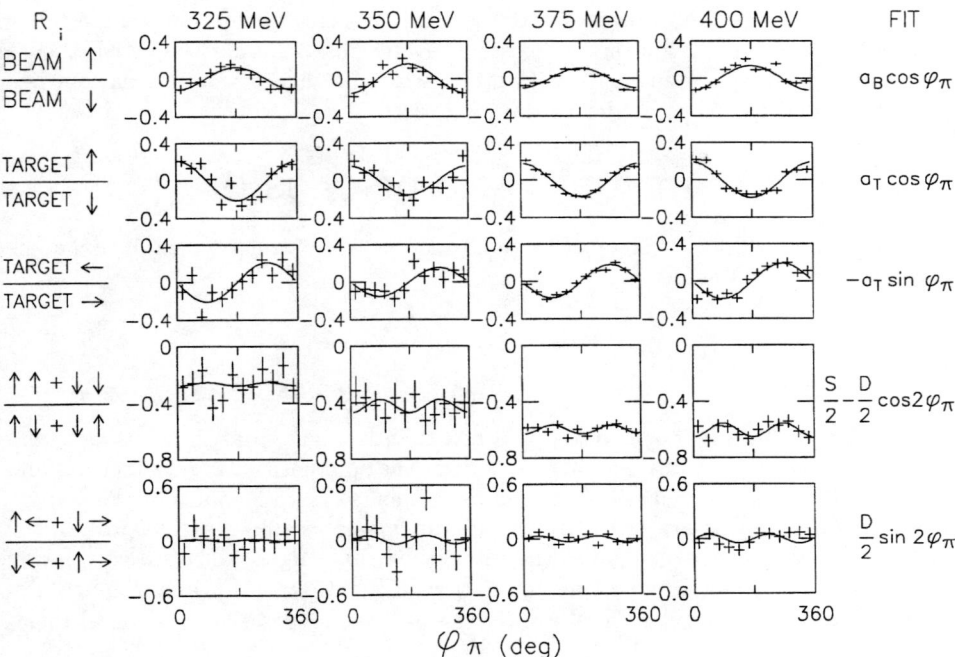

FIGURE 1. Asymmetries for different spin combinations R_i (listed on the left) as a function of the azimuthal angle of the π^+. The solid curves represent a least-square fit using the theoretical ϕ_π dependence (listed on the right), varying a_B, a_T, $S \equiv (A_{xx} + A_{yy})$, and $D \equiv (A_{xx} - A_{yy})$.

however, since target and projectile are identical particles we sample a theta range twice as large as the acceptance suggests. Nevertheless, comparisons with theory may require a corresponding integration range for the theoretical calculation, i.e., the restriction of pion lab angles to the range 6° to 32°.

III RESULTS AND DISCUSSION

Spin-dependent differential cross section as a function of the pion azimuthal angle ϕ_π can be expressed as

$$\sigma(\phi_\pi) = \sigma_{tot}[1 + (p_y A_y^B - q_y A_y^T)\cos\phi_\pi + p_y q_y(\frac{A_{xx}+A_{yy}}{2} - \frac{A_{xx}-A_{yy}}{2}\cos 2\phi_\pi)$$
$$+ terms\ involving\ p_x, p_z, q_x, q_z], \quad (1)$$

where A_y^B and A_y^T are respectively the beam and target analyzing powers, and p_x, p_z, q_x, q_z are very small. The values A_{nn} are the spin correlation coefficients and $p_i(q_i)$ are the beam (target) polarization components. In this equation the observables A_y^B, A_y^T and A_{nn} are integrals over all kinematic variables of the proton, the neutron and the pion except over the pion azimuthal angle ϕ_π. Spin correlation observables are deduced by sorting the data as a function ϕ_π for a specific combination of beam and target spin. We show the ϕ-dependence of the measured ratios in Fig. 1.

The integrated spin correlation coefficient S is related to the spin-dependent cross section by $\Delta\sigma'_T/\sigma_{tot} = -S$, and we find that these two analysis methods agree. The integrated analyzing power is given by $A_y = (-a_B \cdot a_T)^{1/2}$. The resulting polarization observables are shown in Fig. 2 and listed in Table 1. These values, but not those in Fig. 1, are corrected for background contributions. The values for the integrated $\vec{p}\vec{p} \to pn\pi^+$ spin correlation coefficients S and D in Fig. 2 are in remarkably good agreement with recent Hamiltonian predictions following ref. [6].

In our experiment beam and target analyzing powers are related by $A_y^B(\theta) = -A_y^T(\pi - \theta)$ and, therefore, permit a study of the back angles for $A_y(\theta_{c.m.})$ with the present apparatus. Fig. 3 shows the observed $\theta_{c.m.}$ dependence of the analyzing powers $A_y(\theta_{c.m.})$. As mentioned above, the limited size of the detector reduced the usefulness of data near $\theta_{c.m.} = 90°$. However, a large enough range in θ could be covered to demonstrate the change in the angular distributions with energy and the need for partial waves higher than Ss and Sp. The dashed curves show calculations taken from [6].

The solid curves in Fig. 3 represent fits with the relation

$$A_y(x) = (1/\sigma_{unpol}(x)) * (a\ P_1^1(x) + b\ P_2^1(x) + c\ P_3^1(x)), \quad (2)$$

where the expressions $P^1{}_\nu(x)$ are associated Legendre polynomials and $x = cos(\theta)$. $P_3^1(x)$ was not needed or used for the fits shown; however, $P_2^1(x)$, i.e. the inclusion of an $L_\pi = 2$ term, was essential to obtain a reasonable representation of the data at the four energies.

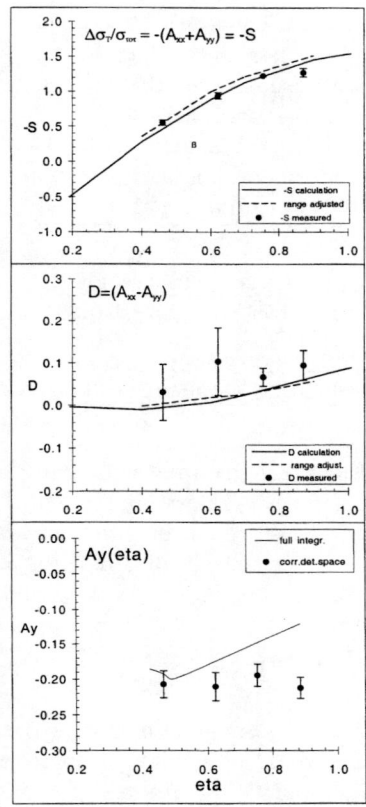
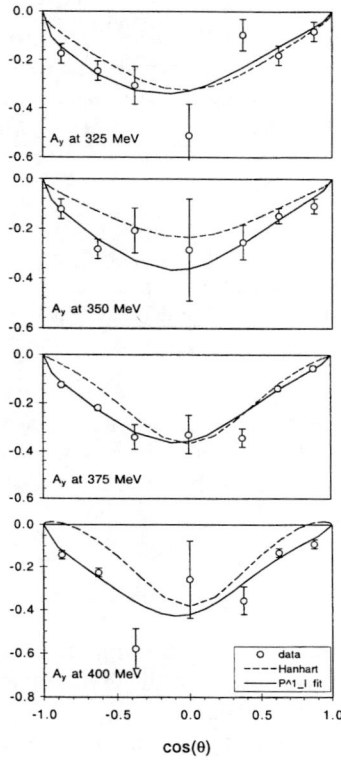

FIGURE 2. The integrated $\vec{p}\vec{p} \to pn\pi^+$ observables $\Delta\sigma'_T/\sigma_{\rm unpol} = -(A_{xx} + A_{yy})$, $D \equiv A_{xx} - A_{yy}$, and A_y as listed in Table 1 are shown as a function of η, the maximum pion center-of-mass momentum in units of the pion mass. The observables A_{xx}, A_{yy} and A_y were integrated over all kinematic variables except the pion azimuthal angle. Solid curves are from Hanhart et al.. These calculations are for the full phase space and are not always directly comparable with the data. The dashed curves approximate the theoretical calculations for the actual phase space observed.

FIGURE 3. The measured $\vec{p}\vec{p} \to pn\pi^+$ analyzing powers $A_y(\theta_{\rm c.m.}, T)$ compared with fits (solid curves) described in the text. The angular distributions suggest some asymmetry around $\theta_{\rm c.m.} = 90°$. Acceptable fits require the inclusion of an associated Legendre polynomial with $L_\pi = 2$. The detector size did not permit reliable measurements for $-0.25 < \cos(\theta_{\rm c.m.}) < 0.25$. The dashed curves are calculations by Hanhart et al.

TABLE 1. Product of beam and target polarization and the deduced integrated spin correlation coefficients. The observables listed are affected differently by the limited observed range for the polar angle θ. Model calculations show that the lack of counts for the range $-0.4 < \cos(\theta_{c.m.}) < 0.4$ makes the listed values for -S and D slightly larger than what a complete integral would have produced. The effect is more pronounced for A_y because $A_y(\theta_{c.m.})$ has a strong minimum near $\theta_{c.m.} = 90°$. The last column provides the full integral over the fits to $A_y(\theta)$ in Fig. 3.

T (MeV)	η	$P \cdot Q$	$-S \equiv -(A_{xx} + A_{yy})$	$D \equiv A_{xx} - A_{yy}$	A_y (full space, Fig.3)
325.6	0.464	0.456±0.003	0.553±0.038	0.031±0.066	-0.207± 0.019
350.5	0.623	0.342±0.004	0.942±0.046	0.104±0.080	-0.210± 0.020
375.0	0.753	0.514±0.004	1.226±0.011	0.067±0.021	-0.194± 0.016
400.0	0.871	0.526±0.006	1.274±0.019	0.095±0.035	-0.212±0.015

We have found that the recent calculations of the Jülich group are generally successful in predicting the integrated spin correlations for $\vec{p}\vec{p} \to pn\pi^+$ in spite of their difficulties with $\vec{p}\vec{p} \to pp\pi°$ [8]. However, an indication of remaining shortcomings is found in the difference between calculated and measured angular distributions for $A_y(\theta_{c.m.})$ in Fig. 3.

ACKNOWLEDGEMENTS

We acknowledge the support of Drs. M. Dzemidzic, F. Sperisen and D. Tedeschi in the early stages of the experiment. We thank Dr. W. Haeberli for his advice and continuing interest and J. Doskow for technical support. We are grateful to the authors of ref. [6] for making available to us calculations for $\vec{p}\vec{p} \to pn\pi^+$ obtained with their model.

REFERENCES

1. H.O. Meyer et al., Nucl. Phys. **A539** (1992) 633.
2. E. Korkmaz, Jin Li, D.A. Hutcheon, R. Abegg, J.B. Elliott, L.G. Greeniaus, D.J. Mack, C.A. Miller, N.L. Rodning, Nucl. Phys. **A535** (1991) 637. 1012. 454.
3. W.W. Daehnick et al., Phys. Rev. Lett. **74** (1995) 2913.
4. J.G. Hardie et al., Phys. Rev. **C56** (1997) 20. 213.
5. R.W. Flammang et al., Phys. Rev. **C58** (1998) 916.
6. C. Hanhart Ph.D. thesis, University of Bonn 1997, and C. Hanhart, J. Haidenbauer, O. Krehl, J. Speth, Phys. Lett. **B444** (1998) 25.
7. T. Wise, A.D. Roberts, and W. Haeberli, Nucl. Inst. Meth. **A 336** (1993) 410.
8. H.O. Meyer et al., Phys. Rev. Lett. **81** (1998) 3096.

Pion Production in pN Collisions near Threshold

Józef Złomańczuk

representing the PROMICE/WASA collaboration[1]

Department of Radiation Sciences, Uppsala University,
S-751 21 Uppsala, Sweden

Abstract. Measurements of the pp→ppπ^0 reaction at 310, 320, 340, 360, 400 and 425 MeV, and quasi-free pn→ppπ$^-$ production in pd collisions at 320 MeV have been carried out at the PROMICE/WASA facility at CELSIUS. The pp→ppπ^0 differential cross sections have been parametrised and used to deduce the poorly known σ_{01} total cross section through the relation: $\sigma_{01}=2\sigma(pn\to pp\pi^-)-\sigma(pp\to pp\pi^0)$. Preliminary results show the σ_{01} to be of the order of 50% of the σ_{11} for the ppπ$^-$ excess energy between 8 and 40 MeV.

In this contribution we report on new measurements of the quasi-free pn→ppπ$^-$ reaction in pd collisions at 320 MeV and of the pp→ppπ^0 reaction at 310, 320, 340, 360, 400 and 425 MeV carried out at the CELSIUS storage ring in Uppsala. The measurements were done using an electron-cooled proton beam and the PROMICE-WASA experimental set-up [1]. For the pp→ppπ^0 reaction, both protons were measured by the forward scintillator hodoscope and the tracker (FD), and the π^0 was reconstructed through the missing mass. Careful calibration of the FD elements resulted in narrow, nearly background-free, missing mass peaks, with a FWHM ranging from 2.2 MeV/c^2 at 310 MeV to about 7 MeV/c^2 at 425 MeV.

In the case of the quasi-free pn→ppπ$^-$ reaction, in addition to the two protons detected in the FD, a third charged particle was measured, either in the FD or in the CsI arrays covering angles from 30^0 to 90^0 on both sides of the target.

Once a pd→ppπ$^-$(p) event had been selected, it was possible to calculate the 4-momenta of the π$^-$ and spectator proton using the momenta of the fast protons and the π$^-$ angle. This allowed us to calculate the invariant mass of the ppπ$^-$ system, needed to find the excitation function, and reject events with large energy for the undetected proton. The latter reduced the contribution from a pd→pppπ$^-$ reaction with all three nucleons involved.

To evaluate the detector acceptance, we require a reasonable phenomenological description of the differential cross sections. In this work we have followed approach of Handler [2] to describe the matrix element of the pp→ppπ^0 and pn→ppπ$^-$ reactions. However, we had to add two extra states, Sd and Ds, to the Ss, Sp, Ps and Pp states used by Handler, in order to fit the experimental distributions. While the Sd state in

---

[1] http://www.tsl.uu.se/wasa

the pp→ppπ^0 reaction has already been observed [3], the Ds state has not been seen at near threshold energies. To account for the proton-proton final state interaction, we have used scattering wave functions calculated for the Paris potential [4] at a radius of 1 fm.

The results obtained for the pp→ppπ^0 reaction are shown in Figs. 1-2. The distributions have been normalised using existing data on the pp→pp total cross section. The parameters describing the matrix element were found at each beam energy independently by fitting the experimental distributions with Monte Carlo predictions. As one can see a fairly good agreement has been reached.

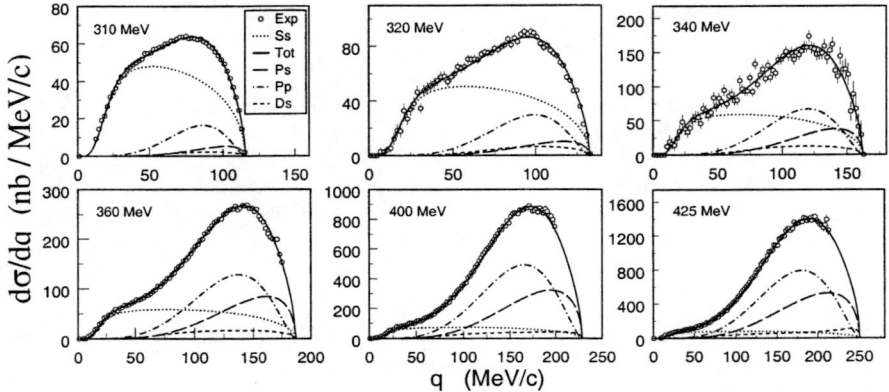

FIGURE 1. Acceptance-corrected distributions in the proton-proton relative momentum q at six beam energies for the pp→ppπ^0 reaction.

The distribution of the two-proton relative momentum (see fig. 1) is composed of contributions from various partial states having different shapes thus it may be used to find those contributions if their shapes are known.

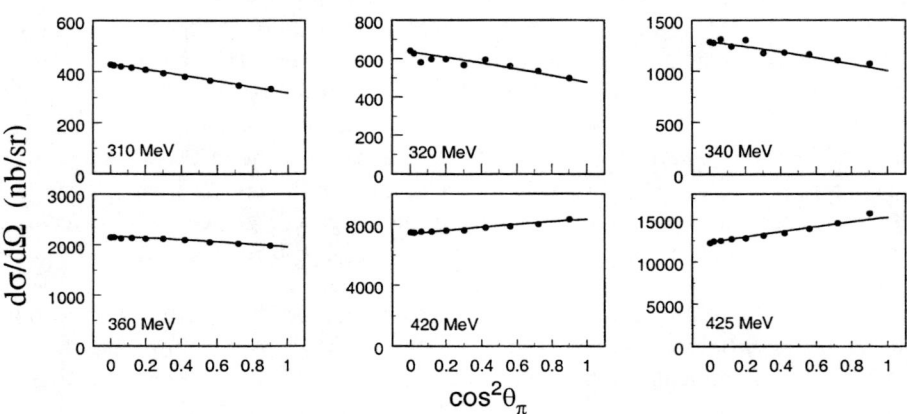

FIGURE 2. Acceptance-corrected distributions in the cosine^2 of the c.m. π^0 angle at six beam energies for the pp→ppπ^0 reaction.

The acceptance-corrected c.m. pion angle distributions presented in Fig. 2 clearly show a change from a negative slope at 310 MeV to a positive one at higher energies.

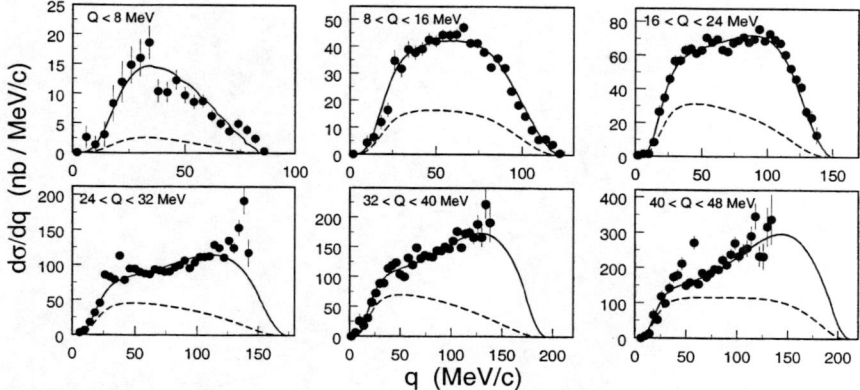

FIGURE 3. Acceptance-corrected distributions in the proton-proton relative momentum q at six c.m. energies for the pn→ppπ⁻ reaction.

The relative momentum distributions obtained for the quasi-free pn→ppπ⁻ reaction are shown in Fig. 3 for six bins of the ppπ⁻ excess energy (invariant mass − rest mass of the ppπ⁻ system). The solid lines represent $1/2(\sigma_{11}+\sigma_{01})$ and the dashed lines the $\sigma_{01}/2$ term. Since σ_{11} is given by the parametrisation of the pp→ppπ⁰ data and the experimental points are normalized with the pd elastic scattering, the values of σ_{01} can be deduced through the fit to the experimental data. As seen, the σ_{01} term provides about 30% of the cross section.

It should be stressed that the results presented here are preliminary and, especially at large values of q, there is some model dependence due to the acceptance. As a cross-check of the pn→ppπ⁻ results, we shall analyse the quasi-free data obtained at 310 and 340 MeV; values deduced for different Q-bins should be independent of the beam energy.

REFERENCES

1. Calén H. *et al.*, Nucl. Instrum. Methods, **A379**, 1996, p. 57.
2. Handler, R, Phys. Rev. **B138**, 1230 (1965).
3. Złomańczuk, J., Phys. Lett. **B436**, 251 (1998).
4. Loiseau, B., and Mathelitsch, L., Z. Pys. **A358**, 435 (1997); Lacombe, M. *et al.*, Phys. Rev. **C21**, 861 (1980).

MESON PRODUCTION FROM NUCLEI AND PRODUCTION OF HEAVY MESONS

Pion Production and Entropy Experiments at Storage Rings

B. Jakobsson[1] and the CHIC Collaboration[2]

[1]Department of Physics, Lund University, Box 118, SE 221 00 Lund, Sweden
[2]Bari-Catania-Copenhagen-Cracow-Lund-St.Petersburg-Uppsala

Abstract. The perspective for future experiments with large detector systems at storage rings, operating in slow ramping mode, is discussed. Data from slow ramping proton-nucleus and heavy ion collision experiments with more modest detector systems, on entropy estimation and charged pion emission from the absolute threshold to the energy region dominated by delta production, are presented and discussed.

INTRODUCTION

The perspective for so called excitation function experiments at storage rings, operating in slow ramping mode is exciting. Detector systems that can provide not only information on limited parts of the emission processes but provide at the same time a useful event characterization, are necessary to match this beam operation properly.

Slow Ramping

The CHIC collaboration has introduced a slow ramping operation mode of the CELSIUS storage ring, both for experiments with proton and heavy ion beams (1,2). This means that after a first injecton phase and a fast acceleration phase up to the energy corresponding to the requested starting point, a slower increase in beam frequency is adopted and data taking starts. Fig. 1 (left) shows that during this energy increase the number of stored ions are decreasing, both due to useful collisions in the target and due to non-useful losses, whereas the luminosity may either increase or decrease depending on the balance between these losses and the increase of the beam frequency.

When running in this mode we found it quite possible to flag the time in the beam cycle and thus the frequency when our aquisition signals an event trigger. The error in the collision energy determination is actually always < 1 MeV (per nucleon). This allowed us to measure "excitation function data" and a number of experiments were performed in this way utilizing slowly ramped beams of protons, ^{12}C, ^{14}N, ^{16}O and ^{20}Ne in the energy region 150 - 500 MeV (per nucleon). During the cause of these experiments we learned that nearly any kind of cycle can be produced at CELSIUS with very high degree of reproduceability and in Fig. 1 (right) we show examples of both normal and more exclusive ramping. It should be mentioned that in all these experiments we introduced gas-jet targets of thickness 10^{12} - 10^{14} atoms/cm^2.

Matching Detectors

The experiments which are discussed below have been performed with fairly simple external detectors. In both cases inclusive data are requested. It is however obvious that even if "excitation function data" still can provide interesting information, both hadron-nucleus and nucleus-nucleus collision data normally require an event selection at least with respect to impact parameter. In particular studies aiming at a better understanding of the nuclear equation of state or more specifically of the liquid-gas phase transition one needs to select a large enough sample of emission sources of similar size and with well determined excitation energy. Thus a detector as close as possible to a 4π one and with as large as possible coverage in momentum-space is ideal. This excludes an external system

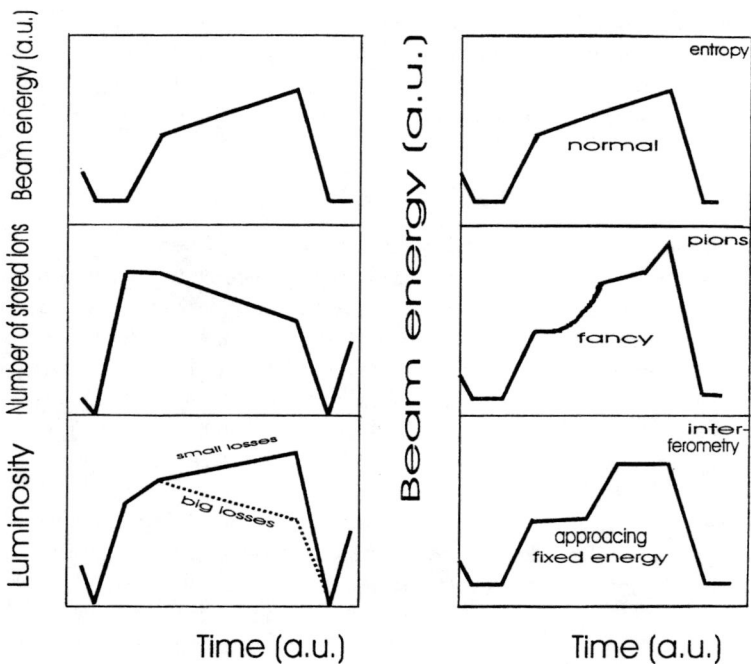

FIGURE 1. Beam energy, flux and luminosity as a function of time in a typical slow ramping cycle (left) and different kinds of slow ramping cycles (right).

since the gas-jet target environment is very occupied by pumps etc. An internal system has better spatial access, possibly 3π, but it must fulfil very delicate vacuum requirements at $\sim 1 \cdot 10^{-10}$ torr.

A high granularity detector of this kind, the CHICSi detector (3,4) has been constructed by the CHIC collaboration and it is now in assembling phase. This detector consisting of 570 Si and Si + GSO telescopes is presented in the report by Pavel Golubev (5).

ENTROPY EXPERIMENTS

Since the pion production experiments, discussed later, were performed with a set of plastic scintillator range telescopes, which registered p, d and t at the same time, unless hardware rejection was explicitly introduced, we could get entropy data "for free". This means a determination of the t/d/p emission ratios. The angular positions and the momentum coverage of the telescopes are shown in the $p_\parallel$ - $p_\perp$ plots of Fig. 2, that also presents simulations of protons in p + Kr and O + Kr experiments at 300 MeV (per nucleon). These impact parameter integrated event simulations are performed within a standard NMD, nuclear molecular model (6). The figures show that the data do represent well emission from an intermediate, strongly interacting source with small contributions from target (or projectile) evaporation.

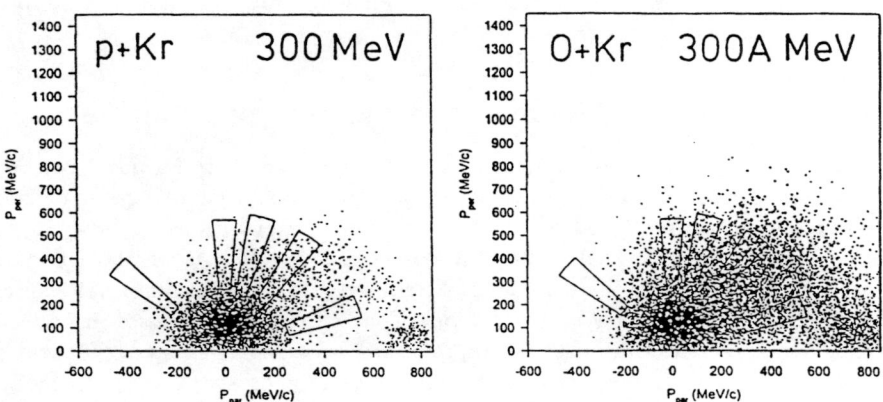

FIGURE 2. NMD simulations of protons in momentum space produced in 300 MeV p + Kr collisions (left) and 300 MeV/nucleon O + Kr collisions (right). The experimental coverage by the detector telescopes is indicated.

Entropy in nuclear collisions

The entropy(S), as it is defined by the equation of state in a classical, system in thermal equilibrium (temperature, T)

$$\frac{1}{T} = (\frac{dS}{dE})_{N,V} \tag{1}$$

or more general in a quantum system,

$$S = ln\rho(E, V, N) \tag{2}$$

where E, V and N stand for energy, volume and nucleon number of the closed system, is of course not directly measurable. A first attempt to calculate the entropy from particle ratios was made by Siemens and Kapusta (7),

$$\frac{S}{A} = \frac{5}{2} - ln(2^{2/3} <\rho_N>) = 3.95 - ln(R_{dp}) \tag{3}$$

where R_{dp} is the ratio of the total number of deuterons to protons from the emission source. This result, valid for an expanding Fermi-Dirac system, was however found to overestimate the entropy vastly in heavy ion collisions (8) and indeed also our d/p results in p + Kr show strong overestimation. Actually the entropy according to formula (3) gives a value of S increasing with beam energy from 6 to 6.5 if E > 30 MeV particles are selected and from 7 to 9 if the more extreme high energy d production is considered by an, E > 30 MeV/nucleon condition. The necessity to go to complete model calculations, containing both dynamics and particle production, was first pointed out by Gelbke (9) and we have now adopted this concept in an extended way.

Complete models for light particle production

We compare our "excitation function data" on p, d and t production to various models where nucleons interact both with a mean-field and with other nucleons individually. The basic model, which would be the most relevant one for determining entropy directly provided that freeze-out conditions can be set, is the EES (expanding evaporative statistical) model of Friedman (10,11). This contains sequential evaporation of classical Weisskopf type affected by expansion and in addition secondary evaporation from unstable fragments. Also a more classical evaporation model (12) is introduced for comparison. The second (microscopic) type of models is represented by the Gudima, Toneev intranuclear cascade (INC) approach (13,14). This contains first pre-equilibrium emission through the exciton prescription, then successive quasi-free NN scattering in a local Thomas-Fermi potential and finally evaporation from excited residues. Mean-field models are represented by the NMD model (6), that contains classical molecular dynamics + residue evaporation. The detailed description of the potentials and equation of state in this prescription are given in ref. (15).

It should be stressed that the statistical models provide composite particles in a completely "natural" way, whereas it has been suggested that additional final state interaction may be necessary for the latter two kinds of models. The coalescence prescription is suggested, i.e. to combine nucleons into d or t whenever they fall inside a momentum space sphere given by a characteristic radius (p_o) for strong interaction,

$$\frac{d^3\sigma}{dp^3} = \frac{4\pi}{3}\frac{\gamma p_o^3}{\sigma_r}(\frac{d^3\sigma_N}{dp^3})^2 \qquad (4)$$

where $d^3\sigma_N/dp^3$ stands for the differential nucleon cross section. Originally this was taken to be the proton cross section but later it was found that for deuterons it is necessary to use the product of the proton and neutron cross sections since they differ substantially.

Light particle production data

In Fig. 3 we present as histograms the experimental d/p sideways emission ratio in p + Kr reactions (left) and Ne + Ar reactions (right). In all cases a smooth increase in the d/p ratio with increasing beam energy is observed which does not call for a dramatic change of the emission processes in the interval 100 - 500 MeV (per nucleon). The p + Kr figure shows a very large difference in deuteron production between the energy cut, E > 30 MeV and the "velocity" cut, E > 30 MeV/nucleon. All curves refer to INC calculations. It should however immediately be mentioned

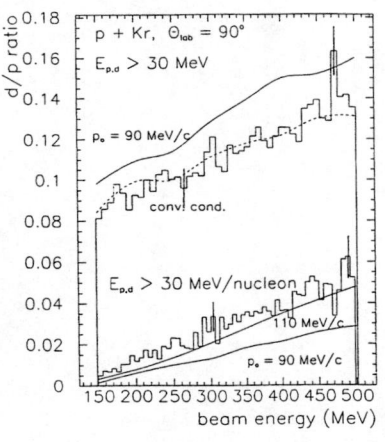
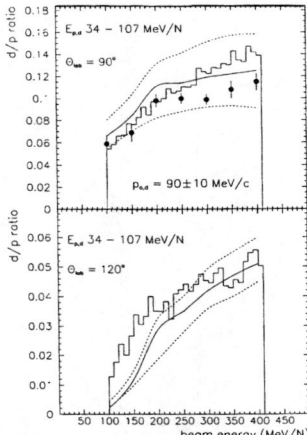

FIGURE 3. The d/p emission ratio in p + Kr collisions (left) and in Ne + Ar collisions (right). Data are represented by the histograms. The curves represent calculations, described in the text.

that these contain coalescence as an artificial extra final-state interaction. Without coalescence the INC predictions show a strong underestimation of the deuteron production. Before going into details about the comparison with the INC calculations it should be mentioned that:

i) the EES model cannot reproduce the p + Kr data whereas the Ne + Kr can be reproduced rather well.

ii) the pure evaporation model fails to predict also the Ne + Kr data.

iii) the NMD model strongly underestimates the deuteron production until coalescence is introduced.

The result iii) thus confirms that even if there is a natural fast deuteron production mode, as in the molecular dynamics prescription, some very strong extra final state interaction is necessary to reproduce the data. The EES result can be interpreted in terms of the difficulties in selecting correctly the preformed source size and excitation. In the p + Kr reactions, excitation energy from 2 to 6 MeV/nucleon is available and this forces the selected production mode to become evaporation from compound nuclei with no energy for compression and therefore nearly no expansion. This is consequently not the process to produce high energy deuterons. In the Ne + Ar case both expanding sources ($E^*/A > 10$ MeV) and pure evaporative sources ($E^*/A < 5$ MeV) are produced as well as the whole region in between and this appears to be the right kind of mixture.

We now return to coalescence. If the fundamental formula (4) gives a proper description of the composite particle production, then the d/p ratio and the t/d ratio should be the same. Spin degeneracy factors should however enter and create a difference through the difference between the $(2S_d + 1)^2$ term and the $(2S_p + 1) \cdot (2S_t + 1)$ term. Thus the d/p ratio should ideally be 9/4 times larger than the t/d ratio. Experimental results from O + Kr collisions in the 80 - 300 MeV/nucleon beam energy interval, are presented, again as histograms in Fig. 4.

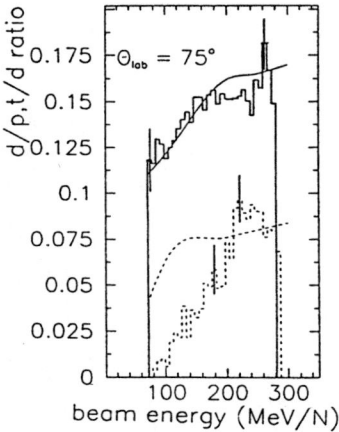

FIGURE 4. The d/p and t/d emission ratio at 75° in O + Kr collisions. Histograms show data and curves calculations, described in the text.

The curves show again INC calculations that include coelescence with a momentum radius, $p_o = 90$ MeV/c. The agreement with the d/p ratio is again excellent but now we observe that although the t/d ratio falls at the right level it appears as if high energy tritons are more difficult to produce in the low collision energy region, say up to 200 MeV/nucleon. For the moment this remains unexplained but it is a good example of the strong predicting power of "excitation function data".

PION EMISSION EXPERIMENTS

Methods of measuring π^+ by means of stopping them in sandwich range telescopes and then identifying either the fast $\pi\mu$ decay or the slow μe decay are classical (16,17). In our experiments we used the fast decay and observed by pulse shape discrimination technique the delayed muon energy signal in the stop detector of the plastic scintillator telescope. This method has high efficiency but can be used only for a limited range of pion energies, between ~ 10 and 80 MeV. It is therefore best adopted for a beam energy region in between the extreme threshold region, where the lowest pion energies dominate, and the high collision energy region, where the contribution from pions with $E > 80$ MeV starts to dominate. It should immediately be said that when comparing total yields we adopted the process of "empirically normalized BUU estimations" of the lowest and highest energy pions that is described in ref (18). In order to measure π^-, the decay method is impossible since these pions are rapidly absorbed by the detector nuclei, leading to decay through evaporation. In order to obtain the π^- yield one must instead use the prompt energy signals in all detectors preceeding the stop detector (three detectors required for a trigger) to identify charged pions and then subtract the π^+ contribution. This gives substantially larger uncertainties for π^- cross sections and it also restricts the useful data to those stop detectors where the $\Delta E - \Delta E$ correlations give acceptable resolution. Fig. 5 exploits both these deficiencies in that only the 44 - 70 MeV pion energy interval is accepted and that the fluctuations are substantially larger for π^- than for π^+ (error bars are not introduced because they make the figure very diffuse).

Excitation function data

Fig. 5 shows the expected π^+ excess for all emission angles between 55° and 150°. The excess is substantially smaller than a straightforward weighting of isospin decomposed $NN \rightarrow NN\pi$ contributions would introduce. However the Coulomb interaction with the proper mean-field accounts for this reduced excess. The absolute normalization, which was not necessary in the previous experiment where we always measured a ratio between yields of two particles, is delicately dependent on the possibility of registrating a monitor particle with well known cross section. Details about how to normalize can be found in ref. (2). The spectrometer data at 580 MeV (19) confirm to some extent our results even if the backward points indicate a larger π^+ to π^- ratio. A detailed estimate of the expected differences between

π^+ and π^- requires however a complicated three-body Coulomb calculation which has not yet been made.

"Excitation function data" of the kind we discuss here put very strong constraints on the models that want to explain them. This is e.g. demonstrated by the comparison in Fig. 6 between our collected differential, $d\sigma/d\Omega$ π^+ production data for proton induced collisions and the QMD calculations (6,15). The QMD approach is that of classical Ehrenfest's equation of motion describing the wave packets moving in a soft Skyrme potential,

$$H = \sum_{i=1}^{A} \frac{\vec{p}_i^2}{2m} + \frac{1}{2} \sum_{i=1}^{A} V_{Skyrme}(\vec{r}_i) + \frac{1}{2} \sum_{i=1}^{Z} \sum_{k=1, i \neq k}^{Z} \frac{e_1 e_2}{|\vec{r}_i - \vec{r}_k|} \tag{5}$$

When two nucleons are closer to each other than $\sqrt{\sigma_{eff}/\pi}$ they scatter isotropically. Absorption is introduced through a simple optical mean-free-path adopted to Coulomb trajectories of pions in the field of the nucleus. With these ingredients the points in Fig. 6 are obtained. The agreement is surprisingly good with one exception - the backward emission (150°, only measured for p + Kr) is grossly underestimated. There have been many speculations about the reason, one being the necessity to modify the isotropic scattering assumption, but so far no satisfactory explanation has been found. Other attempts to describe these "excitation function data" have been made with BUU (Boltzmann, Uhling-Uelenbeck equations) models (18) and INC models (13,14). From these attempts it is worth mentioning

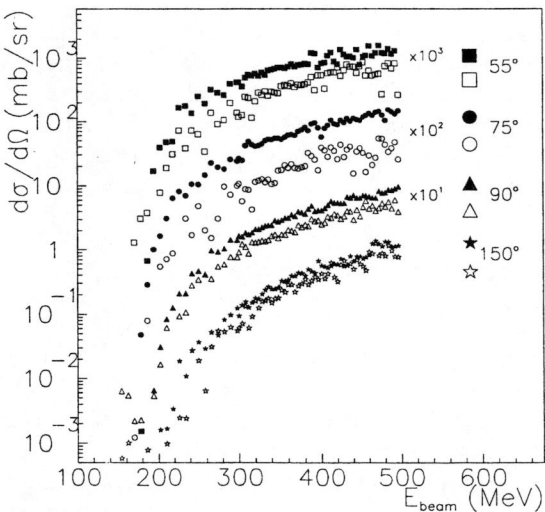

FIGURE 5. Differential cross sections, $d\sigma / d\Omega$, of π^+ and π^- with energies 44 - 70 MeV from p + Kr collisions. The points at 580 MeV beam energy are from ref. (19).

noindent that the BUU calculations have proven the necessity to introduce a proper (degenerate) Fermi distribution for the nucleons in the target nucleus in order to explain the (sub)threshold pion emission. The INC is very succesful in explaining the 90°, p + Ar data (20) but it has not yet been extended to all data and especially not to the backward data where QMD fails.

As mentioned before, total yields must be estimated via semi-empirical extrapolations both below 10 MeV and above 80 MeV. Examples of such yields, from p + Ar and p + N collisions are shown in Fig. 7. Possible comparisons to previous data are scarce but some attempts to extract useful points are made in this figure. Thus it appears as if the agreement is quite good for all higher beam energies and possibly also for the lowest energies. It should be stressed that comparing p + N with p + C should result in some 12% lower yields of π^+ for the latter reaction for natural reasons. There is however a region between, say 200 and 250 MeV where the spectrometer data appear to be significantly lower. Actually it has recently also been pointed out that there is for 200 MeV proton induced reactions a general systematic inconsistency in the available data of the three charge states of pions (21). Even though Coulomb effects are important the order of the yields should from fundamental isospin considerations be π^+ largest, then π^0 and π^- smallest. It appears as if systematic errors in the different measurements methods still are too large even to agree on this order in the data which of course calls for more experiments with larger efforts to set the absolute normalization correctly.

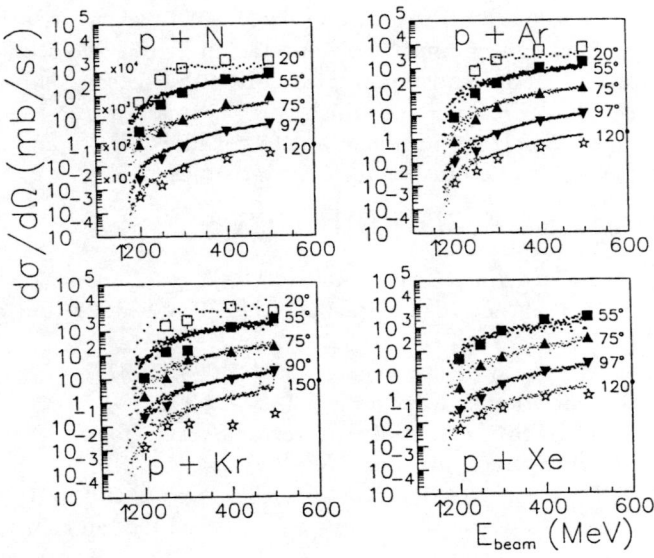

FIGURE 6. Differential cross sections, $d\sigma / d\Omega$, of π^+ and π^- with energies 10 - 80 MeV from p + N, Ar, Kr, Xe collisions. The points come from QMD calculations (6,15).

PERSPECTIVES

We have presented "excitation function data" on entropy and pion production in proton-nucleus and heavy ion collisions from 100 to 500 MeV (per nucleon), collected during slow ramping operation of the CELSIUS ring. The constraints on the models, claiming to explain completely the emission processes, are much higher than in conventional fixed target experiments and in some cases there are even indications of a necessary change from one model to another.

These data are obtained with fairly simple external detectors. The need for a high granularity detector system with maximum coverage in space and momentum space, in order to perform event characterization is obvious and such an internal telescope system, the CHICSi detector (4,5), is under assembly.

The slow ramping cycle could be of quite different nature depending on the physics question that a CHICSi experiment is asking. One could for instance imagine that an interferometry experiment is to be set up in order to study the space-time evolution of the emitting sources in a heavy ion collision. Such experiments require very large statistics in order to analyse the lowest relative momentum region of the two-particle correlation functions, especially if simultaneous nn, np and pp correlations are to be measured as in standard CHIC interferometry experiments (22). Thus it is hardly possible to cover a wide range in energy but it is still interesting to compare data from a few energies. In this case step-like ramp of the kind shown in Fig. 1 (lower, left) should be utilized. A completely different kind of ramp, like the one presented in Fig 1 (upper, left) should be asked for if e.g. a phase-transition like behaviour is expected. A search for the liquid-gas transition may be undertaken by measuring either event-by-event mass distributions of all fragments or by determining the connection between two statistical parameters, like excitation energy and temperature. In either case it is important that the initial source which breaks up or which is in thermal and/or chemical equilibrium is well defined. The combination of a detector like CHICSi and slow ramping operation of a storage ring seems ideal for such studies.

REFERENCES

1. Fokin, A. et al., *Phys. Rev.* **C60**, 024601 (1999).
2. Martensson, J. et al., *Lund U. preprint* **LUIP9804**, 1998 and submitted to *Phys. Rev. C*.
3. Evensen, L. et al., *IEEE Trans. Nucl. Sci.* **44**, 629 (1997).
4. Jakobsson, B., Nucl. Phys. News Int. **9:2**, 22 (1999).
5. Golubev, P., "CHICSi - a 3π Multi Detector System for Studying Heavy Ion Interactions at Storage Rings", in *these Proceedings*.
6. Bondorf, J.P., Idier, D. and Mishustin, I., *Phys. Lett* **B359**, 261 (1995).
7. Siemens, P. J.,and Kapusta, J. I., *Phys. Rev. Lett* **43**, 1486 (1979).
8. Nagamiya, S. et al., Phys. Rev. **C24**, 971 (1981).
9. Gelbke, C. K., in *Proceedings of the Int. Conf. on Nucleus-Nucleus Collisions*

Michigan Strate U., 1982, pp 473c.
10. Friedman, W., *Phys. Rev. Lett* **60**, 2125 (1988).
11. Friedman, W., *Phys. Rev.* **C40**, 2055 (1988) and **C42**, 667 (1990).
12. Duran, D., Private Communication.
13. Toneev, V. D. and Gudima K. K. *GSI Preprint* **GSI-93-52**, 1993.
14. Gudima, K.K., Mashnik, S. G. and Toneev, V. D., *Nucl. Phys* **A401**, 329 (1983).
15. Fokin, A., Thesis, U. Lund, 1998.
16. Sanouillet, G. et al., *CEN Saclay Preprint* **CEA-N-248**, 1986.
17. Norén, B. et al., *Nucl. Phys.* **A489**, 763 (1988).
18. Jakobsson, B., et al., *Phys. Rev. Lett.* **78**, 3828 (1997).
19. Crawford, J. F. et al., *Phys. Rev.* **C22**, 1184 (1980).
20. Gudima, K. K. and Ploszajczak, M., *GANIL Preprint* **P9817**, 1998.
21. Bellini, V. et al., *Z. Physik* **A333**, 393 (1989).
22. Ghetti., R. et al., *Nucl. Instr. and Meth.* **A335**, 156 (1993).

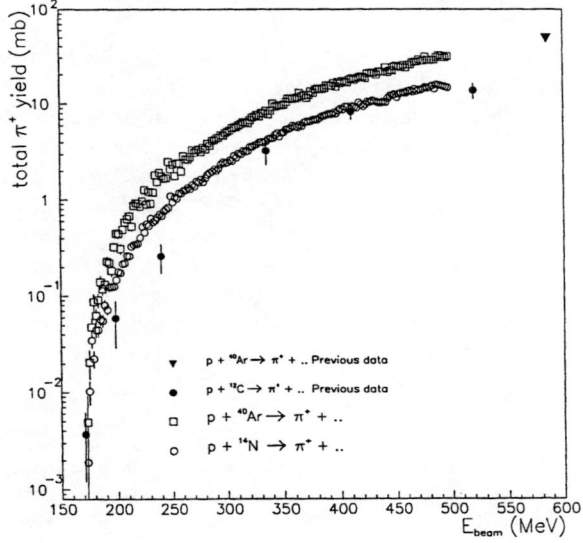

FIGURE 7. Total cross sections of π^+ from p + N and p + Ar collisions. Open points and squares refer to the slow ramping experiments, the closed triangles to spectrometer p + Ar data and the solid points to p + C spectrometer data.

Meson Production in p+d Reactions[1]

GEM Collaboration

M. Betigeri [i], J. Bojowald [a], A. Budzanowski [d], A. Chatterjee [i],
J. Ernst [g], L. Freindl [d], D. Frekers [h], W. Garske [h], K. Grewer [h],
A. Hamacher [a], J. Ilieva [a,e], L. Jarczyk [c], K. Kilian [a],
S. Kliczewski [d], W. Klimala [a,c], D. Kolev [f], T. Kutsarova [e],
J. Lieb [j], H. Machner [a], A. Magiera [c], H. Nann [a,2], L. Pentchev [e],
H. S. Plendl [k], D. Protić [a], B. Razen [a], P. von Rossen [a], B. J. Roy [i],
R. Siudak [d], J. Smyrski [c], R. V. Srikantiah [i], A. Strzałkowski [c],
R. Tsenov [f], K. Zwoll[b]

a. Institut für Kernphysik, Forschungszentrum Jülich, Jülich, Germany
b. Zentrallabor für Elektronik, Forschungszentrum Jülich, Jülich, Germany
c. Institute of Physics, Jagellonian University, Krakow, Poland
d. Institute of Nuclear Physics, Krakow, Poland
e. Institute of Nuclear Physics and Nuclear Energy, Sofia, Bulgaria
f. Physics Faculty, University of Sofia, Sofia, Bulgaria
g. Institut für Strahlen- und Kernphysik der Universität Bonn, Bonn, Germany
h. Institut für Kernphysik, Universität Münster, Münster, Germany
i. Nuclear Physics Division, BARC, Bombay, India
j. Physics Department, George Mason University, Fairfax, Virginia, USA
j. Physics Department, Florida State University, Tallahassee, Florida, USA

Abstract. Pion and η production on the deuteron are studied at energies in the vicinity of the absolute threshold. The data are expected to be sensitive to high momentum components in the deuteron wavefunction as well as to two step processes.

The study of meson production on the deuteron is the first step towards an understanding of meson production on nuclei and the related in–medium effects. In a first approximation, the reaction can be assumed to be mainly a nucleon–nucleon reaction with the second nucleon in the deuteron being a mere spectator [1]. However, the struck nucleon has a Fermi motion and thus an effective mass. In addition rescattering on the spectator nucleon may take place. All these processes

---

[1] presented by H. Machner, e-mail: h.machner@fz-juelich.de
[2] on leave from IUCF, Bloomington, Indiana, USA

do not take place reactions on the nucleon.

Here we will concentrate on π and η meson production, i. e. the reactions

$$p + d \to {}^3He + \pi^0, \qquad (1)$$
$$p + d \to {}^3He + \eta. \qquad (2)$$

The underlying elementary process for both reactions is the $p + n \to d + meson$ reaction. In both reactions, resonances play an important role: in case of the η it is the $N^*(1535)$ which is an s-wave, while in the π channel it is the $\Delta(1232)$, a p-wave. From a comparison of both reactions, one may therefore learn how different boundary conditions influence the reaction mechanism. In the experiments for reaction 1, the isospin related

$$p + d \to {}^3H + \pi^+ \qquad (3)$$

was simultaneously measured. The experiment was performed at the external pro-

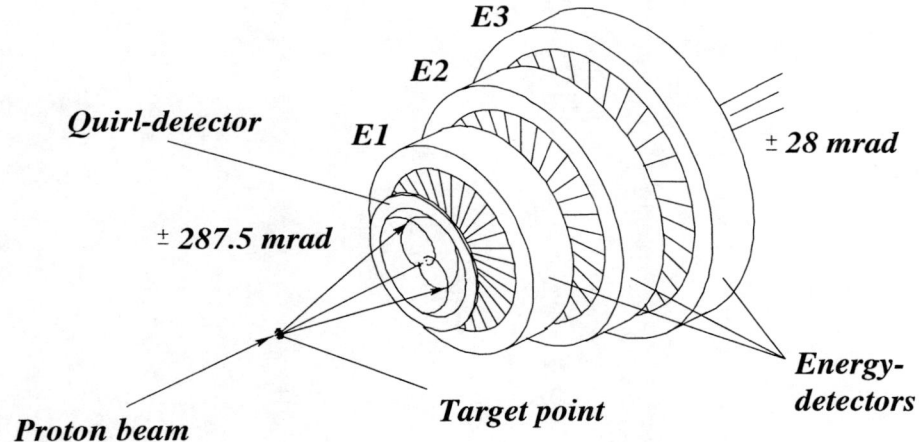

FIGURE 1. The detector system "germanium wall". In the present measurement, detector E2 was removed.

ton beam of the COSY accelerator. Because of its storage mode, one can extract the beam by a stochastic method in time periods over several seconds, thus avoiding pile-up as well as dead time. The beam was focused to the center of a target cell, containing liquid deuterium with dimensions 6 mm in diameter and thicknesses from 2 to 6.5 mm in the different runs, respectively. Recoiling baryons were detected in an angular range in the laboratory system between 30 to 250 mrad by a high purity germanium stack detector called the germanium wall (see Fig. 1). Particles emitted into smaller angles were identified with a magnetic spectrograph. This setup has full acceptance for threshold measurements in the center of mass system.

The first detector has 40000 pixels formed by 200 Archimedes' spirals bent to the left and right on the front and rear side, respectively. The calorimeter detectors which follow have structures of 32 wedges on one side. More details on the detector can be found elsewhere [3]. Due to its high granularity, the germanium wall is well suited for high luminosities. However, in the early runs for reaction 2, a large halo prevented making use of this feature. The halo has now been reduced to the 1% level with the beam not cooled. Such were the conditions for the measurements of the 1 and 3 reactions. data were taken at beam momenta ranging from 750 Mev/c up to 1050 MeV/c in steps of 50 MeV/c, thus covering the full range of the Δ-excitation. For η production a momentum of 1673 MeV/c was chosen which corresponds to the pole position of the N^* resonance. The information deduced

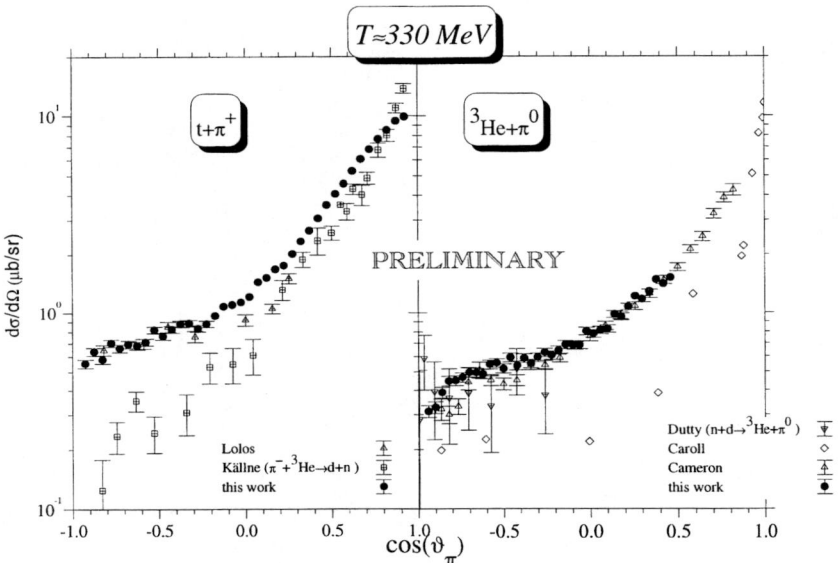

FIGURE 2. Angular distribution for pion from the two reactions. The presently obtained cross sections are compared with previous data for approximately the same beam energy ([4–6]), or deduced from pion absorption employing detailed balance (Ref. [7]). For the data from Dutty et al. [8] charge symmetry is assumed.

from the germanium wall are energy, emission vertex and particle type. The energy information for a given exit channel can be transformed into the emission angle in the centre of mass system without making use of the additional measured quantities. However, possible background can not be subtracted by this method. Therefore, the missing mass of the unobserved pion is extracted by making use of all measurements together with the knowledge of the initial state and applying conservation of momentum, energy, charge and baryon number.

The measured differential cross sections in the centre of mass system for pion production at a beam momentum of 850 MeV/c are shown in Fig. 2, together with earlier measurements close to that beam momentum. The angular distributions for both reactions show a backward peaking of the $A = 3$ nuclei which corresponds to a forward peaking of the pion. This is also found in the other data to which the present results are compared with. The flat part seen in the range $\cos(\theta) \leq 0$ is not found in the time–reversed and isospin–related data from Källne et al. [7]. It is again this region for the $^3He + \pi^+$ channel where the data from Ref. [5] do not agree with the present results. The enhancement is predicted by a two–step process [11,12], where the resultant pion scatters on the spectator nucleon. It is obvious that none of the earlier measurements has the good statistics of the present results. We have fitted Legendre polynomials to the angular distributions. Polynomials up to 4th order were found necessary to achieve reasonable χ^2 values.

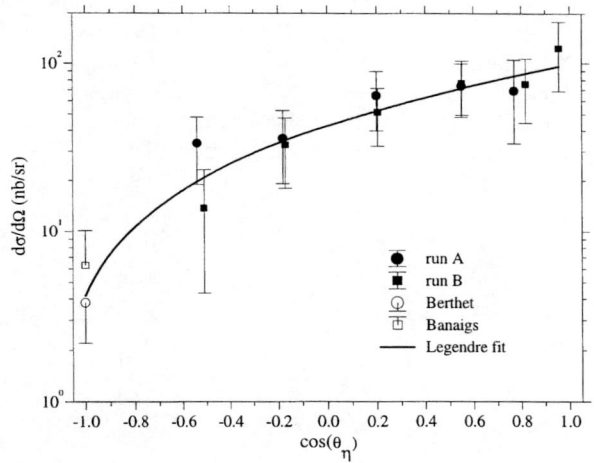

FIGURE 3. Angular distribution of the reaction $p + d \to {}^3He + \eta$. Added are two points from Ref's [9,10]. A Legendre polynomial fit is shown as solid curve.

In Fig. 3 are the results for the η production shown. Results from two different runs agree nicely with each other. Two data points for backward emission from Ref.'s [9,10] are added. The total cross section is extracted from a 2. order Legendre polynomial fit, also shown in the Figure. It should be mentioned that this measurement is still subthreshold in the nucleon–nucleon system, although quite far above the absolute threshold. The large momentum transfer of $\Delta p = 706$ MeV/c, which is nearly double than in the pion reaction, corresponds to a distance of only 0.27 fm. The reaction is, therefore, extremely sensitive to the high momentum components in the deuteron wave function.

We are grateful to the COSY operation crew for their efforts making a good beam. Support by BMBF Germany (06 MS 568 I TP4), Internationales Büro des BMBF (X081.24 and 211.6), SCSR Poland (2P302 025 and 2P03B 88 08), and COSY Jülich is gratefully acknowledged.

REFERENCES

1. M Ruderman, Phys. Rev. **87** (1952) 383.
2. V. Jaeckle, K. Kilian, H. Machner, Ch. Nake, W. Oelert, P. Turek, Nucl. Instruments and Methods in Physics Research A 349. (1994) 15
3. M. Betigeri et al., Nuclear Instr. Methods in Physics Research **421** (1999) 447.
4. G. J. Lolos et al., Nucl. Phys. **A 386** (1982) 477.
5. J. Carol et al., Nucl. Phys. **A 305** (1978) 502.
6. J. M. Cameron et al., Nucl. Phys. **A 472** (1987). 718
7. J. Källne et al., Phys. Rev. **C 24** (1981) 1102.
8. W. Dutty, Diploma thesis, Freiburg (1981), E. Rössle et al., Pion Production and Absorption in Nuclei, R. D. Bent (ed.) AIP Conf. Proc. **79** (1982) 171.
9. J. Banaigs et al., Phys. Lett. **45B** (1973) 394.
10. P. Berthet et al., Nucl. Phys. **A443** (1985) 589.
11. M. P. Locher and H. J. Weber, Nucl. Phys. **B76** (1974) 400.
12. K. Kilian and H. Nann, AIP Conf. Proc. No. 221 (1990) 185

Heavy Meson Production at COSY - 11

P. Moskal[a], H. -H. Adam[b], J. T. Balewski[c,d,e],
A. Budzanowski[c], C. Goodman[e], D. Grzonka[d], L. Jarczyk[a],
M. Jochmann[f], A. Khoukaz[b], K. Kilian[d], P. Kowina[g],
M. Köhler[f], T. Lister[b], W. Oelert[d], C. Quentmeier[b],
R. Santo[b], G. Schepers[b,d], U. Seddik[h], T. Sefzick[d], S. Sewerin[d],
J. Smyrski[a], A. Strzałkowski[a], M. Wolke[d], P. Wüstner[f]

[a] Institute of Physics, Jagellonian University, PL–30059 Cracow, Poland
[b] IKP, Westf. Wilhelms-Universität, Wilhelm-Klemm-Straße 9, D–48149 Münster, Germany
[c] Institute of Nuclear Physics, ul. Radzikowskiego 152, PL-31-342 Cracow, Poland
[d] Institut für Kernphysik, Forschungszentrum Jülich, D–52425 Jülich, Germany
[e] IUCF, Milo B. Samson Lane, Bloomington, IN 47405, USA
[f] ZEL, Forschungszentrum Jülich, D–52425 Jülich, Germany
[g] Institute of Physics, Silesian University, PL-40-007 Katowice, Poland
[h] Egyptian Atomic Energy Authority, 101 Sharia Kaser El-Aini, 13759 Cairo, Egypt

Abstract. The COSY-11 collaboration has measured the total cross section for the $pp \rightarrow pp\eta'$ and $pp \rightarrow pp\eta$ reactions in the excess energy range from Q = 1.5 MeV to Q = 23.6 MeV and from Q = 0.5 MeV to Q = 5.4 MeV, respectively. Measurements have been performed with the total luminosity of 73 nb^{-1} for the $pp \rightarrow pp\eta$ reaction and 1360 nb^{-1} for the $pp \rightarrow pp\eta'$ one. Recent results are presented and discussed.

INTRODUCTION

The word heavy used in the title requires a short explanation. The reason is rather historical, and it seems now that heavy are all mesons but not pions. The talk will concern the production of the η and η' mesons, and since η' is even heavier than η the discussion concerning this meson will constitute the major part of the presentation.

Last year, for the first time total cross sections for the production of the η' meson in the collision of protons close to the reaction threshold have been published [1,2]. Two independent experiments performed at the accelerators SATURNE and COSY have delivered consistent results.

The first remarkable inference derived from these experiments was that the total cross sections for the $pp \rightarrow pp\eta'$ reaction are by about a factor of fifty smaller than the cross sections for the $pp \to pp\eta$ reaction at the corresponding values of the excess energy. Trying to explain this large difference Hibou et al. [1] showed that the one-pion-exchange model with the parameters adjusted to fit the total cross section for the $pp \to pp\eta$ reaction underestimate the η' data by about a factor of two. This discrepancy suggests that short-range production mechanisms as for example heavy meson exchange, mesonic currents [3], or more exotic processes like the production via a fusion of gluons [4] may contribute significantly in the creation of η and η' mesons [5]. Especially that the momentum transfer required to create these mesons is much larger compared to the pion production, and already in case of pions a significant contribution from the short-range heavy meson exchange is necessary in order to obtain agreement with the experiments [6,7].

The second interesting observation was that the energy dependence of the total cross section for the $pp \to pp\eta$ and $pp \to pp\eta'$ reactions does not follow the predictions based on the phase space volume and the proton-proton final state interaction, which is the case in the π^0 meson production [8,9]. Moreover, for η and η' mesons the deviation from this prediction were qualitatively different. Namely, the close to threshold cross sections for the η meson are strongly enhanced compared to the model comprising only the proton-proton interaction [10] in contrary to the observed suppression in the case of the meson η'. The energy dependence of the total cross section for the $pp \to pp\eta$ reaction can be, however, explained when the η-proton attractive interaction is taken into account [11,12]. Albeit η-proton interaction is much weaker than the proton-proton one (compare scattering length $a_{p\eta} = 0.751$ fm $+ i$ 0.274 fm [13] with $a_{pp} = -7.83$ fm [14]) it becomes important through the interference terms between the various final pair interactions [12]. By analogy, the steep decrease of the total cross section when approaching a kinematical threshold for the $pp \to pp\eta'$ reaction could have been explained assuming a repulsive η'-proton interaction [15,16]. This interpretation, however, should rather be excluded now in view of the new COSY-11 data which will be presented in the next chapters.

POSSIBLE PRODUCTION MECHANISMS

The theoretical studies of the mechanisms accounting for the π^0 and η mesons creation in the close to threshold $pp \to pp\pi^0(\eta)$ reactions have shown that the short-range component of the N-N force and the off-shell pion rescattering dominate the production process of the π^0 meson [6,7,17], whereas the η meson is predominantly produced through the excitation of the intermediate baryonic resonance [18–21,10]. However, the comparison of the experimentally determined η and η' total cross section ratio with the predictions based on the one-pion-exchange model indicates that we are still far from the full understanding of the dynamics of the discussed processes. In particular, at present there is not much known about the relative contribution of the possible reaction mechanisms to the production of the meson η'. It is expected that similarly as in the case of pions the η' meson can be produced as depicted in Figures 1a,b,c,d. However, because of the much larger four-momentum transfer, short-range mechanisms, like heavy meson exchange (Figure 1c) or depicted in Figure 1e production via a mesonic current, where the η' is created in a fusion of exchanged virtual ω, ρ, or σ mesons shell contribute even more significantly. Recently Nakayama et al. [3], studied contributions from the *nucleonic* (Fig. 1b), *nucleon resonance* (Fig. 1d), and *mesonic* (Fig. 1e) currents and found that each one separately could describe the absolute values and energy dependence of the close to threshold η' data points [1,2] after an appropriate adjustment of the ratio of the pseudoscalar to the pseudovector coupling. This rather pessimistic conclusion means that it is not possible to judge about the mechanisms responsible for the η' meson production from the total cross section alone.

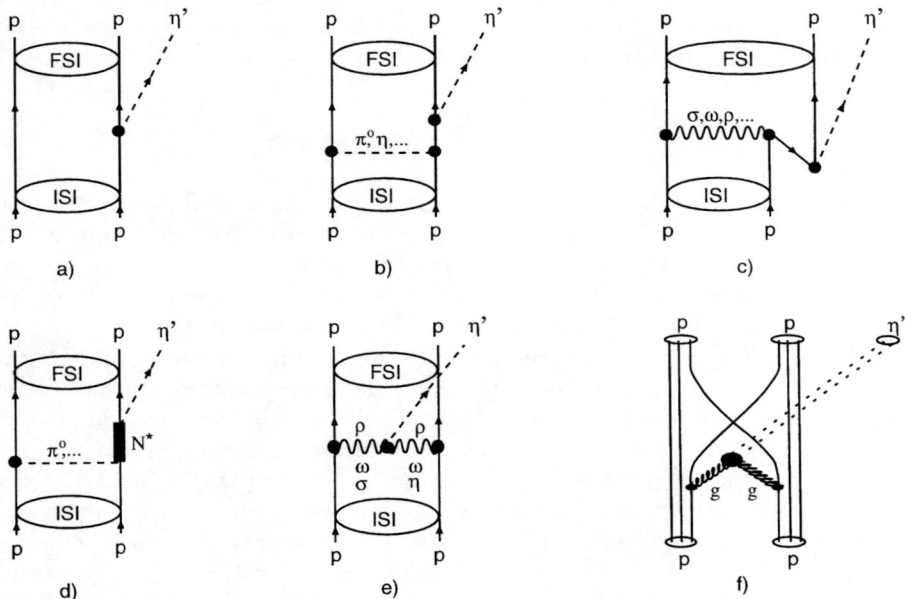

FIGURE 1. Diagrams for the $pp \rightarrow pp\eta'$ reaction near-threshold: (a)— η'-bremsstrahlung (nucleonic current) (b)— "rescattering" term (nucleonic current) (c)— production through the heavy meson exchange (d)— excitation of an intermediate resonance (nucleon resonance current) (e)— emission from the virtual meson (mesonic current) (f)— production via a fusion of gluons (gluonic current).

Moreover, the possible gluonium admixture in the meson η' makes the study even more complicated but certainly also more interesting. Figure 1f depicts appropriate short-range mechanism which may lead to the creation of the flavour singlet state via a fusion of gluons emitted from the exchanged quarks of the colliding protons [22]. Albeit the quark content of η and η' mesons is very similar, this manner of the production should contribute primarily in the creation of the meson η'. This is due to the small pseudoscalar mixing angle ($\Theta_{PS} \approx -15°$) [23] which implies that the η' meson is predominantly a flavour singlet state and is expected to contain a significant admixture of gluons. Further, it is almost two times heavier than η and hence its creation requires much larger momentum transfer which is more probable to be realized in the short-range interactions. Unfortunately, at present there are no theoretical calculations concerning this mechanism. Now, since the effective coupling constant describing the η'-proton-proton vertex is not known, it is even not possible to determine the contribution from the simplest possible production mechanism where the η' is supposed to be emitted as a bremsstrahlung radiation from one of the colliding protons as it is shown in Figure 1a. Therefore, investigations of the η' production have to deal with a few problems at the same time. Namely: unknown reaction mechanism, unknown coupling constant, and unknown proton-η' interaction. In the next section the present status of the knowledge about the effective NNη' coupling constant will be given.

$NN\eta'$ coupling constant

In the effective Lagrangian approach [24,25] the strength of the nucleon-η' coupling is driven by the the $NN\eta'$ coupling constant $g_{NN\eta'}$, which comprises the information about the structure of the η' meson and the nucleon. The knowledge of the coupling constant is necessary in the calculation of the production cross section if one considers the Feynman diagrams as illustrated in Figure 1.

The main difficulty in the determination of this quantity is due to the fact that usually the direct production on the nucleon is either associated with the production through baryonic resonances, as in the case of the $\gamma p \to \eta' p$ reaction [26], or with the exchange of other mesons. Therefore, if the direct production mechanism is not dominant it is not possible to extract the $NN\eta'$ coupling without the clear understanding of the other mechanisms. However, it would be very interesting to determine the $g_{NN\eta'}$ coupling constant and to compare it with the calculations performed on the quark level assuming the η' meson structure. First theoretical considerations concerning this issue have been published last year [27].

Assuming that the η and η' mesons are mixtures of the SU(3) singlet and octet states, one can relate the $NN\eta$ and $NN\eta'$ coupling constants by the following equation [24,28]:

$$g_{NN\eta'} = \frac{sin\Theta + \sqrt{2}cos\Theta}{cos\Theta - \sqrt{2}sin\Theta} \cdot g_{NN\eta} \xrightarrow{\Theta=-15.5°} 0.82 \cdot g_{NN\eta}. \qquad (1)$$

The measurements of the $\gamma p \to p\eta$ [29,30] reaction have yielded that: $0.2 \leq g_{NN\eta} \leq 6.2$, whereas the comparison of the $\pi^- p \to \eta n$ and $\pi^- p \to \pi^0 n$ reaction cross sections implies [29]: $5.7 \leq g_{NN\eta} \leq 9.0$. The above inequalities and equation 1 lead to the following range for the $g_{NN\eta'}$ value: **$4.7 \leq \mathbf{g_{NN\eta'}} \leq 5.1$**, which is to be compared to the η' coupling determined from the fits to low energy nucleon-nucleon scattering in the one-boson-exchange models amounting to $g_{NN\eta'} = 7.3$ [31].

On the other hand, the $g_{NN\eta'}$ coupling constant determined via dispersion methods [32] turns out to be smaller than 1, $g_{NN\eta'} < 1$, which is in contradiction to the above estimations.

The $g_{NN\eta'}$ coupling constant is also related to the issue of the total quark contribution to the proton spin ($\Delta\Sigma$). The approximate equation derived in reference [33] reads: $\Delta\Sigma = \Delta u + \Delta d + \Delta s = \frac{\sqrt{3} f_{\eta'}}{2M} g_{NN\eta'}$, where, $f_{\eta'} \approx 166$ MeV [33] denotes the η' decay constant and M the proton mass. Δu, Δd and Δs are the contributions from up, down and strange quarks, respectively [1]. The total contribution of the quarks to the proton spin amounts to $\Delta\Sigma = 0.38^{+0.09}_{-0.10}$ [34]. Applying this value in the above equation one obtains $g_{NN\eta'} = 2.48^{+0.59}_{-0.65}$, which is consistent with the upper limit ($g_{NN\eta'} \leq 2.5$) set from the comparison of the measured total cross section values for the $pp \rightarrow pp\eta'$ reaction with the calculations based on the effective Lagrangian approach, where only a direct production has been considered [2].

The present estimations for $g_{NN\eta'}$ inferred from different experiments are widely spread from 0.2 to 7.3 and are not consistent with each others. Therefore more effort is needed on experimental as well as theoretical side to fix this important parameter.

THE COSY - 11 EXPERIMENT

The experiments were performed at the cooler synchrotron COSY-Jülich [42] which accelerates protons up to a momentum of 3500 MeV/c. The threshold momenta for the $pp \rightarrow pp\eta$ and $pp \rightarrow pp\eta'$ reactions are equal to **1981.6 MeV/c** and **3208.3 MeV/c**, respectively. About $2 \cdot 10^{10}$ accelerated protons circulate in the ring passing $1.6 \cdot 10^6$ times per second through the H_2 cluster target [43,44] installed in front of one of the dipole magnets, as depicted schematically in Figure 2. The target is realized as a beam of H_2 molecules grouped inside clusters of up to 10^5 atoms.

At the intersection point of the cluster beam with the COSY proton beam the collision of protons may result for example in the production of the η' meson. The ejected protons of the $pp \rightarrow pp\eta(\eta')$ reaction, having smaller momenta than the beam protons, are separated from the circulating beam by the magnetic field. Further they leave the vacuum chamber through a thin exit foil and are registered by the detection system consisting of drift chambers and scintillation counters as depicted in Figure 2.

---

[1] Contribution of quarks heavier than the *strange* quark are usually not considered, but I. Halperin and A. Zhitnitsky suggested [41] that the intrinsic charm component of proton may also carry a significant amount of the proton spin. The quark and gluon contributions to the proton spin are widely discussed in the literature [33–40] based on measurements of the spin asymmetries in deep-inealstic scattering of polarised muons on polarised protons.

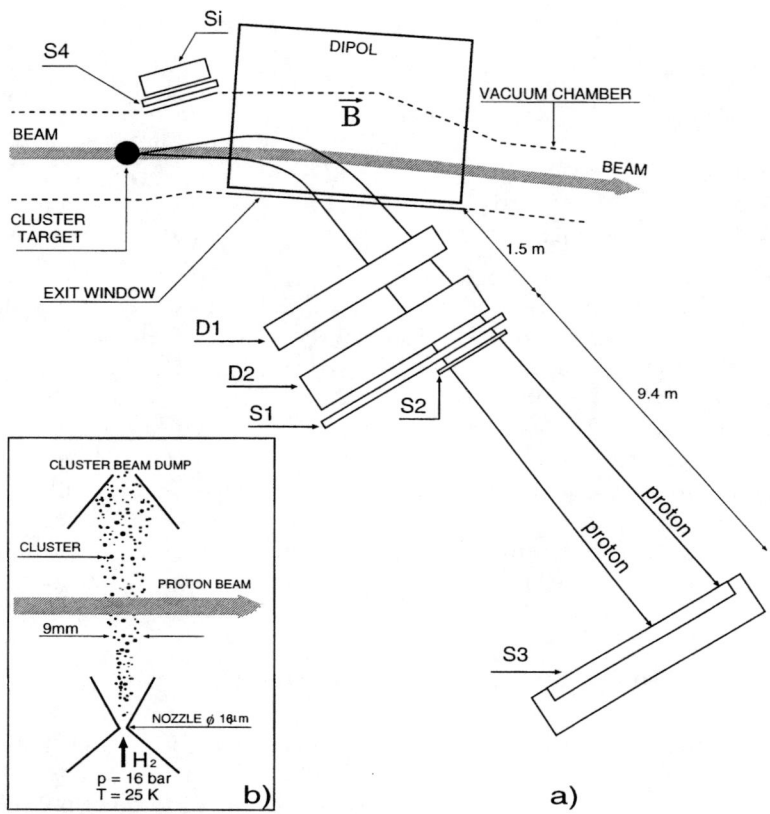

FIGURE 2. a) Schematic view of the COSY-11 detection setup [45]. Only detectors used for the measurements of the $pp \to pp\eta(\eta')$ reactions are shown.
The cluster target is located in front of the accelerator dipole magnet. Protons from the $pp \to pp\eta(\eta')$ reaction are bent by the magnetic field of the dipole magnet, whereas the beam particles keep circulating in the COSY ring. The decay products of the η or η' mesons are not shown, since the analysis is based on the measurement of the four-momenta of the outgoing protons, which leave the vacuum chamber through the thin exit foil and are detected: i) in two drift chamber stacks D1, D2, ii) in the scintillator hodoscopes S1, S2, iii) and in the scintillator wall S3. For the measurement of the elastically scattered protons, additionally, scintillation detector S4 and silicon pad detector Si are used in coincidence with the S1, D1 and D2 detectors.
b) Schematic view of the cluster target.

The measurement of the track direction by means of the drift chambers, and the knowledge of both the dipol magnetic field and the target position allow to reconstruct the momentum vector for each registered particle. The time of flight measured between the S1 (S2) - and the S3 scintillators gives the particle velocity.

Having momentum and velocity for each particle one can calculate its mass, and hence identify it.

In the first step of the data analysis events with two tracks in drift chambers were preselected, and the mass of each particle was evaluated. Figure 3 shows the squared mass of two simultaneously detected particles. A clear separations is seen into groups of events with two protons, two pions, proton and pion and also deuteron and pion. Thus, this spectrum allowed for a software selection of events with two registered protons.

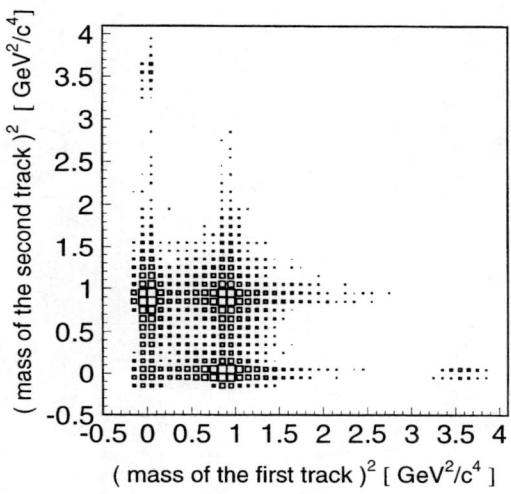

FIGURE 3. Squared masses of two positively charged particles measured in coincidence. Pronounced peaks are to be recognized when two protons, proton and pion, two pions, or pion and deuteron were registered. Note that the number of events is shown in logarithmic scale.

The knowledge of the momenta of both protons before and after the reaction allows to calculate the mass of an unobserved particle or system of particles created in the reaction. Figure 4a depicts the missing mass spectrum obtained for the $pp \to ppX$ reaction at the excess energy value of $Q = 5.8$ MeV above the η' meson production threshold. Most of the entries in this spectrum originate in the multi-pion production, forming a continuous background to the well distinguishable peaks accounting for the ω and η' mesons production, which can be seen at mass values of 782 MeV/c^2 and 958 MeV/c^2, respectively. The signal of the $pp \to pp\eta'$ reaction is better to be seen in the Figure 4b, where the missing mass distribution only in the vicinity of the kinematical limit is presented. Figure 4c shows the missing mass spectrum for the measurement at $Q = 7.6$ MeV together with the multi-pion background as combined from the measurements at different excess energies [46]. Subtraction of the background leads to the spectrum with a clear peak at the mass of the meson η' as shown by the solid line in Figure 4d. The dashed histogram in this figure corresponds to the Monte-Carlo simulations where the beam and target conditions were deduced from the measurements of the elastically scattered protons [46].

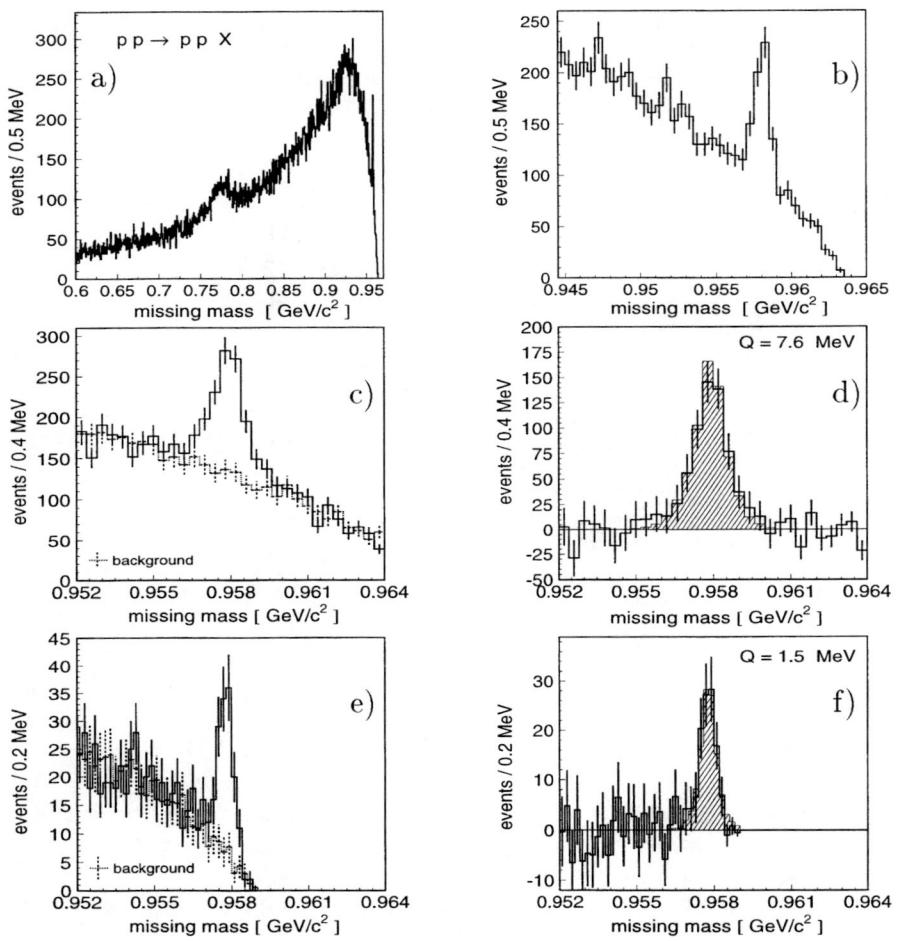

FIGURE 4. Missing mass of the unobserved particle or system of particles; **upper row:** measurements at Q = 5.8 MeV above the η' production threshold; **middle row:** at Q = 7.6 MeV; and **lower row:** at Q = 1.5 MeV. Background shown as dotted lines is combined from the measurements at different energies shifted to the appropriate kinematical limits and normalized to the solid-line histogram.

The scale of the simulated distribution was adjusted to fit the data, but the consistency of the widths is a measure of understanding of the detection system and the target-beam conditions. Histograms from a measurement at Q = 1.5 MeV shown in Figures 4e,f demonstrate the achieved accuracy at the COSY-11 detection system. The width of the missing mass distribution (Fig. 4f), which is now close to the natural width of the η' meson ($\Gamma_{\eta'} = 0.203$ MeV [47]), is again well reproduced by the Monte-Carlo simulations.

RESULTS

Total cross section

Determination of (a) number of the produced η' events from the presented above missing mass distributions, (b) luminosity from the simultaneous measurements of the elastically scattered protons, and (c) detection system acceptance by means of the Monte-Carlo simulations allows for the calculations of the total cross section for the $pp \rightarrow pp\eta'$ reaction. The total cross section for the $pp \rightarrow pp\eta$ reaction was determined by the same method, however, with a much bigger signal to background ratio (40/1) due to the larger total cross section values [48].

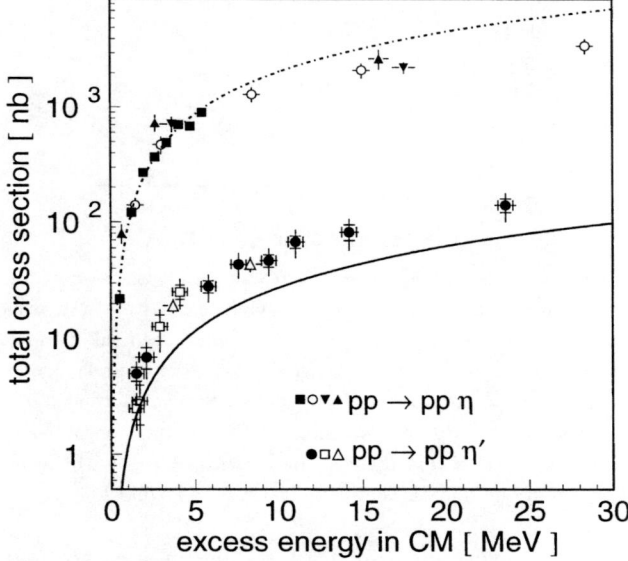

FIGURE 5. Total cross section for the reactions $pp \rightarrow pp\eta$ (upper points) and $pp \rightarrow pp\eta'$ (lower points). Solid squares and solid circles corresponds to the yet unpublished COSY - 11 results. Triangles indicate measurements performed at SATURNE [1,49], open circles at CELSIUS [10], and open squares at COSY [2]. The curves are explained in the text.

Figure 5 shows the compilation of the total cross sections for the η and η' meson production together with the new COSY-11 data shown as filled squares (η) and filled circles (η'). The COSY-11 data on the η production were taken changing continously during the measurement cycle a momentum of the uncooled proton beam. This technique allowed for the precise determination of the total cross section energy dependence near the kinematical threshold. The obtained result confirmed the enhancement of the close to threshold total cross section values compared to the predictions based on the phase space factors and the proton-proton FSI which was earlier observed by the PINOT [49], WASA [10], and SPES III [1,50] collaborations.

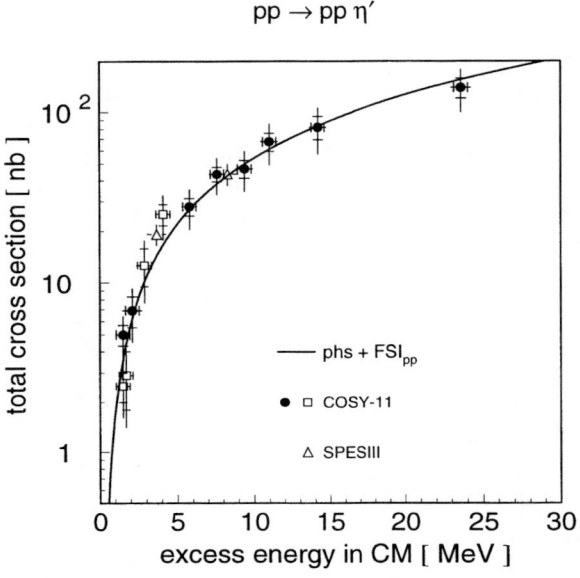

FIGURE 6. Total cross section for the $pp \to pp\eta'$ reaction as a function of the center of mass excess energy. Open squares and triangles are from references [2] and [1], respectively. Filled circles indicate the results of the analysis of the COSY - 11 measurements performed in February 1998. The solid line depicts calculations of the total cross section under assumption that the primary production amplitude is constant and that only a proton-proton interaction take place in the exit channel. The proton-proton scattering amplitude was computed according to the formulas from reference [53]. The obtained energy dependence agrees within a few line thickness with the Fäldt and Wilkin model [52] presented in Figure 5 as a solid line.

New COSY - 11 data concerning η' meson production are shown in Figure 5 as filled circles. These measurements on the $pp \to pp\eta'$ reaction were performed with the stochastically cooled proton beam and the integrated luminosity of 1360 nb^{-1}. Statistical and systematical errors are separated by dashes. The systematical error of the energy equals to 0.44 MeV constitutes of the 0.3 MeV due to the uncertainty in the detection system [51] and 0.14 MeV due to the uncertainty in the η' meson mass [47]. The systematical error of the cross section values, including the overall normalization uncertainty, amounts to 15 % [2,28]. It is worth to stress again, that

SPES III and COSY - 11 results obtained at different laboratories are in a perfect agreement.

The dashed-dotted line in Figure 5 shows the energy dependence predicted by Fäldt and Wilkin [52] normalized now to the COSY-11 data points, and the solid line corresponds to the predictions based on a one-pion-exchange model adjusted to fit the close to threshold $pp \to pp\eta$ data (dashed line) [1]. The factor two discrepancy suggests that the short-range mechanisms may play a prominent role in the production of these mesons [5,4]. However, recent calculations performed by Nakayama et al. [3] indicate that the determination of the total cross section close to threshold is surely not sufficient to establish the contributions from different mechanisms to the overall production amplitude. Specifically, the primary production amplitude for processes studied by these authors (Fig. 1b,d,e) does not change significantly within the present experimental accuracy for the excess energies below Q = 30 MeV. Therefore, the energy dependence of the total cross section for $Q \leq 30$ MeV should be quite well described by the integral of the phase space volume weighted by the squared amplitude of the final state interaction among the outgoing particles. And indeed, as shown in Figure 6, the data are in a good agreement with this model even without considering the η'-proton interaction. This leads to the conclusion that the η'-proton interaction is too weak to influence considerably, within the experimental error bars, the total cross section energy-dependence.

Primary production amplitudes

The cross section for the reaction $pp \to ppX$ can be expressed as:

$$\sigma_{pp \to ppX} = \frac{\int phase\ space \cdot |M_{pp \to ppX}|^2}{flux\ factor}, \qquad (2)$$

where, $M_{pp \to ppX}$ denotes the transition matrix element for the $pp \to ppX$ reaction, and X stands for π^0, η or η' mesons. In analogy with the *Watson-Migdal* approximation [54] for two body processes, it can be assumed that the complete transition amplitude of a production process $M_{pp \to ppX}$ factorizes approximately as [12]:

$$M_{pp \to ppX} \approx M_0 \cdot M_{FSI} \qquad (3)$$

where, M_0 accounts for all possible production processes, and M_{FSI} describes the elastic interaction of protons and X meson in the exit channel. Making further assumptions that only the proton-proton interaction is present in the exit channel ($M_{FSI} = M_{pp \to pp}$) and that the primary production amplitude does not change with the excess energy, it is possible to calculate $|M_0|$. The enhancement from the proton-proton interaction, $|M_{pp \to pp}|^2$, was estimated as an inverse of the squared Jost function, with Coulomb interaction being taken into account [53]. The

$|M_{pp\to pp}|^2$ is a dimensionless factor which turns to zero with vanishing relative protons momentum k, peaks sharply at k≈25 MeV/c and approaches asymptotically unity for large proton-proton relative momenta.

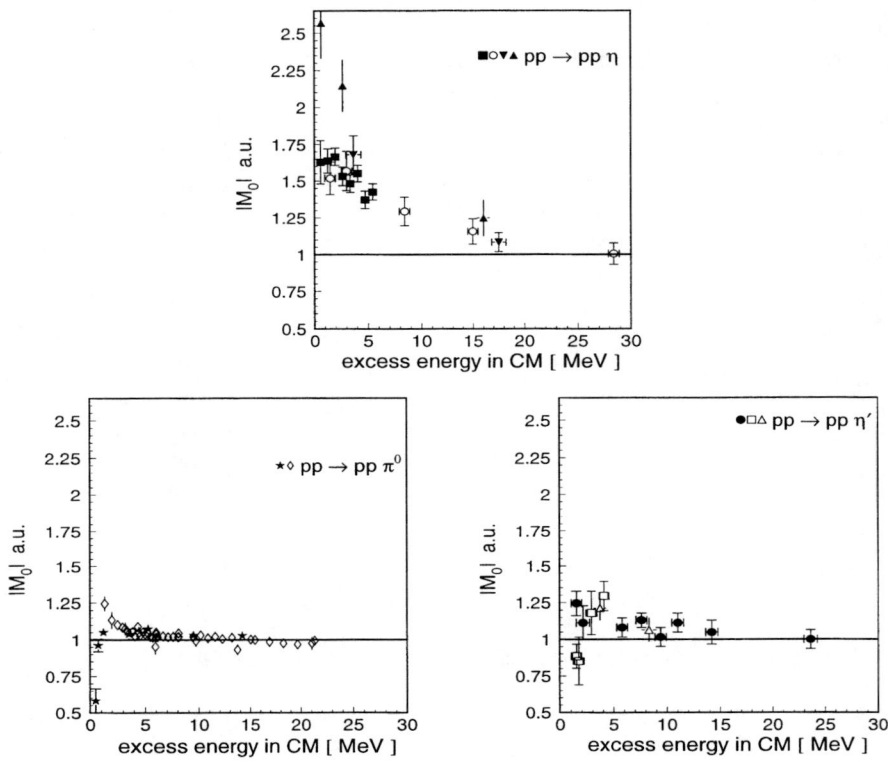

FIGURE 7. Quantity $|M_0|$ extracted from the experimental data for the reactions $pp \to pp\eta$ — upper picture; $pp \to pp\pi^0$ — left picture; $pp \to pp\eta'$ — right picture;

Figure 7 compares the extracted absolute values for the modulus of the primary production amplitude for the near-threshold production of the η, π^0 and η' mesons. The quantity $|M_0|$ is normalized to unity at the point of highest excess energy, for each meson separately. If the performed assumptions in the derivation of $|M_0|$ were fulfilled the obtained values would be equal to one as depicted by the solid line. It can be seen, however, that in the case of the η meson, $|M_0|$ grows with decreasing excess energy reflecting attractive η-proton interaction. In the data for the π^0 production, apart from the two closest-to-threshold points[2], one can notice a tiny grow of $|M_0|$ when the excess energy decreases from Q = 20 MeV to Q = 2 MeV.

---

[2] Due to the steep falling of the total cross section near-threshold already a small change of the energy (0.2 MeV) lifts the points significantly up. Moreover, for the very low energies nuclear

This may be cause by the small π-proton interaction. The deviation from the constant is much smaller than in the η meson case since, the S-wave π-proton interaction is much weaker than the η-proton one.

Similarly, neglecting the two lowest points for the η' meson, one observes about 20 % increase of $|M_0|$ when approaching the threshold. This may indicate a small attractive η'-proton interaction. Anyhow, with the new COSY - 11 data points the possible η'-proton repulsive interaction must be excluded.

Instead of conclusion the article of Bernard et al. [56] is recommended where the threshold matrix-elements for the $pp \to pp\pi^0(\eta, \eta')$ reactions were evaluated in a fully relativistic Feynman diagrammatic approach as reported by N. Kaiser at this conference.

ACKNOWLEDGMENTS

P.M. acknowledges the hospitality and financial support from the Forschungszentrum Jülich and the Foundation for Polish Science.

REFERENCES

1. F. Hibou et al., *Phys. Lett.* **B 438**, 41 (1998).
2. P. Moskal et al., *Phys. Rev. Lett.* **80**, 3202 (1998).
3. K. Nakayama et al., preprint nucl-th/9908077
4. S. D. Bass, preprint hep-ph/9907373
5. C. Wilkin, preprint nucl-th/9810047 and references therein
6. J. Haidenbauer, Ch. Hanhart, J. Speth, *Acta Phys. Pol.* **B 27**, 2893 (1996).
7. C. J. Horowitz, H. O. Meyer, D. K. Griegel, *Phys. Rev.* **C 49**, 1337 (1994).
8. H. O. Meyer et al., *Phys. Rev. Lett.* **65**, 2846 (1990).
9. H. O. Meyer et al., *Nucl. Phys.* **A 539**, 633 (1992).
10. H. Calèn et al., *Phys. Lett.* **B 366**, 39 (1996).
11. U. Schuberth, Ph.D. dissertation at Uppsala University, Acta Universitatis Upsaliensis **5**, 1995
12. A. Moalem et al., *Nucl. Phys.* **A 589**, 649 (1995).
13. A. M. Green, S. Wycech, *Phys. Rev.* **C 55**, R2167 (1997).
14. J. P. Naisse, *Nucl. Phys.* **A 278**, 506 (1977).
15. P. Moskal et al., *Acta Phys. Pol.* **B 29**, 3091 (1998).
16. V. Baru et al., *Eur. Phys. J.* **A 6** (1999) in press.
17. E. Hernández, E. Oset, *Phys. Lett.* **B 350**, 158 (1995).
18. G. Fäldt, C. Wilkin, *Z. Phys.* **A 357**, 241 (1997).
19. J. F. Germond and C. Wilkin, *Nucl. Phys.* **A 518**, 308 (1990).

---

and Coulomb scattering are expected to compete. The limit is approximately at 0.8 MeV of the proton energy in the rest system of the other proton, where the Coulomb penetration factor C^2 is equal to 0.5 [55]. Thus, one should be careful, at small excess energies, where the approximately treated Coulomb interaction dominates.

20. J.M. Laget, F. Wellers, J. F. Lecolley, *Phys. Lett.* **B 257**, 254 (1991).
21. A. Moalem et al., *Nucl. Phys.* **A 600**, 445 (1996).
22. N. Nikolaev, CosyNews **No. 3** May 1998, Published by the Forschungszentrum Jülich in Cooperation with CANU, the COSY User Organisation of the Universities
23. A. Bramon, R. Escribano, M.D. Scadron, *Eur. Phys. J.* **C 7**, 271 (1999).
24. J.-F. Zhang et al., *Phys. Rev.* **C 52**, 1134 (1995).
25. N.C. Mukhopadhyay, J.-F. Zhang, M. Benmerrouche, Talk given at the Workshop on the Structure of the η' Meson. New Mexico State University and CEBAF, Las Cruces, New Mexico, March 8, 1996, World Scientific, 1996, p. 111.
26. R. Plötzke et al., *Phys. Lett.* **B 444**, 555 (1998).
27. D. Lehmann, preprint hep-ph/9808210;
    N. I. Kolev et al., preprint hep-ph/9905438; preprint hep-ph/9903279
28. P. Moskal, PhD Thesis, Jagellonian University, Cracow 1998,
    IKP FZ Jülich, Jül-3685, August 1999, http://ikpe1101.ikp.kfa-juelich.de/
29. M. Benmerrouche, N.C. Mukhopadhyay, *Phys. Rev.* **D 51**, 3237 (1995).
30. M. Benmerrouche, N.C Mukhopadhyay, *Phys. Rev. Lett.* **67**, 1070 (1991).
31. M.M. Nagels et al., *Nucl. Phys.* **B 147**, 189 (1979).
    M.M. Nagels et al., *Few body systems and nucl. forces I*, 17 (1978).
32. W. Grein, P. Kroll, *Nucl. Phys.* **A 338**, 332 (1980).
33. A.V. Efremov et al., *Phys. Rev. Lett.* **64**, 1495 (1990).
34. B. Adeva et al., *Phys. Rev.* **D 58**, 112002 (1998).
35. B. Adeva et al., *Phys. Rev.* **D 58**, 112001 (1998).
36. D. Adams et al., *Phys. Rev.* **D 56**, 5330 (1997).
37. J. Ellis, M. Karliner, hep-ph/9601280 and references therein
38. G.M. Shore, G. Veneziano, *Phys. Lett.* **B 244**, 75 (1990).
39. V.W. Hughes et al., *Phys. Lett.* **B 212**, 511 (1988).
40. J. Ashman et al., *Phys. Lett.* **B 206**, 364 (1988).
41. I. Halperin, A. Zhitnitsky, hep-ph/9706251
42. R. Maier, *Nucl. Instr. & Meth. in Phys. Res.* **A 390**, 1 (1997).
43. H. Dombrowski et al., *Nucl. Instr. & Meth. in Phys. Res.* **A 386**, 228 (1997).
44. A. Khoukaz, PhD Thesis, Westfälische Wilhelms-Universität Münster (1996).
45. S. Brauksiepe et al., *Nucl. Instr. and Meth. in Phys. Res.* **A 376**, 397 (1996).
    D. Grzonka, W. Oelert, KFA-IKP(I)-1993-1 http://ikpe1101.ikp.kfa-juelich.de/
46. P. Moskal, Ann. Rep. 1999, IKP FZ-Jülich, to be published.
47. C. Caso et al., *Eur. Phys. J.* **C 3**, 1 (1998).
48. J. Smyrski et al., FZJ-IKP(I)-1999-1, submitted for publication in Phys. Lett.
49. E. Chiavassa et al., *Phys. Lett.* **B 322**, 270 (1994).
50. A. M. Bergdolt et al., *Phys. Rev.* **D 48**, R2969 (1993).
51. P. Moskal et al., Ann. Rep. 1997, IKP FZ Jülich, Jül-3505, Feb. 1998, p. 41.
52. G. Fäldt and C. Wilkin, *Phys. Lett.* **B 382**, 209 (1996).
53. B. L. Druzhinin, A. E. Kudryavtsev, V. E. Tarasov, *Z. Phys.* **A 359**, 205 (1997).
54. K.M. Watson, *Phys. Rev.* **88**, 1163 (1952).
55. J. D. Jackson, J. M. Blatt, *Rev. of Mod. Phys.* **22**, 77 (1950).
56. V. Bernard, N. Kaiser, Ulf-G. Meißner, *Eur. Phys. J.* **A 4**, 259 (1999).

THEORETICAL ISSUES IN MESON PRODUCTION

Meson exchange models for Meson production

C. Hanhart[*][1]

*Department of Physics, University of Washington, Seattle, WA 98195-1560, USA.

Abstract. The production of mesons in nucleon–nucleon collisions is reviewed from the viewpoint of the meson–exchange picture. In the first part various possible production mechanisms and their relative importance are discussed. In addition, general features of meson production are described.

In the second part special emphasis is put on pion production. Implications of chiral perturbation theory are discussed. Results based on a specific meson-exchange model for pion production in nucleon–nucleon collisions are presented. The model is utilized to calculate several spin-dependent observables of the reactions $pp \to pp\pi^0$ and $pp \to pn\pi^+$ such as spin-correlation parameters. This allows us to study the role of the Delta resonance at close to threshold energies.

The talk closes with a brief discussion of the production of ϕ mesons as an example for the production of heavier mesons. A strategy is outlined that allows to extract information on the structure of the nucleon from the reaction $NN \to NN\phi$.

---

[1]) Supported by USDOE and the Alexander-von-Humboldt Foundation

INTRODUCTION

With the advent of new accelerator technology measurements with very high accuracy became possible for meson production close to threshold [1]. This improvement opens the possibility for the investigation of a lot of interesting physics phenomena. It is the goal of this talk to highlight a few of those. This will we done using the meson exchange picture.

The largest part of the presentation will concentrate on pion production. As this is the first and most important inelasticity of the nucleon–nucleon system it shall deal as a guideline for models of the production of heavier mesons. In addition, for this channel there are not only analyzing powers but also double polarization data available. A study of this system will therefore allow us to get a feeling about what can be learned from meson production. In particular it will be demonstrated that using polarization observables will allow to study resonances that couple too weakly to significantly influence the unpolarized observables – in this example it will be the Δ (1235) far below the resonance position. However, the same kind of sensitivity should be found for resonances that have only a minor effect on the total cross section even at their resonance position. In addition: in the unpolarised observables almost all the effects of the Delta–isobar are driven by the $^1D_2 \to {}^5S_2$ $NN \to N\Delta$ transition leading to p–wave π–production[2]. Polarization observables allow to study less dominant transition amplitudes selectively and thus allow direct insight into the NN interaction.

There is one special feature to the production of pions: since they are pseudo Goldstone bosons of the spontaneous broken chiral symmetry, there is the possibility of studying pion production in nucleon–nucleon collisions with chiral perturbation theory. I will briefly compare the features of the meson exchange approach to those of chiral perturbation theory.

As an example for the production of heavier mesons I will continue with a short discussion of how one can use data on the production of vector mesons close to threshold to deduce information on the structure of the nucleon.

The presentation will close with a brief outlook.

GENERAL FEATURES OF MESON PRODUCTION IN NUCLEON–NUCLEON COLLISIONS

To produce a meson in nucleon nucleon collisions the initial kinetic energy needs to be large enough to put the outgoing meson on its mass shell. The initial center of mass momentum required to produce a meson of mass m turns out to be at

---

[2] This transition is prohibited for the $pp \to pp\pi^0$ reaction and this is why the Delta is less important for the neutral pion production (see below).

least $p = \sqrt{Mm + \frac{m^2}{4}}$, where M denotes the nucleon mass. If the kinematics is chosen close to the threshold the outgoing particles are (almost) at rest. Thus, large characteristic for the meson production close to threshold. This leads to a big momentum mismatch between the final wave-function and the initial one leading to a suppression of the direct production mechanisms (c.f. Fig. 1a and d for pion production). Therefore short range processes dominate the production (c.f. Fig. 1b [3] and c). Note, however, once one moves away from threshold the range of allowed final momenta increases. Thus moderate momentum transfers become more and more likely and higher partial waves come into play. Thus, one should expect very different production mechanisms to be dominant as a function of the excess energy. The range of energies just described can be identified with the range of $0 \leq \eta \leq 1$, where η denotes the maximum meson momentum in units of the meson mass. Therefore, in this presentation we will mainly concentrate on this regime.

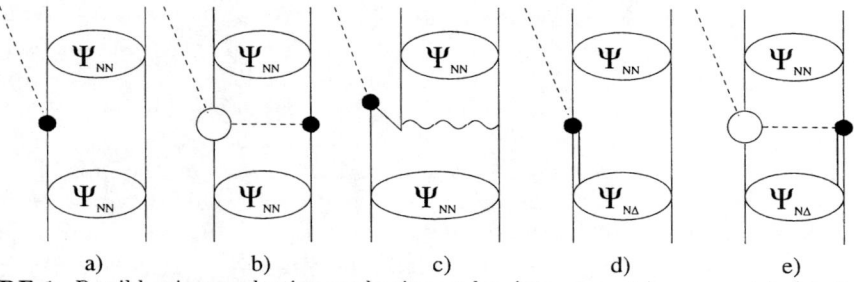

FIGURE 1. Possible pion production mechanisms taken into account in our model: (a) direct production; (b) pion rescattering; (c) contributions from pair diagrams; (d) and (e) production involving the excitation of the $\Delta(1232)$ resonance. Note that diagrams where the Δ is excited after pion emission are also included.

The goal of the meson exchange picture is to find a description of meson production that is both transparent enough to be able to identify the relevant physics of the process and still contains all the relevant dynamics. In this context it is important to understand the special role that the nucleon–nucleon system plays in the initial as well as the final state.

As early as 1952 Watson pointed out the significance that the final state nucleon–nucleon interaction should play in meson production processes close to threshold [2]. To see this let us consider meson absorption on a two nucleon system, which is related to meson production through detailed balance. The cross section is a measure of the volume in which all three particles must be simultaneously in order to make the absorption process possible. However, since the elastic scattering cross section for the two nucleons is very much larger than the production cross section and the NN interaction close to threshold is strongly attractive, a two step process is favored in which first the two nucleons approach at close range – driven

---

[3] For large space like momentum transfers, the πN T–matrix enters far off–shell and thus also the pion exchange contribution should be considered as a short range process.

by the NN interaction – before the meson comes into play. Therefore the energy dependence of the absorption (production) matrix element should be given by the energy dependence of the initial (final) nucleon–nucleon interaction. If we neglect the effect of the meson–nucleon interaction and use the effective range expansion for the nucleon–nucleon interaction the energy dependence of the total production cross–section close to threshold should be given by

$$\sigma_{NN \to NNx}(\eta) \propto \int_0^{m_x \eta} dq' q'^2 \frac{p'}{1 + a(a + r_o)p'^2} , \qquad (1)$$

where $a(r_o)$ denotes the nucleon–nucleon scattering length (effective range) in the relevant channel and p' and q' denote the relative momenta of the two nucleons and of the meson with respect to the two nucleons respectively. It was shown in ref. [3] that the inclusion of the Coulomb effect is important, if for a two proton final state. The inclusion of the Coulomb effect in eq. 1 is straightforward [4]. It turns out that the energy dependence of the total cross section that follows from eq. 1 is in nice agreement with almost all experimental evidence on meson production near threshold [1]. The only exception is the production of η mesons, since the ηN interaction is too strong to be neglected (see also ref. [5]).

The above argument shows that it is important to include the final state interactions of the produced particles in order to describe the energy dependence of the total cross sections correctly. Reversing this statement indicates, that it is possible to relate the energy dependence of the near threshold production cross sections to low energy parameters of the dominant interactions. Examples for this are given in ref. [6] for the ΛN interaction and in refs. [7,8] for the ηN interaction.

Care should be taken, however, when trying to deduce a general formula like eq. 1 not only to describe the energy dependence but also to set the normalization of the effect of the FSI independent of the production operator. This statement is illustrated by different groups using different prescriptions for the normalization of the FSI factor that differ by more than an order of magnitude, although both use the meson exchange picture to calculate the production operator (c.f. e.g. refs. [5] and [9]). Based on quite general arguments one can show, that there is no model independent way to fix the normalization introduced by the FSI factor [11] [4].

A similar reasoning that lead to the successful description of the energy dependence of the total cross section leads to an expression that describes the effect of the initial state interaction [11]: In order to allow the production to proceed, the two incoming nucleons have to come very close to each other. There are two mechanisms that potentially prevent this, namely elastic and inelastic scattering. If we focus on the meson exchange picture, it can be argued that the principal value integral that introduced the scheme dependence for the FSI is actually small for

---

[4] The method applied in refs. [5,9] was also critizised in ref. [10] from a different point of view, namely, that the inclusion of the FSI through using a plane wave that is modified in strength according to the on–shell scattering data misses the short range correlations introduced by the FSI and thus leads to wrong relative weights of the individual contributions.

high energies[5]. Based on this observation one can derive a rough estimate for a suppression factor induced by the ISI (for a discussion of the range of applicability see [11])

$$\lambda = \frac{1}{4}(1+\eta_L)^2 - \eta_L \sin^2(\delta_L) ,\qquad(2)$$

where η_L and δ_L denote the inelasticity and the phaseshifts of the partial wave in the initial state relevant for the meson production under consideration. Using the numbers given by the SAID database [12] we find for the η and η' production values of $\lambda = \frac{1}{5}$ and $\lambda = \frac{1}{3}$ respectively. Note, that the former number covers the bulk effect of the initial state interaction calculated within a meson exchange model [13]. Thus, the ISI leads to a non negligible suppression of the total cross section and should be taken into account.

PION PRODUCTION AND THE ROLE OF THE Δ CLOSE TO THRESHOLD

As this is meant to be an overview presentation, this section starts with a brief history of the field.

Essentially all recent theoretical investigations on pion production near threshold are based on the model proposed by Koltun and Reitan in 1966 [14]. In this model two production mechanisms are considered: direct production (Fig. 1a), and pion rescattering (Fig. 1b) – the latter, however, in on–shell approximation, where the πN–T–matrix is replaced by the scattering length. It is important to note that – as a consequence of chiral symmetry – the isoscalar scattering length[6] almost vanishes. This model was utilized by Miller and Sauer in 1991 [3] to analyze the first set of high precision data of the reaction $pp \to pp\pi^0$ near threshold that became available from IUCF [15]. It turned out that such a model grossly underestimates the empirical cross section [3]. Note, that isovector rescattering is prohibited in this reaction channel.

In 1992 Niskanen extended this model by including the Δ (1235) isobar (cf. Fig. 1d and e) [16]. Furthermore, he allowed for an energy dependence in the s-wave rescattering term, however, still keeping the on–shell approximation [17]. These improvements roughly doubled the predicted $pp \to pp\pi^0$ cross section, but Niskanen's results still underestimate the IUCF data by a factor of 3.6.

Another new production mechanism was introduced by Lee and Riska in 1993 [18]. These authors considered effects from meson-exchange currents due to the exchange of heavy mesons that excite a nucleon–antinucleon pair, as shown in Fig. 1c. It was found that the resulting contributions (in particular the one of the

---

[5] Remember: there is no such thing as a model independent splitting of the NN interaction and the production operator.
[6] The πN–T–matrix can be written as a linear combination of an isoscalar and an isovector part.

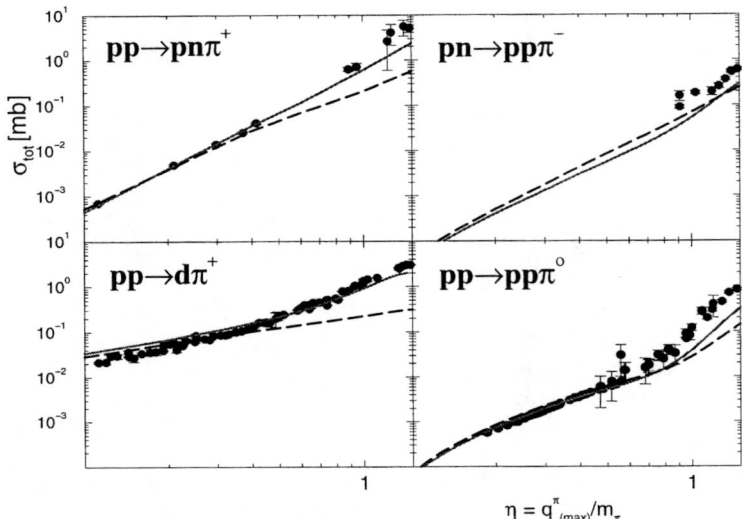

FIGURE 2. *Total cross sections for the different pion production channels. The solid line represents the result of the full model of ref. [32], whereas for the dashed line the contributions involving the Δ where omitted. For a compilation of the experimental data see ref. [1].*

ω meson) enhance the pion production cross section by a factor of 3-5 [18,19] and thus eliminate most of the under prediction found in earlier investigations.

However, in 1995 Hernández and Oset presented an alternative explanation for the missing strength in the π^0 production close to threshold [20]. These authors took into account the off-shell properties of the πN amplitude in the evaluation of the rescattering diagram. It turned out, that this enhances the rescattering contribution sufficiently to describe the empirical $pp \to pp\pi^0$ cross-section without the inclusion of additional short ranged diagrams. In the same year our group performed a calculation using a microscopic model for the πN T-matrix [21]. Although the effect of the rescattering was not as large as reported in [20] its enhancement due to the off-shell part of the T-matrix was quite significant.

Since the pion is the Goldstone Boson of the chiral symmetry one might hope that the framework of chiral perturbation theory allows more insight into this fundamental reaction. In the literature there is a series of papers available, that use tree level chiral perturbation theory to calculate the cross sections close to threshold for both neutral pion production [22–25] and charged pion production [26,27]. In addition there are two calculations available that calculate the π^0 production to one loop order [28,29] – these, however, use heavy baryon chiral perturbation theory which is inappropriate in the present kinematics [5]. The advantage of chiral perturbation theory compared to the meson exchange picture is, that one has an organizing principle at hand. Although the expansion parameter is quite large

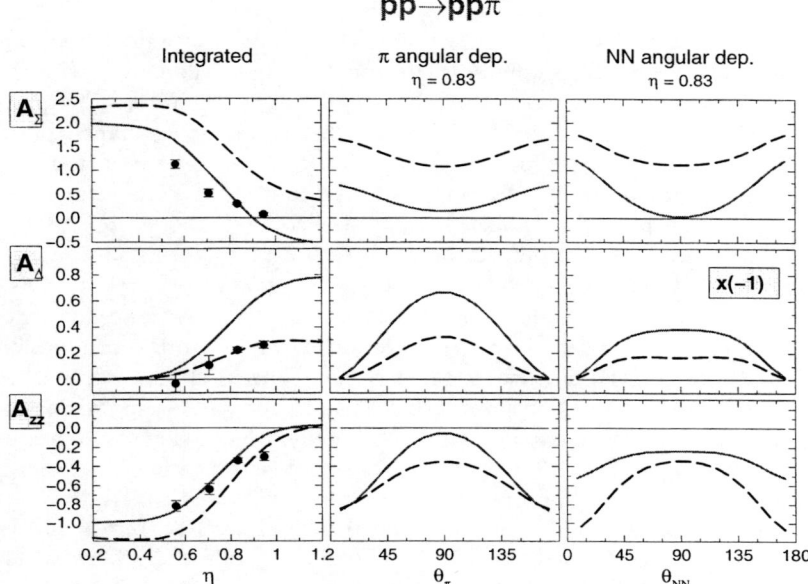

FIGURE 3. *Spin correlation parameters for the reaction* $pp \to pp\pi^0$, *where* $A_\Sigma = A_{xx} + A_{yy}$ *and* $A_\Delta = A_{xx} - A_{yy}$. *Curves are the same as in Fig. 2. The experimental data are taken from Refs. [35,36].*

in case of the $NN \to NN\pi$ reactions, namely $Q = \sqrt{m/M} \simeq \frac{1}{3}$ [23], it is very illuminating to study the lowest order contributions and to compare those to the diagrams depicted in Fig. 1.

Based on the counting rules employed in ref. [23] the lowest order contributions to pion production close to threshold ($O(Q)$) are for the neutral pion production the direct production from the nucleon as well as from the delta (Fig. 1a and d), whereas in case of the charged pion production at the same order there appears a rescattering diagram (Fig. 1b) involving the isovector πN interaction in addition [26,27]. As it was argued above, the large momentum transfer characteristic of meson production close to threshold leads to a suppression of the direct production mechanisms in favor of rescattering and short range processes. Therefore, one expects the production of charged pions to be under better control than the one of neutral pions – that is indeed what we experience: the lowest order calculation for the π^+ production deviates from the data by a factor of 2 [26,27] whereas the one for π^0 production deviates by more than an order of magnitude (the latter occurs because the individual contributions are small and there is a cancelation between the Δ contribution (Fig. 1d) and the nucleonic one (Fig. 1a) [23]).

At the next order ($O(Q^3)$) in both channels loops enter[7] as well as contact interactions (the effective field theory analog of the short range contributions of the meson exchange picture) and rescattering diagrams. The strength of the contact interactions is not constrained by chiral symmetry. Therefore the authors of refs. [23,24,27] tried to estimate their strength by means of resonance saturation: the contact interactions were identified with short ranged diagrams (Fig. 1c) that were evaluated using parameters from realistic NN potentials. However, it is not clear a priori, that this procedure makes sense for the contact interactions. The parameters relevant for the rescattering can be related to πN scattering [30].

Taking all this together it should not come as a surprise that the s–wave pion production is not well under control theoretically. As it was argued before the typical momentum transfers decrease as we move away from the production threshold and it is the large momentum transfer that causes all the trouble. A more appropriate approach to pion production close to threshold thus seems to be to first understand the higher partial waves and then to move down in energy. This procedure should allow one to study the onset of new physics in a more controlled way what highlights the importance of looking at higher partial waves. It should be noted that one expects the chiral expansion to work better, when the pions are produced in higher partial waves [31]. Therefore in what follows we will concentrate on polarization observables for here the higher partial waves show up most clearly.

To be more concrete in what follows we will compare the results of a particular meson exchange model [32] to the data. In the literature there is only one additional model that considers all the different pion production channels as well as higher partial waves, namely the one by the Osaka group. This model was described in a different presentation in some detail [33], and thus we will not discuss it here.

In the Jülich model [32] all standard pion-production mechanisms (direct production (Fig. 1a), pion rescattering (Fig. 1b), contributions from pair diagrams (Fig. 1c) are considered – the latter, however, are treated as a parametrisation of all the missing short range mechanisms. In addition, production mechanisms involving the excitation of the $\Delta(1232)$ resonance (cf. Fig. 1d,e) are taken into account explicitly. All NN partial waves up to orbital angular momenta $L_{NN} = 2$, and all states with relative orbital angular momentum $l \leq 2$ between the NN system and the pion are considered in the final state. Furthermore all πN partial waves up to orbital angular momenta $L_{\pi N} = 1$ are included in calculating the rescattering diagrams in Fig. 1b,e. Thus, our model includes not only s-wave pion rescattering but also contributions from non resonant p-wave rescattering. For details about the model we refer the reader to ref. [32].

Results of this model for total cross sections for the reactions channels $pp \rightarrow pp\pi^0$, $pp \rightarrow pn\pi^+$, $pn \rightarrow pp\pi^-$, and $pp \rightarrow d\pi^+$ are displayed in Fig. 2. Results for analyzing powers were presented in Refs. [32,34]. It was found that the model yields a very good overall description of the data from the threshold up to the Δ resonance region. In fact, a nice quantitative agreement with basically all experimental infor-

---

[7] This is true only when using the counting scheme of ref. [23].

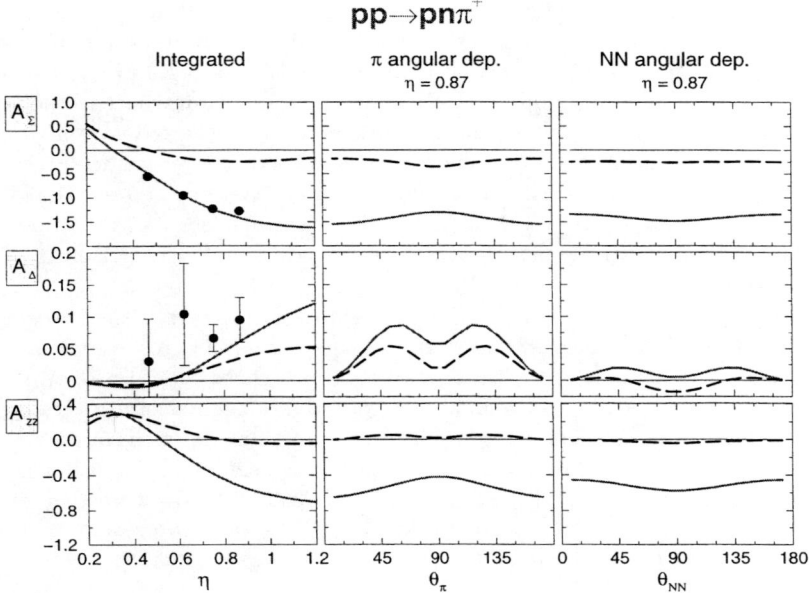

FIGURE 4. *Spin correlation parameters for the reaction $pp \to pn\pi^+$. Same description of the curves as in Fig. 2. The experimental data are taken from Ref. [37].*

mation (then available) was observed over a wide energy range. Thus, this model is very well suited as a starting point for a detailed analysis of the forthcoming spin-dependent observables of the reaction $NN \to NN\pi$.

Looking at the results for the total cross sections one observes that the role of the Δ is very different in the different production channels: it dominates the π^+ production at values of $\eta > 1$ but it appears much less relevant anywhere else. As we will see in the following, this is no longer true for polarization observables. Here the Δ turns out to be very important in all production channels in the whole energy range if one wants to describe the spin observables as well. This clearly demonstrates the power of polarization observables to investigate resonances even if their effect is too weak to show up in unpolarized observables!

Predictions for the spin correlation coefficient combinations $A_\Sigma = A_{xx} + A_{yy}$, $A_\Delta = A_{xx} - A_{yy}$ and A_{zz} are shown in Figs. 3 (for $pp \to pp\pi^0$) and 4 (for $pp \to pn\pi^+$). The polar integrals of these observables are displayed in the left panels as a function of η. The other two panels contain the results for the different observables as a function of the pion angle (middle panel) and of the angle between the nucleons (right panel). Data for these observables are currently being analyzed [38].

Again the solid line shows the results for the full model, whereas the dashed line are the results when all contributions involving the Δ are omitted. As can be seen clearly from the figure, the inclusion of the isobar is essential for all the polarization observables displayed even at energies as low as $\eta = 0.4$!

Our model is not able to describe $A_{xx} - A_{yy}$ in the reaction $pp \to pp\pi^0$ (c.f. Fig. 3). The numerator of this observable is sensitive to Pp partial waves only [39] (Here capital letters denote the NN relative momentum whereas the small letters denote the pion angular momentum with respect to the NN system). Does figure 3 therefore tell us that all but the Pp piece is correctly reproduced by our model[8]? The answer is no. There is a possible other explanation, namely that it is not the numerator of $A_{xx} - A_{yy}$ but the denominator that causes the deviation. Note, that our model underestimates the total cross section for π^0 production by a factor of 2 at energies of $\eta = 1$ (c.f. Fig. 2).

As it was pointed out in ref. [36], it is possible to relate the integrated double polarization observables to the spin cross sections (the cross sections one gets for a given initial spin state). At moderate energies these should be dominated by the lowest partial waves. A possible way to get better insights could therefore be to look at the spin cross sections directly. However, in their determination the total cross section enters besides the double polarization observables. The former, however, is badly known at energies around $\eta = 1$. Better data is needed to accurately constrain the spin cross sections in the energy region of interest. Fortunately those data will soon be available [40].

VECTOR MESON PRODUCTION

In this section we will briefly discuss, how the investigation of vector meson production in nucleon–nucleon collisions can improve our knowledge on the structure of the nucleon. In particular we will outline a method on how to deduce information on the $NN\phi$ coupling constant from data on differential cross sections recently measured by the DISTO collaboration [41].

This information can be regarded as complementary to recent experiments on $\bar{p}p$ annihilation at rest, where the unexpectedly large cross section ratios $\sigma_{\bar{p}p \to \phi X}/\sigma_{\bar{p}p \to \omega X}$ (cf. Ref. [42] for a compilation of data) were interpreted by some authors as a clear signal for an intrinsic $\bar{s}s$ component in the nucleon [42]. However, in an alternative approach based on two-step processes these data were explained without introducing any strangeness in the nucleon and any explicit violation of the OZI rule [43].

In this context ϕ production in nucleon-nucleon collisions is of specific interest. Here one does not expect any significant contributions from competing OZI-allowed two-step mechanisms. Therefore cross section ratios $\sigma_{pp \to pp\phi}/\sigma_{pp \to pp\omega}$ should provide a clear indication for a possible OZI violation and the amount of hidden strangeness in the nucleon – note, that data recently presented by the DISTO collaboration indicate that this ratio is about 8 times larger than the OZI estimate [41]. Based on a model calculation in this section it will be investigated, if the

---

[8]) This was given as a possible explanation of this deviation in ref. [32].

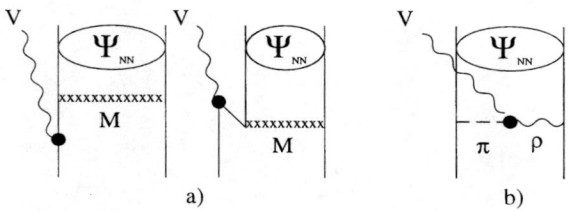

FIGURE 5. ϕ and ω-meson production currents included in the present study: (a) nucleonic current (the diagrams with the production after the meson exchange are also included), (b) meson exchange current. $v = \omega, \phi$ and $M = \pi, \eta, \rho, \omega, \sigma, a_o$.

observed enhancement over the OZI estimate in the cross section implies a $g_{NN\phi}$ that is likewise enhanced and therefore at variance with the OZI rule. It should be emphasized that the investigation is by far not as ambitious as the study of pion production, for the individual ingredients are much less known. The major goal of this study is to extract constraints on $g_{NN\phi}$ from the ω to ϕ ratio.

As before we also describe the $pp \to pp\phi$ reaction within a relativistic meson-exchange model, where the transition amplitude is calculated in DWBA in order to take the NN final state interaction into account. (See Ref. [44] for the details of the formalism.) For the NN interaction we employ the model Bonn B [45]. The effect of the ISI is accounted for via an appropriate adjustment of the (phenomenological) form factors at the hadronic vertices.

In a previous study of the reaction $pp \to pp\omega$ [44] it was found that the dominant production mechanisms are the nucleonic and $\omega\rho\pi$ mesonic currents, as depicted in Fig. 5. Note, that in contrast to the pion production here we take the Born diagram for the $\pi N \to \omega N$ T-matrix multiplied with a phenomenological form-factor and not a microscopic T-matrix. However, this is not a serious draw back for the current investigation, since this formfactor is anyhow adjusted to the data. The reason why it is still possible to extract important information from the vector meson production is, that the angular distribution of the produced vector meson provides a unique and clear signature of the magnitude of these currents, thus allowing one to disentangle these two reaction mechanisms. Therefore, it is possible to fix uniquely the magnitudes of the nucleonic and the meson-exchange current by analyzing the angular distribution of the ϕ meson measured by the DISTO collaboration [41]. Furthermore, since the $NN\phi$ coupling constant enters only in the nucleonic current it is possible to extract its value from such an analysis. It is determined by the requirement of getting the proper contribution of the nucleonic current needed to reproduce the angular distributions.

The parameters of the model (coupling constants, cutoff masses of the vertex form factors) are mostly taken over from the employed NN model. The $\phi\rho\pi$ coupling constant is obtained from the measured decay width of $\phi \to \rho + \pi$. However, besides the $g_{NN\phi}$ that we want to extract from the analysis, there are still some

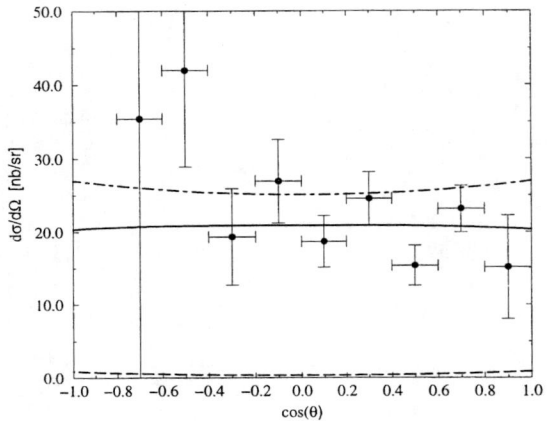

FIGURE 6. Angular distribution for the reaction $pp \to pp\phi$ at an incident energy of $T_{lab} = 2.85$ GeV. The dashed-dotted curve corresponds to the mesonic current contribution, the dashed curve to the nucleonic current contribution. The solid curve is the total contribution. The experimental data are from Ref. [41].

more free parameters: The cutoff mass of the $\phi\rho\pi$ vertex form factor of the meson-exchange current, and the form factor and tensor- to vector coupling constant ratio $\kappa_\phi \equiv f_{NN\phi}/g_{NN\phi}$ of the nucleonic current. It is possible to fix most of them by performing a combined analysis of the ω and ϕ data as well as other sources (cf. Ref. [46] for details). The remaining unknown is the tensor coupling. Here we assume that $\kappa_\phi = \kappa_\omega$, as also suggested by SU(3) symmetry, and $-0.5 \leq \kappa_\omega \leq 0.5$.

After the above considerations, we are now prepared to apply the model to the reactions $pp \to pp\omega$ and $pp \to pp\phi$. The angular distribution for ϕ-meson production measured at $T_{lab} = 2.85$ GeV is shown in Fig. 6. We observe that the angular distribution is fairly flat. Recalling the results we obtained for ω production [44] this tells us that ϕ-meson production should be almost entirely due to the $\phi\rho\pi$ meson-exchange current. Only a very small contribution of the nucleonic current is required if the angular distribution drops at forward and backward angles, as indicated by the data[9].

Details about our strategy for fixing the various parameters and information about the data used in the analysis can be found in Ref. [46]. We get a set of values which range from

$$-1.4 \leq g_{NN\phi}^{model} \leq -0.163 \:. \tag{3}$$

Note, that the spread is not an estimate of the theoretical uncertainty of the model but reflects the remaining uncertainty within the model – this will be reduced

---

[9] Note, that since there are identical particles in the initial state the angular distribution has to be symmetric around 90 degrees.

sufficiently as soon as more experimental information is available. Nevertheless, it is encouraging to see that the extracted values all lie within fairly narrow bounds. This clearly indicates to us that the dependence on the model parameters is not very strong, and that the magnitude of $g_{NN\phi}$ is primarily determined by the experimental information used.

The values of $g_{NN\phi}$ obtained may be compared with those resulting from SU(3) flavor symmetry considerations and imposition of the OZI rule,

$$g_{NN\phi}^{OZI} = -3g_{NN\rho}\sin(\alpha_v) \cong -(0.60 \pm 0.15) ,$$

where the factor $\sin(\alpha_v)$ is due to the deviation from the ideal $\omega - \phi$ mixing. The numerical value is obtained using the values of $g_{NN\rho} = 2.63 - 3.36$ [47] and $\alpha_v \cong 3.8°$. Comparing this value with the ones extracted from our model analysis, we conclude that the preliminary data presently available can be described with using a $NN\phi$ coupling constant that is compatible with the OZI rule. This clearly indicates that a dynamical model is needed for drawing any conclusion about the validity of the OZI rule.

SUMMARY AND OUTLOOK

The main points of the presentation can be summarized as

- it is important to take into account the nucleon–nucleon interaction in the final as well as in the initial state to get a quantitative understanding of meson production close to threshold,

- the interesting energy regime for meson production is $0 \leq \eta \leq 1$, for this regime includes both large and moderate momentum transfers,

- polarization observables allow to detect resonances, even if their effect is too small to show up in the total cross section. Note, that this statement is true only if there is a spin dependent coupling of the resonance to the produced meson.

It should be made clear that theory lacks far behind experiment. However, the large amount of experimental information that is going to be available within the next years will help us to get a better understanding of meson production reactions and will therefore guide us to better models and insights into the phenomenology of the nucleon–nucleon interaction.

ACKNOWLEDGMENTS

The author is very grateful to J. Haidenbauer for comments on the manuscript and to G. A. Miller and U. van Kolck for useful discussions. He also thanks K. Nakayama, J. W. Durso, J. Haidenbauer, J. Speth and O. Krehl for the collaboration that lead to the results presented here.

REFERENCES

1. For a short review and references to experimental as well as theoretical papers see, e.g., H. Machner and J. Haidenbauer, J. Phys. G **25**, R231 (1999).
2. K. Watson, Phys. Rev. **88**, 1163 (1952).
3. G. A. Miller and P. Sauer, Phys. Rev. **C44**, 1725 (1991).
4. H.O. Meyer et al., *Nucl. Phys.* A **539**, 633 (1992).
5. V. Bernard, N. Kaiser and Ulf-G. Meißner, Eur. Phys. J A4(1999) 259 and N. Kaiser, these proceedings.
6. J.T. Balewski et al., *Eur.Phys.J.* **A2**, 99 (1998).
7. A.M. Green, J.A. Niskanen and S. Wycech, Phys. Rev. C **54**.
8. V.Y. Grishina et al., nucl-th/9905049.
9. A. Sibirtsev and W. Cassing, Eur. Phys. J. **A2**, 333 (1998)
10. J. A. Niskanen, Phys.Lett. **B456**, 107 (1999).
11. C. Hanhart and K.Nakayama, Phys. Lett. B **454**, 176 (1999).
12. extracted from the VIRGINIA TECH PARTIAL-WAVE ANALYSES ON-LINE (http://clsaid.phys.vt.edu/ CAPS/said_branch.html)
13. M. Batinic, A. Svarc and T.-S.H. Lee, *Phys.Scripta* **56**, 321 (1997).
14. D. Koltun and A. Reitan, Phys. Rev. **141**, 1413 (1966).
15. H.O. Meyer et al., *Phys. Rev. Lett.* **65**, 2846 (1990);
16. J. Niskanen, Phys. Lett. **B289**, 227 (1992).
17. J. Niskanen, Phys. Rev. **C49**, 1285 (1994).
18. T.-S. Lee and D. Riska, Phys. Rev. Lett. **70**, 2237 (1993).
19. C.J. Horowitz, H.O. Meyer and D.K. Griegel, Phys. Rev. **C49**, 1337 (1994).
20. E. Hernández and E. Oset, Phys. Lett. **B350**, 158 (1995).
21. C. Hanhart et al., Phys. Lett. **B358**, 21 (1995).
22. T.Y. Park et al., Phys. Rev. **C53**, 1519 (1996)
23. T.D. Cohen et al., *Phys. Rev.* C **53**, 2661 (1996).
24. U. van Kolck, G.A. Miller and D.O. Riska, Phys. Lett. **B388** (1996) 679.
25. T. Sato, T.-S.H. Lee, F. Myhrer and K. Kubodera, *Phys. Rev.* C **56**, 1246 (1997).
26. C. Hanhart et al., *Phys. Lett.* B **424**, 8 (1998).
27. C. da Rocha, G. A. Miller, and U. van Kolck, nucl-th/9904031
28. By E. Gedalin, A. Moalem, L. Razdolskaya, *Phys. Rev.* C **60**, 31 (1999).
29. V. Dmitrasinovic, K. Kubodera, F. Myhrer and T. Sato, nucl-th/9902048
30. V. Bernard, N. Kaiser and Ulf-G. Meißner, *Nucl. Phys.* B **457**, 147 (1995) and *Nucl. Phys.* A **615**, 483 (1997).
31. U. van Kolck, private communication.
32. C. Hanhart, J. Haidenbauer, O. Krehl, and J. Speth, *Phys. Lett.* B **444**, 25 (1998).
33. Y. Maeda et al., πN Newsletter **13**, 326 (1997) and K. Tamura, these proceedings.
34. C. Hanhart, J. Haidenbauer, and J. Speth, *Acta Phys. Pol.* B **29**, 3047 (1998).
35. H.O. Meyer et al., *Phys. Rev. Lett.* **81**, 3096 (1998).
36. H.O. Meyer et al., nucl-ex/9907017.
37. K. Swapan et al., nucl-ex/9907016.
38. J. Balewski, these proceedings.
39. H.O. Meyer, in *Baryons '98. Proceedings of the 8th International Conference on the*

Structure of Baryons, edited by D.W. Menze and B.Ch. Metsch (World Scientific, Singapore 1999), pp. 493.
40. J. Zlomaczuk, these proceedings.
41. F. Balestra et al., Phys.Rev.Lett.81(1998)4572
42. J. Ellis et al., *Phys. Lett.* B **353**, 319 (1995).
43. See, e.g., V.E. Markushin and M.P. Locher, *Eur. Phys. J.* A **1**, 91 (1998).
44. K. Nakayama et al., *Phys. Rev.* C **57**, 1580 (1998).
45. R. Machleidt, *Adv. Nucl. Phys.* **19**, 189 (1989).
46. K. Nakayama et al., Phys. Rev. **C60** (1999) 055209.
47. G. Höhler and E. Pietarinen, *Nucl. Phys.* B **95**, 210 (1975).
48. Collaboration CRN Strasbourg, IPN Orsay, LNS Saclay, in *Nouvelles de Saturne no. 19*, p.51 (Saclay 1995)

Diagrammatic Approach to Meson Production in Proton-Proton Collisions near Threshold

Norbert Kaiser

TU München, Physik Dept. T39, D-85747 Garching, Germany

Abstract

We evaluate the threshold T-matrices for the reactions $pp \to pp\pi^0$, $pn\pi^+$, $pp\eta$, $pp\omega$, $p\Lambda K^+$ and $pn \to pn\eta$ in a relativistic Feynman diagram approach. We employ an effective range approximation to take care of the strong S-wave pN and $p\Lambda$ final state interaction. We stress that the heavy baryon formalism is not applicable in the NN-system above π-production threshold due to the large external momentum, $|\vec{p}| \simeq \sqrt{Mm_\pi}$. The magnitudes of the experimental threshold amplitudes extracted from total cross section data, $\mathcal{A} = (2.7 - 0.3\,i)\,\text{fm}^4$, $\mathcal{B} = (2.8 - 1.5\,i)\,\text{fm}^4$, $|\mathcal{C}| = 1.32\,\text{fm}^4$, $|\Omega| = 0.53\,\text{fm}^4$, $\mathcal{K} = \sqrt{2|K_s|^2 + |K_t|^2} = 0.38\,\text{fm}^4$ and $|\mathcal{D}| = 2.3\,\text{fm}^4$ can be reproduced by (long-range) one-pion exchange and short-range vector meson exchanges, with the latter giving the largest contributions. Pion loop effects in $pp \to pp\pi^0$ appear to be small. The presented diagrammatic approach requires further tests via studies of angular distributions and polarization observables.

INTRODUCTION

With the advent of the proton cooler synchrotrons at Bloomington, Jülich and Uppsala, high precision data for meson production in pp-collisions near threshold have become available. The first data on the process $pp \to pp\pi^0$ were a big surprise because the experimental cross sections turned out to be a factor of five larger than theoretical predictions based on direct pion production and neutral pion rescattering fixed from on-shell πN-data. Subsequently, it was argued that heavy meson exchanges might be able to remove this discrepancy. For the status of calculations in dynamical models see the contributions of S. Coon, C. Hanhart and K. Tamura.

We will consider here a rather simple approach to meson production in NN-collisions near threshold. It is primarily based on (relativistic) Feynman diagrams and therefore it avoids the (potentially ambiguous) off-shell effects of NN-potential models. In order to make transparent the underlying physics, we will use some simplifying assumptions. First, we approximate the T-matrix by its value at threshold which is parametrized in terms of one (or two) constant complex amplitudes. Secondly, we employ an effective range approximation for the NN S-wave final-state interaction. Clearly, this simple approach needs refinement (to justify the assumptions made) and further tests. In particular, one has to find out whether the meson-exchange diagrams which nicely explain the magnitude of the total cross sections for various reactions, are also able to reproduce the more exclusive data given by angular distributions and polarization observables.

One important feature of the reaction $NN \to NN\pi$ at threshold should be kept in mind, namely the large initial center-of-mass momentum, $|\vec{p}|^2 = m_\pi(M+m_\pi/4)$. We demonstrate that the heavy baryon formalism is not applicable at such extreme kinematics where the external momentum $|\vec{p}| \simeq \sqrt{Mm_\pi}$ diverges in the heavy nucleon limit $M \to \infty$. In order to avoid these problems it seems mandatory to perform fully relativistic calculations. In this spirit we calculate various tree (and one-pion loop) diagrams contributing to $pp \to pp\pi^0$, $pn\pi^+$, $pp\eta$, $pp\omega$, $p\Lambda K^+$ and $pn \to pn\eta$ at threshold.

THRESHOLD AMPLITUDE AND FINAL STATE INTERACTION

Consider the reaction $p_1(\vec{p}) + p_2(-\vec{p}) \to p+p+\pi^0$ in the center-of-mass (cm) frame at threshold. The invariant T-matrix can be expressed in terms of one complex-valued (constant) amplitude, denoted by $\mathcal{A}$,

$$\mathcal{T}_{\text{cm}}^{\text{th}}(pp \to pp\pi^0) = \mathcal{A}\,(i\,\vec{\sigma}_1 - i\,\vec{\sigma}_2 + \vec{\sigma}_1 \times \vec{\sigma}_2) \cdot \vec{p} \quad . \tag{1}$$

$\vec{\sigma}_{1,2}$ are the spin-operators of the two protons. The value of the proton cm momentum to produce a neutral pion at rest is $|\vec{p}| = \sqrt{m_\pi(M+m_\pi/4)} = 362.2$ MeV with $M = 938.27$ MeV being the proton mass and $m_\pi = 134.97$ MeV the neutral pion mass. Obviously, $|\vec{p}|$ vanishes in the chiral limit of zero pion mass and thus the soft-pion theorem which requires a vanishing threshold T-matrix in the chiral limit $m_\pi = 0$ is trivially fulfilled. In the threshold region, the final-state di-proton system as well as the neutral pion are dominated by angular momentum zero states, thus we are dealing with a $^3P_0 \to {}^1S_0 s$ transition. Consequently, one deduces from unitarity $\mathcal{A} = |\mathcal{A}|\,e^{i\delta(^3P_0)}$ with the $^3P_0$ pp-phase shift to be taken at the threshold energy in the lab frame, $T_{\text{lab}}^{\text{th}} = m_\pi(2 + m_\pi/2M) = 279.65$ MeV, where $\delta(^3P_0) = -6.3°$. The pseudoscalar threshold T-matrix eq.(1) incorporates the Pauli exclusion principle for the (indistinguishable) outgoing protons, since left multiplication with the spin exchange operator $(1+\vec{\sigma}_1 \cdot \vec{\sigma}_2)/2$ leads to a minus sign. Diagrams with crossed proton lines are therefore automatically included. Approximating the near threshold T-matrix by the T-matrix exactly at threshold one would get for the total cross section $\sigma_{\text{tot}}(T_{\text{lab}}) = const \times (T_{\text{lab}} - T_{\text{lab}}^{\text{th}})^2$. In the case of $pp \to pp\pi^0$ near threshold, however, such an approximation is not sufficient as can be seen from Fig. 1. The deviation from the pure 3-body phase space behavior has to do with the very strong pp final-state interaction in the $^1S_0$ partial wave.

We follow here a procedure derived by Watson and in this approach the total cross section for $pp \to pp\pi^0$ including final-state interaction takes the form

$$\sigma_{\text{tot}}(T_{\text{lab}}) = |\mathcal{A}|^2 \left(\frac{M}{4\pi}\right)^3 \frac{2\sqrt{T_{\text{lab}}}}{(2M+T_{\text{lab}})^{3/2}} \tag{2}$$
$$\times \int_{2M}^{W_{\text{max}}} dW \sqrt{(W^2-4M^2)\,\lambda(W^2, m_\pi^2, 4M^2+2MT_{\text{lab}})}\, F_p(W) \quad ,$$

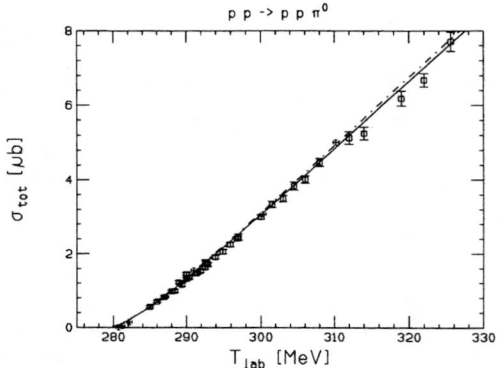

Figure 1: One-parameter fit (solid line) to $\sigma_{\text{tot}}(pp \to pp\pi^0)$ as described in the text.

T_{lab} [MeV]	285	290	295	300.3	308	314	319	325.6
σ_{tot} [μb]	0.56	1.31	2.06	3.07	4.47	5.25	6.18	7.71
$|\mathcal{A}|$ [fm^4]	2.73	2.69	2.64	2.70	2.71	2.64	2.66	2.73

Table 1: Values of $|\mathcal{A}|$ extracted from the total cross sections for $pp \to pp\pi^0$.

with $F_p(W)$ the correction factor due the pp final-state interaction. We evaluate it in the effective range approximation

$$F_p(W) = \left\{1 + \frac{a_p}{4}(a_p + r_p)(W^2 - 4M^2) + \frac{a_p^2 r_p^2}{64}(W^2 - 4M^2)^2\right\}^{-1}, \quad (3)$$

with W the final-state pp-invariant mass. $W_{\max} = \sqrt{4M^2 + 2MT_{\text{lab}}} - m_\pi$ is its kinematical endpoint and $\lambda(x,y,z) = x^2 + y^2 + z^2 - 2yz - 2xz - 2xy$ denotes the Källen function. Furthermore, $a_p = 7.81$ fm and $r_p = 2.77$ fm are the scattering length and effective range parameter for elastic pp-scattering including electromagnetic effects. We have fixed the normalization of $F_p(W)$ such that in the limit $a_p = 0$ (i.e. vanishing final-state interaction), it becomes identical to 1. Furthermore, the condition $F_p(2M) = 1$ ensures that there is no final-state interaction effect exactly at threshold, as it must be according to the definition of $\mathcal{A}$ in eq.(1). In Appendix C of ref.[1] a derivation of the correction factor $F_p(W)$ in scattering length approximation ($r_p = 0$) is given using effective field theory methods. With the help of eqs.(2,3), we extract the nearly constant values of $|\mathcal{A}|$ shown in Table 1. The empirical value of $\mathcal{A}$ is thus

$$\mathcal{A}^{(\exp)} = (2.7 - 0.3\,i) \text{ fm}^4 \,. \quad (4)$$

In Fig. 1 we show by the solid line the resulting total cross sections in comparison to the data from IUCF and CELSIUS using eqs.(2,3,4).

RELATIVISTIC MESON EXCHANGE DIAGRAMS

Tree Level Goldstone Boson Contributions

We consider tree level and one-loop pion exchanges contributing to the T-matrix for $pp \to pp\pi^0$ at threshold. Due to the large proton momentum even at threshold, $|\vec{p}| \simeq \sqrt{Mm_\pi}$, it is not appropriate to employ the frequently used heavy baryon formalism. Let us demonstrate the failure of the heavy baryon formalism for a simple (but generic) example. Consider diagram b) in Fig. 2 which involves a propagating nucleon after emission of the real π^0. We show that this nucleon propagator cannot be expanded in powers of $1/M$ in the usual way. Let $v^\mu = (1, \vec{0})$ be the four-vector which selects the center-of-mass frame. The four-vector of the propagating nucleon is $Mv + k$ with $k^\mu = (-m_\pi/2, \vec{p})$ and $k^2 = -Mm_\pi$. We start on the left hand side with the correct relativistic result and then perform the usual $1/M$-expansion of the heavy baryon formalism

$$-\frac{1}{m_\pi} = \frac{1}{v \cdot k + k^2/2M} = \frac{1}{v \cdot k} \sum_{n=0}^{\infty} \left(\frac{-k^2}{2Mv \cdot k} \right)^n = -\frac{2}{m_\pi} \sum_{n=0}^{\infty} (-1)^n \ . \quad (5)$$

One sees that infinitely many terms of the $1/M$-expansion contribute to the same order and the resulting series does not even converge. The source of this problem is the extreme kinematics of the reaction $NN \to NN\pi$. The heavy baryon formalism cannot cope with external momenta scaling as $|\vec{p}| \simeq \sqrt{Mm_\pi}$. This problem is related to trivial kinematics and immediately overcome in fully relativistic calculations. We thus turn to the relativistic description of the chiral pion-nucleon

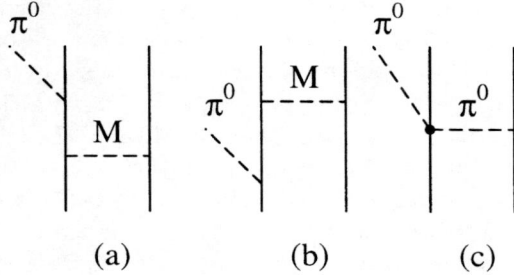

Figure 2: Feynman graphs for neutral pion production, $(M = \pi^0, \eta, \omega, \rho^0)$.

system. We mention only the for our purpose relevant pion-nucleon vertices. The πNN-vertex (and also the ηNN-vertex) is of pseudovector type as required by chiral symmetry. The second order isoscalar chiral $\pi\pi NN$-contact interaction, $\pi^0(q_1) + N(p_1) \to \pi^0(q_2) + N(p_2)$, reads

$$\frac{i}{f_\pi^2}\Big\{ -4c_1 m_\pi^2 + 2c_3 q_1 \cdot q_2 + \frac{c_2'}{2M}[(p_1+p_2) \cdot q_1 \not{q}_2 + (p_1+p_2) \cdot q_2 \not{q}_1]$$
$$+ \frac{c_2''}{2M^2}(p_1+p_2) \cdot q_1 \, (p_1+p_2) \cdot q_2 \Big\} \ . \quad (6)$$

This form is unique since on mass-shell it gives the most general second order $(s \leftrightarrow u)$ crossing symmetric polynomial contribution to the invariant isoscalar πN-amplitudes. The low-energy constants c_1, c'_2, c''_2, c_3 have already been determined (at tree level) from low–energy πN-data and we list their values for completeness: $c_1 = -0.64$, $c'_2 = -5.63$, $c''_2 = 7.41$, $c_3 = -3.90$, in units of GeV^{-1}. Consider first the one-pion (π^0) exchange, shown in Figs. 2a,b (i.e. M = π^0). We stress that these are relativistic Feynman graphs, i.e. in the intermediate states they contain the full relativistic Dirac propagator which sums up several time-orderings. We find

$$\mathcal{A}^{(\pi,\mathrm{dir})} = \frac{g_{\pi N}^3}{4M^4(1+\mu)(2+\mu)} = 0.48\,\mathrm{fm}^4 \,, \tag{7}$$

with $\mu = m_\pi/M = 0.144$ and $g_{\pi N} = 13.4$ the strong pion-nucleon coupling constant. Next, there is the so-called π^0-rescattering, as shown in Fig. 2c,

$$\mathcal{A}^{(\pi,\mathrm{res})} = \frac{g_{\pi N}\,\mu}{f_\pi^2 M(1+\mu)}\left[\frac{c_3}{2} + \left(1 + \frac{\mu}{4}\right)c'_2 + \left(1 + \frac{\mu}{4}\right)^2 c''_2 - 2c_1\right] = 0.46\,\mathrm{fm}^4 \,, \tag{8}$$

with $f_\pi = 92.4$ MeV the weak pion decay constant. Again, there is a marked difference to the heavy baryon case. To leading order (in μ), the relativistic couplings c'_2, c''_2 combine to give the $c_2 = c'_2 + c''_2$ term in the heavy baryon approach. In previous HBChPT calculations the c_2-term was found with an incorrect prefactor $1/2$. The relative factor of two in the relativistic calculation comes from the fact that products of nucleon and pion four-momenta are not dominated anymore by the term nucleon mass M times pion energy. For the reaction $NN \to NN\pi$ the product of the nucleon and pion three-momenta can be equally large, since $|\vec{p}| \simeq \sqrt{Mm_\pi}$. Interestingly, the combination of low-energy constants in eq.(8) is dominated (to about 90%) by the last term $\sim -2c_1$, which is related to the pion-nucleon sigma-term, $\sigma_{\pi N}(0) = -4c_1 m_\pi^2 = (45 \pm 8)$ MeV (to leading order). Therefore the strength of the π^0-rescattering (at threshold) is almost entirely due to this particular chiral symmetry breaking term. The effects from the $\Delta(1232)$-resonance encoded in the low-energy constants c'_2, c''_2, c_3 turn out to be very small. The next Goldstone boson which can contribute is the $\eta(547)$-meson. We find

$$\mathcal{A}^{(\eta)} = \frac{g_{\pi N}\,g_{\eta N}^2\,\mu}{4M^2(m_\eta^2 + M^2\mu)(2+\mu)} = 0.02\,\mathrm{fm}^4 \,, \tag{9}$$

where we have employed the SU(3)-value for the ηN-coupling constant together with the simplified ratio of the octet axial vector coupling constants $D/F = 1.5$, which leads to $g_{\eta N} = \sqrt{3}g_{\pi N}/5 = 4.64$.

Pion Loop Effects

We do not attempt a full one-loop calculation here, but rather consider certain (simple) classes of loop graphs. We use dimensional regularization and minimal subtraction to eliminate divergences and set the renormalization scale equal to the proton mass M. For estimating the genuine size of pion loop effects, such a procedure ignoring renormalization via counterterms should be sufficient. Consider first a certain class of pion-loop diagrams which involve the $\pi\pi$-interaction, shown in Fig. 3. Notice that only the full class of diagrams is independent of the choice of interpolating pion field. One finds a term proportional to the pion-loop contribution to the nucleon scalar form factor evaluated at a momentum transfer $t = -Mm_\pi$,

$$\mathcal{A}^{(\mathrm{loop},1)} = \frac{g_{\pi N}(2+\mu)(\mu-1)}{12 M f_\pi^2 m_\pi^2 (1+\mu)} \sigma_{\pi N}(-M^2 \mu)_{\mathrm{loop}} = -0.10 \,\mathrm{fm}^4 \;. \tag{10}$$

This amounts to a very small -4% correction in comparison to the empirical value of $\mathcal{A}$. Next, we consider the entire class of loop graphs proportional to $g_{\pi N}/f_\pi^4$ and get

$$\mathcal{A}^{(\mathrm{loop},2)} = (-0.01 + 0.22 - 0.25)\,\mathrm{fm}^4 = -0.04\,\mathrm{fm}^4 \;, \tag{11}$$

a contribution which actually vanishes in the chiral limit $\mu \to 0$. For explicit expressions, see ref.[1]. We observe sizeable cancelations between various sets of loop graphs and conclude therefore that these seem not to play a significant role for explaining the magnitude of the threshold amplitude $\mathcal{A}$.

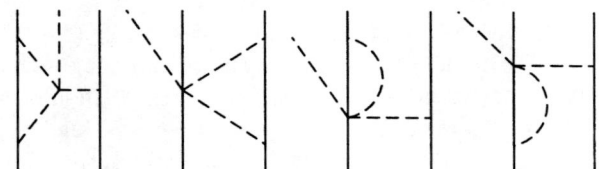

Figure 3: Class of one-pion loop graphs involving the $\pi\pi$-interaction.

Heavy Meson Exchanges

The heavy meson exchanges can play a significant role in $NN \to NN\pi$. Consider first $\omega(782)$-exchange. There is some uncertainty about the coupling constant $g_{\omega N}$. In conventional boson-exchange models of the NN-force one can nowadays work with a coupling constant $g_{\omega N}$ that is compatible with the value $g_{\omega N} = 10.1 \pm 0.9$ derived from forward NN-dispersion relations. There is agreement that the tensor-to-vector coupling ratio of the ω-meson κ_ω is very small. If we set $\kappa_\omega = -0.16$ and use the coupling constant $g_{\omega N} = 10.3$, we get

$$\mathcal{A}^{(\omega)} = \frac{g_{\pi N} g_{\omega N}^2}{2M^2(m_\omega^2 + M^2\mu)(2+\mu)}\left[2 + \mu(\kappa_\omega^2 - \kappa_\omega - 1) + \mu^2 \kappa_\omega\left(1 + \frac{9}{8}\kappa_\omega\right)\right] = 1.45\,\mathrm{fm}^4 \;, \tag{12}$$

which is quite large. Similarly, we evaluate the $\rho(770)$-exchange contribution. We use the coupling constant $g_{\rho N} = 3.04$ (obtained via ρ-universality, $g_{\rho N} = g_\rho/2$, with $g_\rho = 6.08$ fixed from the $\rho^0 \to \pi^+\pi^-$ decay width $\Gamma_\rho = 152.4\,\text{MeV}$) and the tensor-to-vector coupling ratio $\kappa_\rho = 6.1$, which leads to

$$\mathcal{A}^{(\rho)} = \frac{g_{\pi N}\, g_{\rho N}^2}{2M^2(m_\rho^2 + M^2\mu)(2+\mu)} \left[2 + \mu(\kappa_\rho^2 - \kappa_\rho - 1) + \mu^2 \kappa_\rho \left(1 + \frac{9}{8}\kappa_\rho\right)\right] = 0.51\,\text{fm}^4 . \tag{13}$$

Note that due to the large value $\kappa_\rho = 6.1$ the terms proportional to μ and μ^2 in the square bracket of eq.(13) are most important. Another mechanism is the emission of the neutral pion from the anomalous $\omega\rho\pi$-vertex, with the $\rho(770)$ coupling to one and the $\omega(782)$ to the other proton. The pertinent interaction vertex is given by $\mathcal{L}_{\omega\rho\pi} = -(G_{\omega\rho\pi}/f_\pi)\,\epsilon^{\mu\nu\alpha\beta}\,(\partial_\mu \omega_\nu)\,\vec{\rho}_\alpha \cdot \partial_\beta \vec{\pi}$. Using $G_{\omega\rho\pi} = -1.14$ (consistent with ω- and ϕ-decays) the contribution of the anomalous $\omega\rho\pi$-vertex to $\mathcal{A}$ follows as

$$\mathcal{A}^{(\omega\rho\pi)} = \frac{g_{\omega N}\, g_{\rho N}(1+\kappa_\rho)(1+\kappa_\omega) G_{\omega\rho\pi} M\mu^2}{2f_\pi(m_\omega^2 + M^2\mu)(m_\rho^2 + M^2\mu)} = -0.06\,\text{fm}^4 , \tag{14}$$

which is quite small. However, one expects this vertex to be of larger importance in the neutral pion P-wave amplitudes. Combining the various meson exchange contributions evaluated so far, we get

$$\mathcal{A}^{(\text{thy})} = 2.72\,\text{fm}^4 , \tag{15}$$

which compares well with the empirical value given in eq.(4). Of course, taken the uncertainty in some of the coupling constants, the 1% agreement between our theoretical prediction for $\mathcal{A}$ and its empirical value eq.(4) should not be taken too seriously. We only want to make the point that these well-known boson-exchange diagrams, when evaluated fully relativistically, can explain the near threshold data for $pp \to pp\pi^0$.

CHARGED PION PRODUCTION IN PP-COLLISIONS

Here we will discuss charged pion production using the same approach. The T-matrix for charged pion production in proton-proton collisions, $p_1(\vec{p}) + p_2(-\vec{p}) \to p + n + \pi^+$, at threshold in the center-of-mass frame reads

$$\mathcal{T}_{\text{cm}}^{\text{th}}(pp \to pn\pi^+) = \frac{\mathcal{A}}{\sqrt{2}}\,(i\vec{\sigma}_1 - i\vec{\sigma}_2 + \vec{\sigma}_1 \times \vec{\sigma}_2) \cdot \vec{p} - \sqrt{2}\mathcal{B}\,i(\vec{\sigma}_1 + \vec{\sigma}_2)\cdot\vec{p} . \tag{16}$$

Since the pn-system at rest can be in a spin-singlet or in a spin-triplet state, two different transitions are possible at threshold. The threshold amplitude for the singlet transition $^3P_0 \to {}^1S_0 s$ is by isospin symmetry proportional (with a factor $1/\sqrt{2}$) to the threshold amplitude $\mathcal{A}$ for the reaction $pp \to pp\pi^0$ introduced in eq.(1). The new threshold amplitude for the triplet transition $^3P_1 \to {}^3S_1 s$ is called $\mathcal{B}$. Again one deduces from unitarity, $\mathcal{B} = |\mathcal{B}|e^{i\delta(^3P_1)}$, with the $^3P_1$ pp-phase shift to be taken at $T_{\text{lab}}^{\text{th}} = 292.3\,\text{MeV}$, where $\delta(^3P_1) = -28.1°$. Notice, that the abovementioned unitarity relation neglects the (small) inelasticity due to the $pp \to d\pi^+$ channel which opens 4.8 MeV lower at $T_{\text{lab}} = 287.5\,\text{MeV}$.

Extraction of the Threshold Amplitudes

Employing the same method as before to correct for the strong S-wave pn final-state interaction, the total cross section for $pp \to pn\pi^+$ reads

$$\sigma_{\text{tot}}(T_{\text{lab}}) = \left(\frac{M}{4\pi}\right)^3 \frac{2\sqrt{T_{\text{lab}}}}{(2M + T_{\text{lab}})^{3/2}} \int_{M+M_n}^{W_{\text{max}}} \frac{dW}{W} \sqrt{\lambda(W^2, M^2, M_n^2)} \qquad (17)$$

$$\times \sqrt{\lambda(W^2, m_{\pi^+}^2, 4M^2 + 2MT_{\text{lab}})} \left\{ |\mathcal{A}|^2 F_s(W) + 2|\mathcal{B}|^2 F_t(W) \right\} .$$

The correction factors from the pn singlet and triplet S-wave final-state interaction are given in effective range approximation by

$$F_{s,t}(W) = \left\{ 1 + (a_{s,t} + r_{s,t}) \frac{a_{s,t}}{4W^2} \lambda(W^2, M^2, M_n^2) + \frac{a_{s,t}^2 r_{s,t}^2}{64W^4} \lambda^2(W^2, M^2, M_n^2) \right\}^{-1}, \qquad (18)$$

with W the final-state pn invariant mass and $W_{\text{max}} = \sqrt{4M^2 + 2MT_{\text{lab}}} - m_{\pi^+}$. M still denotes the proton mass and $M_n = 939.57$ MeV stands for the neutron mass. The singlet and triplet scattering lengths and effective range parameters for elastic np-scattering have the values: $a_s = 23.75$ fm, $a_t = -5.42$ fm, $r_s = 2.75$ fm and $r_t = 1.76$ fm.

T_{lab} [MeV]	294.3	295.1*	289.0*	299.3	306.3	314.1	319.2
$\sigma_{\text{tot}}^{\text{exp}}$ [μb]	0.71	1.1*	3.84*	4.81	13.91	25.5	41.1
$\sigma_{\text{tot}}^{\text{fit}}$ [μb]	0.57	1.07	3.46	4.81	13.82	25.76	34.14

Table 2: Fit of the total cross sections for $pp \to pn\pi^+$.

The data base of total cross sections for the process $pp \to pn\pi^+$ in the 30 MeV region above threshold consists at present of five data points measured at IUCF and of two data points measured at COSY (marked in Table 2 by an asterisk). We leave out the data point at the highest energy $T_{\text{lab}} = 319.2$ MeV where the π^+ angular distributions are no more isotropic and thus P-waves start to become important. We also found it important to ignore in the fit the data point at the lowest energy $T_{\text{lab}} = 294.3$ MeV. Using eq.(17,18) for the total cross section and the value of $|\mathcal{A}| = 2.72$ fm^4 as determined from the $pp \to pp\pi^0$ data, one finds in a best fit of the remaining three IUCF data points for the modulus of the triplet threshold amplitude

$$|\mathcal{B}|^{(\text{exp})} = 3.16 \text{ fm}^4 , \qquad (19)$$

with a very small total $\chi^2 = 0.044$ (see also Table 2). An unconstrained fit of the same three data points gives $|\mathcal{A}| = 3.00$ fm^4 and $|\mathcal{B}| = 3.15$ fm^4 with a marginally smaller total $\chi^2 = 0.042$. It is quite remarkable that $|\mathcal{A}|$ is found to be in 10% agreement with the value obtained from fitting the large set of precise near threshold $pp \to pp\pi^0$ data. Due to the very strong pn final-state interaction in the $^1S_0$

exit channel ($a_s = 23.75$ fm) the singlet transition contributes a factor 30 to 40 less to the total cross section than the triplet transition. Using the information from the $^3P_1$ phase shift, one gets the following experimental value of the triplet threshold amplitude $\mathcal{B}$,

$$\mathcal{B}^{(\text{exp})} = (2.8 - 1.5\,i)\,\text{fm}^4 \;. \qquad (20)$$

This number should be considered as indicative since the systematic error is not under control when only three data points are fitted.

Diagrammatic Approach

Next, we turn to the evaluation of the relativistic Feynman diagrams contributing to $pp \to pn\pi^+$ at threshold. In addition to the ones considered for $pp \to pp\pi^0$, there is now the possibility for isovector pion-rescattering. The chiral πN-Lagrangian contains such vertices at leading order (the so-called Weinberg-Tomozawa vertex) and at next-to-leading order (a vertex proportional to the low-energy constant $c_4 = 2.25$ GeV^{-1}). One finds

$$\mathcal{B}^{(\pi,\text{iv})} = \frac{g_{\pi N}(c_4 m_\pi - 1)}{2M^2 f_\pi^2 (1+\mu)} = -0.82\,\text{fm}^4 \;. \qquad (21)$$

We note that a naive application of the heavy baryon formalism would give the contribution from the Weinberg-Tomozawa vertex with a wrong prefactor 3/4. From the other pseudoscalar meson (π and η) exchange diagrams, one finds the following contributions to the triplet amplitude $\mathcal{B}$,

$$\mathcal{B}^{(\pi,\text{dir})} = \frac{g_{\pi N}^3 (3+2\mu)}{4M^4(1+\mu)(2+\mu)} = 1.58\,\text{fm}^4 \;, \qquad (22)$$

$$\mathcal{B}^{(\pi,\text{res})} = \mathcal{A}^{(\pi,\text{res})} = 0.46\,\text{fm}^4 \;, \qquad (23)$$

$$\mathcal{B}^{(\eta)} = -\mathcal{A}^{(\eta)} = -0.02\,\text{fm}^4 \;. \qquad (24)$$

Note that the (direct) 1π-exchange contribution $\mathcal{B}^{(\pi,\text{dir})}$ is rather large due to an enhancement factor $3+2\mu$ in comparison to $\mathcal{A}^{(\pi,\text{dir})}$ given in eq.(7). From the vector meson (ρ and ω) exchange diagrams one finds the following contributions to $\mathcal{B}$,

$$\mathcal{B}^{(\omega)} = \frac{g_{\pi N} g_{\omega N}^2}{M^2(m_\omega^2 + M^2\mu)(2+\mu)} \left(1 - \frac{\mu}{4}\kappa_\omega\right)^2 = 1.56\,\text{fm}^4 \;, \quad \mathcal{B}^{(\omega\rho\pi)} = 0 \;, \quad (25)$$

$$\mathcal{B}^{(\rho)} = \frac{g_{\pi N} g_{\rho N}^2 (\mu \kappa_\rho - 4)}{4M^2(m_\rho^2 + M^2\mu)(2+\mu)} \left[3 + \mu\left(2 + \frac{\kappa_\rho}{4}\right)\right] = -0.38\,\text{fm}^4 \;. \qquad (26)$$

One observes that the sizeable ω-exchange contribution $\mathcal{B}^{(\omega)}$ is approximately equal to $\mathcal{A}^{(\omega)}$ given in eq.(12). Note also that the expression for ω-exchange cannot be recovered by simply substituting the vector meson mass and coupling constants in the expression for ρ-exchange. The reason for that are certain diagrams with (charged) ρ^+-exchange which have no analogy in the case of the (neutral) ω-meson.

In the spirit of vector meson dominance one could also think of an isovector $\pi\rho NN$-contact vertex of the form $\mathcal{L}_{\pi\rho N} = (g_\rho g_{\pi N}/2M)\bar{N}\gamma^\mu\gamma_5\vec{\tau}\cdot(\vec{\pi}\times\vec{\rho}_\mu)N$. The form of this vertex and the coupling constant in front are copied from the Kroll-Ruderman term for charged pion photoproduction, replacing the charge e by the universal ρ-coupling $g_\rho = 6.08$ and the photon field by the isotriplet ρ-meson field $\vec{\rho}_\mu$. The respective ρ^0 and ρ^+-exchange diagrams give rise to the following contribution to the triplet amplitude

$$\mathcal{B}^{(\rho,\mathrm{KR})} = \frac{g_{\pi N}g_\rho g_{\rho N}}{M^2(m_\rho^2 + M^2\mu)}\left(1 - \frac{\mu}{4}\kappa_\rho\right) = 0.46\,\mathrm{fm}^4 \;. \tag{27}$$

Admittedly, this contribution is somewhat speculative since the Kroll-Ruderman vertex for ρ-mesons presumes a particular realization of vector meson dominance. Summing up the various tree level contributions given in eqs.(21–27), we get

$$\mathcal{B}^{(\mathrm{thy})} = 2.84\,\mathrm{fm}^4 \;, \tag{28}$$

which is very close the real part of the experimental value in eq.(20), $\mathrm{Re}\,\mathcal{B}^{(\mathrm{exp})} = 2.8$ fm^4. We thus conclude that also the dominant real part of the triplet amplitude $\mathrm{Re}\,\mathcal{B}$ can be understood in terms of these well-known tree-level meson exchange diagrams when evaluated fully relativistically. Of course, the relatively large imaginary part $\mathrm{Im}\,\mathcal{B}^{(\mathrm{exp})} = -1.5$ fm^4 remains unexplained in tree approximation. For a discussion of $\mathrm{Im}\,\mathcal{B}$ see appendix B in ref.[1]. As mentioned earlier this imaginary part originates (because of unitarity) from the fact that the $^3P_1$ pp-phase shift is rather large at the pion production threshold. The large imaginary part $\mathrm{Im}\,\mathcal{B}$ thus reflects the strong initial-state interaction in the $^3P_1$ entrance channel.

ETA-MESON PRODUCTION IN PP-COLLISIONS

In this section we will discuss η-production using a similar approach. The T-matrix for η-production in proton-proton collisions, $p_1(\vec{p}) + p_2(-\vec{p}) \to p + p + \eta$, at threshold in the center-of-mass frame reads

$$\mathcal{T}_{\mathrm{cm}}^{\mathrm{th}}(pp \to pp\eta) = \mathcal{C}\,(i\,\vec{\sigma}_1 - i\,\vec{\sigma}_2 + \vec{\sigma}_1\times\vec{\sigma}_2)\cdot\vec{p} \;. \tag{29}$$

with $\mathcal{C}$ the (complex) threshold amplitude for η-production. The η-production threshold is reached at a proton laboratory kinetic energy $T_{\mathrm{lab}}^{\mathrm{th}} = m_\eta(2 + m_\eta/2M) = 1254.6$ MeV, where $m_\eta = 547.45$ MeV denotes the eta-meson mass.

Extraction of the Threshold Amplitude

In the case of η-production near threshold it is also important to take into account the ηp final-state interaction, since the ηN-system interacts rather strongly near threshold. In fact recent coupled-channel analyses of the $(\pi N, \eta N)$-system find for the real part of the ηN-scattering length $\mathrm{Re}\,a_{\eta N} = 0.72$ fm. For comparison, this value is a factor 5.7 larger than the $\pi^- p$ scattering length, $a_{\pi^- p} = 0.125$

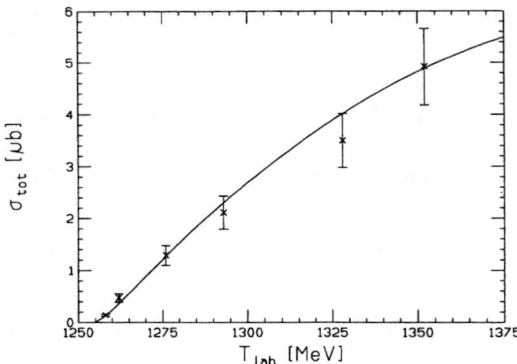

Figure 4: The eta-production total cross section $\sigma_{\text{tot}}(pp \to pp\eta)$ versus T_{lab}.

fm, measured in pionic hydrogen. We assume that the correction due to the S-wave ηp final-state interaction near threshold can be treated in effective range approximation analogous to the S-wave pp final-state interaction. We furthermore make the assumption that the final-state interactions in the pp-subsystem and in the two ηp-subsystems do not influence each other and that they factorize. Using such a factorization ansatz, the total cross section for $pp \to pp\eta$ reads

$$\sigma_{\text{tot}}(T_{\text{lab}}) = |\mathcal{C}|^2 \left(\frac{M}{4\pi}\right)^3 \frac{2\sqrt{T_{\text{lab}}}}{(2M+T_{\text{lab}})^{3/2}} \int_{2M}^{W_{\text{max}}} dW\, W F_p(W)$$

$$\times \int_{s_\eta^-}^{s_\eta^+} ds_\eta\, F_\eta(s_\eta)\, F_\eta(6M^2 + 2MT_{\text{lab}} + m_\eta^2 - W^2 - s_\eta) \ . \quad (30)$$

with $W_{\text{max}} = \sqrt{4M^2 + 2MT_{\text{lab}}} - m_\eta$ the endpoint of the di-proton invariant mass spectrum and $F_p(W)$ given by eq.(3). The variable s_η is the invariant mass squared of the first ηp-pair and $\tilde{s}_\eta = 6M^2 + 2MT_{\text{lab}} + m_\eta^2 - W^2 - s_\eta$ is the invariant mass squared of the second ηp-pair. The correction factor $F_\eta(s_\eta)$ due to the S-wave ηp final-state interaction reads in effective range approximation

$$F_\eta(s_\eta) = \left| 1 - \frac{i\, a_{\eta N}}{2\sqrt{s_\eta}} \sqrt{\lambda(s_\eta, m_\eta^2, M^2)} + \frac{a_{\eta N}\, r_{\eta N}}{8 s_\eta} \lambda(s_\eta, m_\eta^2, M^2) \right|^{-2} . \quad (31)$$

The (complex) ηN-scattering length $a_{\eta N} = (0.72 + 0.26\, i)$ fm and the (complex) ηN-effective range parameter $r_{\eta N} = (-1.50 - 0.24\, i)$ fm are taken from recent analyses. Using eq.(30,31) for the total cross section one finds in a least square fit of the six data points from CELSIUS for the modulus of the threshold amplitude

$$|\mathcal{C}|^{(\text{exp})} = 1.32\, \text{fm}^4 \ , \quad (32)$$

with a total $\chi^2 = 3.8$. The resulting energy dependent cross section from threshold up to $T_{\text{lab}} = 1375$ MeV is shown in Fig. 4 by the solid line together with the data from CELSIUS. It is rather astonishing that one can describe the total cross section data up to 100 MeV above threshold with a constant threshold amplitude $\mathcal{C}$ and a simple factorization ansatz for the three-body final-state interaction.

Diagrammatic Approach

Next, we turn to the evaluation of the relativistic Feynman diagrams contributing to $pp \to pp\eta$ at threshold. The resulting expressions can be simply copied from the case $pp \to pp\pi^0$ making only the substitution $(g_{\pi N}, m_\pi) \to (g_{\eta N}, m_\eta)$. One finds from $\pi^0, \eta, \omega, \rho^0$-exchange

$$\mathcal{C}^{(\pi^0)} = \frac{g_{\eta N} g_{\pi N}^2 m_\eta}{4M^2(m_\pi^2 + Mm_\eta)(2M + m_\eta)} = 0.17\,\text{fm}^4\,, \tag{33}$$

$$\mathcal{C}^{(\eta,\text{dir})} = \frac{g_{\eta N}^3}{4M^2(M + m_\eta)(2M + m_\eta)} = 0.02\,\text{fm}^4\,, \tag{34}$$

$$\mathcal{C}^{(\omega)} = \frac{g_{\eta N} g_{\omega N}^2}{2M(m_\omega^2 + Mm_\eta)(2M + m_\eta)}$$
$$\times \left[2 + \frac{m_\eta}{M}(\kappa_\omega^2 - \kappa_\omega - 1) + \frac{m_\eta^2}{M^2}\kappa_\omega\left(1 + \frac{9}{8}\kappa_\omega\right)\right] = 0.24\,\text{fm}^4\,, \tag{35}$$

$$\mathcal{C}^{(\rho^0)} = \frac{g_{\eta N} g_{\rho N}^2}{2M(m_\rho^2 + Mm_\eta)(2M + m_\eta)}$$
$$\times \left[2 + \frac{m_\eta}{M}(\kappa_\rho^2 - \kappa_\rho - 1) + \frac{m_\eta^2}{M^2}\kappa_\rho\left(1 + \frac{9}{8}\kappa_\rho\right)\right] = 0.51\,\text{fm}^4\,. \tag{36}$$

Note that the ρ^0-exchange has become dominant because of the large tensor-to-vector coupling ratio $\kappa_\rho = 6.1$ and the larger ratio $m_\eta/M = 0.58$. Interestingly, recent measurements of the η-angular distributions suggest the dominance of vector meson exchange. Besides these diagrams with η-emission before and after meson exchange between the protons, one has to account for the strong ηp-rescattering. Microscopically, the strong ηN S-wave interaction originates (among other things) from the nucleon resonance $S_{11}(1535)$ which is supposed to have a very large coupling to the ηN-channel. Instead of introducing this resonance together with several parameters (mass, width, coupling constant), we introduce merely a local $NN\eta\eta$-contact vertex of the form $\mathcal{L}_{\eta N} = K \bar{N}N\eta^2$. The interaction strength K is then determined by the real part of the ηN-scattering length. This means that the pseudovector Born graphs plus the contact vertex sum up to give the empirical value of Re $a_{\eta N} = 0.72$ fm. This leads to the equation

$$4\pi\left(1 + \frac{m_\eta}{M}\right) \text{Re}\, a_{\eta N} = 2K - \frac{g_{\eta N}^2 m_\eta^2}{M(4M^2 - m_\eta^2)}\,, \tag{37}$$

which results in a value of $K = 7.39$ fm. The η-rescattering graph (analogous to Fig. 2c) leads to the following contribution to the threshold amplitude

$$\mathcal{C}^{(\eta,\text{res})} = \frac{g_{\eta N} K}{Mm_\eta(M + m_\eta)} = 0.38\,\text{fm}^4\,. \tag{38}$$

Evidently, all contributions to $\mathcal{C}$ scale with the (empirically not well determined) ηN-coupling constant $g_{\eta N}$. The numbers given above which add up to the empirical value of $|\mathcal{C}| = 1.32\,\text{fm}^4$ follow with $g_{\eta N} = 5.13$. Such a value of $g_{\eta N}$ is close to the SU(3)-prediction and it is consistent with all existing empirical information on $g_{\eta N}$. The main point we want to make here is that even the $pp \to pp\eta$ threshold amplitude can be understood in terms of these well-known meson exchange diagrams when evaluated relativistically. With rather mild assumptions on the coupling constant $g_{\eta N}$ and the form of the ηN-rescattering one can easily reproduce the empirical value $|\mathcal{C}| = 1.32\,\text{fm}^4$.

Results for $pn \to pn\eta$

The threshold T-matrix for η-production in NN-collisions takes the general form

$$T_{\text{cm}}^{\text{th}}(NN \to NN\eta) = \frac{\mathcal{C}}{4}(i\vec{\sigma}_1 - i\vec{\sigma}_2 + \vec{\sigma}_1 \times \vec{\sigma}_2) \cdot \vec{p}\,(3 + \vec{\tau}_1 \cdot \vec{\tau}_2)$$
$$+ \frac{\mathcal{D}}{4}(i\vec{\sigma}_1 - i\vec{\sigma}_2 - \vec{\sigma}_1 \times \vec{\sigma}_2) \cdot \vec{p}\,(1 - \vec{\tau}_1 \cdot \vec{\tau}_2)\,. \quad (39)$$

Here, the amplitude $\mathcal{C}$ (introduced in eq.(29)) belongs to the transition $^3P_0 \to {}^1S_0 s$ with total isospin $I_{\text{tot}} = 1$ and the amplitude $\mathcal{D}$ belongs to the transition $^1P_1 \to {}^3S_1 s$ with total isospin $I_{\text{tot}} = 0$. The latter amplitude $\mathcal{D}$ is accessible only through the reaction $pn \to pn\eta$. For the sake of completeness, we give the contributions of the one-boson exchange diagrams to the triplet amplitude $\mathcal{D}$,

$$\mathcal{D}^{(\pi)} = -3\,\mathcal{C}^{(\pi^0)} = -0.51\,\text{fm}^4\,, \quad \mathcal{D}^{(\eta)} = \mathcal{C}^{(\eta,\text{dir})} + \mathcal{C}^{(\eta,\text{res})} = 0.40\,\text{fm}^4\,, \quad (40)$$

$$\mathcal{D}^{(\omega)} = \frac{g_{\eta N} g_{\omega N}^2}{2M(m_\omega^2 + Mm_\eta)(2M + m_\eta)}$$
$$\times \left[2 + \frac{m_\eta}{M}(1 - \kappa_\omega - \kappa_\omega^2) - \frac{m_\eta^2}{M^2}\kappa_\omega\left(1 + \frac{7}{8}\kappa_\omega\right)\right] = 0.44\,\text{fm}^4\,, \quad (41)$$

$$\mathcal{D}^{(\rho)} = \frac{3g_{\eta N} g_{\rho N}^2}{2M(m_\rho^2 + Mm_\eta)(2M + m_\eta)}$$
$$\times \left[-2 + \frac{m_\eta}{M}(\kappa_\rho^2 + \kappa_\rho - 1) + \frac{m_\eta^2}{M^2}\kappa_\rho\left(1 + \frac{7}{8}\kappa_\rho\right)\right] = 1.54\,\text{fm}^4\,, \quad (42)$$

which add up to $\mathcal{D}^{(\text{thy})} = 1.87\,\text{fm}^4$. The expression for the total cross section of the reaction $pn \to pn\eta$ is obtained from eq.(30) replacing the factor $2|\mathcal{C}|^2 F_p(W)$ by $|\mathcal{C}|^2 F_s(W) + |\mathcal{D}|^2 F_t(W)$. Recently, total cross sections for $pn \to pn\eta$ have been extracted from measurements of the process $pd \to pp n \eta$ at CELSIUS. A best fit of the two data points closest to threshold: $\sigma_{\text{tot}}(T_{\text{lab}} = 1295\,\text{MeV}) = (11 \pm 1.6)\,\mu\text{b}$ and $\sigma_{\text{tot}}(T_{\text{lab}} = 1321\,\text{MeV}) = (20 \pm 1.9)\,\mu\text{b}$, gives

$$|\mathcal{D}|^{(\text{exp})} = 2.3\,\text{fm}^4\,. \quad (43)$$

Compared to this value the theoretical prediction, $\mathcal{D}^{(\text{thy})} = 1.87\,\text{fm}^4$, is only about 20% too small.

OMEGA-MESON PRODUCTION IN PP-COLLISIONS

In this section we discuss briefly omega-production using the same approach. The T-matrix for $\omega(782)$-production in pp-collisions, $p_1(\vec{p}) + p_2(-\vec{p}) \to p + p + \omega$, at threshold in the cm frame reads

$$\mathcal{T}^{\text{th}}_{\text{cm}}(pp \to pp\omega) = \Omega\,(i\,\vec{\sigma}_1 - i\,\vec{\sigma}_2 + \vec{\sigma}_1 \times \vec{\sigma}_2) \cdot (\vec{\epsilon} \times \vec{p}) \quad , \tag{44}$$

with Ω the (complex) threshold production amplitude and $\vec{\epsilon}$ the ω-meson polarization vector. Using a formula analogous to eq.(2) for the total cross section one finds in a best fit of the five points near threshold measured recently at SATURNE,

$$|\Omega|^{(\text{exp})} = 0.53\,\text{fm}^4 \quad . \tag{45}$$

The resulting energy dependent total cross section is shown in Fig. 5 by the full line. If one furthermore takes into account the finite ω-decay width $\Gamma_\omega = 8.4\,\text{MeV}$ by smearing the cross section over an ω-mass distribution of Breit-Wigner shape one obtains the dashed curve in Fig. 5. In this case the best fit value is $|\Omega|^{(\text{exp})} = 0.46\,\text{fm}^4$, which does not differ much from eq.(45). The contributions of (tree-level)

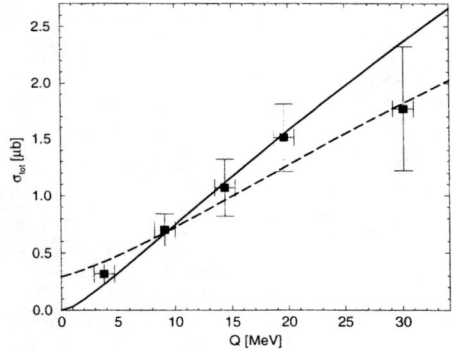

Figure 5: The total cross section $\sigma_{\text{tot}}(pp \to pp\omega)$ versus the cm excess energy Q.

meson-exchange diagrams to the threshold amplitude Ω have been calculated in ref.[2]. It is essential that ω-emission from proton lines interferes destructively with ω-emission from the anomalous $\omega\rho\pi$-vertex. The contribution of scalar σ-meson exchange turns out to be negligibly small. Most interestingly, it is found in ref.[2] that the empirical value $|\Omega| = 0.53\,\text{fm}^4$ can be reproduced by $(\pi, \eta, \omega, \rho)$-exchange diagrams and ω-emission from the anomalous $\omega\rho\pi$-vertex using the same coupling constants as for the other reactions $NN \to NN\pi, NN\eta$. The set of coupling constants used here gives $\Omega^{(\text{thy})} = (0.40 + 0.04 + 0.26 + 0.72 - 0.89)\,\text{fm}^4$. Due to the mentioned large cancelations this agreement also holds for the somewhat smaller value $|\Omega|^{(\text{exp})} = 0.46\,\text{fm}^4$ after a slight decrease of $G_{\omega\rho\pi}$. With regard to the simplicity of the approach presented here this is indeed very remarkable.

KAON-PRODUCTION IN PP-COLLISIONS

Finally, we consider kaon and lambda-hyperon production in pp-collisions, $pp \to p\Lambda K^+$, near threshold using the same approach. The corresponding T-matrix at threshold in the cm frame reads

$$\mathcal{T}_{\rm cm}^{\rm th}(pp \to p\Lambda K^+) = \frac{K_s}{\sqrt{3}}(i\vec{\sigma}_1 - i\vec{\sigma}_2 + \vec{\sigma}_1 \times \vec{\sigma}_2) \cdot \vec{p} + \frac{K_t}{\sqrt{3}} i(\vec{\sigma}_1 + \vec{\sigma}_2) \cdot \vec{p} \quad , \qquad (46)$$

with the amplitudes K_s and K_t belonging to the singlet transition $^3P_0 \to {}^1S_0 s$ and the triplet transition $^3P_1 \to {}^3S_1 s$. Using a formula analogous to eq.(17) for the total cross section with final state interaction set equal in $^1S_0$ and $^3S_1$ $p\Lambda$-states, one finds in a best fit of the seven data points near threshold measured at COSY,

$$\mathcal{K}^{(\exp)} = \sqrt{2|K_s|^2 + |K_t|^2} = 0.38\,{\rm fm}^4 \quad . \qquad (47)$$

The corresponding energy dependent total cross section is shown in Fig. 6. In ref.[3]

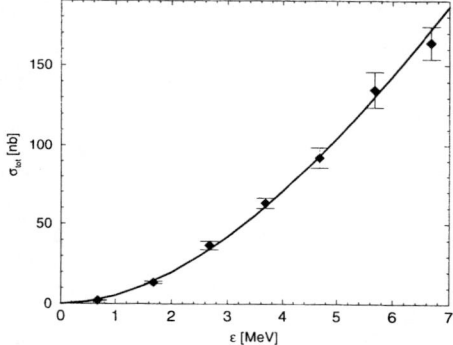

Figure 6: The total cross section $\sigma_{\rm tot}(pp \to p\Lambda K^+)$ versus the cm excess energy ϵ.

the contributions of various vector and pseudoscalar meson exchange diagrams to K_s and K_t were calculated using SU(3)-symmetry. It was found that subprocesses like ω- or K^+-exchange or the $S_{11}(1650)$-excitation alone can explain the value $\mathcal{K}^{(\exp)} = 0.38\,{\rm fm}^4$. With reasonable assumptions on couplings and the cutoff $\Lambda_c = 1.5\,{\rm GeV}$ entering a common meson-nucleon form factor this also holds for the total sum of many processes. This is possible because of cancelations of terms with different sign and the freedom to shift strength between K_s and K_t when only the combination $\mathcal{K} = 0.38\,{\rm fm}^4$ is fixed by the data. For further details on the subject see ref.[3] and the references cited therein.

REFERENCES

[1] V. Bernard et al., *Eur. Phys. J.* **A4** (1999) 259; nucl-th/9806013 and refs. therein
[2] N. Kaiser, *Phys. Rev.* **C** (1999) in print; nucl-th/9907114 and refs. therein
[3] N. Kaiser, *Eur. Phys. J.* **A5** (1999) 105; nucl-th/9902046 and refs. therein

Off-Mass-Shell πN Scattering and $pp \to pp\pi^0$

M. T. Peña[*,**], S. A. Coon[†], J. Adam Jr.[‡], and A. Stadler[*,⋆]

[*] *Centro de Física Nuclear, 1699 Lisboa, Portugal*
[**] *CFIF,Instituto Superior Técnico, 1096 Lisboa, Portugal*
[†] *Physics Department, New Mexico State University, Las Cruces, NM 88003, USA*
[‡] *Institute of Nuclear Physics, Řez n. Prague, CZ-25068, Czech Republic*
[⋆] *Departamento de Física, Universidade de Évora, 7000 Évora, Portugal*

Abstract. We adapt the off-shell πN amplitude of the Tucson-Melbourne three-body force to the half-off-shell amplitude of the pion rescattering contribution to $pp \to pp\pi^0$ near threshold. This *pion* rescattering contribution, together with the impulse term, provides a good description of the data when the half-off-shell amplitude is linked to the phenomenological invariant amplitudes obtained from meson factory πN scattering data.

The precise measurements of $pp \to pp\pi^0$ [1,2] could be used to calibrate or constrain the πN scattering amplitude F^+ underlying 2π exchange three-nucleon forces, in a manner complementary to the standard constraints of on-mass-shell πN data and the implementation of chiral symmetry [3,4]. To see this, consider a two-pion-exchange three-body-force diagram and strip off one of the outer nucleons so that the emerging pion is on its mass shell. The result is the pion "rescattering" diagram found to be tiny if assumed to be proportional to the tiny isospin even s-wave πN scattering length. However, the off-mass-shell πN amplitudes of PCAC-current algebra [3,4] have s-wave terms which are of the same magnitude as the p-wave terms familiar from Δ-isobar models. A qualitative estimate of $pp \to pp\pi^0$ due to the impulse diagram plus half-off-mass-shell pion rescattering diagram was given a long time ago by Hachenberg and Pirner [5]. Later calculations with half-off-shell amplitudes appear to confirm the Hachenberg-Pirner findings of an enhancement of the cross section via *s*-wave pion rescattering [6,7]. In contrast, the pion rescattering diagram calculated with chiral perturbation theory appears to decrease the theoretical cross section rather far below the data [8].

We calculate in momentum space the non-relativistic impulse term plus half-off-shell pion rescattering term. The T-matrix which enters into the latter is

$$T_\pi^{TM} = \frac{-i}{(2\pi)^3} \frac{g}{2m} \boldsymbol{\sigma}_2 \cdot \boldsymbol{k} \frac{1}{\mu^2 - k^2}(-\bar{F}^+ - \Delta F^+) \qquad (1)$$

where k is the four-momentum of the pion exchanged between protons 1 and 2 ($\boldsymbol{k} = \boldsymbol{p}_2' - \boldsymbol{p}_2$), and F represents the appropriate invariant amplitude of $\pi(k) + N(p_1) \to \pi(q) + N(p_1')$, (proton 1 emitting the real pion). The Tucson-Melbourne (TM) Z-graph contribution (labeled ΔF^+) is given in Refs. [3,4] and the (covariant nucleon pole removed) non-spin flip even current algebra πN amplitude for general pion four momenta q, k is

$$\bar{F}^+(\nu, t, q^2, k^2) = [(1-\beta)(\frac{q^2+k^2}{\mu^2} - 1) + \beta(\frac{t}{\mu^2} - 1)]\frac{\sigma}{f_\pi^2} + C^+(\nu, t, q^2, k^2) \quad (2)$$

where σ is the pion-nucleon σ term, $f_\pi \approx 93$ MeV, and C^+ contains the higher order Δ isobar contribution calculated dispersively [9]. The latter amplitude must have the simple form [3,9]

$$C^+(\nu, t, q^2, k^2) = c_1 \nu^2 + c_2 q \cdot k + O(q^4). \quad (3)$$

On the other hand, the assumed form of the multiplier of σ/f_π^2 (adapted [3,10] for πN scattering from the $SU(3)$ generalization of the Weinberg low energy expansion for $\pi\pi$ scattering) is such that $\bar{F}^+$ satisfies the soft pion theorems. The c_2 and β constants in the coefficient of the $q \cdot k$ term can be eliminated in favor of the on-shell (measurable) quantity $\bar{F}^+(0, \mu^2, \mu^2, \mu^2)$ [3,4]. We expand $\bar{F}^+$ in powers of q, k and drop terms of $\mathcal{O}(\mu^2/m^2)$ to get a nonrelativistic amplitude with which we do quantum mechanics. In the kinematics of the Tucson-Melbourne 2π exchange three-body force, the quantity ν^2 is of $\mathcal{O}(\mu^4/m^2)$ and $c_2\nu^2$ is therefore dropped from the (two pions off-mass-shell) amplitude. It is easy to see, however, that, exactly at pion production threshold, the needed values in $\bar{F}^+$ of (1) are $\nu = \mu$, $t = -m\mu$, $q^2 = \mu^2$, $k^2 = -m\mu$, so the quantity $c_1\nu^2$ must be retained in a realistic calculation (which, by the way, should *not* "freeze" the amplitude at the threshold values). The retention of $c_1\nu^2$ and placing of q^2 on-shell for the produced pion are the only changes from the structure of the TM amplitude in the three-nucleon force. We follow Ref. [11] and remove a spurious term from $(-\bar{F}^+ - \Delta F^+)$ in Eq. 1; it corresponds to a pion produced directly from a four-fermion (contact) interaction and should not be present in a rescattering diagram.

The three parameters, $\sigma/f_\pi^2 \approx 1.35\mu^{-1}$, $\bar{F}^+(0, \mu^2, \mu^2, \mu^2) \approx -0.08\mu^{-1}$, and and $c_1 \approx +1.23\mu^{-3}$ of the present TM πN amplitude are found from a recent interior dispersion analysis [12] of the SP98 phase shift solution to meson factory data. The monopole $\pi^0 NN$ vertex function of the exchanged pion reproduces the 2% Goldberger-Treiman discrepancy [10] suggested by the current πN and NN data. Our results are shown in Figure 1 and compared with the data points labeled IUCF and Celsius from Refs. [1,2] respectively and with our calculation of the ChPT [8] treatment of pion rescattering.

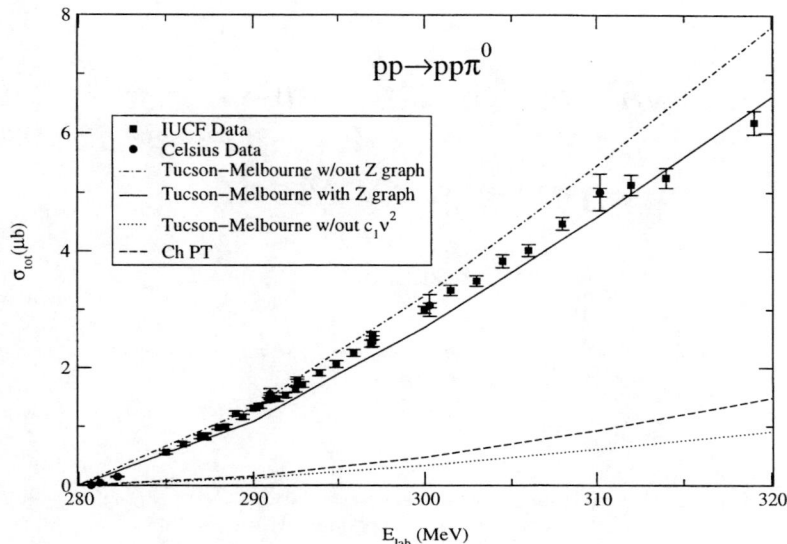

FIGURE 1. Cross section for $pp \to pp\pi^0$ using the Bonn-B NN potential for the initial and final state interaction of the two protons. All calculations include both impulse and pion rescattering diagrams. The "frozen kinematics" approximation is not used.

REFERENCES

1. H. O. Meyer, et al., Nucl. Phys. **A539**, 633(1992).
2. A. Bondar, et al., Phys. Lett. B **356**, 8 (1995).
3. S. A. Coon, et al., Nucl Phys. **A317**, 242 (1979).
4. S. A. Coon and M. T. Peña, Phys. Rev. C **48**, 2559 (1993).
5. F. Hachenberg and H. J. Pirner, Ann. Phys. (N. Y.), **112**, 401 (1978).
6. V. P. Efrosinin, D. A. Zaikin, and I. I. Osipchuk, Z. Phys. **A322** 573 (1985); Phys. Lett. **B246**, 10 (1990).
7. E. Hernández and E. Oset, Phys. Lett. **B350**, 158 (1995).
8. B. Y. Park, et al., Phys. Rev. C **53**, 1519 (1996); T. D. Cohen, et al., Phys. Rev. C **53**, 2661 (1996).
9. M. D. Scadron and L. R. Thebaud, Phys. Rev. D **9**, 1544 (1974); M. D. Scadron, *Few Body Dynamics*, editors A. N. Mitra, et al., North Holland(1976), p. 325.
10. S. A. Coon, lectures at 11th Indian-Summer School, Prague, Sept. 1998; Czech. J. Phys. **49** 1235 (1999); nucl-th/9906011.
11. J. L. Friar, D. Hüber, and U. van Kolck, Phys. Rev. C **59**, 53 (1999).
12. W. B. Kaufmann, G. E. Hite, and R. J. Jacob, πN Newsletter, No. 13, (1997) ed. D. Drechsel et al., p. 16; W. B. Kaufmann and G. E. Hite, Phys. Rev. C **60**, 055204 (1999).

Observables for the $pd \to {}^3\text{H}\pi^+$ and $pd \to {}^3\text{He}\pi^0$ Reactions in a $pp \to d\pi^+$ Model

W. R. Falk

Department of Physics and Astronomy, University of Manitoba
Winnipeg, MB, Canada R3T 2N2

Abstract. Differential cross sections and spin observables A_y, iT_{11}, T_{20}, and T_{22} are calculated for the $pd \to {}^3\text{H}\pi^+$ reaction in a $pp \to d\pi^+$ model at energies near threshold. The results are compared with experimental data for the reactions $\vec{p}d \to {}^3\text{He}\pi^0$ and $\vec{d}p \to {}^3\text{He}\pi^0$. Good agreement of these predictions with the data for the proton analyzing powers is obtained, and for most of the other observables satisfactory agreement is found. Effects of various assumptions in the model are investigated and discussed.

INTRODUCTION

Understanding pion production in the three nucleon $p+d$ system is important since it represents a level of complexity next to that of the elementary $NN \to NN\pi$ reaction. Effects of the nuclear environment may begin to manifest themselves at this level which has major implications for dealing with $A(p,\pi^+)B$ reactions.

In general, theoretical descriptions of $pd \to X\pi$ reactions have not been particularly successful, despite great efforts by many groups over the years [1]. In order to circumvent many of the problems that beset fully microscopic model calculations a number of investigators have used experimental data from $NN \to NN\pi$ reactions as input to the three nucleon sector [2,3]. Following this approach, we apply a phenomenological $pp \to d\pi^+$ model [4] developed for the general case of $A(p,\pi^+)B$ reactions, for predicting the numerous experimental observables measured for the $pd \to {}^3\text{He}\pi^0$ reaction near threshold. Although the model specifically calculates observables for (p,π^+) reactions, via the $pp \to d\pi^+$ process, it can also be used to calculate observables for the π^0 reaction, provided that the elementary $pn \to (pn)\pi^0$ process occurs primarily through the NN isospin transition $1 \to 0$. This should be a valid assumption near threshold.

TABLE 1. Low energy amplitudes for the $pp \to d\pi^+$ reaction.

l_π	Amplitude ($mb^{1/2}$)	Phase (deg)
0	$\|a_1\| = 0.86\eta^{1/2}$	-6.0
1	$\|a_0\| = 0.086\eta^{3/2}$	-26.8
1	$\|a_2\| = 2.15\eta^{3/2}$	9.6
2	$\|a_6\| = 0.52\eta^{5/2}$	-2.9

TABLE 2. Bombarding energies (lab) and pion (c.m.) momenta.

$\vec{p}+d \to {}^3\text{He}+\pi^0$		$\vec{d}+p \to {}^3\text{He}+\pi^0$		$p+d \to {}^3\text{H}+\pi^+$	
T_p (MeV)	η_{π^0}	T_d (MeV)	η_{π^0}	η_{π^+}	T_p (MeV)
200.5	0.127	400.75	0.126	0.145	209.2
205.0	0.240	409.75	0.239	0.256	214.2
210.0	0.323	419.75	0.322	0.337	219.5

MODEL

The model is described in detail in Ref. [4]. It uses as input experimental amplitudes for the $pp \to d\pi^+$ reaction. Low energy parameters for this reaction have been measured by a number of investigators [5]. A representative set of amplitudes extracted from these results for the lowest pion partial waves is shown in Table 1.

In the model the dynamical parameter that defines the energy at which the $pp \to d\pi^+$ reaction amplitudes are evaluated is specified by setting the c.m. pion momentum in the $pp \to d\pi^+$ reaction equal to that in the $A(p,\pi^+)B$ reaction. Furthermore, in comparing $pn \to d\pi^0$ and $pp \to d\pi^+$ reactions, account must be taken of the the different masses of the particles involved, and the Coulomb effects present in the π^+ reaction. This was implemented by setting $C_0^2 \eta_{\pi^+} = \eta_{\pi^0}$, where C_0^2 is the Coulomb correction factor. The resulting energies for comparing the various reactions are shown in Table 2. Experimental measurements for the $\vec{p}d \to {}^3\text{He}\pi^0$ reaction [6] have been made at the proton energies shown in column 1; measurements for the $\vec{d}p \to {}^3\text{He}\pi^0$ reaction [7] have been made at the deuteron energies shown in column 3. The corresponding proton bombarding energies used in the calculations for the $pd \to {}^3\text{H}\pi^+$ reaction are shown in column 6.

RESULTS AND DISCUSSION

The reference calculations used the $pp \to d\pi^+$ reaction amplitudes given in Table 1, a D state component in the wave function for the deuteron target, and an S state only ${}^3\text{H}$ final nucleus wave function [4]. Results of these calculations are shown by the solid lines in Figs. 1-2 for A_y and T_{20}, respectively. The number displayed in each panel represents the proton bombarding energy shown in the last column of

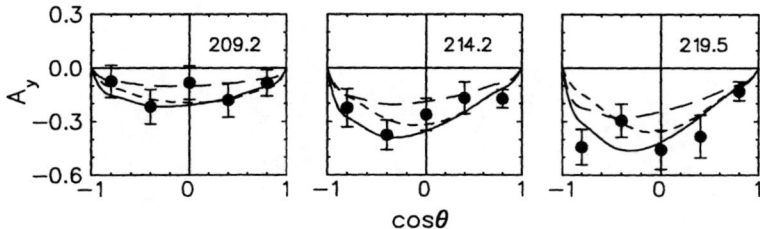

FIGURE 1. Angular distributions of A_y for the $pd \to X\pi$ reaction. The data are from Ref. [6].

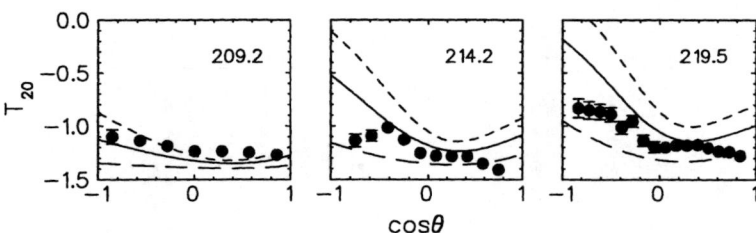

FIGURE 2. Angular distributions of T_{20} for the $pd \to X\pi$ reaction. The data are from Ref. [7].

Table 2. The long dashes represent calculations using only $pp \to d\pi^+$ reaction amplitude a_1 and the a_2 amplitude reduced to half its value. Finally, the short dashes represent calculations where the ^{2}H target wave function is taken to be a pure S state. Analyzing power angular distributions fit the data very well using the full amplitude set and a D state in ^{2}H. Some reduction in the p-wave amplitude would provide an improved fit to the T_{20} distributions. Angular distributions of the differential cross sections follow the trend of the data but exhibit a somewhat greater forward-backward asymmetry. Experimental values for iT_{11} and T_{22} are consistent with zero at all angles and energies. The calculated results for T_{22} are very small, consistent with experiment, but for iT_{11} values as large as 0.25 are obtained. These observations suggest that using the on-shell $pp \to d\pi^+$ amplitudes affect different observables in different ways.

REFERENCES

1. L. Canton, et al., Phys. Rev. C **57**, 1588 (1998); Phys. Rev. C **56**, 1231 (1997).
2. Jean-Francois Germond and Colin Wilkin, J. Phys. G. **16** 381 (1990).
3. H.O. Meyer and J.A. Niskanen, Phys. Rev. C **47**, 2474 (1993).
4. W.R. Falk, Phys. Rev. C **50**, 1574 (1994).
5. E. Korkmaz, et al., Nucl. Phys. **A535**, 637 (1991); M. Drochner et al., Phys. Rev. Lett. **77**, 454 (1996); P. Heimberg et al., Phys. Rev. Lett. **77**, 1012 (1996).
6. L.K. Warman, Ph.D thesis, Indiana University, 1998.
7. V.N. Nikulin et al., Phys. Rev. C **54**, 1732 (1996).

Pion production mechanism in nucleon-nucleon collisions

K. Tamura*, Y. Maeda† and N. Matsuoka†

* Physics Division, Fukui Medical University, Fukui 910-1193, Japan
† Research Center for Nuclear Physics, Osaka University, Osaka 567-0047, Japan

Abstract. The theoretical analysis for the threshold pion production has been performed by using the models of heavy-meson exchange and pion-rescattering including the effect of nucleon resonances. The d-wave pion production amplitude is discussed about the angular distribution of π^0 production. The spin correlation observables give information about the partial wave amplitudes.

INTRODUCTION

The behavior of the pion production near threshold provides us with considerable information on the nature of low-energy strong-interaction physics. Recently several measurements of the reaction $NN \to NN\pi$ near the pion-production threshold have been performed. The cross section of $pp \to pp\pi^0$ was measured very precisely [1,2]. It has been pointed out that large contribution of the s-wave pion-production amplitude is necessary to reproduce the total cross section of the reaction $pp \to pp\pi^0$. Lee and Riska found that the short range effect for the two-body pion-production operator gave enhancement of the cross section by a factor 3~5 (heavy-meson exchange model) [3,4]. Hernández and Oset took into account the off-shell properties of the πN amplitude in the pion-rescattering process (pion-rescattering model) [5]. They found constructive interference between the amplitudes of the direct pion-production process and the pion-rescattering process. Hanhart et al. examined the off-shell πN amplitude with a realistic meson-theoretical model [6]. The works with chiral perturbation method [7] predict the s-wave pion-production amplitude with the opposite sign to the results of the heavy-meson exchange model and the pion-rescattering model.

In this paper, theoretical calculations are performed by using the heavy-meson exchange model (HM) and the pion-rescattering model (RES). Recently spin correlation measurement has been done at IUCF [9]. Large probability of s-wave amplitude is necessary to explain the experimental data.

THEORETICAL MODEL

The parameters of the heavy-meson-exchange model are taken from Bonn potential. We adopted Hamilton model for the pion-rescattering model [5]. The effects of the Δ and $N(1440)$ resonance and the meson dissociation mechanism ($\rho\pi\pi$, $\omega\rho\pi$) are also taken into account. The partial waves for the initial and final states of the NN system are summed up to $\ell = 6$ incorporating the Coulomb interaction. In Fig.1 are shown the calculated results of the total cross section. The hatched regions show results with the models of heavy-meson exchange and pion rescattering using the NN wave function given by the various nuclear forces (Paris, Bonn, Argonne V14 and Reid soft-core). The calculated results give good fit to the experimental data. Our results also show good explanation for $d\sigma/d\Omega$ and A_y of the reactions $pp \to pn\pi^+$ and $pp \to d\pi^+$.

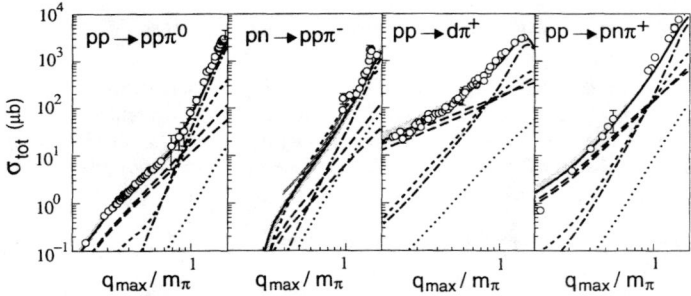

FIGURE 1. Total cross section of pion production. Curves show the result with Paris potential. Solid curve is the result of HM. Other lines show each mechanism: long dash (HM, RES), short dash (direct), dash-dot (Δ and N^*), dot (meson dissociation).

D-WAVE PION PRODUCTION MECHANISM

The angular distribution of π^0 production was measured at RCNP and TSL. The experimental data show convex structure for the angular distribution. This structure mainly comes from the d-wave amplitude $^3P_2$-$^3F_2 \to {}^1S_0 +$ d-wave.

The NN state $^3P_2$-$^3F_2$ has strong coupling to the ΔN state. By this reason the angular dependence of the π^0 production is sensitive to the short range structure of the Δ excitation mechanism. The pion production operator of Δ excitation contains the factor $\vec{q}^{\,2}/(\vec{q}^{\,2} + \mu^2)$. Here $\vec{q}$ and μ are the momentum and mass of the exchanged meson. If one write this factor as $\vec{q}^{\,2}/(\vec{q}^{\,2} + \mu^2) = 1 - \mu^2/(\vec{q}^{\,2} + \mu^2)$,

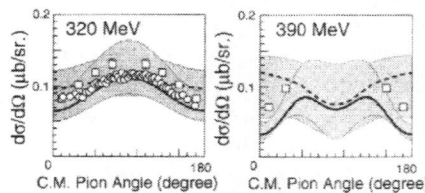

FIGURE 2. Angular distribution of π^0 production. Solid (Dashed) line is the calculation without (with) the contact interaction. Data are from RCNP(box) and TSL(circle) [8].

then the term 1 gives a contact interaction. Because we introduced the vertex form factor ($\Lambda_{\pi NN}$=700 MeV), the contact interaction is smeared out and gives sizable effect at short range region ($r \leq 0.5$ fm). In Fig.2, the calculations including the contact interaction (dashed line) failed to explain the convex structure of experimental data. The results without the contact interaction are shown as solid lines which give good fit to the experimental data.

SPIN CORRELATION OBSERVABLES ($\vec{P}\vec{P} \to PP\pi^0$)

Recently Meyer et al. [9] measured the spin correlation observables with the beam and target polarization. By using the partial wave amplitude Pp, Ps and Ss, we can see the global structure of the observables as

$$A_{xx} + A_{yy} = 2Ss^2/\sigma, \qquad A_{xx} - A_{yy} = 4Pp^2/15\sigma,$$
$$A_{zz} = -Ss^2 - Ps^2 + Pp^2/3, \quad A_y = \pi\sqrt{5}Pp \times Ps \sin\delta/10\sigma$$

Here σ means the total cross section and δ is given by the phase shift of NN states. Fig.3 show the experimental data and calculated results. The combination $A_{xx} + A_{yy}$ can be explained by our model satisfactorily. Therefore the model gives reasonable magnitude of the Ss amplitude. But in the case of the combination $A_{xx} - A_{yy}$, our model gives overestimation. Furthermore the calculated result of analyzing power is very small. Analyzing power A_y is proportional to the product of p-wave amplitude Pp and s-wave amplitude Ps in which nucleons of the final state couple to P-wave. In the case of Ss amplitude, the main contribution comes from the transition $^3P_0 \to {}^1S_0 + s$. Because the $^1S_0$ state has large strength at short range region ($r \leq 0.5$ fm), the wave function has good overlap with the Yukawa function of the heavy meson exchange. Contrary to this, in the Ps-amplitude case, the overlap between the wave function and the Yukawa function becomes small because the P-wave NN state has no strength at short range region. As the result the Ps amplitude becomes very small. In other word, we do not have s-wave pion production mechanism which might be effective around 1 fm region in our model.

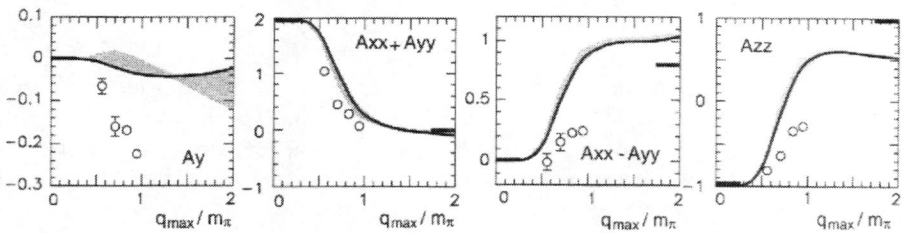

FIGURE 3. Spin correlation observables of $pp \to pp\pi^0$. Solid line is the calculation of HM with Paris potential. The data are from Ref. [9].

REFERENCES

1. H.O. Meyer et al., Phys. Rev. Lett. 65 (1990) 2846.
2. A. Bondar el al., Phys. Lett. B356 (1995) 8 .
3. T.-S.H. Lee and D. Riska, Phys. Rev. Lett. 70 (1993) 2237.
4. C.J. Horowitz et al., Phys. Rev. C49 (1994) 1337.
5. E. Hernández and E. Oset, Phys. Lett. B350 (1995) 158.
6. C. Hanhart, et al., Phys. Lett. B358 (1995) 21.
7. T. Sato et al., Phys. Rev. C56 (1997) 1246, and references therein.
8. J. Zlomanczuk et al., Phys. Lett. B436 (1998) 251.
9. H.O. Meyer et al., Phys. Rev. Lett. 81 (1998) 3096; nucl-ex/9907017.

Near threshold Λ and Σ^0 production in pp collisions

A. Gasparian[a,b], J. Haidenbauer[a], C. Hanhart[c], L. Kondratyuk[b], and J. Speth[a]

[a]*Institut für Kernphysik, Forschungszentrum Jülich GmbH, D-52425 Jülich, Germany*
[b]*Institute of Theoretical and Experimental Physics, 117258, B.Cheremushkinskaya 25, Moscow, Russia*
[c]*Nuclear Theory Group and INT, Dept. of Physics, University of Washington, Seattle, WA 98195-1560, USA*

Abstract. The reactions $pp \to p\Lambda K^+$ and $pp \to p\Sigma^0 K^+$ are studied near their thresholds. The strangeness production process is described by the π- and K exchange mechanisms. Effects from the final state interaction in the hyperon-nucleon system are taken into account rigorously. It is shown that the experimentally observed strong suppression of Σ^0 production compared to Λ production can be explained by a destructive interference between π and K exchange in the reaction $pp \to p\Sigma^0 K^+$.

Recently the total cross sections for the reactions $pp \to p\Lambda K^+$ and $pp \to p\Sigma^0 K^+$ were measured for the first time near their thresholds, and specifically at the same excess energies [1,2]. It was found that the cross section for the Σ^0 production is about a factor of 30 smaller than the one for the Λ production [2]. This is rather surprising because data at higher energies [3] indicate that the cross section for Λ production exceeds the one for Σ^0 production only by a factor of around 2.5.

In this paper we present results of an exploratory investigation of the origin of the observed strong suppression of the near-threshold Σ^0 production in comparison to the Λ channel. In particular we want to examine a possible explanation that was suggested in Ref. [2], namely effects from the strong ΣN final state interaction (FSI) leading to a $\Sigma N \to \Lambda N$ conversion. We treat the associated strangeness production in the standard distorted wave Born approximation. We assume that the strangeness production process is governed by the π- and K exchange mechanisms as depicted in Fig. 1. In order to have a solid basis for our study of possible conversion effects we employ a microscopic YN interaction model developed by the Jülich group (specifically model A of Ref. [4]). This model is derived in the meson-exchange picture and takes into account the coupling between the ΛN and ΣN channels.

The vertex parameters (coupling constants, form factors) appearing at the πNN

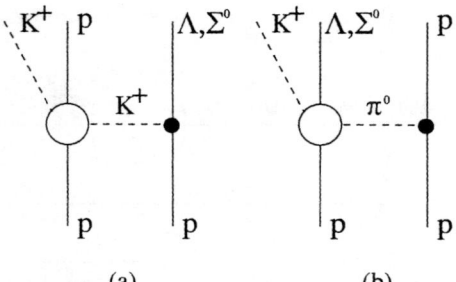

FIGURE 1. Mechanisms for the reactions $pp \to p\Lambda K^+, p\Sigma^0 K^+$ considered in the present investigation: (a) kaon exchange; (b) pion exchange.

and KNY vertices in the production diagrams in Fig. 1 are taken over from the Jülich YN interaction. The elementary amplitudes T_{KN} and $T_{\pi N \to KY}$ are taken from microscopic models of KN scattering [5] and of the reaction $\pi N \to K\Lambda, K\Sigma$ [6] that were developed by our group. However, for simplicity reasons we use the scattering length and on-shell threshold amplitudes, respectively, instead of the full (off-shell) KN and $\pi N \to KY$. The off-shell extrapolation of the amplitudes is done by multiplying those quantities with the same form factor that is used at the vertex where the exchanged meson is emitted. Only s-waves are considered.

We do not take into account the initial state interaction (ISI) between the protons. Therefore we expect an overestimation of the cross sections in our calculation [7]. But since the thresholds for the Λ and Σ^0 production are relatively close together (at $T_{lab} = 1582$ MeV and at $T_{lab} = 1796$ MeV) and, moreover, the energy dependence of the NN interaction is relatively weak in this energy region the ISI effects should be very similar for the two strangeness production channels and therefore should roughly drop out when ratios of the cross sections are taken.

The cross section ratio for K exchange alone and based on the Born diagram is 16, cf. Table 1. Including the YN FSI, i.e. possible conversion effects $\Sigma N \to \Lambda N$, leads to a strong enhancement of the cross section in the Λ channel but only to a moderate enhancement in the Σ^0 channel. As a consequence, the resulting cross section ratio becomes significantly larger than the value obtained from the Born term and, in fact, exceeds the experimental value. In case of pion exchange the Born diagram yield a cross section ratio of 0.9. Adding the FSI increases the cross section ratio somewhat, but it remains far below the experiment. (Note that some values in Table 1 differ from those given in the talk, due to an error in the isospin coefficients found in the meantime.)

Thus, it's clear that, in principle, K exchange alone could explain the cross section ratio - especially after inclusion of FSI effects. However, we also see from Table 1 that π exchange is possibly the dominant production mechanism for the Σ^0 channel and therefore it cannot be neglected. In fact, the two production mechanisms play quite different roles in the two reactions under consideration, cf. Table 1. K exchange yields by far the dominant contribution for $pp \to p\Lambda K^+$. Here the cross

TABLE 1. Contributions of different diagrams to the total cross section of the reactions $pp \to p\Lambda K^+, p\Sigma^0 K^+$. The experimental results for Λ (Σ^0) production are for the excess energy of 13.2 MeV (13.0 MeV). The Jülich YN model A [4] is employed for the final-state interaction.

diagrams		$\sigma_{pp \to p\Lambda K^+}$ [nb]	$\sigma_{pp \to p\Sigma^0 K^+}$ [nb]	$\frac{\sigma_{pp \to p\Lambda K^+}}{\sigma_{pp \to p\Sigma^0 K^+}}$
K exchange	Born term	706	45	16
	with FSI	2310	56	41
π exchange	Born term	68	76	0.9
	with FSI	109	103	1.1
"$K + \pi$"	with FSI	2360	247	9.5
"$K - \pi$"	with FSI	2490	72	35
experiment [2]		505 ± 33	20.1 ± 3.0	25 ± 6

section obtained from π exchange is more than order of magnitude smaller. In case of the reaction $pp \to p\Sigma^0 K^+$, however, π- and K exchange give rise to contributions of comparable magnitude. This feature becomes very important when we now add the two contributions coherently and consider different choices for the relative sign between the π and K exchange amplitudes. In one case (indicated by "$K + \pi$" in Table 1) the π and K exchange contributions add up constructively for $pp \to p\Sigma^0 K^+$ and the resulting total cross section is significantly enhanced over the individual results. On the other hand, the cross section for $pp \to p\Lambda K^+$ changes very little as compared with the result based on K exchange alone, simply because the contribution from π exchange is very small. Therefore, the cross section ratio becomes smaller and is now a factor of 3 below the experimental value. If we choose the other sign between the π and K exchange amplitudes (indicated by "$K - \pi$") the results for $pp \to p\Lambda K^+$ remain basically unchanged. However, in case of $pp \to p\Sigma^0 K^+$ we get a strongly destructive interference between the amplitudes which, in turn, leads to a drastically reduced total cross section. As a consequence, also the cross section ratio changes considerably and is now in rough agreement with the experiment (cf. Table 1).

REFERENCES

1. J.T. Balewski et al., Phys. Lett. B **420**, 211 (1998).
2. S. Sewerin et al., Phys. Rev. Lett. **83**, 682 (1999).
3. V. Flaminio at al., Compilation of cross sections, CERN-HERA report 79-03 (1979).
4. B. Holzenkamp et al., Nucl. Phys. **A500**, 485 (1989).
5. M. Hoffmann et al., Nucl. Phys. **A593**, 341 (1995).
6. M. Hoffmann, Jülich report, No. 3238 (1996); M. Hoffmann, in *IKP/COSY Annual Report 1996*, Jül-3365, p. 131.
7. M. Batinić, A. Švarc, and T.-S. H. Lee, Phys. Scripta **56**, 321 (1997).

STRANGENESS
PRODUCTION

Theoretical Issues in Strangeness Production

Jean-Marc Laget

CEA/Saclay, DAPNIA/SPhN
F91191 GIF-sur-Yvette Cedex, France

Abstract. After pioneering works on hypernuclei, strangeness production mechanisms have been studied in hadron collisions and photoreactions in the sixties. Recent experiments at SATURNE and COSY, in the hadronic sector, as well as ELSA and JLab, in the electromagnetic sector, have confirmed our basic ideas on the reaction mechanisms. In the near future, strangeness production at JLab, HERMES and COMPASS may prove to be a powerful tool to study hadronic matter.

INTRODUCTION

Since the strange quark is not a normal building block of nucleons and nuclei, strangeness production is a powerful tool to study properties of Hadronic Matter. As in the study of any complex system, the inclusion of an "impurity" and the study of its subsequent propagation provides us with a way to reveal configurations or states that can not be reached otherwise. In the fifties, pioneering studies of hypernuclei led to a first estimate of the Nucleon-Hyperon scattering lengths, while more recent studies revealed the inner shell structure of heavy hypernuclei (see Ref [1] for a review). The world data set on strangeness production in pp scattering was very meager [2], until SATURNE was able to accurately determine cross sections and spin observables. Now COSY is in the process of producing a comprehensive set of data near and just above threshold.

In the electromagnetic sector, the sparse data which have been collected in the fifties and early sixties are going to be superseded by works ongoing at ELSA and JLab.

In the written version of this talk, I will concentrate on new (theoretical and experimental) developments, referring the reader to a lecture [3] and a recent paper [4] for a general background and reference to older experiments.

I will start with the Hadronic Sector, deal with the Electromagnetic Sector, discuss what can we learn in the Few Body Systems and end my talk discussing issues in the Strangeness Response of Hadronic Matter.

HADRONIC SECTOR

Physics Landscape

The world set of total cross-sections of the three channels $pp \to pK^+Y$, $(Y = \Lambda, \Sigma^0, \Sigma^+)$, is very well reproduced over the energy range $2 \leq T_P \leq 6$ GeV by the simplest model [5] which assumes π and K meson exchange, supplemented by Nucleon Hyperon rescattering (FSI). The model relies on the experimental amplitudes of each elementary subprocess (elastic Kaon scattering, Hyperon production by pion beams), Hyperon Nucleon coupling constants consistent with SU3, and the Hyperon Nucleon scattering amplitudes of the Nijmingen group [6]. The only free parameter is the cut off mass of the dipole form factor which is used at each meson baryon vertex: the chosen value, $\Lambda_m^2 = 1$ GeV, acts as an overall normalization factor and falls within the range of accepted values in this kind of game. Above 2 GeV, the variation with energy and the relative strength of each of the three channels are well accounted for: more particularly the predicted ratio of the cross sections of the Λ and the Σ^0 production channel is about 3, in good accord with experiment. Also the Σ production channels are found to be dominated by π meson exchange, while the Λ production channel is found to get contribution from both π and K meson exchanges.

This simple model leads also to a excellent agreement with the spectra of the Kaons, emitted at different angles in the reaction $pp \to K^+X$, accurately measured at SATURNE [7]. Above the Plane Wave contribution, the low energy ΛP scattering reproduces the characteristic enhancement, near the end point of the spectrum, while the coupling between the Λ and Σ channels is responsible for the narrow structure near the Σ production threshold. The model reproduces also the angular distribution of these spectra.

Recent Experiments

At COSY, precise determination of the cross sections of strangeness production channels are possible over a wider part of the phase space than at SATURNE. Fig. 1 shows the distribution of the square of the mass of the ΛP system over the full phase space at $T_P = 1.97$ GeV. It exhibits the same enhancements at the Λ and Σ production thresholds, and resembles to the first spectra recently measured at COSY [8]. However, the analysis of the full statistics is needed for a meaningful comparison. When integrated over the full phase space these data lead to accurate value of the total cross-section of the Λ production channel at $T_P = 1.73$ and 1.97 GeV.

Accurate data have also been recorded, both for the Λ [9] and Σ^0 [10] production channels, very close to their respective threshold. Fig. 2 summarizes these results. Without any adjustment, the simple model [5] reproduces fairly well not only the

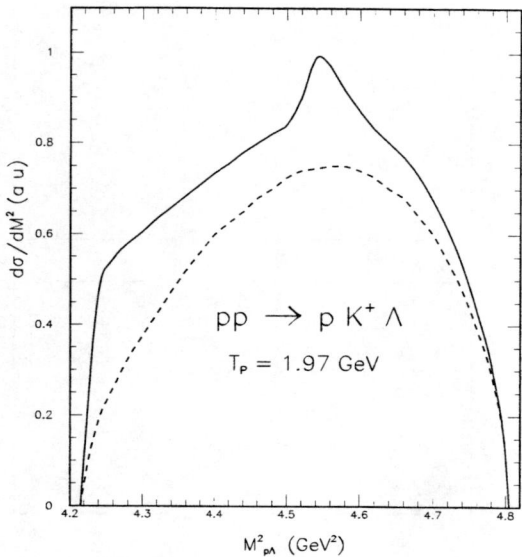

FIGURE 1. The squared mass distribution of the ΛP system in the COSY kinematics. Dashed line: Plane Wave. Full line: FSI.

shape and magnitude (over four decades), but also the ratio between the cross-sections of the Λ and Σ^0 production channels. This ratio varies from about 27 near threshold to about 3 at high energy (above 2 GeV) ! Near threshold the model underestimates the data by about 50%: This can be easily accommodated by adjusting the cut-off mass of the hadronic form factors. I did not play this game, since care has to be taken not to spoil the good agreement with the SATURNE data [7].

The balance between π and K exchange has been beautifully confirmed by the last experiment performed at SATURNE, DISTO [11]. Among other things, the polarization transfer coefficient, D_{NN}, between the incident proton and the Λ was determined in the reaction $\vec{p}p \rightarrow pK^+\vec{\Lambda}$. It is plotted in Fig. 3, against the Feynman variable $X_F = P_\Lambda/(P_\Lambda)_{max}$. When $X_F = 1$, i.e when the Λ is emitted in the direction of the incident beam with the highest momentum, π exchange leads to $D_{NN} = +1$, while K exchange leads to $D_{NN} = -1$. Without any adjustment, the simple model [5] predicts -0.3, in close agreement with the data.

Issues

These findings show that the basic mechanisms are understood, and that the simplest model works. However, discrepancies of the order of 50% still remain in the energy variation of the cross sections. This may be due to the neglect of a

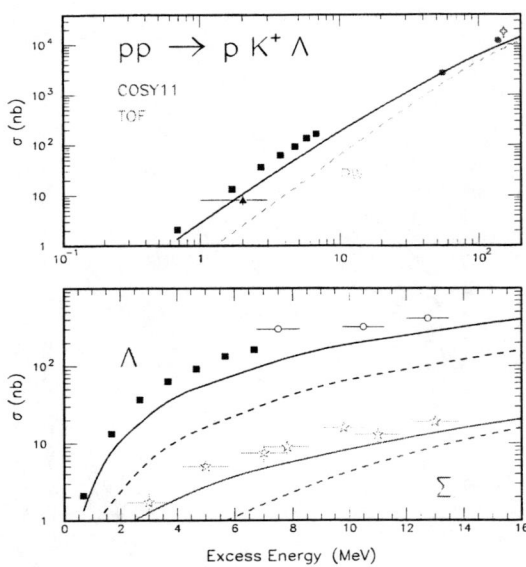

FIGURE 2. The total cross-sections recently determined at COSY. Dashed lines: Plane Wave. Full lines: FSI.

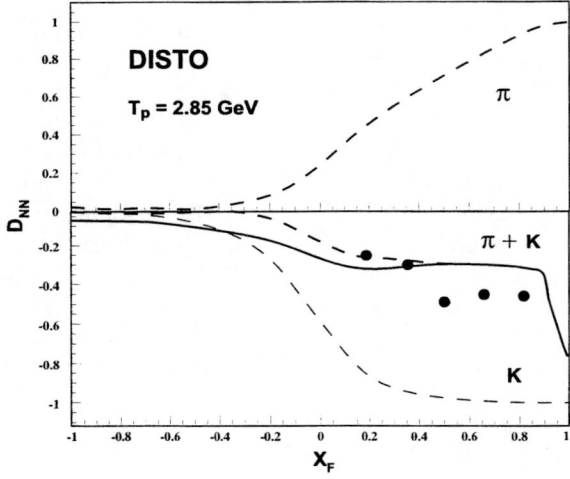

FIGURE 3. Spin transfer coefficient D_{NN} as measured at Saturne. Dashed lines: Plane Wave. Full line: FSI.

possible small contribution of ρ and K^* exchange. This may also reflect the effects of Unitarity constraints. The reason why the simple model [5] works is that the exchange of spin zero mesons saturate the cross section. The corresponding Born Plane Wave amplitudes do not diverge when the energy increases: it is therefore easy to simulate Unitarity constraints by using form factors at the various meson baryon vertices, and supplement the Born term with Nucleon- Hyperon rescattering. This is not the case for the spin 1 (or higher) exchange amplitudes, which diverges with the energy: The use of form factors does not help to cure the problem, and this is the reason why Regge Poles have been introduced. Indeed a Regge Pole description [12] leads to a fair understanding of both the angular distribution and the Λ polarization in the $pp \to pK + \Lambda$ reaction at 400 GeV, again in terms of the elementary amplitudes of the production of Hyperons by mesons.

The implementation of Regge Poles remains to be done at low energy. As I shall discuss now, this has been achieved in the Electromagnetic Sector.

Near the Hyperon production thresholds, hadronic cross-sections are sensitive to Hyperon- Nucleon amplitudes. However a better understanding of the reaction mechanism has to be achieved. The Electromagnetic Sector is in better shape in this respect.

ELECTROMAGNETIC SECTOR

Effective Lagrangian Models

Following the successful description of pion photoproduction in the Δ and the second resonance region, Effective Lagrangian Models start from Born Terms in the tree approximation. However, when Kaon Nucleon coupling constants consistent with SU3 are used, they badly overestimate the experimental cross sections of Hyperon electroproduction channels. The reason is that, contrary to the pion production channels, the energy is higher, more coupled channels are open and unitarity constraints are more difficult to implement. It is possible to accommodate the models with the sparse available data, at the expenses of many additional contributions, which interfere destructively: baryonic resonance exchanges in the s as well as in the u channels, strange mesonic resonances in the t channel, form factors at the various meson baryon vertices.

This is illustrated in Fig. 4 where typical results from Effective Lagrangian Models [13–15] are compared to recent data form ELSA [16] up to 2 GeV, and older data [17] at higher energies. While each reproduces the data in the 1–2 GeV range (they have been tailored for that) they diverge above 2 GeV. The use of form factors at the hadronic vertices helps, but does not cure the disease since it only postpones the divergence at higher energy. The reason is that, in order to fit the data (and since many degree of freedom are available), these models rely heavily on the exchange of particles (mesons in the t channel, baryons in the s and u channels) with non zero spin. The corresponding Born amplitudes are known to diverge with

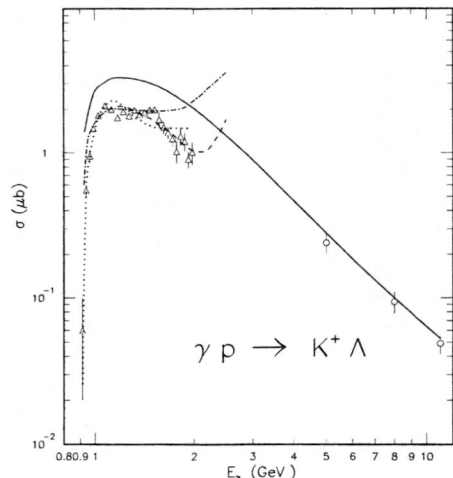

FIGURE 4. The total cross-section of the $\gamma p \to K^+ \Lambda$ reaction. Dot-dashed line: Lagrangian model, without hadronic form factors and off-shell effects [13]. Dashed line: with off-shell effects [14]. Dotted line: with form factors [15]. Full line: Regge model [4].

energy. This is the reason why Regge Models have been introduced. Indeed, a recent application [4] reproduces both the energy variation and the magnitude of the cross sections for strangeness production off the nucleon above 2 GeV. Below, it overestimates the ELSA data, but here it is meant to average over the contributions of the various possible resonances (duality). It is also worthwhile to note that the recent CLAS preliminary data [18] are higher than the ELSA ones, and closer to the Regge model, in the range $E_\gamma = 1.5 - 2.$ GeV.

Regge Models

They provides us with an economical and elegant way to describe meson photoproduction at high energies. Analyticity of the scattering amplitude is built in. The exchange of families (Regge Trajectories) of mesons, in the t channel, and baryons, in the u channel, ensures unitarity, and describes the off-shell behavior of the amplitudes far from the poles. The striking feature is that they exhibit the right energy and momentum dependencies.

Taking advantage of a better knowledge of coupling constants, and of a larger data set, we have revived a Regge Model which reproduces, without free parameters, cross sections and polarization observables, available at high energies, in pion and kaon photoproduction channels [4] as well as electroproduction channels [19,20]. It relies on the full spin structure of the elementary vertices, and is based on the exchange of K and K^* trajectories. I refer the reader to these works for a

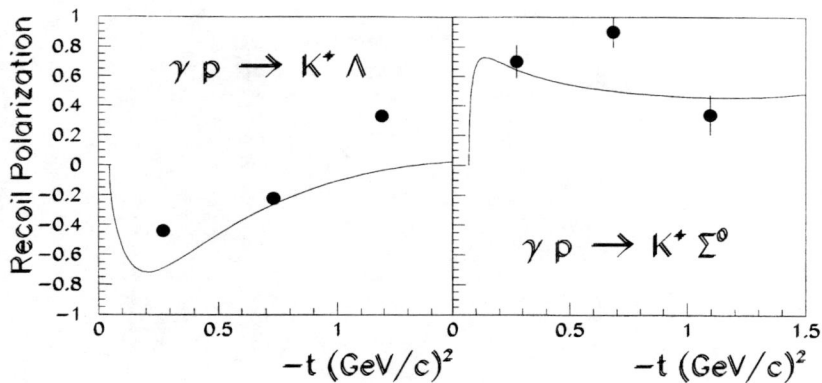

FIGURE 5. Recoil Polarization of the Λ or the Σ emitted in the reaction $\gamma p \to K^+ Y$.

comprehensive review. Let me only flash two recent results: they concern spin observables and constitute a more stringent check of the model.

Fig. 5 shows the recoil polarization of the hyperon recoiling in the reaction $p(\gamma, K^+)Y$, recently determined at ELSA [16]. At low t the Regge model [20] reproduces fairly well the data, and predicts the right sign in both the Λ and Σ channels. This spin observable is particularly sensitive to the modeling of the reaction, since it depends on the imaginary part of the amplitude, and is usually difficult to reproduce. It comes out naturally in the Regge model.

Fig. 6 shows the Transverse and Longitudinal cross sections of the $p(e, e'K^+)\Lambda$ reaction, which have been recently determined at JLab [21] through a Rosenbluth separation. While Lagrangian Models [13,22] predict a linear rise of their ratio with the four momentum transfer Q^2 of the virtual photon, the Regge model [20] correctly reproduces the saturation and the decrease of the ratio at higher virtuality of the photon.

Issues

At low energies (let say below 2 GeV), many resonances contribute. Only a comprehensive experimental determination of the reaction amplitudes (partial wave analysis) will allow to decipher the various contributions and constrain Lagrangian models. On the theoretical side, a serious effort has to be done in order to implement Unitarity constraints in highly coupled channels.

At high energies (above 2 GeV), Regge Model provides us with a successful, simple and economic way to describe data at forward angles ($t \leq 1$ GeV) and backward angles ($u \leq 1$ GeV). In between (around 90°) there is room for hard scatterings, which can be described by saturating Regge trajectories [4].

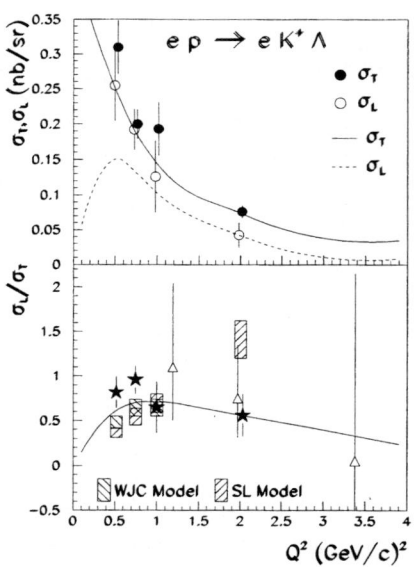

FIGURE 6. The Transverse and Longitudinal cross-sections, and their ratio, as recently measured at JLab in the reaction $p(e, e'K^+)\Lambda$. The curves are the predictions of the Regge model. The hashed areas are the predictions of Effective Lagrangian Models.

FEW BODY SYSTEMS

Such a Regge model of the elementary amplitude can be easily implemented in Few Body systems, following the method [23] which has been successfully used in the pion photoproduction sector. Among others, the determination of the Hyperon-Nucleon scattering amplitudes appears to be particularly appealing at low as well as high energy. The typical diagram is depicted in Fig. 7: the elementary Kaon production amplitude and the deuteron wave function are well under control. It provides the most direct way to access Hyperon-Nucleon scattering, in the absence of Hyperon beams. Furthermore the determination of the polarization of the emitted Hyperon provides us with a strong additional constraint.

Low Energy YN scattering

Low energy YN rescattering leads to characteristic enhancements in the spectrum of the Kaons, emitted in the reaction $D(\gamma, K^+)Yn$, near the Λ and Σ production thresholds. This is clearly apparent in Fig. 8, which shows the latest prediction for a kinematics accessible at CEBAF [24]. This picture is confirmed by a recent measurement shown in Fig. 9. However, the effects are much less dramatic than in proton scattering [5]. The reason is that the dominant mechanism is the quasi-free

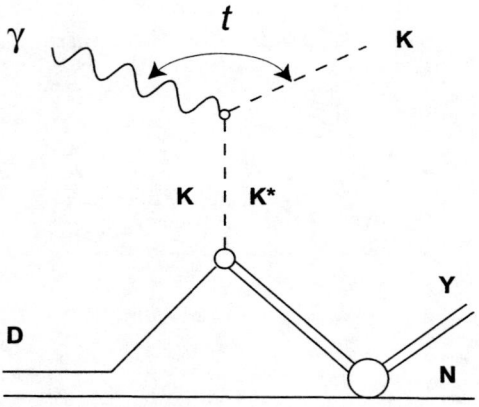

FIGURE 7. The rescattering graph.

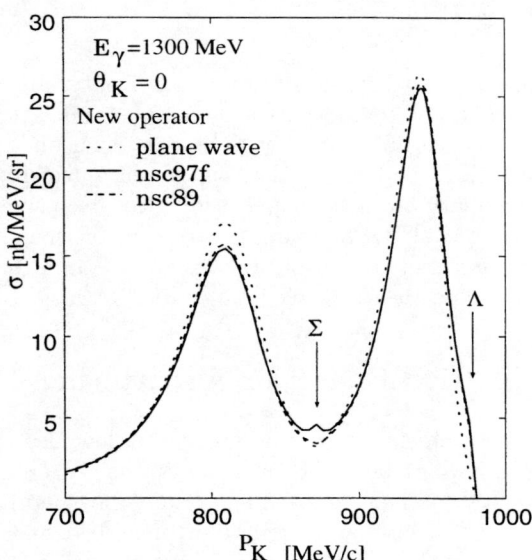

FIGURE 8. Theoretical cross-section [24] of the $D(\gamma, K^+)Yn$ reaction.

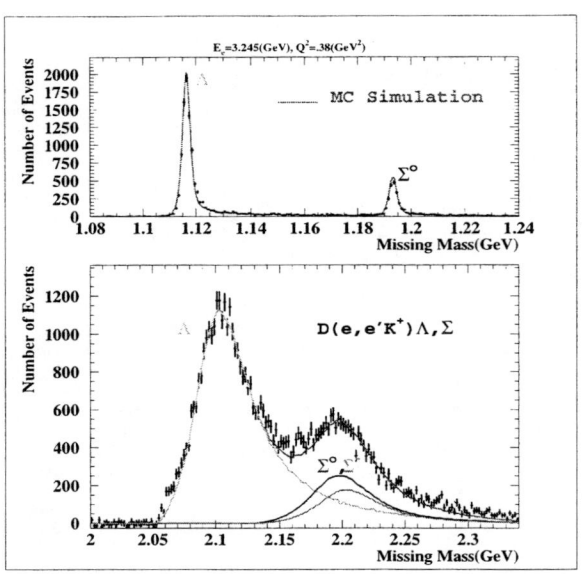

FIGURE 9. The yield of the $D(e,e'K^+)Yn$ reaction at CEBAF [25]. Top panel: proton target.

production of Kaons on a nucleon almost at rest in Deuterium. This mechanism is responsible for the two broad peaks in Figs. 8 and 9 (Λ and Σ production), of which the width is due to the Fermi motion of the target nucleon inside the Deuteron. Λn and $\Sigma^\circ n$ rescatterings appear on the tails of the quasi free peaks. The way to overcome this difficulty consists in performing an exclusive experiment where one select high momenta of the recoiling neutron, in order to suppress quasi-free mechanisms. This has become possible thanks to the high duty factor, the high luminosity of CEBAF and the large acceptance of CLAS.

High Energy YN scattering

Such exclusive measurements, at higher energies, allow also to reach an interesting kinematical regime where both the nucleon and the Hyperon propagate on-shell in the intermediate state. The corresponding triangular singularity induces a narrow peak on top of the quasi-free contribution, in well defined part of the phase space. I have no quantitative estimate for strangeness photoproduction, but the cross sections and kinematical range are very similar to pion photoproduction. As an example, Fig. 10 illustrate this idea in the $D(\gamma, p\pi^-)p$ reaction, in a kinematics accessible at CEBAF. The momentum of the slowest proton is kept high enough (400 MeV/c) in order to kill the quasi-free process. Two peaks, respectively due to πp and pp scattering, are clearly apparent. The rescattering amplitude depends only on on-shell amplitudes and on the low momentum components of the Deuteron

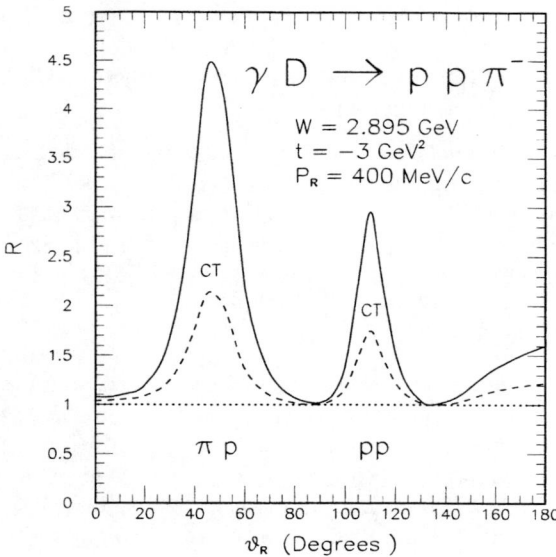

FIGURE 10. The cross-section of the $D(\gamma, p\pi^-)p$.

wave function. I refer the reader to Refs. [26,27] for a detailed discussion of these issues.

The pion production sector provides us with a solid testing ground of this conjecture. In the kaon production sector, this particular kinematics gives access to the on-shell Hyperon-Nucleon scattering amplitude. Two regime are interesting. At low momentum transfer t, between the incoming photon and the outgoing Kaon, the propagating Hyperon is not too much affected and one access the "normal" on-shell Hyperon-Nucleon scattering amplitude. At high momentum transfer, the constituents of the Hyperon are forced to stay in the small interaction volume and a "small" Hyperon propagates experiencing Color Transparency. To what extent the presence of a strange quark will affect Color Transparency is a fascinating open question.

STRANGENESS RESPONSE OF HADRONIC MATTER

The strange quark can be exploited to decipher the inner structure of Hadronic Matter. Let me end my talk with three examples at higher energies:

- *Deep Inelastic Scattering.* When detected in coincidence with the scattered electron, the Kaon gives access to the strange quark structure function $s(x)$ in unpolarized as well as in polarized nucleons. The light quark structure functions $\Delta u(x)$ and $\Delta d(x)$ have already been determined at SMC [28] and HERMES [29] by recording π^+ or π^- emitted off polarized nucleons. The implementation of Ring Imaging Cherenkov Detector (RICH) at HERMES will soon allow to determine $\Delta s(x)$.

- *High P_T Kaon pair production and the gluon distribution.* Electroproduction of pair of mesons with high transverse momentum P_T is thought to be sensitive to the gluon distribution in the nucleon. A first attempt has been performed at HERMES [30] by recording pion pairs emitted off of polarized nucleon targets. In near future, the detection of kaon pairs at HERMES and at COMPASS will allow to get rid of possible competing background processes and may lead to a more reliable determination of $\Delta g(x)$.

- *Polarized Λ production and the Transversity Structure Function $h_1(x)$.* The determination of the polarization of the Λ produced in the interaction of unpolarized electrons with transversally polarized nucleon is one of the few ways to access the Transversity Structure Function $h_1(x)$. The Λ is used as a "quark polarimeter". Such an experiment appears to be feasible with an electron beam around 30 GeV, but the luminosity of the present facility should be increased by orders of magnitude. I refer the reader to Ref. [31] for a comprehensive discussion.

Exploratory studies of these channels are underway at HERMES and COMPASS. Exploiting the full benefit of the creation and propagation of the strange quark in Hadronic Matter requires to increase the luminosity by three order of magnitudes. This the goal of the ELFE project in Europe [32], or of the upgrade of the energy of CEBAF up to 24 GeV in USA. But the future could well be a world facility!

CONCLUSION

The production and the propagation of the strange quark provides us with an "intruder" which probes the inner structure of hadronic matter. Such a opportunity has not been fully exploited in the past, due to the low intensity of the available beams. This is changing with the advent of new facilities (COSY, JLab, HERMES, COMPASS), where strangeness production reactions are undergoing a renewed interest.

On the theoretical side, the main players in the game are identified. At intermediate energies (COSY, JLab), a Regge description is best suited to the description of reaction mechanisms. It has already been implemented in the electromagnetic sector; this remains to be done in the hadronic sector. At high energies, the detection of Kaons or Λ tags the interaction of the probe with a strange quark. However a good understanding of the hadronization chain must be achieved.

I am pretty confident that many new, experimental as well theoretical, results will be reported at the next Conference. Surprises may even happen in the mean time!

REFERENCES

1. R. Bertini, *Europhysics News* **22**, 115 (1991).
2. V. Flamino et al., *Compilation of Cross Sections* **CERN-HERA 79-03** (1979).
3. J. M. Laget, *Journal of the Korean Physical Society* **26**, S244 (1993).
4. M. Guidal, J. M. Laget and M. Vanderhaeghen, *Nucl. Phys.* **A627**, 645 (1997).
5. J. M. Laget, *Phys. Lett.* **B259**, 24 (1991).
6. M. M. Nagels, T. A. Rijken and J. J. de Swart, *Phys. Rev.* **D15**, 2547 (1977); **D20**, 1633 (1979).
7. R. Siebert et al., *Nucl. Phys.* **A567**, 819 (1994).
8. R. Bilger et al., *Phys. Lett.* **B420**, 217 (1998).
9. J. T. Balevsky et al., *Phys. Lett.* **B420**, 211 (1998).
10. S. Severin et al., *Phys. Rev. Rev.* **83**, 682 (1999).
11. F. Ballestra et al., *Phys. Rev. Lett.* **83**, 1534 (1999).
12. J. Soffer anfd Törnquist, *Phys. Rev. Lett.* **68**, 907 (1992).
13. J. C. David et al., *Phys. Rev.* **C53**, 2613 (1996).
14. T. Mizutani et al., *Phys. Rev.* **C58** 75 (1998).
15. H. Haberzettl et al., *Phys. Rev.* **C58**, R40 (1998).
16. M. Q. Tran et al., *Phys. Lett.* **B445**, 20 (1998).
17. A. M. Boyarski et al., *Phys. Rev. Lett.* **22** 1131 (1969).
18. R. Schumacher et al., Communication to PANIC99.
19. M. Vanderhaeghen, M. Guidal and J.-M. Laget, *Phys. Rev.* **C57**, 1454 (1998).
20. M. Guidal, J.-M. Laget and M. Vanderhaeghen, *Phys. Rev.* **C61**, in press (2000).
21. G. Niculescu et al., *Phys. Rev. Lett.* **81**, 1805 (1998).
22. R. A. Williams et al., *Phys. Rev.* **C46**, 1617 (1992).
23. J.-M. Laget, *Phys. Rep.* **69**, 1 (1969).
24. H. Yamamura et al., *Phys. Rev.* **C**, in press (2000); e-print nucl-th/9907029 v2 (July 1999).
25. D. Abbott et al., *Nucl. Phys.* **A639**, 197c (1999); private communication.
26. J.-M. Laget, *Proceedings of the Worskshop "Physics and Instrumentation with 6–12 GeV Beams"*. Eds. S. Dytman et al., JLab User Production, 1998, pp. 57.
27. J.-M. Laget, *Proceedings of the Workshop "Exclusive and Semi-Exclusive Processes at High Momentum Transfer"* (JLab, USA, May 1999). Eds. C. Carlson and A. Radyushkin, World Scientific, 1999, in press.
28. D. Adeva et al., *Phys. Lett.* **B420**, 180 (1998); D. Adams et al., *Phys. Lett.* **B369**, 93 (1996).
29. K. Akerstaff et al., *Phys. Lett.* **B464**, 123 (1999).
30. A. Airapetian et al., e-print hep-ex/9907020 (July 1999).
31. R. A. Kunne *Conference Proceedings* **44**. Eds J. Arvieux et al., Italian Physical Society, 1993, pp. 401.
32. K. Aulenbacher et al., Report CERN 99-10 (1999).

First Results on Strangeness Production from the ANKE Facility

S. Barsov[a], U. Bechstedt[b], G. Borchert[b], W. Borgs[b], M. Büscher[b], M. Debowski[c], M. Drochner[b], W. Erven[b], R. Eßer[d], P. Fedorets[e], D. Gotta[b], M. Hartmann[b], H. Junghans[b], A. Kacharava[f,g], B. Kamys[h], F. Klehr[b], H. R. Koch[b], V. I. Komarov[f], V. Koptev[a], P. Kulessa[b], A. Kulikov[f], V. Kurbatov[f], G. Macharashvili[f,g], R. Maier[b], S. Mikirtytiants[a], S. Merzliakov[f], H. Müller[c], A. Mussgiller[i], M. Nioradze[g], H. Ohm[b], A. Petrus[f], D. Prasuhn[b], K. Pysz[j], F. Rathmann[k], B. Rimarzig[c], Z. Rudy[h], R. Schleichert[b], Chr. Schneider[b], H. Schneider[b], O. W. B. Schult[b], H. Seyfarth[b], K. Sistemich[b], H. J. Stein[b], H. Ströher[b], and P. Wüstner[b] for the ANKE collaboration*

[a]*Petersburg Nuclear Physics Institute, 188350 Gatchina, Russia*
[b]*Forschungszentrum Jülich, 52425 Jülich, Germany*
[c]*Forschungszentrum Rossendorf, 01314 Dresden, Germany*
[d]*Universität zu Köln, 50923 Köln, Germany*
[e]*Institute of Theoretical and Experimental Physics, 117259 Moscow, Russia*
[f]*Joint Institute for Nuclear Research, 141980 Dubna, Russia*
[g]*Tbilisi State University, 380086 Tbilisi, Georgia*
[h]*Jagellonian University, 30059 Cracow, Poland*
[i]*Fachhochschule München, 80335 München, Germany*
[j]*Institute of Nuclear Physics, 31342 Cracow, Poland*
[k]*Universität Erlangen-Nürnberg, 91058 Erlangen, Germany*

*For a complete collaboration list see (8)

Abstract. ANKE is a second generation experimental facility at the internal beam of the accelerator COSY at the Forschungszentrum Jülich (Germany). It has recently (May 1998) been installed in the ring, commissioned and immediately after this been exploited for first experiments. Using circulating beams of protons with energies between 1.0 and 2.3 GeV, sub- and near threshold production of K^+ mesons in collisions with carbon-strip targets has been investigated. The detection system of ANKE used for the first measurements is capable of detecting positively charged ejectiles in the momentum range between about 100 and 600 MeV/c. In particular, it has been optimized for K^+ identification in the presense of an intense background of protons and pions. The final results of the experiments (cross sections and momentum spectra) will be used to obtain more detailed information on the subthreshold production mechanism.

INTRODUCTION

Subthreshold production of hadrons, i.e. production below the corresponding threshold energy in free NN-collisions has been studied since many years for antiprotons (1), pions (2) and kaons (3). All investigations aim at identifying effects of the nuclear environment on the elementary production process, so-called medium effects. Proton-nucleus collisions are the intermediate step between NN- and nucleus-nucleus collisions, and, in fact, the results are necessary input for the interpretation of the heavy-ion experiments. Subthreshold K^+ production is particularly well suited for the following reasons: a) the kaon mass is large so that its production requires high internal momenta, and b) kaons have a large mean free path in nuclear matter, which minimizes final state interaction of the outgoing meson.

The possibility to study nuclear reactions on internal targets with reasonably high luminosity makes ANKE at COSY an ideal tool for a detailed investigation of subthreshold K^+ production (in fact, ANKE is based on a corresponding proposal (4)). Kaon momentum- and angular distributions and finally also coincidences between K^+ and accompanying particles (p, d) will be measured in order to make as exclusive experiments as possible.

THE ANKE SPECTROMETER

ANKE is a magnetic spectrometer, installed in one of the straight sections of the race-track shaped Cooler Synchrotron (COSY) at the Forschungszentrum Jülich, Germany. It currently consists of 3 dipole magnets (D1, D2, and D3), a target station and a dedicated detection system for positively charged particles, which are produced in proton beam – target interactions in forward and backward direction (see fig. 1; more details see ref.(5)).

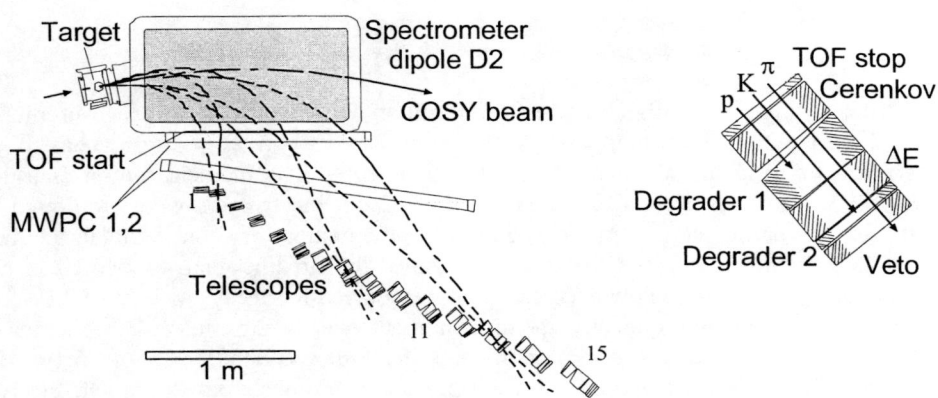

FIGURE 1. Schematic drawing of the ANKE set up at COSY (lhs) and details of one telescope (rhs).

The targets which have been used up to now, were made of thin strips of carbon (~ 50 µg/cm^2) or diamond (~ 400 to 20.000 µg/cm^2). They were fixed about 30 cm in front of dipole D2 inside the COSY beam pipe. After injection and acceleration, the COSY proton beam was steered onto these targets for periods between 30 and 60 seconds to initiate the interactions under investigation.

The kaons are detected and identified by a set of detectors, based on time-of-flight and energy-loss measurements using plastic scintillators. In addition, lucite Cherenkov- and plastic veto detectors are exploited as well as MWPC's, which determine the particle trajectories. In order to maximize the difference in energy loss between pions and kaons, a degrader (Cu-plate of variable thickness) is installed in front of the ΔE-counters. A second Cu-degrader is inserted between ΔE- and veto detectors to stop kaons, so that its decay into muons or pions can be exploited for identification via a delayed coincidence (mean lifetime τ = 12.4 ns). For the deepest subthreshold measurement at T_p = 1.0 GeV, this option has been used in the on-line trigger.

Subthreshold K$^+$ Production

While the threshold energy for K$^+$Λ production in free nucleon-nucleon collisions is 1.58 GeV, production of kaon-hyperon pairs is possible in proton-nucleus interactions at much smaller incident proton energies; the absolute threshold is around 0.7 GeV (i.e. the Q-value of the reaction pp → pKΛ; for p + C, the absolute threshold is 0.75 GeV). Approaching this threshold, one is faced with two general problems: a) the cross section decreases rapidly, and b) the signal-to-background ratio becomes very small. Therefore, in order to perform such experiments, one needs a) high luminosity, b) a large acceptance detector with short flight paths to minimize the effects of kaon decay, and c) a very efficient kaon identification. ANKE at COSY has been designed to fulfill these requirements as much as possible.

K$^+$ Production in pA Collisions at ANKE

After installation of ANKE in the COSY ring in May '98, a series of measurements on K$^+$-production in proton-induced reactions on carbon strip targets has been performed during last year and the first half of 1999. The incident proton energies were 1.0, 1.2, 1.5, 2.0, and 2.3 GeV. Dedicated on-line triggers were employed at different projectile energies in order to optimize the number of kaon events in the data sample written to tape. For the lowest energies, the off-line analysis, which is not completed yet, has up to now been based on the delayed K$^+$ decay.

To be more specific: typically, the measurements were performed at luminosities of (1-2) 10^{32} cm^{-2} s^{-1}, resulting in a count rate of all start counters of about 6·10^5 Hz. About 80% of the observed particles are pions and protons from the target, the rest mainly consists of beam protons scattered from vacuum chamber and the material of D2. These background events are effectively reduced by a good-trajectory requirement, using the MWPC information. The target protons and pions can be discriminated from kaons of the same momentum by time-of-flight. Finally, p, π^+ and K^+ of a given momentum have different ranges in matter. The thickness of scintillation counters and Cu-degraders were chosen such that protons are stopped before they reach the ΔE-counters. Kaons are stopped in the second degrader or at the far end of the ΔE-counters, while pions (and also fast scattered beam protons) transverse the whole telescope. The first degrader has been tapered to compensate the momentum spread in one telescope such that the difference in energy-loss signal of the ΔE-counter between pions and kaons is maximized. Finally, in the analysis performed so far, only kaon candidate events have been accepted with signals in the veto counters delayed by at least 1.3ns with respect to the stop- and ΔE-counters of the same telescope. This condition enhances $K^+ \rightarrow \mu^+\nu_\mu$ and $K^+ \rightarrow \pi^+\pi^0$ events ($\tau = 12.4$ ns) with respect to the backgrounds mentioned above, but at the same time reduces the detection efficiency for kaons by at least a factor of two. More details concerning the analysis can be found in refs. (5,6,7).

In figure 2, preliminary results of the measurements of subthreshold K^+ production are presented:

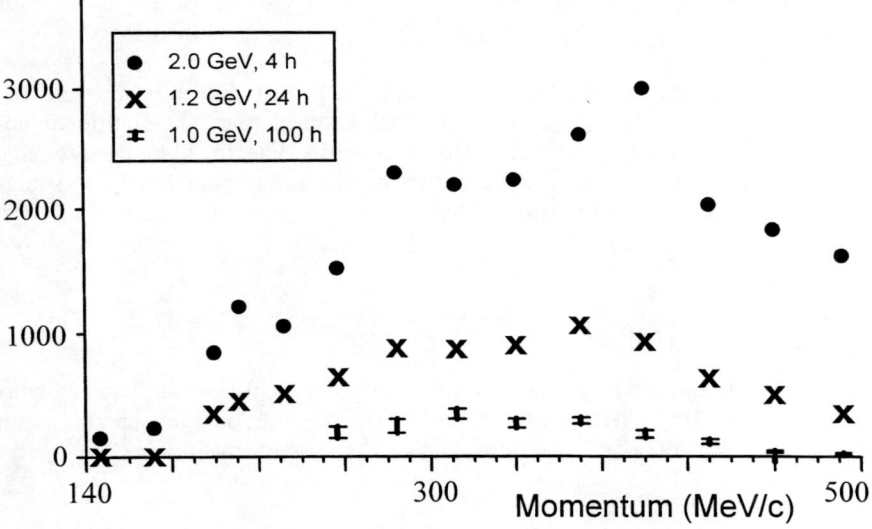

FIGURE 2. Preliminary results for K^+ momentum spectra at 1.0, 1.2 and 2.0 GeV. The number of kaons in each telescope is not yet corrected for kaon detection efficiencies of the different telescopes.

Numbers of kaons produced during the indicated measuring times are plotted for three different incident proton energies as a function of kaon momentum: the bin sizes correspond to the central values of the individual telescopes. As seen from the figure the spectra cover the momentum range from about 150 MeV/c up to about 500 MeV/c, except for the lowest momentum bites for 1 GeV, where kaons have not been identified unambigiously up to now.

In order to convert these kaon numbers into double differential cross sections, the kaon detection efficiencies of the different telescopes, the number of target nuclei and the number of incident protons have to be taken into account; this is currently being done - final results can be expected in the near future.

FUTURE INVESTIGATIONS

In future, it is planned to continue the subthreshold K^+-measurements by studying the A-dependence (i.e. using other targets like Al, Cu, and Pb) and by searching for correlated emission of kaons and light particles (proton, deuteron); (K^+d)-events could be a clear indication for two-step processes with a pion produced in an intermediate step.

For a next phase, ANKE has been equipped with a cluster target to study proton-proton and proton-deuteron interactions. A pellet target is being built for ANKE to be able to perform high luminosity measurements on a variety of targets (H, D, N, noble gases). Finally, a polarized storage-cell gas target is being developed for single and double polarization experiments.

ANKE will be supplemented by forward and backward detectors as well as a detection system for negatively charged particles, in particular K^-. Additional new detector systems are also planned: a spectator- and vertex detector to be used together with the ABS/storage-cell target has been started, and a compact large solid angle electromagnetic calorimeter is being considered.

ACKNOWLEDGMENTS

We would like to acknowledge the help of our colleagues at IKP for the installation and commissioning of ANKE in the COSY ring as well as the support of the infrastructure departments (ZAT, ZEL) of the Forschungszentrum Jülich.

REFERENCES

(1) Schröter, A. et al., *Zeit. f. Physik* **A350** (1994) 101
(2) Balada, A. et al., *Phys. Rev.* **C46**, 604 (1992)
(3) Debowski, M. et al., *Zeit. f. Physik* **A356**, 313 (1997)
(4) Borgs, W., COSY proposal # 18, unpublished (1991)
(5) Sistemich, K. et al., *Nucl. Instr. Meth.*, to be published (1999)
(6) Junghans, H., PhD Thesis, University of Cologne (Germany), 1999
(7) Büscher, M. et al., *Zeit. f. Physik* **A355,** 93 (1996)
(8) http://ikpd15.ikp.kfa-juelich.de:8085/doc/ANKE.html

Hyperon and Charged Kaon Pair Production Close to Threshold

M. Wolke[†], H.-H. Adam[*], J.T. Balewski[†,‖,¶], A. Budzanowski[‖],
C. Goodman[¶], D. Grzonka[†], L. Jarczyk[§], M. Jochmann[‡],
A. Khoukaz[*], K. Kilian[†], M. Köhler[‡], P. Kowina[††], T. Lister[*],
P. Moskal[†,§], N. Lang[*], W. Oelert[†], C. Quentmeier[*], R. Santo[*],
G. Schepers[*,†], U. Seddik[**], T. Sefzick[†], S. Sewerin[†], M. Siemaszko[††],
J. Smyrski[§], A. Strzałkowski[§], P. Wüstner[‡], W. Zipper[††]

[†] *Institut für Kernphysik, Forschungszentrum Jülich, D-52425 Jülich, Germany*
[*] *Institut für Kernphysik, Westfälische Wilhelms-Universität, D-48149 Münster, Germany*
[‖] *Institute of Nuclear Physics, PL-31-342 Cracow, Poland*
[¶] *Indiana University Cooler Facility, Bloomington, IN 47405, USA*
[§] *Institute of Physics, Jagellonian University, PL-30-059 Cracow, Poland*
[‡] *Zentrallabor für Elektronik, Forschungszentrum Jülich, D-52425 Jülich, Germany*
[††] *Institute of Physics, University of Silesia, PL-40-007 Katowice, Poland*
[**] *Egyptian Atomic Energy Authority, 13759 Cairo, Egypt*

Abstract. Close-to-threshold data on the elementary kaon and antikaon production channels in the proton–proton interaction have been taken using the COSY-11 installation at the cooler synchrotron COSY Jülich.

The experimental technique applied at the internal COSY-11 facility — designed for meson production studies at small excess energies — is outlined. The threshold excitation functions for the kaon–hyperon production via the reactions $pp \to pK^+\Lambda$ and $pp \to pK^+\Sigma^0$ are presented. The magnitude of the production amplitudes is compared at equal excess energies and physical implications of the observed Σ^0 suppression in the threshold region are discussed. In addition, within a Dalitz plot analysis the spin-averaged S-wave scattering parameters could be extracted for the Λ–p channel.

With the possibility of detecting all final state particles the elementary antikaon production in the reaction $pp \to ppK^+K^-$ has been investigated. Results on the exclusive total cross section fix the scale of the strangeness dissociation into two kaons.

INTRODUCTION

The associated strangeness production in proton–proton collisions close to threshold provides an elucidative probe for studying both the structure of the nucleon and the baryon–baryon interaction.

As reaction dynamics are expected to depend on the strangeness content in the initial state, strangeness production directly addresses the strangeness content of the nucleon on the sea quark level. The strange–antistrange quark pair creation at threshold requires high momentum transfers between the colliding nucleons. Consequently, the short range part of the nucleon–nucleon interaction is probed.

Close to the reaction threshold only few partial waves contribute, simplifying the theoretical interpretation of the data. With vanishing relative momenta in the exit channel, from extended interaction times the experimental observables are influenced by strong final state interaction effects. In turn, the elastic interaction in the final state may be used to extract low energy scattering parameters.

Both the strangeness dissociation into a kaon–hyperon and a kaon–antikaon system have been studied at the COSY-11 facility in the nucleon–nucleon collision. Close-to-threshold cross section values of elementary kaon and antikaon production are an essential input for transport model calculations to describe the strangeness production in heavy ion collisions.

EXPERIMENTAL TECHNIQUE

The internal COSY-11 facility at the proton synchrotron and storage ring COSY Jülich [1] has been designed for threshold meson production studies. From the experimental point of view, due to small transverse momenta close to threshold, ejectiles are limited to a narrow forward cone in the laboratory. Thus, the concept of an internal magnetic spectrometer with 4π geometrical acceptance at threshold is attainable with a comparatively small detector setup.

A cluster target [2] is installed in front of a regular C–shaped COSY dipole which both separates reaction products from the circulating beam and analyzes their momenta (fig. 1). Two sets of drift chamber stacks (D1,D2 in fig. 1) allow to measure the direction of positively charged reaction products, which are diverted towards the inside of the storage ring. Particle momenta are reconstructed from tracing back of the trajectories through the well known dipole field towards the target position. A particle identification is achieved by combining the momentum information with time–of–flight measurements.

For the ppK^+K^- and pK^+Y final states — where Y denotes a hyperon — start and stop timings for the proton identification are generated by two scintillator arrangements S1/S2 and S3 placed with a distance of 9.1 m. Due to their short lifetime for the identification of kaons an indirect time–of–flight measurement is used: Tracing back proton trajectories towards the target position intrinsically determines the time of the event at the target as a start signal, the stop timing being provided by the "start" scintillator hodoscope S1.

Negatively charged particles, which are deflected towards the yoke of the dipole, are detected by a dedicated part of the detection system consisting of a scintillator and a high granularity array of silicon pad detectors (Si_{dip}), both mounted inside the dipole gap. Short–lived neutral mesons or baryons are identified us-

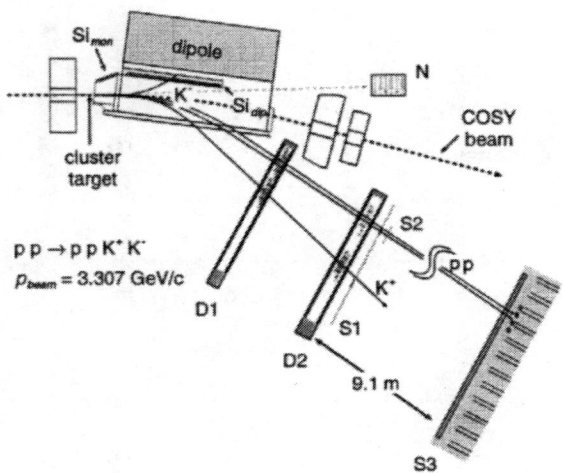

FIGURE 1. Schematic view of the internal COSY-11 installation and Monte Carlo generated particle tracks for the reaction pp → ppK$^+$K$^-$.

ing the missing mass technique with the complete four–momentum information of positively charged ejectiles. The detection system is described in more detail in [3].

As an absolute normalization proton–proton elastic scattering is recorded simultaneously during data taking. Measured differential elastic counting rates are related to high precision cross section data available in the COSY energy range from the results of the EDDA collaboration [4]. Considering the integrated luminosity and the overall acceptance, i.e. both geometrical coverage and detector efficiencies studied in Monte Carlo simulations, total cross section values are derived from the observed counting rates. At present, typical luminosities at COSY-11 are in the range of $2 - 4 \times 10^{30}$ cm^{-2} s^{-1}.

HYPERON PRODUCTION IN PROTON–PROTON COLLISIONS

Final State Identification

For events with an identified proton in the final state the invariant mass of a second charged particle — determined from the indirect time–of–flight method — is plotted versus the missing mass with respect to the assumed (pK$^+$)–subsystem at an excess energy of 12.9 MeV above the Σ^0 threshold in figure 2. Enhancements at both the Λ and the Σ^0 mass are clearly visible within the invariant mass resolution for the charged kaon mass. Kaon signals at missing mass values between the hyperon masses are explained by the misinterpretation of protons originating from the

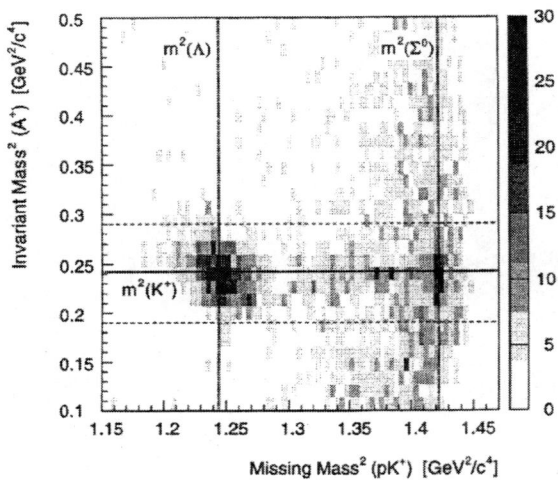

FIGURE 2. Invariant mass squared of a second charged particle in addition to an identified proton versus the missing mass squared of the assumed (pK$^+$) subsystem at an excess energy of 12.9 MeV with respect to the Σ^0 threshold. The range of identified kaons is limited by a cut equivalent to three standard deviations of the K$^+$ invariant mass distribution.

hyperon decay as protons from the primary reaction. Events outside the indicated range of the kaon mass distribution are reproduced in Monte Carlo simulations by scattering of beam particles on the rest gas or secondary reactions of elastically scattered protons in the beam pipe, i.e. by events not originating from the target. Thus, background contributions to the missing mass spectrum — the projection of events within the kaon mass resolution onto the missing mass axis in figure 3a — can be derived from equivalent projections outside the range of identified kaons (for details see [7]).

The missing mass distribution after background subtraction is depicted in figure 3b. A missing mass resolution of about 1 MeV/c^2 is obtained at an excess energy of 12.9 MeV with respect to the (pK$^+\Sigma^0$) threshold. The broadening of the Λ peak is understood as a kinematical effect as close to threshold the missing mass resolution approximately scales with the square root of the excess energy.

The pp $\to$ pK$^+\Lambda/\Sigma^0$ Total Cross Section

The reactions pp $\to$ pK$^+\Lambda$ and pp $\to$ pK$^+\Sigma^0$ have been measured at equal excess energies up to 13 MeV at the COSY-11 installation [5–7], no further data are available in this threshold energy range.

For both hyperon production channels the energy dependence of the total cross section is much better described by a phase space behaviour modified by the (pY)

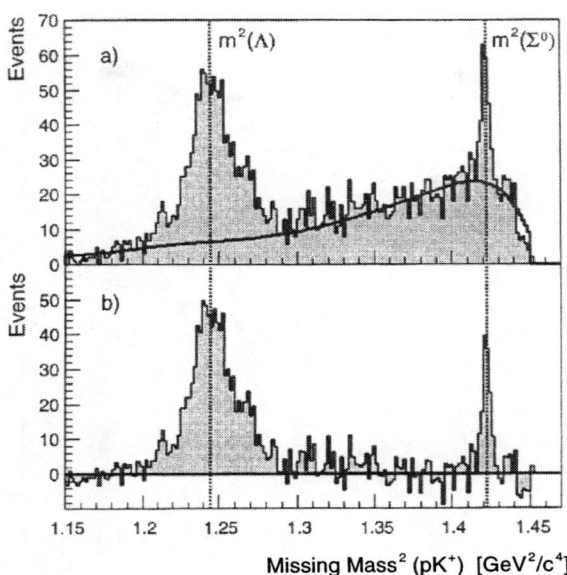

FIGURE 3. Missing mass squared with respect to the (pK$^+$) subsystem at an excess energy of 12.9 MeV above the Σ^0 threshold before (a) and after (b) subtraction of background (solid line).

final state interaction [8] than by a pure phase space calculation (fig. 4).

The most striking feature of the data, however, is the observed Σ^0 suppression in the close–to–threshold energy range with

$$\frac{\sigma\left(\mathrm{pp} \to \mathrm{pK}^+\Lambda\right)}{\sigma\left(\mathrm{pp} \to \mathrm{pK}^+\Sigma^0\right)} = 28^{+6}_{-9}, \qquad (1)$$

whereas at excess energies $\geq$ 300 MeV this ratio is known to be about 2.5 [9]. The large production ratio can be discussed qualitatively against the background of a description of possible production mechanisms within the framework of the meson exchange picture:

From π–exchange, relating the production in nucleon–nucleon scattering to $\pi \mathrm{p} \to \mathrm{K}\Lambda(\Sigma)$ data [10], a cross section ratio of $\approx$ 0.4 underestimates the experimental results by about two orders of magnitude. Considering K–exchange the ratio in proton–proton collisions is essentially given by the ratio of coupling constants $g^2_{N\Lambda K}/g^2_{N\Sigma K}$. Despite the existing uncertainty in the literature for the coupling, the choice of the SU(6) prediction [11] results in a cross section ratio of 27 in good agreement with the data although neglecting both initial and final state interactions. Inclusive measurements of K$^+$ production at higher excess energies (Q $\approx$ 170 MeV) obtained at SPES 4 [12] show enhancements at both the Λp and ΣN thresholds of similar strength. Obviously, both the SATURNE data and the

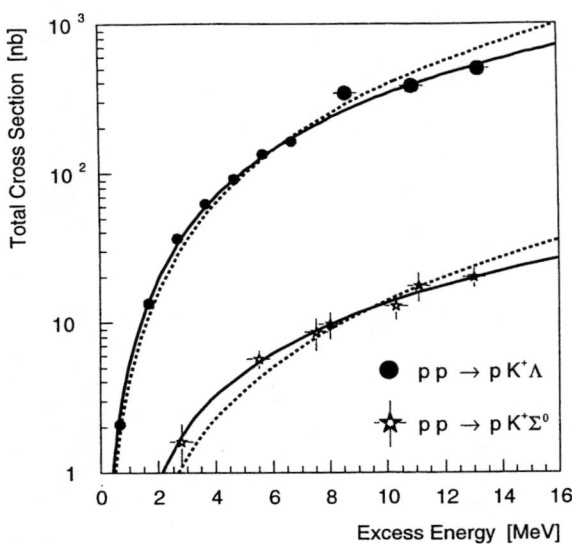

FIGURE 4. Data on the total cross section for the reactions pp → pK$^+$Λ and pp → pK$^+$Σ^0 obtained at COSY–11 [5–7]. The statistical error is included or is smaller than the symbol size. Horizontal error bars denote the uncertainty of the excess energy. The energy dependence is much better described by a phase space behaviour modified by the proton–hyperon final state interaction (solid line [8]) than by a pure phase space calculation (dashed line).

depletion of the Σ^0 signal close to threshold observed at COSY–11 might be explained by a strong final state conversion ΣN → Λp, an effect known for example from K$^-$ absorption on deuterium [13]. In turn, the final state conversion implies an initially larger Σ yield compared to the SU(6) prediction.

First theoretical investigations within the framework of the Jülich meson exchange model [14] taking into account both π– and K–exchange and rigorously treating the final state interaction disfavour a final state conversion as the origin of the observed large production cross section ratio. While Λ production is dominated by kaon exchange, both pion and kaon exchange are found to contribute in case of Σ production with similar strength and a destructive interference of these amplitudes is concluded to be one possible explanation of the Σ^0 suppression. It should be noted that contributions from direct production and heavy meson exchange, i.e. ρ and K* exchange, have so far been neglected but might influence the Λ/Σ^0 production ratio [15–17].

Low Energy (Λp) Scattering Parameters

With the elastic scattering in the final state significantly influencing physical observables — as it has already been noted in case of the energy dependence of the total cross section for hyperon production (fig. 4) — the effect of final state interaction has been used to extract low energy Λ–p scattering parameters [18]. With both unpolarized beam and unpolarized target, only spin averaged values for scattering length $\bar{a}$ and effective range $\bar{r}$ can be deduced from the data. However, the spin dependence is expected to be small [21, 22].

Compared to direct Λ–p scattering experiments, which are difficult to perform due to the short Λ lifetime, the complementary Λ–p final state interaction approach covers an energy range down to threshold, whereas direct data [19, 20] are available at excess energy values above 3.8 MeV (fig. 5).

An appropriate tool to study two–body correlations in a three-particle final state is provided by a Dalitz plot analysis depicting the double differential cross section

$$\frac{d^2\sigma}{dS_{pK}\,dS_{\Lambda K}} = \int f_C\left(q_{pK}\right) f_{FSI}\left(q_{\Lambda p}\right)\,d\rho^3, \qquad (2)$$

where S and q denote the two particle total energy and the momentum in the centre-of-mass of the respective subsystem. The s-wave ansatz in equation 2 assumes

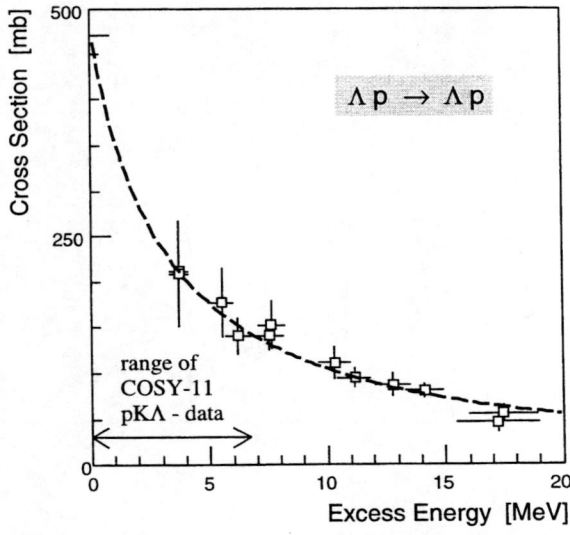

FIGURE 5. Total cross section for Λ–p elastic scattering close to threshold [19,20]. The energy range of the COSY-11 data indicated by the arrow expands the range of available data down to threshold. A combined fit (dashed line) allows to determine spin averaged low energy Λ–p scattering parameters.

the production cross section to be dominated by the available three–body phase space volume $d\rho^3$ modified by variations from a uniform distribution due to the significant contributions to the final state interaction, i.e. the strong Λ–p interaction $f_{FSI}(q_{\Lambda p})$ and the p–K^+ Coulomb repulsion described by the Coulomb penetration factor $f_{coul}(q_{pK})$ [18,23]. Parameterizing the attractive Λ–p interaction according to the Watson formalism [24]

$$f_{FSI}(q_{\Lambda p}) = \frac{1}{\overline{a}^2 q^2 + \left(\frac{1}{2}\overline{r}\,\overline{a}\,q^2 - 1\right)^2} \quad (3)$$

the ansatz for the Dalitz plot contains only the spin averaged scattering length $\overline{a}$ and effective range $\overline{r}$ as free parameters, which significantly influence its differential occupation (fig. 6a). Monte Carlo generated Dalitz plot distributions for different values of scattering parameters $(\overline{a},\overline{r})$ folded with the acceptance of the detection system are compared to the experimental data (fig. 6b) with about 2400 events in total in a maximum likelihood fit.

In the analysis scattering parameters turn out to be strongly correlated constraining only the combination of scattering length and effective range in the $(\overline{a},\overline{r})$ plane

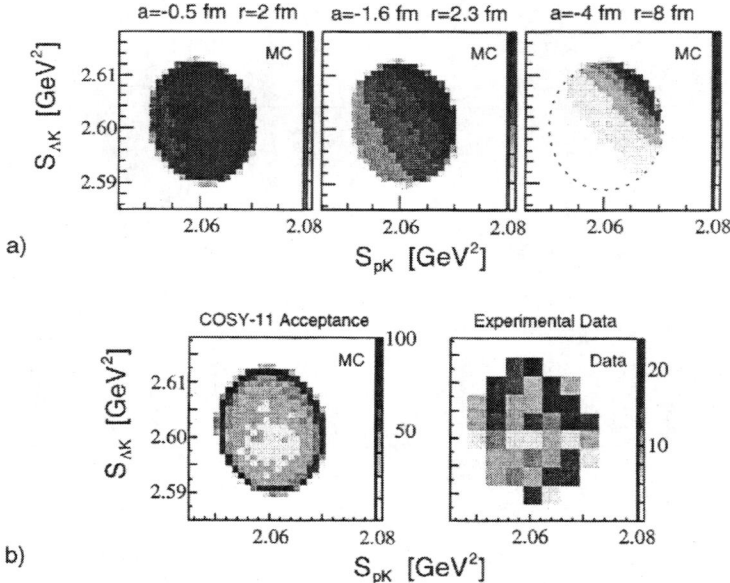

FIGURE 6. Dalitz plots for the reaction $pp \to pK^+\Lambda$ from a measurement at an excess energy of 7 MeV with respect to the $(pK^+\Lambda)$ threshold. Monte Carlo generated distributions demonstrate the sensitivity on low energy Λ–p scattering parameters (a). After being folded with the acceptance of the detection system, the distributions are compared to the experimental data covering the complete kinematically allowed range (b).

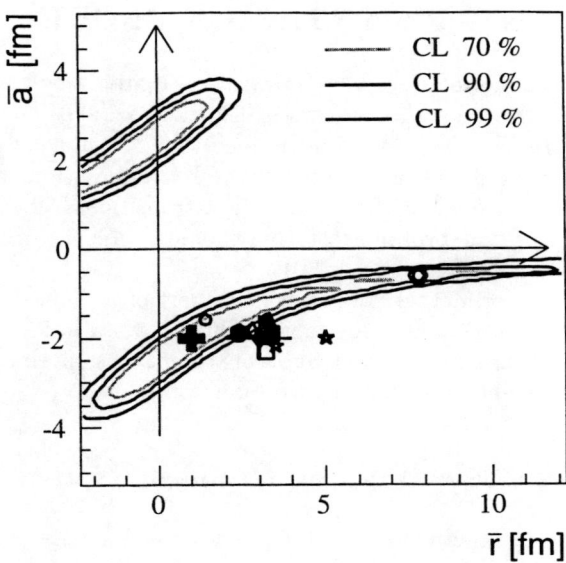

FIGURE 7. Constraints on low energy Λ–p scattering parameters from the logarithm of a likelihood function in the $(\bar{a},\bar{r})$ plane based on close-to-threshold COSY-11 data. Unphysical solutions with a positive scattering length implying a (Λp) bound state are to be excluded. In comparison, available literature values [13,19–22,25] are included. A combined fit with Λ–p elastic scattering data results in spin averaged values of $\bar{a} = -2.0$ fm and $\bar{r} = 1.0$ fm for the Λ–p scattering length and effective range, respectively, as indicated by the filled cross.

(fig. 7). In comparison, available experimental and theoretical values [13,19–22,25] are located close but for the most part slightly outside the 99 % confidence level contour.

Separate values of scattering length and effective range are accessible using the complementary additional experimental information from low energy Λ–p elastic scattering [19, 20]. A fit of the elastic cross section parameterized in terms of the scattering parameters is constrained according to the maximum likelihood ridge from the COSY-11 data (fig. 5). Although due to the strong correlation of the parameters errors are difficult to derive, the combined fit allows a new determination of spin–averaged Λ–p scattering parameters (fig. 7)

$$\bar{a}_{\Lambda p} = -2.0 \,\text{fm}, \quad \bar{r}_{\Lambda p} = 1.0 \,\text{fm}. \tag{4}$$

ELEMENTARY ANTIKAON PRODUCTION

Studies of $K\overline{K}$ production have been motivated originally by the ongoing discussions on the nature of the scalar objects f_0 (980) and a_0 (980) [26–29]. Within the framework of the Jülich meson exchange model [29] a $K\overline{K}$ molecule structure of the f_0 (980) is closely related to the strength of the $K\overline{K}$ interaction, which determines both shape and absolute value of the $\pi\pi \to K\overline{K}$ transitions [30]. Similar effects in $K\overline{K}$ production in proton–proton scattering should probe the $K\overline{K}$ interaction and thus a possible $K\overline{K}$ nature of the f_0 (980).

At present, the experimental approach falls short of its initial objective, due to unexpectedly low cross sections and, consequently, low statistics. However, despite the vicinity of comparatively broad hyperon resonances, preliminary data show clear signals of the elementary antikaon production.

Event Selection and Preliminary Results

Analogously to the experimental technique described in case of the hyperon production channels, (ppK$^+$) final states are identified from two protons and an additionally identified kaon using the indirect time–of–flight method.

First data showed a large scattering of events on the scale of the missing mass in the order of 50 MeV/c^2 with respect to the identified (ppK$^+$) subsystem [31] contrasting sharply with typical values known from hyperon production at COSY–11 (fig. 3). However, with no detection of a negatively charged kaon being required,

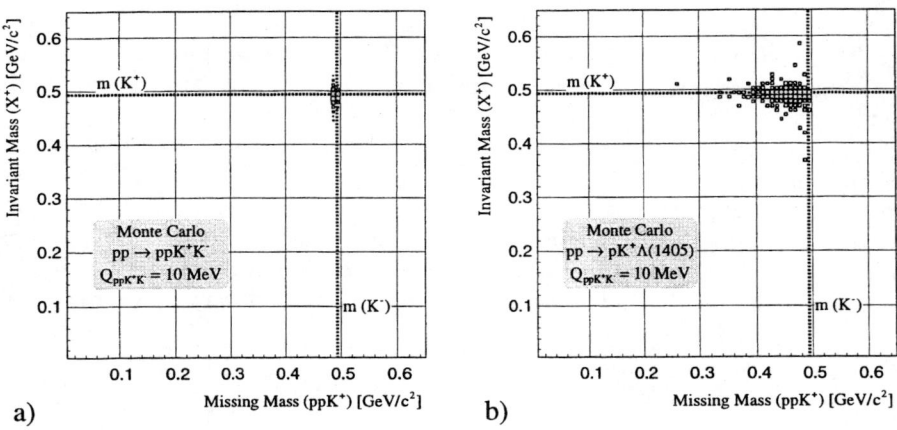

FIGURE 8. Invariant mass of a third positively charged ejectile in addition to two identified protons versus missing mass with respect to the assumed (ppK$^+$)–subsystem at an excess energy of 10 MeV above the (K$^+$K$^-$) threshold for the reactions pp $\to$ ppK$^+$K$^-$ (a) and pp $\to$ pK$^+\Lambda$ (1405) (b) from Monte Carlo simulations.

the reaction type pp → ppK$^+$X, with X denoting an undetected negatively charged meson system, remains ambiguous concerning the underlying reaction mechanism: One of the identified protons may originate from the weak decay of an initially produced hyperon via pp → pK$^+$Y → pK$^+$pX. A dominant influence of the hyperon resonances $\Lambda(1405)$ and $\Sigma(1385)$ is expected due to an acceptance of the detection system decreasing with increasing excess energy. In consequence, the missing mass with respect to the identified (ppK$^+$) system may shift to values too low for a (ppK$^+$K$^-$) hypothesis as demonstrated by the simulations in figure 8, which give a strong hint on the importance of hyperon resonance channels in view of the large scattering with respect to the (ppK$^+$) missing mass observed in the data (see also fig. 9).

From the above discussion, to discriminate the (ppK$^+$K$^-$) final state from hyperon production channels, the K$^-$ identification seems to be experimentally crucial. Preliminary results from data taken at an excess energy of 17.1 MeV with respect to the (ppK$^+$K$^-$) threshold and at a significantly increased integrated luminosity in September 1998 are presented in figure 9a [32]. For events within the experimental resolutions for the charged kaon mass on both the scale of the invariant mass of a third charged ejectile and the missing mass of the assumed (ppK$^+$) subsystem a K$^-$ consistent hit in the negative particle detection system is required: From the measured momenta of positively charged ejectiles and momentum conservation the K$^-$ momentum is calculated. Thus, the negatively charged

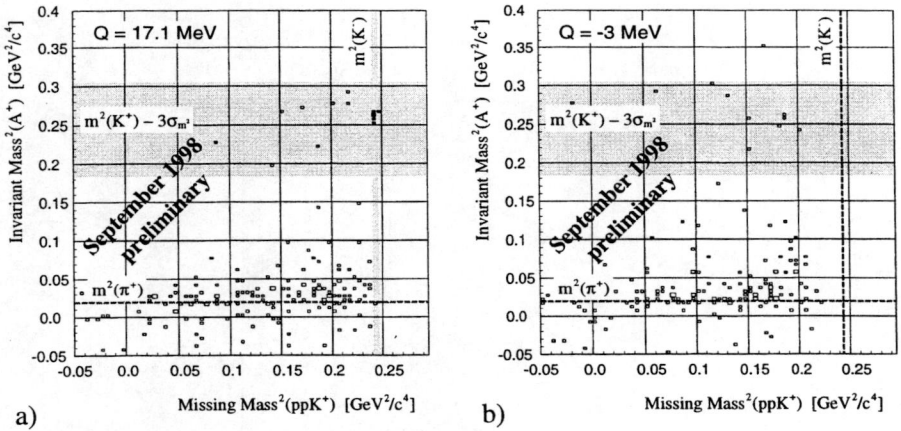

FIGURE 9. Preliminary results for the invariant mass of a third positively charged ejectile in addition to two identified protons versus the missing mass with respect to the assumed (ppK$^+$)–subsystem at an excess energy of 17.1 MeV above (a) and -3 MeV (b) below the (K$^+$K$^-$) threshold, respectively. Experimental resolutions corresponding to three standard deviations are indicated by the shaded regions. For events within the limits of both the invariant mass and missing mass resolutions a K$^-$ consistent hit in the negative particle detection system is required [32].

kaon can be traced through the dipole field to the dedicated detection system and the reconstructed hit position is compared to the registered signals in both silicon pad detector and scintillator. Within the resolution for the invariant mass of the assumed positively charged kaon a group of six events clearly stands out at a missing mass corresponding to the K^- mass. In a measurement below threshold (fig. 9b) at a comparable value of the integrated luminosity the region expected for a (ppK^+K^-) final state remains completely empty.

To confirm this result corresponding to a first preliminary cross section value for K^+K^- production in the order slightly below 1 nb, additional data have been taken at the same excess energy of 17.1 MeV in 1999. Figure 10 depicts a very preliminary result for the missing mass with respect to an identified (ppK^+) subsystem at a further increased luminosity from an increase in beam intensity [32]: In the missing mass spectrum both a broad structure in agreement with expectations for a $pK^+\Sigma(1385)/\Lambda(1405)$ final state and a clearly separated distinct signal corresponding to a (ppK^+K^-) hypothesis are visible.

In conclusion, strong experimental evidence is deduced from the data for an exclusive identification of the elementary antikaon production channel in proton-proton scattering. Final values of the total cross section will be available in the near future.

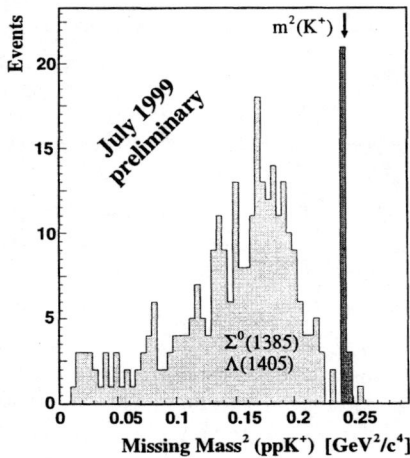

FIGURE 10. Missing mass with respect to an identified (ppK^+)-subsystem at an excess energy of 17.1 MeV above the (K^+K^-) threshold from July 1999 (very preliminary [32]).

FROM PRESENT TO FUTURE

Besides the production of isoscalar (η, η', see also [33]) and vector mesons (ω, Φ), both kaon and antikaon production channels in proton–proton scattering will be part of the experimental programme at the COSY-11 facility in future.

The strong suppression of the Σ^0 signal close to threshold observed in the exclusive pp $\rightarrow$ pK$^+\Lambda/\Sigma^0$ measurements as compared to the ratio at high energies might be explained by a destructive interference of kaon and pion exchange. A measurement of the energy dependence of the production ratio both at COSY-11 and the complementary TOF spectrometer at COSY [34] will further constrain the interpretation of the close-to-threshold data. An extension of the COSY-11 detection system by a dedicated Cerenkov detector for a unique kaon identification independent of the indirect time–of–flight method will provide access to final states of different isospin configuration, i.e. the pp $\rightarrow$ nK$^+\Sigma^+$ channel.

Strong experimental evidence for an exclusive identification of the elementary antikaon production channel in the nucleon–nucleon interaction — the (ppK$^+$K$^-$) final state — will lead to a first total cross section value in the near future. The use of additional drift chambers at the exit window of the momentum analyzing dipole will further increase the acceptance for K$^+$ detection by a factor of two. Thus, already at present luminosity values, a determination of the close–to–threshold excitation function for K$^+$K$^-$ production is feasible.

ACKNOWLEDGEMENTS

The COSY-11 collaboration is grateful to numerous people who have supported the experimental facility. The COSY operating team significantly contributed to the presented results by providing the excellent quality proton beam. Explicitly, we would like to thank J. Haidenbauer, C. Hanhart and C. Wilkin for many fruitful discussions.

REFERENCES

1. Maier, R., *Nucl. Instr. & Meth.* **A 390** 1–8 (1997) and Maier, R., *Nucl. Phys. News Int.* **7** No. 4 5–13 (1997).
2. Dombrowski, H. et al., *Nucl. Instr. & Meth.* **A 386** 228–234 (1997).
3. Brauksiepe, S. et al., *Nucl. Instr. & Meth.* **A 376** 397–410 (1996).
4. Albers, D. et al., *Phys. Rev. Lett.* **78** 1652–1655 (1997)
5. Balewski, J.T. et al., *Phys. Lett.* **B 388** 859–865 (1996).
6. Balewski, J.T. et al., *Phys. Lett.* **B 420** 211–216 (1998).
7. Sewerin, S., Schepers, G. et al., *Phys. Rev. Lett.* **83** 682–685 (1999).
8. Fäldt, G. and Wilkin, C., *Z. Phys.* **A 357** 241–243 (1997).
9. Flaminio, V., Moorhead, W.G., Morrison, D.R.O. and Rivoire, N., *Compilation of Cross Sections*, CERN-HERA 84-01 (1984).

10. Jones, J. et al., *Phys. Rev. Lett.* **26** 860–863 (1972); Baker, R.D. et al., *Nucl. Phys.* **B 145** 402–408 (1978); Saxon, D.H. et al., *Nucl. Phys.* **B 162** 522–546 (1980); Hart, J.C. et al., *Nucl. Phys.* **B 166** 73–83 (1980); Candlin, D.J. et al., *Nucl. Phys.* **B 226** 1–28 (1983); Candlin, D.J. et al., *Nucl. Phys.* **B 311** 613–629 (1988).
11. De Swart, J.J., *Rev. Mod. Phys.* **35** 916–939 (1963); Dover, C.B. and Gal., A., *Prog. Part. Nucl. Phys.* **12** 171–239 (1985).
12. Siebert, R. et al., *Nucl. Phys.* **A 567** 819–843 (1994).
13. Tan, T.H., *Phys. Rev. Lett.* **23** 395–398 (1969).
14. Gasparian, A., Haidenbauer, J., Hanhart, C., Kondratyuk, L. and Speth, J., e–Print Archive nucl-th/9909017 (1999) and proceedings of this conference.
15. Tsushima, K., Sibirtsev, A. and Thomas, A.W., *Phys. Lett.* **B 390** 29–35 (1997).
16. Tsushima, K., Sibirtsev, A., Thomas, A.W. and Li, G.Q., *Phys. Rev.* **C 59** 369–387 (1999).
17. Kaiser, N., *Eur. Phys. J.* **A 5** 105–110 (1999).
18. Balewski, J.T. et al., *Eur. Phys. J.* **A 2** 99–104 (1998).
19. Alexander, G. et al., *Phys. Rev.* **173** 1452–1460 (1968).
20. Sechi-Zorn, B. et al., *Phys. Rev.* **175** 1735–1740 (1968).
21. Holzenkamp, B., Holinde, K. and Speth, J., *Nucl. Phys.* **A 500** 485–528 (1989); Reuber, A., Holinde, K. and Speth, J., *Nucl. Phys.* **A 570** 543–579 (1994).
22. Nagels, M.M., Rijken, T.A. and de Swart, J.J., *Phys. Rev.* **D 20** 1633–1645 (1979).
23. Hanhart, C. et al., *Phys. Lett.* **B 358** 21–26 (1995).
24. Goldberger, M.L. and Watson, K.M., *Collision Theory*, New York: John Wiley & Sons, 1964.
25. Maessen, P.M.M., Rijken, T.A. and de Swart, J.J., *Phys. Rev.* **C 40** 2226–2245 (1989).
26. Jaffe, R.L., *Phys. Rev.* **D 15** 267–280 (1977).
27. Weinstein, J. and Isgur, N., *Phys. Rev.* **D 41** 2236–2257 (1990).
28. Morgan, D. and Pennington, M.R., *Phys. Rev.* **D 48** 1185–1204 (1993).
29. Lohse, D., Durso, J.W., Holinde, K. and Speth, J., *Nucl. Phys.* **A 516** 513–548 (1990).
30. Krehl, O., Rapp, R. and Speth, J., *Phys. Lett.* **B 390** 23–28 (1997).
31. Lister, T., PhD thesis: Westfälische Wilhelms–Universität Münster, 1998; Lister, T., Quentmeier, C. and Wolke, M., *Annual Report 1998 IKP/COSY*: Forschungszentrum Jülich, Jül–3640, 1999.
32. Quentmeier, C., PhD thesis: Westfälische Wilhelms–Unisversität Münster, in preparation.
33. Moskal, P. et al., proceedings of this conference.
34. Wirth, S. et al., proceedings of this conference.

Determination Of The Lifetime Of Heavy Λ-Hypernuclei At COSY Jülich

H. Ohm[a], W. Borgs[a], W. Cassing[b], M. Hartmann[a], L. Jarczyk[c], B. Kamys[c], H.R. Koch[a], P. Kulessa[a,c], H.J. Maier[d], R. Maier[a], M. Matoba[e], D. Prasuhn[a], K. Pysz[f], Z. Rudy[c], O. Schult[a], H. Ströher[a], A. Strzalkowski[c], Y. Uozumi[e] and I. Zychor[g]

[a] *Institut für Kernphysik, Forschungszentrum Jülich, D-52425 Jülich, Germany*
[b] *Institut für Theoretische Physik, Universität Giessen, D-35392 Giessen, Germany*
[c] *M. Smoluchowski Institute of Physics, Jagellonian University, PL-30059 Cracow, Poland*
[d] *Technologielaboratorium, LMU München, D-85748 Garching, Germany*
[e] *Department of Nuclear Engineering, Kyushu University, Fukuoka 812, Japan*
[f] *H. Niewodniczanski Institute of Nuclear Physics, PL-31342 Cracow, Poland*
[g] *The Andrzej Soltan Institute for Nuclear Studies, PL-05400 Swierk, Poland*

Abstract. The lifetime of heavy hypernuclei has been determined at COSY Jülich making use of the (p,K)-reaction and the recoil-shadow technique based on the detection of fragments from the Λ-decay induced fission. Hypernuclei were produced by bombarding extremely thin U and Bi targets with protons of 1.5 GeV and 1.9 GeV kinetic energy. High luminosity was achieved due to the multi-pass technique with the target being exposed to the internal beam of COSY. Position sensitive low pressure MWPCs permitted clear identification of fission fragments based on energy loss and TOF information. The lifetime of hypernuclei was deduced from the position distributions of fission events in the shadow region and from the recoil momentum calculated within the BUU model. Values of 240±60 ps and 161±7(stat.)±14(syst.)ps were found with U and Bi-targets, respectively. The deduced production cross sections for hot hypernuclei prior to the evaporation process of $\sigma = 150^{+150}_{-80}$ μb for U and 350±140 μb for Bi are in agreement with the theoretical predictions.

1. INTRODUCTION

The free Λ particle which consists in the simple quark picture of u, d and s quarks decays in vacuum with a lifetime of τ = 263 ps according to Λ→p+π⁻+38 MeV and Λ→n+π⁰+41 MeV. Bound in a nucleus the Λ-particle occupies its ground state. Then Pauli blocking due to the occupation of low-lying proton and neutron states happens so that the nonmesonic decay Λ+p→n+p+177 MeV becomes dominant since it is almost unaffected by Pauli blocking due to the larger energy release. It is the main decay channel already for nuclei like $^{12}_\Lambda$C and in very heavy hypernuclei the mesonic decay width is predicted to be of the order of only 10^{-3} of the total decay width so that it can usually be neglected for A>200. The existing data for lighter nuclei indicate saturation of the lifetime of hypernuclei already around mass number 12. Recent theoretical studies deduce from the mass dependence of the lifetime of the heaviest hypernuclei that the ratio R_n/R_p of the neutron-induced to the proton-induced nonmesonic decay rate is larger than 2 which suggests a possible violation of the $\Delta I=1/2$ rule (1).

Up to mass number 56 there is an increasing amount of data on the lifetime of hypernuclei, see e.g. (2). For heavier masses data on the lifetime of hypernuclei are only available above A=200 and these data suffer partly from large experimental uncertainties or they are inconsistent. All of these are from Λ-decay induced delayed fission studies. The reaction of uranium and bismuth targets with stopped antiprotons and observation of fission fragments using a recoil-shadow technique resulted in lifetimes of 180 ± 40(stat.) ± 60(syst.) ps and 130 ±30(stat.) ± 30(syst.) ps (3), respectively. From the Bi+e$^-$ reaction at 1.2 GeV and observation of position distributions in mica detectors a hypernucleus lifetime of (2.7 ± 0.5) ns (4,5) was obtained. In view of this situation we have decided to produce heavy hypernuclei at COSY Jülich with the (p,K)-reaction and apply the recoil shadow technique to fragments from the Λ-decay induced fission. The production reaction is well suited for this technique because of the large and well defined momentum transfer and fission fragments can easily be identified even in the expected intense background of lighter particles. COSY Jülich makes possible to achieve high luminosities even with very thin targets which are needed for this technique. In addition, with its flexibility COSY allows to measure excitation functions which facilitates unambiguous identification of the production process.

2. THE METHOD

In order to extract the time scale for delayed fission events measured with the recoil shadow technique the velocity of the cold hypernuclei recoiling out of the target has to be known. Since the experimental setup does not allow to determine this momentum distribution experimentally we make use of coupled channel Boltzmann-Uehling-Uhlenbeck (CBUU) transport calculations (6,7). Theoretical momentum distributions of residual nuclei from the reaction ^{238}U+p have been compared with experimental data and found to be in good agreement up to T_p = 3 GeV. The same agreement is expected when events with hypernucleus formation are selected. A wide variety of simulations for the recoil momentum distribution has been performed using various models for the ΛN scattering, but the resulting recoil momentum distribution only changed on the few % level. This leads to a contribution to the systematical error for the Λ lifetime of 2-3 ps.

In Figure 1 we show schematically the predicted cross sections and branching ratios of the processes induced by 1.9 GeV protons colliding with Au, Bi and U. The cross section σ_{HN} for the formation of hot hypernuclei depends only slightly on the target mass. The survival probability P_s against prompt fission, however, decreases with that mass while the fissility P_f of hypernuclei increases strongly. The relevant parameter for the lifetime measurements is thus the product of the survival probability P_s, the Λ

induced fission probability P_f and the Λ-formation cross section σ_{HN}. This gives 16 µb for p+Au, 24 µb for p+Bi and 42 µb for p+U.

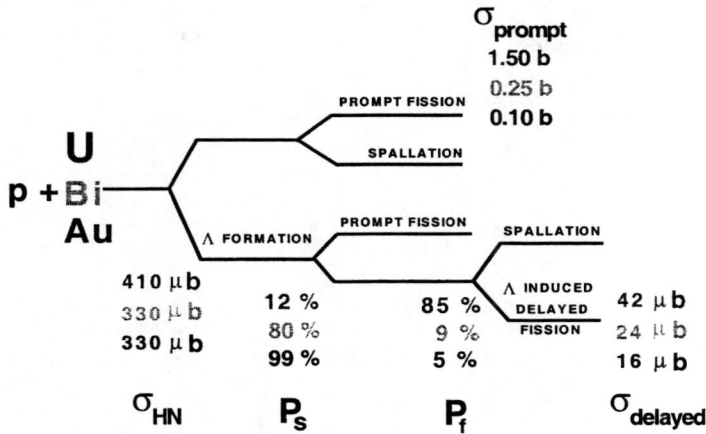

Figure 1. Schematic representation of contributions from different competing processes in p+Au, Bi, U reactions at $T_p = 1.9$ GeV as calculated in the CBUU model. The measured prompt fission cross sections are taken from ref. (8,9).

3. THE EXPERIMENT

The principle of the setup is presented in Figure 2. The target is suspended such that it is not touched by the beam during injection and acceleration. When the beam is moved onto the target by means of steerer magnets fission fragments are recorded with two position-sensitive multiwire proportional chambers (MWPC). The target and its support structure establish a "shadow" region at the MWPCs which is accessible only for fragments which originate downstream of the target and are due to delayed fission induced by the Λ-decay of hypernuclei recoiling out of the target. The "bright" region of the MWPCs is partly screened by a diaphragm (with a 1.15 mm wide slit cut in beam direction) placed downstream of the target. This reduces the load of prompt fission fragments on the detectors by two orders of magnitude thus diminishing dead time losses and multiple-hit effects. The detection of both prompt and delayed fission fragments serves for monitoring and for normalization of the cross section. Background determination is performed at a proton energy of 1.0 GeV which is sufficiently low to avoid any significant hypernucleus production.

Very thin targets are used which are based on a double layer of $2*15$ µg/cm^2 carbon. Typically they are 13 mm long and 3 mm wide and carry 15 µg/cm^2 target material on either side (Bi or UF$_4$) at their lower end. These thin strips have to be flat and stay in a fixed position within <0.1 mm, a condition which is fulfilled only with the help of electrostatic forces from 2 electrodes below the target onto which up to 7 kV are applied. By varying the voltage between the electrodes one can in addition move the target precisely in order to optimize the position of the projection of the shadow edge on the detectors. During the experiment the target is continously monitored by a TV-camera.

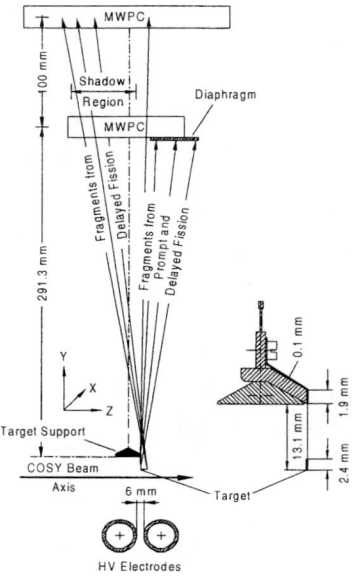

Figure 2. Recoil-shadow method used at COSY for measurements of the lifetime of heavy Λ hypernuclei. The target arrangement is shown on an expanded scale to the right. The position of the target material at the lower end of the strip defines together with the edge of the holder right above the target the boundary between the shadow region and that part of the detector which is open for prompt fission fragments.

Two low-pressure MWPCs serve for selective detection of fission fragments for track determination and time-of-flight measurements. Each chamber consists of a central cathode foil sandwiched between two wire planes with a wire spacing of 1 mm. The orientation of the wires is parallel and perpendicular to the beam. Position information is read out via delay lines for which the time reference is taken from the cathode signals. In addition, the cathode signals supply ΔE-information and define start and stop for the TOF measurement with a resolution better than 1 ns FWHM. The sensitivity of the chambers for minimal ionizing particles was found to be $< 4*10^{-11}$.

3.1 Tests And Data Analysis

The detection system was tested and calibrated with fragments from the spontaneous fission of ^{252}Cf and the α-particles emitted from the same source were used in order to test the rejection of light particles. Figure 3 shows the event distribution on the lower detector from which it can be deduced that the response is homogeneous over the full area apart from a region very close to the left edge. The curved boundary is due to a flange.

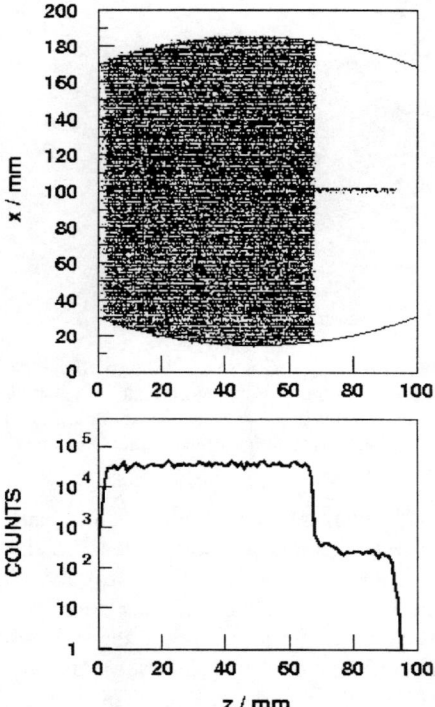

Figure 3. Hit distribution of ^{252}Cf fission fragments (upper part) and projection onto the axis which is parallel to the beam. The curved boundaries are due to the shadow of a flange while the blank area to the right is shadowed by the diaphragm the slit of which is visible.

In the data analysis events are selected according to the following conditions:

1. The sum of the delays of the anode signals from both ends of a delay line is constant unless a double hit has occured. Therefore, narrow gates are set in the sum spectra in order to select single hits only.

2. In the ΔE verusus TOF distributions fission fragments show a signature which allows to clearly separate them from lighter particles, see Figure 4. These data were obtained in the Bi+p reaction at 1.0 GeV. The intersecting straight lines shown are gates defined on the basis of ^{252}Cf fission events.

3. Finally, the trajectory for each event is calculated from the position information of both detectors and traced back to the target. Only those events are accepted which originate from a narrow region around the target.

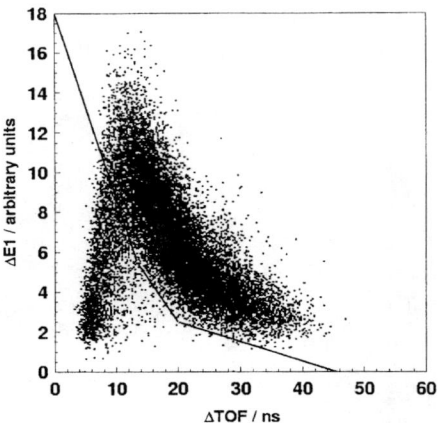

Figure 4. Plot of the energy loss in the lower detector versus TOF between the two chambers measured with a Bi target and 1.0 GeV protons. Events in the right upper part show the behaviour expected for for fission fragments while the events in the left lower part are due to light particles for which the energy loss increases with decreasing velocity. The intersecting straight lines are gates for the data analysis defined on the basis of ^{252}Cf fission events.

Events accepted with these conditions are sorted into position spectra. Background events are subtracted from the hypernucleus data and then the distribution is fitted with the exponential decay law folded with the velocity distribution of the hypernuclei obtained from BUU calculations. The systematic uncertainties include as main components target deformation, small angle scattering, uncertainty of the geometrical dimensions of the apparatus, uncertainty of the recoil momentum distribution and ist modification by energy loss in the target and the uncertainty of the average vertical position where the beam hits the target. Assuming that the individual contributions are independent one gets a typical total systematic error of about 14 ps.

3.2 The Operation Of The Accelerator

The optics of the accelerator at the target position of this experiment have been optimized for highest luminosity which implies small horizontal and vertical β-functions and at the same time small beam radii. The focussing properties of the ring are shifted continously during acceleration in order to avoid crossing of the transition energy. Since these changes resulte in inacceptably large beam extensions a rather qick change is made to optical conditions close to injection optics after the goal energy has been reached. After this procedure a vertical steerer bump ramps the beam slowly onto the target so as to obtain a rather smooth rate distribution on the detectors until the beam has been used up. Three individual accelerator cycles with different beam energies can be combined to a supercycle so that background measurements at 1.0 GeV can be performed immediately after production cycles at e.g. 1.5 or 1.9 GeV beam energy, see Figure 2. In this way systematical errors due to slow changes of the target deformation can be minimized when background events measured at 1.0 GeV are subtracted from hypernucleus events.

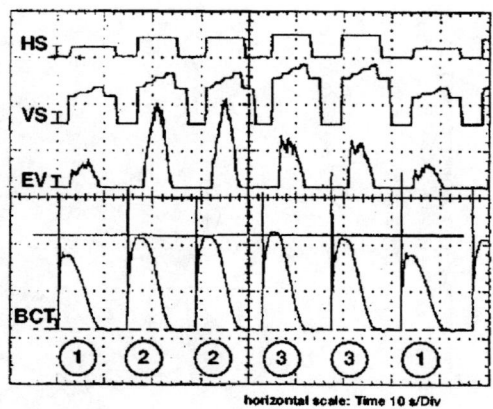

Figure 5. Supercycle for the hypernucleus experiment consisting of 1 cycle at 1.0 GeV, 2 cycles at 1.5 GeV and 2 cycles at 1.9 GeV. The signals shown are the horizontal closed orbit correction (HS), the vertical steerer bump (VS), the event rate of the experiment (EV) and the intensity of the circulating beam (BCT) as function of time.

4. EXPERIMENTAL RESULTS

4.1 The Reaction ^{238}U + p

Hit distributions along the beam axis measured with a uranium target are presented in Figure 6 for 1.0 GeV proton energy (upper part) and 1.5 GeV (lower part) (10). The high number of events seen on the right are due to prompt fission fragments which are recorded for monitoring purposes. The shape of the slope towards the shadow region is mainly governed by the absorption of fission fragments in the target and by small angle scattering at the shadow edge and in a foil separating the UHV of the accelerator from the vacuum of the detector chamber. The 3 events seen in the 1.0 GeV data result in an average background of about 1 event per wire in the 1.5 GeV data when normalized to the same counting time, see Figure 6, lower part. The statistical accuracy of the data is not yet as good as in the later performed Bi experiment since the present low-β ion-optics were not yet available. Thefore the background distribution was assumed to be constant. A fit to the 1.5 GeV data results in a hypernucleus lifetime of (240 ± 60)ps. The comaparison of the number of delayed and prompt events yield after normalization to the known prompt fission cross section for the production of those hypernuclei which undergo delayed fission decay $\sigma_{f,\Lambda}=(20\pm4)\mu b$. Taking into account the survival probability of hypernuclei after the production process and the fission probability after Λ-decay we obtain $\sigma=150^{+150}_{-80}\mu b$ for the production of hot hypernuclei.

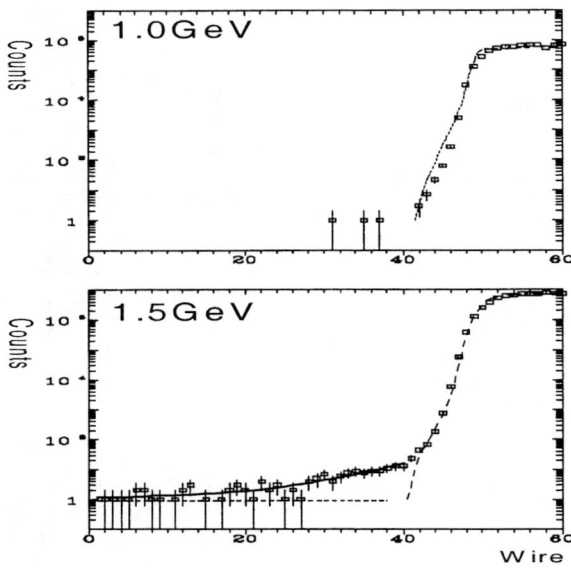

Figure 6. Position distribution (hits/wire) on the lower detector for proton energies of 1.0 GeV (upper part) and 1.5 GeV (lower part). The increasing number of events for wire numbers >60 is due to prompt fission fragments which are emitted in forward direction hit the unshielded part of the chamber.

4.2 The Reaction ^{209}Bi + p

The lifetime of hypernuclei from the ^{209}Bi + p reaction is deduced from the hit distributions shown in Figure 7. In the background spectra measured at 1.0 GeV (left part) there are only few events extending into the shadow region up to wire number ≈55. The rate increases towards higher wire numbers around the shadow edge, reaches a maximum and then decreases again due to the presence of the diaphragm. The flat distribution with the number of counts reduced by 2 orders of magnitude above channel number 70 stems from fragments which pass through the slit of the diaphragm. In the distribution in the right part which was measured at 1.9 GeV within the same supercycle there are significantly more events in the shadow region due to the decay of hypernuclei. After subtraction of the properly normalized background the lifetime was deduced from a fit (solid line in Figure 7) to be

$$\tau = 161 \pm 7(\text{stat.}) \pm 14(\text{syst.}) \text{ ps.}$$

From a comparison of the intensity of these delayed events with the intensity of prompt fission events and by normalizing to the known prompt fission cross section of $\sigma = (255 \pm 40)$ mb (8,9) we were able to deduce the cross section for the production of hot hypernuclei $\sigma = (350 \pm 140)$ μb which agrees well with the BUU prediction of 330 μb. From a fit with a fixed lifetime of 2.7 ns an upper of limit 80 nb was found for the production cross section of such a component.

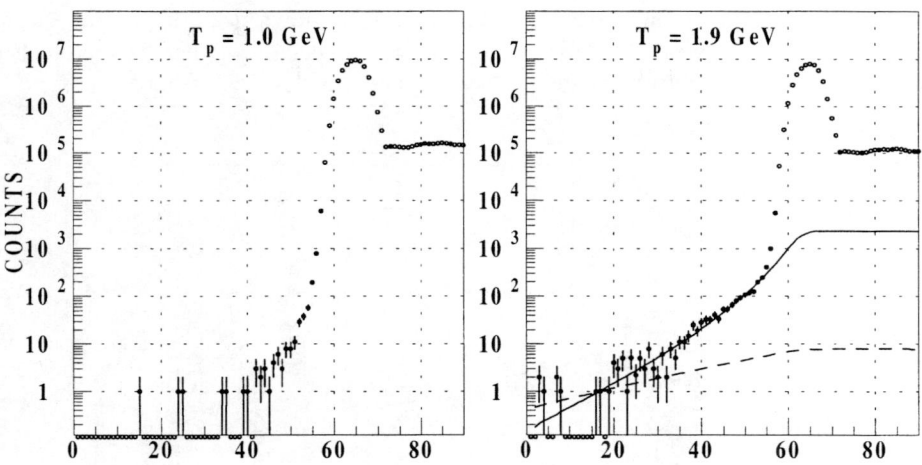

Figure 7. *Left part:* distribution of hits in the lower MWPC for collisions of 1.0 GeV protons with the Bi target. The abscissa is the number of wires (spacing of 1 mm) parallel to the COSY beam. The reduced number of events for wire numbers > 70 in the forward direction is due to the presence of a diaphragm with a narrow slit along the beam direction. *Right part:* the analogous distribution for 1.9 GeV bombarding energy. The full line represents the fit from which the lifetime was obtained. The dashed curve is a fit with a fixed lifetime of 2.7 ns which gives an upper limit of the production cross section of such a component.

5. DISCUSSION

The known lifetime information for heavy hypernuclei is collected in Table 1. It is obvious that our Bi value is the most precise one but there is general agreement within the relatively large error bars among the results obtained with protons and antiprotons. The 2.7 ns component measured in the Bi + e⁻ reaction however, cannot be understood when compared with the other values.

TABLE 1. The lifetime of heavy Λ-hypernuclei measured in different reactions

Experiment		Hypernucleus Lifetime / ps	Reference
U+p	COSY	240 ± 60	(10)
Bi+p	COSY	161 ± 7 (stat.) ± 14 (syst.)	(11)
Bi+$\bar{p}$	CERN	180 ± 40 (stat.) ± 60 (syst.)	(3)
U+$\bar{p}$	CERN	130 ± 30 (stat.) ± 30 (syst.)	(3)
Bi+e⁻	Kharkov	2700 ± 500	(4,5)

When the heavy hypernuclei are compared with the lighter ones there is almost consistency with the assumption that the lifetimes saturate for higher masses, see Figure 8. Only our new Bi value indicates a slight decrease of the lifetime. This has recently been interpreted as being due to a difference in the decay rate for neutron and proton induced nonmesonic decay (1). The conclusion drawn from this finding is that the $\Delta I = \frac{1}{2}$ rule may be violated for the nonmesonic decay of the Λ-hyperon. It is desirable to get further precise data in order to make this statement firmer.

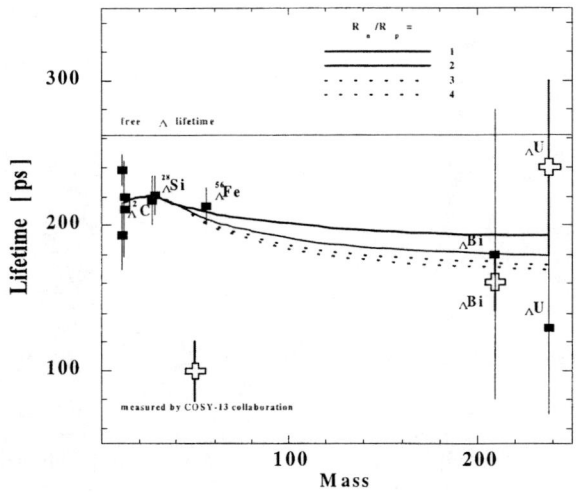

Figure 8. The new lifetime data for very heavy hypernuclei together with published results (2,10,11,12). The lines drawn are results from phase space calculations including different neutron-to-proton ratios of the nonmesonic decay rate.

6. CONCLUSIONS

We have shown that heavy hypernuclei can be produced with the (p,K) reaction and identified with the recoil shadow method. From the hit distribution seen in the shadow region lifetimes have been deduced which are consistent with values reported for hypernuclei produced with anitprotons. The 2.7 ns lifetime reported for the Bi+e reaction could not be confirmed. The present values are the most precise ones in this mass region. Comparison with lighter nuclei indicated that the hypernucleus lifetime does not saturate but slowly decrease for very heavy cases. The cross sections found for the hypernucleus production with U and Bi targets agree well with BUU predictions thus confirming that this theoretical approach gives a good description of the production process.

ACKNOWLEDGEMENTS

The project was supported by the German DLR International Bureau of the BMBF and the Polish Committee for Scientific Research (Grant No. 2 P03B 161 17).

REFERENCES

1. Z. Rudy et al., Eur. Phys.J. A5 (1999) 127
2. H.C. Bang et al., Phys. Rev.Lett. 81 (1998)4321
3. T.A. Armstrong et al., Phys. Rev. C47 (1993) 1957
4. V.I. Noga et al., Yad. Fiz. 43 (1986) 1332
5. V.I. Noga et al., Yad. Fiz. 46 (1987) 1313
6. Z. Rudy et al., Z. Phys. A351 (1995) 217
7. Z. Rudy et al., Z. Phys. A354 (1996) 445
8. J. Hudis and S. Katcoff, Phys. Rev. C13 (1976) 1961
9. E.S. Matusevich et al., Sov. J. Nucl. Phys. 7 (1968) 709
10. H. Ohm et al., Phys. Rev. C55 (1997) 3062
11. P. Kulessa et al., Phys. Lett. B427 (1998) 403
12. J.J. Szymanski et al., Phys. Rev. C43 (1991) 849

Near Threshold Two Meson Production with the $pd \to {}^3\text{He}\pi^+\pi^-$ and $pd \to {}^3\text{He}K^+K^-$ Reactions [1]

F. Bellemann,[1] A. Berg,[1] J. Bisplinghoff,[1] G. Bohlscheid,[1] J. Ernst,[1]
C. Henrich,[1] F. Hinterberger,[1] R. Ibald,[1] R. Jahn,[1] L. Jarczyk,[2]
R. Joosten,[1] A. Kozela,[3] H. Machner,[4] A. Magiera,[2] R. Maschuw,[1]
T. Mayer-Kuckuk,[1] G. Mertler,[1] J. Munkel,[1]
P. von Neumann-Cosel,[5] D. Rosendaal,[1] P. von Rossen,[4]
H. Schnitker,[1] K. Scho,[1] J. Smyrski,[2] A. Strzalkowski,[2] R. Tölle,[4]
and C. Wilkin [6]

(COSY-MOMO Collaboration)

[1] *Institut für Strahlen- und Kernphysik, Universität Bonn, Bonn, Germany*
[2] *Institute of Physics, Jagellonian University, Cracow, Poland*
[3] *Institute of Nuclear Physics, Cracow, Poland*
[4] *Institut für Kernphysik, Forschungszentrum Jülich, Jülich, Germany*
[5] *Institut für Kernphysik, Technische Universität Darmstadt, Darmstadt, Germany*
[6] *Department of Physics and Astronomy, University College London, London WC1E 6BT, United Kingdom*

Abstract. Near threshold two meson production via the reactions $pd \to {}^3\text{He}\,\pi^+\pi^-$ and $pd \to {}^3\text{He}\,K^+K^-$ was measured kinematically complete with the MOMO experiment at COSY. The obtained two pion invariant mass spectra and angular distributions depict a remarkable deviation from phase space. The two kaon data are consistent with phase space topped by a clear signal of the ϕ meson.

The MOMO experiment focusses on near threshold two meson production via the reactions $pd \to {}^3\text{He}\,\pi^+\pi^-$ and $pd \to {}^3\text{He}\,K^+K^-$. It takes advantage of the good quality of the external COSY beam and the high resolution spectrometer BIG KARL. The setup consists of a high granularity scintillating fibers meson detector near the target with a $\pm$ 45 deg. opening angle, and the spectrometer, which is used for ${}^3\text{He}$ identification. The large solid angle and high resolution of this detection method will yield precision data on the low energy (T<90MeV)

---

[1] Work supported by the BMBW und the Forschungszentrum Jülich

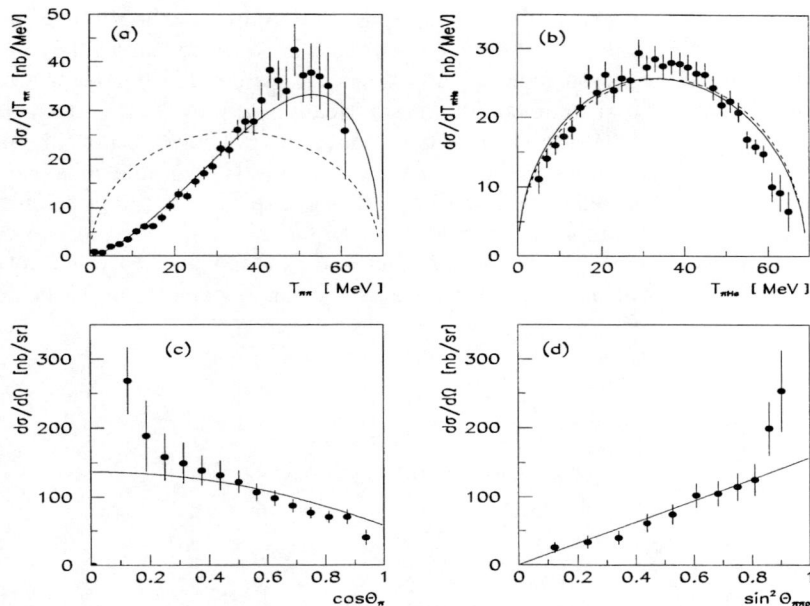

FIGURE 1. Differential cross sections for the $pd \to {}^3\text{He}\,\pi^+\pi^-$ reaction at $Q = 70$ MeV as a function of (a) the pion-pion excitation energy $T_{\pi\pi}$, (b) the excitation energy in the $\pi\text{-}{}^3\text{He}$ system, (c) the angle θ_π between one of the pions and the beam direction in the overall c.m. system, and (d) the angle $\theta_{\pi\pi p}$ between the two-pion relative momentum and the beam axis, also in the c.m. system. In the first two cases the dashed curves represent the predictions of phase space normalized to the data, whereas in all cases the solid curves are predictions assuming that the pion pair emerges in the relative p wave. The linear deviations in $T_{\pi\pi}$ from phase space in (a) and the linearity of the cross section with $\sin^2\theta_{\pi\pi p}$ in (d) are clear indications of the dominance of pion-pion p wave effects.

meson-meson interaction and probe into questions like meson-nucleon resonances and $K\bar{K}$ molecules.

The MOMO vertex detector consists of 672 scintillating fibers (round, 2.5mm diameter) arranged in three planes tilted 60 deg. versus each other. The fibers are read out by 16-fold photomultipliers. The MOMO scattering chamber houses a 4mm LD$_2$ target with extremely thin windows. The reaction $pd \to {}^3\text{He}\,\pi^+\pi^-$ was measured at three different proton beam momenta(1060 MeV/c, 1150 MeV/c, 1200 MeV/c), corresponding to 28 MeV, 70 MeV and 92 MeV center of mass energy above the reaction threshold. Recently, first data of the reaction $pd \to {}^3\text{He}\,K^+K^-$ at 2.62 GeV/c (Q=56 MeV) were obtained. Beam intensities ranged up to some 10^9 protons per second. The ^{3}He particles could be unambigiously identified by time of flight and energy loss measurements. The two-meson hits on the vertex detector could be uniquely identified by their hit patterns. Good events must be coplanar in respect to the total meson momentum axis, which is defined by the beam and the ^{3}He momenta. In total some 30 000 kinamatically complete $\pi^+\pi^-$- events were observed. The obtained two - pion invariant mass spectra show a strong deviation from phase space at all three energies, whereas the π-^{3}He missing mass spectra follow phase space. The pion angular distributions display a remarkable sidewise peaking (in the c.m.s.) and a preferential back to back emission of the two pions. This behaviour can be well described by calculations assuming a p wave between the two pions and s waves in the π-^{3}He systems [1]. Fig. 1 depicts the 1150 MeV/c data.

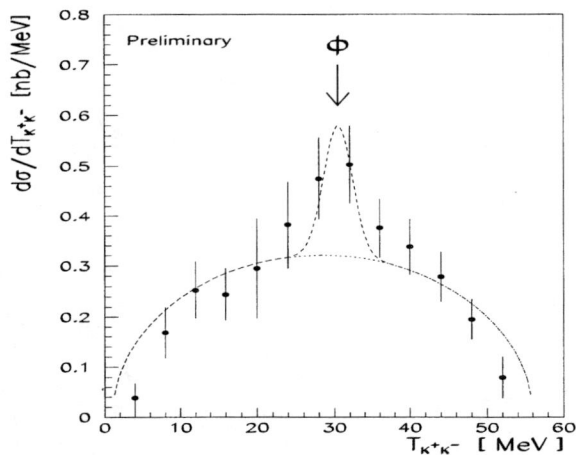

FIGURE 2. Two kaon invariant mass spectrum for the $pd \to$ ^{3}HeK^+K^- reaction at $Q = 56$ MeV.

Figure 2 shows our first results on two kaon production via the reaction $pd \to^3$ HeK^+K^- at a beam momentum of 2.62 GeV/c (Q = 56 MeV). In total, some 1500 two kaon events were observed. In contrast to our two pion data, the obtained invariant mass spectra and angular distributions depict no significant deviation from phase space. A clear signal of the ϕ meson was also observed.

REFERENCES

1. F. Bellemann *et al.*, Phys. Rev. C **60**, 61002 (1999)

Associated Strangeness Production at COSY-TOF

S. Wirth for the COSY-TOF Collaboration

Physikalisches Institut, Universität Erlangen-Nürnberg, D-91058 Erlangen, Germany

Abstract. The associated strangeness production in elementary reactions like pp → KYN close to reaction thresholds is one of the main topics to be investigated at the Time-of-Flight spectrometer TOF located at the Cooler Synchrotron COSY (FZ Jülich). The concept of the event reconstruction is based on the identification of the delayed hyperon and kaon decay ($\Lambda \to p\pi^-$, $\Sigma^+ \to p\pi^0$, $n\pi^+$ and $K^0_s \to \pi^+\pi^-$). Since the highly granulated detector covers almost the full phase space all differential distributions as well as total cross sections and the hyperon polarization can be extracted from the data.

INTRODUCTION

Hyperon production in elementary reactions is studied exclusively at the external COSY beam using the time-of-flight spectrometer TOF. For the associated strangeness production in the reaction pp → $K^+\Lambda p$ total and differential cross sections as well as Dalitz plots, mass spectra of the subsystems and the Λ-polarization are measured. In the actual step the production of Σ-hyperons in the channels pp → $K^0\Sigma^+p$, $K^+\Sigma^+n$, and $K^+\Sigma^0p$ is included.

The data should shed light on the dynamics of the $\bar{s}s$-creation in the threshold region, especially investigating the role of N^* resonances, the influence of final state interactions and the coupling of the different channels. The results should also provide information on the structure of the involved baryons, in the favorable case also on a possible strangeness content of the nucleon.

PRINCIPLE OF MEASUREMENT

The experimental setup of COSY-TOF used up to now for the measurement of the associated strangeness production consisted of a very small liquid hydrogen target (4 mm length, 6 mm diameter), the "Erlangen start detector" and the "Jülich quirl" in a distance of 1 m as stop detector. Furthermore, a large area scintillator wall was added behind TOF to measure primary and decay neutrons appearing in the Σ^+-channel. The start detector system covers the whole phase space of the primary reaction products. As shown in Fig. 1, it consists of the so-called "starttorte" built by two layers of segmented scintillators, a double-sided silicon microstrip detector and two scintillating fiber hodoscopes [1, 2].

The identification of the Λ is done by its delayed decay into two charged particles p and π^-, whereas the signature of the Σ^+-decay is given only by a kink in the track. In this case, due to the short decay length ($c\tau = 2.4$ cm) a precise measurement of the track very close to the target is essential, which is performed by the double-sided microstrip detector. Additional indications are the delayed decay $K^0_s \to \pi^+\pi^-$ and/or a detected neutron, respectively. The measurement of all charged primary and secondary particles (with an accuracy in the vertex reconstruction of better than 1 mm) allows a kinematically complete event reconstruction which leads to a missing mass resolution of about 10 MeV.

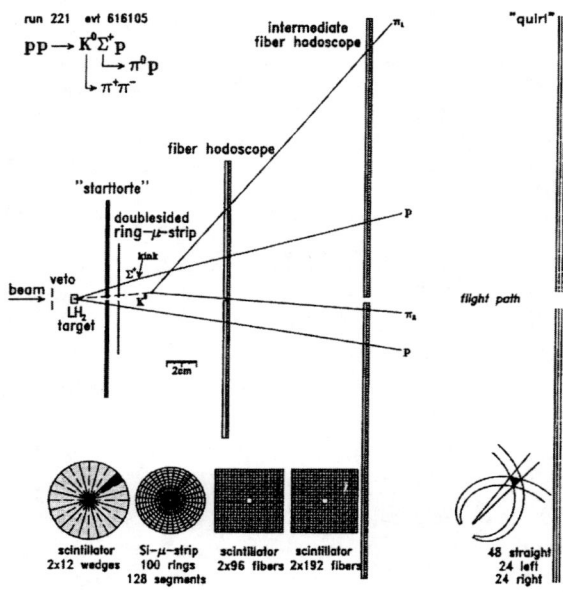

FIGURE 1. Schematic view of the "Erlangen Startdetector" and the "Jülich Quirl" together with a candidate event of the type pp → $K^0\Sigma^+p$

RESULTS

For the Λ-channel clean nearly background free samples could be extracted at beam momenta between 2.50 and 2.85 GeV/c, which allow to evaluate total cross sections, angular distributions, Dalitz plots with mass projections and also the Λ-polarization. In part these results are already published [3]. The total cross section together with the results of COSY-11 [4] just at threshold give a consistent picture allowing already to test different model calculations and their parameters. The Dalitz plots show that the data more or less smoothly cover the full phase space. The projections for the pΛ- and KΛ-subsystems at $p_{Beam} = 2.75$ GeV/c, however, indicate significant deviations from the phase space (see figure 2). The enhancement on the left side of the pΛ-mass mainly corresponds to the pΛ-final-state interaction. It is also interesting to notice that the data point sticking up in the middle of this spectrum corresponds to the pΣ-mass. Whether this has to be interpreted as a coupled channels effect needs further confirmation from more precise data. The shift to higher masses in the the KΛ-spectrum can be explained in resonance model calculations [5] where the influence of the $N^*(1710)$ and $N^*(1720)$ resonances affects the KΛ-spectrum in the observed way.

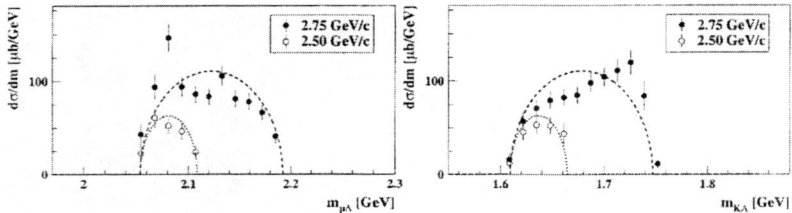

FIGURE 2. Projections of the Dalitz plots on 2-particle subsystems at 2.50 and 2.75 GeV/c

In the channel $pp \rightarrow K^0\Sigma^+p$ for the first time close to threshold a subsample of about 30 events could be extracted. Here a total yield of about 300 events is expected from the full data set at $p_{Beam} = 2.85$ GeV/c.

OUTLOOK

The hyperon program at COSY-TOF is planned as a long term activity. It will include measurements with a polarized beam and a polarized target to get a deeper understanding of the investigated process and the structure of the involved objects. Furthermore, a deuteron target will be used to study strangeness production at the neutron.

ACKNOWLEDGEMENTS

We gratefully acknowledge support from the German BMBF and the FZ Jülich.

REFERENCES

1. W. Eyrich et al., Physica Scripta 48, 88 (1993).
2. W. Eyrich et al., "Scintillating Fiber Hodoscopes for COSY-TOF" in SCIFI 97 Conference on Scintillating Fiber Detectors, edited by A. D. Bross, R. C. Ruchti and M. R. Wayne, AIP Conference Proceedings 450, New York: American Institute of Physics, 1998, pp. 381-384.
3. R. Bilger et al., Phys. Lett. B 420, 217 (1998).
4. J.T. Balewski et al., Phys. Lett. B 420, 211 (1998).
5. K. Tsushima, A. Sibirtsev and A.W. Thomas, Phys. Lett. B 390, 29 (1997).
   A. Sibirtsev, K. Tsushima and A.W. Thomas, Phys. Lett. B 421, 59 (1998).

POLARIZED BEAMS AND POLARIZED TARGETS

Polarization in Meson Production Reactions

L. D. Knutson

University of Wisconsin, Madison, Wisconsin 53706

Abstract.
A comprehensive formalism for describing polarization observables in meson production reactions is presented. Particular attention is given to the complications that arise when the final state contains three particles. A general formula for the partial wave expansion of the polarization observables is presented, and a number of applications of the formalism are discussed.

INTRODUCTION

In the last few years we have seen a substantial amount of new work, both experimental and theoretical, on the subject meson production. As is well known to everyone at this conference, much of the new experimental work has been carried out at storage ring machines. In many respects, storage rings are ideally suited to the study of these reactions, in particular making it possible to map out features of cross sections and polarization observables at energies just above threshold.

Experiments carried out recently at the IUCF Cooler Ring (see for example Refs. [1–3]) have demonstrated the great value of employing polarized beams and targets in the study of meson production reactions, and the present paper will deal mainly with the description of polarization measurements of this kind.

We shall focus here on reactions of the type NN → NNπ, though at least some of the results are easily extended to mesons of arbitrary spin and to reactions such as NN → dπ.

Compared to "ordinary" reactions that have only two particles in the final state, 3-body final state reactions are kinematically much more complex. In ordinary reactions there is effectively only one kinematic variable, the reaction angle θ, but with three particles in the final state the number of independent variables is five. This is a non-trivial complication, because it means that one can no longer simply plot experimental results *vs* θ. Typically what one does is to either restrict the kinematics to fixed angles (and plot the observable of interest as a function of position along a kinematic locus) or else integrate over some variables. In either case, one loses potentially valuable information.

The description of polarization observables is also somewhat more complex for 3-body final state reactions than it is for ordinary reactions. In particular, the rules that tell us which observables may be non-zero when parity is conserved are more complex when the final state contains more than two particles. Further conceptual complexities also arise when there are identical particles in the initial and/or final state.

For all of these reasons, the analysis and interpretation of polarization data for meson production reactions is not at all straightforward. If one has obtained measurements with a polarized beam or target, how are the results most effectively presented? What might one learn about the reaction process from a particular type of measurement? What kind of angular distributions are allowed for a given observable? These are questions that can confound even experts in the field.

Some of the complexity inherent in 3-body final state reactions can be overcome by introducing the idea of a partial-wave analysis. When meson production reactions are initiated at energies just above threshold, the reactions proceed mainly into states of low orbital angular momentum. If one assumes that only a few partial waves are important, it is then possible draw important conclusions. In particular one can make definite predictions about the kinds of angular patterns that are allowed for a given observable. In addition, one can say much about how each partial wave affects a given observable.

In the following sections we first describe how to generalize the definitions of the polarization observables for reactions with three-body final states. We then introduce the partial-wave formalism and discuss the partial wave contributions to the reaction $pp \to pp\pi^0$. The partial wave expansion of the cross section is then obtained and several features of the cross section formula are illustrated. We then move on the polarization observables. We first describe how these quantities may be calculated and we then present a general partial wave expansion formula that can be used for any polarized beam/polarized target spin observable. The paper then concludes with some further illustrations and examples. Additional applications of the formalism presented here are discussed in Ref. [4].

POLARIZATION EXPERIMENTS

Most of us are already somewhat familiar with the general rules that apply to simple polarization experiments. However, as indicated above, these rules do not necessarily carry over to meson production reactions. In order to understand the source of the complications, we will find it useful to begin by reviewing some simple cases.

The simplest polarization experiments involve a beam of polarized spin-$\frac{1}{2}$ particles incident on an unpolarized target, with the reaction leading to a 2-particle final state. It is well known that in this kind of experiment the reaction rate is sensitive only to the component of the polarization vector that is perpendicular to the reaction plane. Sensitivity to the longitudinal and transverse in-plane components is disallowed provided that parity is conserved.

To define the measurable quantities it is useful to introduce a coordinate frame with the z-axis along the beam direction. The usual convention is to then choose the y-axis to be along $\vec{k}_{in} \times \vec{k}_{out}$ where $\vec{k}_{out}$ is the momentum of the detected particle, which means that the reaction product is detected at $\phi = 0$. With this definition, $\hat{y}$ is perpendicular to the reaction plane and, assuming that parity is conserved, we have the result that reaction rate is sensitive only to P_y. It follows that the differential cross section (which we designate by W) may be written in the form

$$W = W_0 \left\{ 1 + P_y A(\theta) \right\}. \tag{1}$$

Here W_0 represents the unpolarized cross section, and $A(\theta)$ is the analyzing power of the reaction.

This particular expression for the polarized cross section is well suited for experiments that employ isolated small detectors. However, if one is using detectors that span a range of azimuthal angles (as is normally the case in storage ring experiments) Eq. (1) is somewhat clumsy to use, since different points on the detector correspond to different choices of $\hat{y}$ and thus to different values of P_y.

For this and for other reasons as well, it will be useful to generalize Eq. (1) somewhat. Rather than defining $\hat{y}$ in terms of the reaction kinematics, the directions of $\hat{x}$ and $\hat{y}$ may be chosen arbitrarily – for example, we might take $\hat{y}$ to be vertical (up) and $\hat{x}$ to be horizontal. In this case, the polarized-beam cross section at angles θ and ϕ may depend on both P_x and P_y, but only in the following well-defined way:

$$W = W_0 \left\{ 1 + P_y A(\theta) \cos \phi - P_x A(\theta) \sin \phi \right\}. \tag{2}$$

One of the properties of this particular cross section formula is that each polarization term has a simple, unique azimuthal form. With the beam polarized along $\hat{y}$ the radiation pattern varies as $\cos \phi$ and when the polarization is along $\hat{x}$ the pattern varies as $-\sin \phi$. As we shall see below, the azimuthal behavior is not so simple for reactions with 3-body final states. In anticipation of this result, it is useful to rewrite Eq. (2) in the form

$$W = W_0 \left\{ 1 + P_y A_y(\theta, \phi) + P_x A_x(\theta, \phi) \right\}, \tag{3}$$

where the $\sin \phi$ and $\cos \phi$ factors have now been absorbed into the definitions of A_x and A_y.

Let us now generalize this result to the case in which the final state has three particles. The first point to note is that when three particles are present, two relative momentum vectors are needed to specify the final-state kinematics. For our purposes it will be most convenient to define $\vec{q}$ to be the momentum of the meson in the 3-body center-of-momentum frame and $\vec{p}$ to be the momentum of nucleon #1 in the NN center-of-momentum frame.

We know that, in general, the reaction observables may be functions of all the independent kinematic variables. Thus, in reactions such as NN $\to$ NNπ, the observables will depend on the directions of both $\vec{p}$ and $\vec{q}$ and also the way in which the total c.m. energy is shared between p and q. Thus, if X is any measurable quantity we may write

$$X = X(\theta_p, \phi_p, \theta_q, \phi_q, \epsilon) \tag{4}$$

where the angles θ_p and ϕ_p (θ_q and ϕ_q) specify the direction of $\vec{p}$ ($\vec{q}$), and where ϵ is some kind of energy-sharing parameter. For example we could take ϵ to be the total kinetic energy in NN c.m. frame.

From this discussion it is already clear that the azimuthal dependences of the observables must be more complex than in the two-particle case. In Eq. (3) we had $A_y \propto \cos\phi$ and $A_x \propto -\sin\phi$, but since there are now two azimuthal angles, this result obviously does not carry over. What we may say in the more general case is that A_y and A_x must transform under rotations about the z-axis as $\cos\phi$ and $-\sin\phi$. All functions of ϕ_p and ϕ_q that satisfy this condition are permitted.

The other new complication for three-body final states is that the cross section may depend on the longitudinal component of the polarization as well as on the transverse components (see Ref. [5]). Parity conservation places certain restrictions on the form of the longitudinal analyzing power, but this quantity is no longer required to be zero. Thus the appropriate cross section formula for the three-body case is

$$W = W_0\left\{1 + P_y A_y + P_x A_x + P_z A_z\right\}. \tag{5}$$

As we have said, one of the new features here is that A_x and A_y no longer vary simply as $\sin\phi$ and $\cos\phi$. On the other hand, it is important to recognize that, as in the two-body case, the observables A_x and A_y are still not independent. Both quantities are measured with transverse beam polarization and therefore the two measurements are completely equivalent. Obviously, the radiation pattern observed with y polarization will be identical to that for x polarization except that the pattern will be rotated by 90° about the beam axis. Therefore, in Eq. (5) there are two independent polarization observables, the longitudinal analyzing power, A_z, and the transverse analyzing power, A_x or A_y.

For experiments in which both the beam and the target are polarized the cross section must include additional terms. Besides the beam analyzing power terms given in Eq. (5), we will now have analogous target analyzing power terms, plus the spin-correlation terms that depend on both beam and target polarization. If the components of the target polarization are represented by Q_x, Q_y and Q_z the full expression can be written in the form

$$W = W_0\{1 + \sum_i P_i A_{i0} + \sum_j Q_j A_{0j} + \sum_{i,j} P_i Q_j A_{ij}\}. \tag{6}$$

where the sums extend over the indices x, y and z.

In Eq. (6) there are three beam analyzing power terms, three target analyzing power terms and nine spin-correlation terms. From the discussion above, we already understand that for both beam and target, only two of the three analyzing powers are independent observables. The question of how many independent spin correlation parameters there are is somewhat more complex and will be addressed later. We shall also come back later to the question of what azimuthal dependences are allowed for each term.

PARTIAL WAVE FORMALISM

At this point let us turn to the question of what one can learn about the polarization observables from a partial wave analysis. The basic idea of a partial wave decomposition is simple and fundamental. If the total kinetic energy in the final state is small, then we expect the reaction to be dominated by transitions to states with low orbital angular momentum values. It follows that an expansion of the outgoing spherical waves would involve mainly low-order spherical harmonics, and therefore the observables should have a relatively simple angular dependence.

For reactions with two-body final states the consequences are well understood. Suppose, for example, that a particular reaction leads only to s- and p-wave final states. Then the outgoing wave will involve spherical harmonics of order $l = 0$ and $l = 1$. Since the observables are quadratic functions of the amplitudes, expansion of the cross sections (including both the polarized and unpolarized cross sections) in spherical harmonics will involve terms only up to $l = 2$. Thus, for example, the unpolarized cross section may vary no more rapidly than $\cos^2 \theta$. Similarly the quantity $W_0 A_{y0}$ must be a simple function, with an angular dependence no more complex than $c_1 \sin \theta + c_2 \sin \theta \cos \theta$.

Let us now see how these same general ideas carry over to meson production reactions. All of the results to be presented here follow from a single formula that expresses the "scattering amplitude" for the reaction in terms of the contributing partial waves.

To begin, one must define the initial and final angular momentum states. In the initial state the nucleons have orbital angular momentum l, total spin s_i, and total angular momentum J. The quantities s_a and s_b are the individual nucleon spin quantum numbers, and the corresponding magnetic quantum numbers are σ_a and σ_b. In the final state the nucleon spin quantum numbers are represented by s_1 and s_2 with spin projections σ_1 and σ_2, and the NN total spin quantum number is s_f. Similarly, we use l_p for the NN relative orbital angular momentum, and j for the NN total angular momentum. For the meson, the spin is represented by quantum numbers s_3 and σ_3. Finally l_q is the angular momentum of the meson relative to the NN center of mass, and j' is the vector sum of s_3 and l_q.

The scattering matrix, $M^{\sigma_1,\sigma_2,\sigma_3}_{\sigma_a,\sigma_b}$, gives the amplitude of the outgoing wave in spin state $\sigma_1, \sigma_2, \sigma_3$ obtained when the nucleons in the initial state have spin projections σ_a and σ_b. The form of this amplitude can be deduced from general principles, and the result is

$$M^{\sigma_1,\sigma_2,\sigma_3}_{\sigma_a,\sigma_b} = i K_i \sum \left[\frac{2l+1}{2J+1}\right]^{\frac{1}{2}} \langle s_a \sigma_a, s_b \sigma_b | s_i \sigma_i \rangle \langle l0, s_i \sigma_i | JM \rangle$$

$$\times \langle s_1\sigma_1, s_2\sigma_2|s_f\sigma_f\rangle\langle s_f\sigma_f, l_p\lambda_p|jm\rangle\langle s_3\sigma_3, l_q\lambda_q|j'm'\rangle$$
$$\times \langle t_a\nu_a, t_b\nu_b|t_i\nu_i\rangle\langle t_1\nu_1, t_2\nu_2|t_f\nu_f\rangle\langle t_f\nu_f, t_3\nu_3|t_i\nu_i\rangle$$
$$\times \langle jm, j'm'|JM\rangle U_\alpha\, Y_{l_p}^{\lambda_p}(\hat{p})\, Y_{l_q}^{\lambda_q}(\hat{q}). \tag{7}$$

In this equation the indices t and ν represent the isospin quantum numbers, with notation that exactly parallels that used for the the ordinary spin. The quantity K_i is a kinematic constant, while U_α is a matrix element which will depend on the quantum numbers in both the initial state (l, s_i, J and t_i) and the final state (s_f, l_p, l_q, j, j', and t_f). The matrix element is also a function of the energy sharing parameter ϵ. The sum in Eq. (7) extends over the initial and final state quantum numbers listed above, and also over the magnetic quantum numbers σ_i, σ_f, M, m, m', λ_p, λ_q, ν_i and ν_f.

The expression given is Eq. (7) is somewhat complex. However, the main point to note is that, as in the two-particle case, the complexity of the angular pattern is determined by the final state orbital angular momentum values, l_p and l_q.

Our formula for the scattering amplitude simplifies considerably if we consider the special case, $pp \to pp\pi^0$. In this case the isospin Clebsch-Gordon coefficients simply become numerical factors. In addition, s_3 is zero, which eliminates one more coefficient. The result is

$$M_{\sigma_a,\sigma_b}^{\sigma_1,\sigma_2} = \sqrt{2}\, i\, K_i \sum_\alpha \sum_{\substack{M,\sigma_i,m \\ \lambda_p,\sigma_f,\lambda_q}} \left[\frac{2l+1}{2J+1}\right]^{\frac{1}{2}} \langle s_a\sigma_a, s_b\sigma_b|s_i\sigma_i\rangle\langle l0, s_i\sigma_i|JM\rangle$$
$$\times \langle s_1\sigma_1, s_2\sigma_2|s_f\sigma_f\rangle\langle s_f\sigma_f, l_p\lambda_p|jm\rangle\langle jm, l_q\lambda_q|JM\rangle U_\alpha\, Y_{l_p}^{\lambda_p}(\hat{p})\, Y_{l_q}^{\lambda_q}(\hat{q}). \tag{8}$$

Here summation over α is shorthand for summation over initial and final state quantum numbers J, l, s_i, j, l_p, s_f and l_q.

One obvious question at this point is whether one needs to modify this formula in some way to account for the fact that there are identical particles in the initial and final states. The answer is that (with one minor exception) no modification is needed. If one follows the obvious procedure of including only initial and final angular momentum states that are consistent with the antisymmetrization requirement, then the resulting formula will satisfy all of the necessary symmetries. The only "special" requirement is that when the initial state has identical particles, an extra factor of 2 must be included to reproduce the appropriate incident flux normalization. This extra factor of 2 is already included in Eq. (8).

PARTIAL WAVES FOR $pp \to pp\pi^0$

The reaction $pp \to pp\pi^0$ is a interesting special case to focus on in part because the number of contributing partial waves is relatively small at energies just above threshold. We shall adopt the standard notation and designate the various matrix elements as Ss, Ps, Pp, etc., where the first letter indicates the final-state pp angular momentum, l_p, and the second the pion angular momentum, l_q.

At energies just above threshold one expects the cross section to be dominated by the single allowed Ss transition that proceeds from $^3P_0$ initial state to the $^1S_0$ pp final state. At higher energies, transitions to the Ps states become important, and one might also expect contributions from Pp, Sd and Ds final states. Table 2 lists all of the allowed $pp \to pp\pi^0$ matrix elements through $l_p + l_q = 2$.

Experimental evidence for Ps and Pp transitions has been seen in polarization observables (see Ref. [3]) and there is evidence for Sd transitions in the differential cross section [6].

TABLE 1. Partial waves for the reaction $pp \to pp\pi^0$.

	Initial			Final			
	l	s_i	J	l_p	s_f	j	l_q
Ss:	1	1	0	0	0	0	0
Ps:	0	0	0	1	1	0	0
	2	0	2	1	1	2	0
Pp:	1	1	1	1	1	0	1
	1	1	1	1	1	1	1
	1	1	0	1	1	1	1
	1	1	2	1	1	1	1
	3	1	2	1	1	1	1
	1	1	1	1	1	2	1
	1	1	2	1	1	2	1
	3	1	2	1	1	2	1
	3	1	3	1	1	2	1
Sd:	1	1	2	0	0	0	2
	3	1	2	0	0	0	2
Ds:	1	1	2	2	0	2	0
	3	1	2	2	0	2	0

EXPANSION OF THE CROSS SECTION

Starting from Eq. (8), it is straightforward (at least in principle) to obtain expressions for the polarization observables. If we are concerned only with measuring polarized beam and target cross sections (with no measurement of final state polarizations), then the observables, Q, are found by taking the trace of an appropriate matrix product; *i.e.*

$$Q = \text{Tr}\left[MTM^\dagger\right], \qquad (9)$$

where T is some spin operator.

The simplest observable is the unpolarized cross section, W_0. In this case the operator T is just proportional to the unit matrix, and one has

$$W_0 = \frac{1}{(2s_a+1)(2s_b+1)} \text{Tr}\left[MM^\dagger\right], \tag{10}$$

The goal at this point is to obtain a formula for the partial-wave expansion of the cross section which shows explicitly the various allowed angular dependences. To obtain this result we first substitute Eq. (8) into Eq. (10). We then combine the spherical harmonics and carry out an angular momentum reduction. In the process, one finds that it is possible to express the angular dependence in terms of a set of functions known as the bipolar harmonics. These quantities are defined by the following formula:

$$B^\Lambda_{L_p,L_q;L}(\hat{p},\hat{q}) = \sum_{\Lambda_p,\Lambda_q} \langle L_p\Lambda_p, L_q\Lambda_q | L\Lambda \rangle Y^{\Lambda_p}_{L_p}(\hat{p}) Y^{\Lambda_q}_{L_q}(\hat{q}). \tag{11}$$

The expansion of the cross section in terms of the bipolar harmonics is given by

$$W_0 = \frac{2K_i^2}{(2s_a+1)(2s_b+1)} \left(\frac{1}{4\pi}\right) \sum_{L_p,L_q;L} \left[\sum_{\alpha,\alpha'} C^{\alpha,\alpha'}_{L_p,L_q;L} U_\alpha U^*_{\alpha'}\right] B^0_{L_p,L_q;L}(\hat{p},\hat{q}). \tag{12}$$

According to this formula, the coefficient of a given bipolar harmonic involves a sum over pairs of matrix elements weighted by numerical factors, C, which are are easily evaluated. The explicit expression for the numerical factors is found by carrying out the angular momentum reduction mentioned above. For the spin-zero mesons, the result is

$$C^{\alpha,\alpha'}_{L_x,L_y;L} = (-)^{l'_p+l'_q+l'} \delta_{s_i,s'_i}\delta_{s_f,s'_f}[(2l+1)(2l'+1)(2l_p+1)(2l'_p+1)(2l_q+1)(2l'_q+1)$$
$$\times (2j+1)(2j'+1)(2J+1)(2J'+1)]^{\frac{1}{2}} \langle l_p 0, l'_p 0 | L_p 0 \rangle \langle l_q 0, l'_q 0 | L_q 0 \rangle \langle l 0, l' 0 | L 0 \rangle$$
$$\times W(j,s_f,L_p,l'_p;l_p,j')\, W(l,s_i,L,J';J,l') \begin{Bmatrix} j & l_q & J \\ j' & l'_q & J' \\ L_p & L_q & L \end{Bmatrix}. \tag{13}$$

In this formula the W's are Racah coefficients and the quantity in brackets is a Wigner nine-j coefficient. Computer codes for evaluating these angular momentum coefficients are available from various sources, and so the C's are easily calculated.

DISCUSSION AND EXAMPLES

Partial wave expansion formulas such as the ones given in the preceding section are useful for a number of purposes. To carry out a full partial wave analysis of measurements one would need to know the actual numerical values of the C coefficients. However, many useful conclusions can be drawn simply from the form of the equations and by understanding which coefficients may be non-zero. In this section we illustrate this point with a few examples.

We note, first of all, that the various cross section terms in Eq. (12) arise from interference between two matrix elements, which in our notation are labelled by the quantum numbers α and α'. Because of the δ_{s_i,s'_i} and δ_{s_f,s'_f} factors in Eq. (13), two partial waves may interfere only if the NN spin quantum numbers match in both the initial and final states. Applying this rule to the states listed in Table 1, we conclude that the dominant Ss transition may interfere with the Sd and Ds terms but not with Ps or Pp. Similarly, there can be no interference between Ps and Pp contributions.

As we shall see later, in our general formula for the polarization observables the factor δ_{s_f,s'_f} is still present. Thus, if we divide the partial waves into two groups, one containing terms with $s_f = 0$ (Ss+Sd+Ds) and the other corresponding to $s_f = 1$ (Ps+Pp), interference will be permitted only within each group. Since Ss is dominant at energies just above threshold, we conclude that the Ds and Sd terms may be more important at low energies than one might otherwise have thought.

From Eq. (13) it is fairly easy to draw conclusions about the "radiation patterns" that are allowed if only a few partial waves are present. The Clebsch-Gordon coefficient $\langle l_p 0, l'_p 0 | L_p 0 \rangle$ requires that L_p not exceed $l_p + l'_p$ and also that $l_p + l'_p + L_p$ be even. There are analogous requirements for L_q. It follows, for example, that if only Ss and Ps transitions are important, the cross section may only involve terms proportional to $B^0_{0,0;0}$ and $B^0_{2,0;2}$; i.e.

$$W_0 = A + B \left(\tfrac{3}{2} \cos^2 \theta_p - \tfrac{1}{2} \right) \tag{14}$$

If Sd transitions are allowed as well, then one may also may have terms involving $B^0_{0,2;2}$:

$$W_0 = A + B \left(\tfrac{3}{2} \cos^2 \theta_p - \tfrac{1}{2} \right) + C \left(\tfrac{3}{2} \cos^2 \theta_q - \tfrac{1}{2} \right), \tag{15}$$

where the C term arises from Ss-Sd interference. It is this last term that is seen in the experiments of Ref. [6].

For our final example we focus on the ϕ-dependences of the cross section. Note, first of all, that the expansion of the cross section involves only angular functions $B^\Lambda_{L_p,L_q;L}$ with $\Lambda = 0$. This result comes out of the angular momentum algebra and is a consequence of the fact that the cross section must be invariant under rotations about the z-axis. Since the bipolar harmonics are constructed from spherical harmonics, we have

$$B^\Lambda_{L_p,L_q;L} \propto e^{i(\Lambda_p \phi_p + \Lambda_q \phi_q)}, \tag{16}$$

where the values of Λ_p and Λ_q are restricted to those satisfying $\Lambda_p + \Lambda_q = \Lambda$. One can show that if parity is conserved, the non-zero C coefficients all have $L_p + L_q + L$ even, and it is easy to demonstrate that the corresponding B functions are purely real for $\Lambda = 0$. As a result the cross section may involve terms that go as $\cos n(\phi_p - \phi_q)$. However, the values of n may not exceed either L_p or L_q.

Returning to our earlier examples, we may conclude that cross section contributions arising from Ss^2, Ps^2 or Ss-Sd interference must be independent of both ϕ angles. However, Pp^2 contributions may give rise to cross section terms that go as $\cos(\phi_p - \phi_q)$ and $\cos 2(\phi_p - \phi_q)$.

POLARIZATION OBSERVABLES

We next turn to the polarization observables. Formulas for these quantities may be obtained by using the appropriate spin operators for T in Eq. (9). For the analyzing powers the operators we want are the Pauli matrices. By employing the density matrix formalism one can demonstrate that

$$W_0 A_{i0} = \frac{1}{(2s_a+1)(2s_b+1)} \, \text{Tr}\left[M \sigma_i M^\dagger\right], \tag{17}$$

where the subscript i can be x, y or z. In a similar way, the spin correlation parameters are obtained by using for T the direct product of Pauli matrices for beam and target particles:

$$W_0 A_{ij} = \frac{1}{(2s_a+1)(2s_b+1)} \, \text{Tr}\left[M \, \sigma_i^{(b)} \otimes \sigma_j^{(t)} \, M^\dagger\right]. \tag{18}$$

It turns out that it is somewhat cumbersome to work with the Cartesian spin operators that appear in Eqs. (17) and (18). Obtaining the partial wave expansions is simplified considerably if one introduces a set of spherical tensor spin operators that are defined, for each particle, by the equations

$$\begin{aligned} \tau_{00} &= I \\ \tau_{10} &= \sigma_z \\ \tau_{1\pm 1} &= \mp \tfrac{1}{\sqrt{2}}(\sigma_x \pm i\sigma_y) \end{aligned} \tag{19}$$

where I is the 2×2 unit matrix.

With these new operators, we may now introduce a set of "spherical tensor" polarization observables. These quantities are defined by the equation

$$T_{k_1 q_1, k_2 q_2} = \frac{1}{(2s_a+1)(2s_b+1)} \, \text{Tr}\left[M \, \tau_{k_1 q_1}^{(b)} \otimes \tau_{k_2 q_2}^{(t)} \, M^\dagger\right]. \tag{20}$$

From the definitions given above, it is straightforward to find simple relationships between the usual Cartesian analyzing powers and spin correlation coefficients and the spherical tensor observables. These relationships are given in Table 2 below. In deriving these formulas, we make use of the identity

$$T_{k_1 -q_1, k_2 -q_2} = (-)^{q_1+q_2} T^*_{k_1 q_1, k_2 q_2}, \tag{21}$$

which follows from the definitions in Eq. (19) and the Hermiticity of the Pauli operators.

TABLE 2. Translation Formulas.

Cartesian Observable	Spherical Observable
Differential cross section:	
W_0	$T_{00,00}$
Beam analyzing powers:	
$W_0 A_{x0}$	$-\sqrt{2}\,\mathrm{Re}\,[T_{11,00}]$
$W_0 A_{y0}$	$-\sqrt{2}\,\mathrm{Im}\,[T_{11,00}]$
$W_0 A_{z0}$	$T_{10,00}$
Target analyzing powers:	
$W_0 A_{0x}$	$-\sqrt{2}\,\mathrm{Re}\,[T_{00,11}]$
$W_0 A_{0y}$	$-\sqrt{2}\,\mathrm{Im}\,[T_{00,11}]$
$W_0 A_{0z}$	$T_{00,10}$
Spin correlation parameters:	
$W_0 A_{zz}$	$T_{10,10}$
$W_0 [A_{xx}+A_{yy}]$	$-2\,\mathrm{Re}\,[T_{11,1-1}]$
$W_0 [A_{xx}-A_{yy}]$	$2\,\mathrm{Re}\,[T_{11,11}]$
$W_0 A_{xz}$	$-\sqrt{2}\,\mathrm{Re}\,[T_{11,10}]$
$W_0 A_{zx}$	$-\sqrt{2}\,\mathrm{Re}\,[T_{10,11}]$
$W_0 A_{yz}$	$-\sqrt{2}\,\mathrm{Im}\,[T_{11,10}]$
$W_0 A_{zy}$	$-\sqrt{2}\,\mathrm{Im}\,[T_{10,11}]$
$W_0 [A_{xy}+A_{yx}]$	$2\,\mathrm{Im}\,[T_{11,11}]$
$W_0 [A_{xy}-A_{yx}]$	$2\,\mathrm{Im}\,[T_{11,1-1}]$

EXPANSION OF THE POLARIZATION OBSERVABLES

As indicated in the preceding section, the introduction of the spherical tensor spin operators greatly simplifies the job of obtaining a single, general purpose formula for the partial wave expansion of the observables. The simplification comes from the fact that the spin operators of Eq. (19) are easily represented in angular momentum language. The relevant expression is [7]

$$\langle \sigma | \tau_{kq} | \sigma' \rangle = (-)^{s-\sigma'} \sqrt{2s+1} \, \langle s\sigma, s\,-\sigma' | kq \rangle. \tag{22}$$

To obtain the partial wave expansion formula we now substitute this expression, along with Eq. (8) for M, into Eq. (20). Carrying out the angular momentum reduction, we obtain the result

$$T_{k_1 q_1, k_2 q_2} = \frac{2K_i^2}{(2s_a+1)(2s_b+1)} \left(\frac{1}{4\pi}\right)$$
$$\times \sum_{L_p, L_q, L} \left[\sum_{\alpha, \alpha'} C^{\alpha, \alpha'; \xi}_{L_p, L_q; L} U_\alpha U^*_{\alpha'} \right] B^Q_{L_p, L_q; L}(\hat{p}, \hat{q}), \tag{23}$$

where $Q = q_1 + q_2$ and where the label ξ is shorthand for the indices $k_1 q_1, k_2 q_2$.

The expansion coefficients, C, are generalized versions of the coefficients given in Eq. (13). The full expression is

$$C_{L_p,L_q;L}^{\alpha,\alpha';\xi} = (-)^{l_p'+l_q'+l'} \delta_{s_f,s_f'} \sum_{K,I} [(2s_a+1)(2s_b+1)(2k_1+1)(2k_2+1)(2K+1)(2I+1)$$
$$\times (2s_i+1)(2s_i'+1)(2l+1)(2l'+1)(2l_p+1)(2l_p'+1)(2l_q+1)(2l_q'+1)$$
$$\times (2j+1)(2j'+1)(2J+1)(2J'+1)]^{\frac{1}{2}} \langle k_1 q_1, k_2 q_2 | KQ \rangle$$
$$\times \langle l_p 0, l_p' 0 | L_p 0 \rangle \langle l_q 0, l_q' 0 | L_q 0 \rangle \langle I0, KQ | LQ \rangle W(j, s_f, L_p, l_p'; l_p, j')$$
$$\times \begin{Bmatrix} s_i & J & l \\ s_i' & J' & l' \\ K & L & I \end{Bmatrix} \begin{Bmatrix} j & l_q & J \\ j' & l_q' & J' \\ L_p & L_q & L \end{Bmatrix} \begin{Bmatrix} s_a & s_a & k_1 \\ s_b & s_b & k_2 \\ s_i & s_i' & K \end{Bmatrix}. \quad (24)$$

The partial-wave expansion formulas given in Eq. (23) and (24) represent our central result. Several authors have previously presented partial wave expansion formulas for pion production reactions (see for example Refs. [8–11]). However, the present equations are considerably more general than any of the previous formulations.

EQUIVALENT OBSERVABLES

We now turn to the issue of which spin observables in Eq. (6) are independent. We already understand that the transverse analyzing powers, A_{x0} and A_{y0}, are equivalent in the sense that the radiation pattern observed with polarization along $\hat{y}$ must be identical to that observed with polarization along $\hat{x}$ except for rotation by 90° about the z-axis. In other words, we may write

$$A_{x0}(\theta_p, \phi_p, \theta_q, \phi_q) = A_{y0}(\theta_p, \phi_p + \tfrac{\pi}{2}, \theta_q, \phi_q + \tfrac{\pi}{2}). \quad (25)$$

The same result holds for the target analyzing powers A_{0x} and A_{0y}.

The corresponding symmetries for the spin-correlation parameters can be discovered by inspecting Table 2. The rule is that two observables are equivalent provided that i) they are given by the real and imaginary parts of the same spherical tensor, and ii) that the spherical tensor has $q_1 + q_2 \neq 0$.

We may demonstrate this result as follows. In Eq. (23) we see that the spherical tensor observables involve only B functions with $\Lambda = Q = q_1 + q_2$. However, from Eq. (16) the B functions transform under rotations according to the rule

$$B^Q(\theta_p, \phi_p + \beta, \theta_q, \phi_q + \beta) = e^{iQ\beta} B^Q(\theta_p, \phi_p, \theta_q, \phi_q) \quad (26)$$

and from this it follows that

$$T_{k_1 q_1, k_2 q_2}(\theta_p, \phi_p + \beta, \theta_q, \phi_q + \beta) = e^{iQ\beta} T_{k_1 q_1, k_2 q_2}(\theta_p, \phi_p, \theta_q, \phi_q) \quad (27)$$

From Eq. (27) we may now obtain the desired symmetry relations. For example, for spin observables with $Q = 1$ we take $\beta = \pi/2$ and obtain

$$T_{k_1q_1,k_2q_2}(\theta_p,\phi_p+\tfrac{\pi}{2},\theta_q,\phi_q+\tfrac{\pi}{2}) = iT_{k_1q_1,k_2q_2}(\theta_p,\phi_p,\theta_q,\phi_q), \tag{28}$$

from which it follows that

$$\operatorname{Im} T_{k_1q_1,k_2q_2}(\theta_p,\phi_p+\tfrac{\pi}{2},\theta_q,\phi_q+\tfrac{\pi}{2}) = \operatorname{Re} T_{k_1q_1,k_2q_2}(\theta_p,\phi_p,\theta_q,\phi_q). \tag{29}$$

Analogous formulas are easily obtained for any parameter with $Q \ne 0$. For example, for $Q = 2$ we have

$$\operatorname{Im} T_{k_1q_1,k_2q_2}(\theta_p,\phi_p+\tfrac{\pi}{4},\theta_q,\phi_q+\tfrac{\pi}{4}) = \operatorname{Re} T_{k_1q_1,k_2q_2}(\theta_p,\phi_p,\theta_q,\phi_q). \tag{30}$$

Applying these general results to the observables listed in Table 2 we obtain the following equivalences:

$$\begin{array}{ll} A_{y0} \Leftrightarrow A_{x0} & A_{0y} \Leftrightarrow A_{0x} \\ A_{yz} \Leftrightarrow A_{xz} & A_{zy} \Leftrightarrow A_{zx} \\ A_{xx}-A_{yy} \Leftrightarrow A_{xy}+A_{yx} & \end{array} \tag{31}$$

Note that there is no corresponding equivalence relationship between $A_{xx}+A_{yy}$ and $A_{xy}-A_{yx}$. These two parameters are proportional to the real and imaginary parts of $T_{11,1-1}$, but for this function the real and imaginary parts are not connected by a rotation since $Q = 0$. Thus, 6 of the 9 spin-correlation parameters that appear in Eq. (6), are independent.

If the two particles in the initial state are identical there are additional symmetries, since measurements with beam and target polarization states interchanged must be equivalent. It is straightforward to show that if parity is conserved, identical particles in the initial state requires

$$A_{ij}(\theta_p,\phi_p,\theta_q,\phi_q,\epsilon) = A_{ji}(\pi-\theta_p,\phi_p+\pi,\pi-\theta_q,\phi_q+\pi,\epsilon). \tag{32}$$

In practice, this equation relates the beam and target analyzing powers and also relates the spin-correlation coefficients A_{xz} and A_{zx}. Thus, for $pp \to NN\pi$ there are two independent analyzing powers and five independent spin-correlation parameters.

FURTHER COMMENTS AND EXAMPLES

The formulas given in Eqs. (23) and (24) are the source of a great deal of useful information about pion production reactions. We will illustrate this point with a few examples.

The first example concerns the nature of the interference contributions to the various measurable quantities. From Eq. (24) one can easily demonstrate that the expansion coefficients $C_{L_p,L_q;L}^{\alpha,\alpha';\xi}$ are either symmetric or antisymmetric under interchange of the states α and α':

$$C^{\alpha,\alpha';\xi}_{L_p,L_q;L} = (-)^{k_1+k_2} C^{\alpha',\alpha;\xi}_{L_p,L_q;L}. \tag{33}$$

Since the $C's$ are real, it follows immediately that the unpolarized cross section ($k_1 + k_2 = 0$) and the spin correlation parameters ($k_1 + k_2 = 2$) depend only on $\text{Re}[U_\alpha U^*_{\alpha'}]$ whereas the analyzing powers ($k_1 + k_2 = 1$) depend only on $\text{Im}[U_\alpha U^*_{\alpha'}]$.

One may learn about the allowed angular patterns for a given observable by determining which specific expansion coefficients are non-zero. This can either be done by noting the various triangle inequalities that must be satisfied in Eq. (24), or else by simply calculating the coefficients directly. Some results for the analyzing power A_{y0} are given in Table 3. To obtain these results we included all of the $pp \to pp\pi^0$ partial waves listed in Table 1. We then keep a list of the various allowed angular patterns along with the partial wave combinations that may contribute to each. In Table 3 we see that many distinct angular patterns are allowed for the analyzing power. Thus, if one has kinematically complete measurements of A_{y0} in 4π geometry, it should be possible (at least in principle) to extract a substantial amount of information from the data.

TABLE 3. Allowed angular patterns for $W_0 A_{y0}$. In all cases, the angular function is $B^1_{L_p,L_q;L}(\theta_p, \phi_p, \theta_q, \phi_q)$. A $\checkmark$ in the table indicates that the corresponding combination of partial waves may contribute.

Angular Function $L_p, L_q; L$	Ps-Pp	$(Pp)^2$	Ss-Sd	Ss-Ds	Sd-Ds	$(Sd)^2$	$(Ds)^2$
0,1;1	$\checkmark$						
0,2;2		$\checkmark$	$\checkmark$			$\checkmark$	
2,0;2		$\checkmark$		$\checkmark$			$\checkmark$
2,1;1	$\checkmark$						
2,1;2	$\checkmark$						
2,1;3	$\checkmark$						
2,2;1		$\checkmark$				$\checkmark$	
2,2;2		$\checkmark$				$\checkmark$	
2,2;3		$\checkmark$				$\checkmark$	
2,2;4		$\checkmark$				$\checkmark$	
4,0;4							$\checkmark$
0,4;4					$\checkmark$		

It is also useful, at this point, to generalize our earlier discussion about the allowed ϕ dependences for the various observables. In Eq. (23) the observable $T_{k_1q_1,k_2q_2}$ is expressed in terms of Bipolar harmonics with azimuthal quantum number $Q = q_1 + q_2$. As shown in Eq. (16), the B functions involve terms proportional to $e^{i(\Lambda_p\phi_p+\Lambda_q\phi_q)}$ where $\Lambda_p + \Lambda_q = Q$, and from Table 2 one can find the value of Q for each observable. If we then incorporate the fact that the term in square brackets in Eq. (23) is purely imaginary for the analyzing powers and purely real for the cross section and spin correlation parameters we can easily obtain formulas that show the allowed ϕ dependences for any observable. For example, for the transverse analyzing powers ($Q = 1$) we have

$$W_0 A_{y0} = \sum_n F_n(\theta_p, \theta_q) \cos[(n+1)\phi_p - n\phi_q] \tag{34}$$

and

$$W_0 A_{x0} = -\sum_n F_n(\theta_p, \theta_q) \sin[(n+1)\phi_p - n\phi_q], \tag{35}$$

where the same functions F_n appear in both A_{x0} and A_{y0}. In a similar way, for the spin correlation parameters $A_{xx} - A_{yy}$ and $A_{xy} + A_{yx}$ (which have $Q = 2$) we obtain

$$W_0 (A_{xx} - A_{yy}) = -\sum_n G_n(\theta_p, \theta_q) \cos[(n+2)\phi_p - n\phi_q] \tag{36}$$

and

$$W_0 (A_{xy} + A_{yx}) = \sum_n G_n(\theta_p, \theta_q) \sin[(n+2)\phi_p - n\phi_q]. \tag{37}$$

Next we will look at some of the properties of two particular polarization observables, A_{z0} and $A_{xy} - A_{yx}$. These are special observables which are required to be zero for reactions with two-body final states (assuming parity is conserved), but which may be non-zero in meson production reactions. One of the complications of trying to measure these new observables is that all of the allowed angular dependences are fairly complex. If one includes all of the partial wave contributions listed in Table 1 and constructs a table analogous to Table 3, one finds that for A_{z0} only three terms are present, $B^0_{2,1;2}$, $B^0_{2,2;1}$ and $B^0_{2,2;3}$. Of the three, the simplest function is $B^0_{2,1;2}$,

$$B^0_{2,1;2} = -i \frac{3\sqrt{10}}{8\pi} \sin\theta_p \cos\theta_p \sin\theta_q \sin(\phi_p - \phi_q). \tag{38}$$

For the spin-correlation parameter $A_{xy} - A_{yx}$ one finds that $B^0_{2,1;2}$ is the only term present. The point to be made here is that, if one wishes to measure either of these quantities, a substantial amount of care must be taken. In particular, we see from Eq. (38) that if one integrates over either θ_p or ϕ_p or ϕ_q, any signal present in the full data set will be lost. To measure these special quantities, it is important to understand ahead of time what kinds of angular patterns are possible.

Finally, we note that it is possible to find special combinations of measurable quantities that are sensitive to specific partial wave combinations. For example, one finds that if we include only the partial waves of Table 1, the combination

$$W = W_0 [1 - A_{xx} - A_{yy} - A_{zz}] \tag{39}$$

may arise only from the (Ps)^2 interference terms. This observation has already been used in Ref. [3] to extract the total (Ps)^2 cross section from measurements of spin-dependent total cross section differences.

ACKNOWLEDGEMENTS

The work described in this paper are the result of a fruitful collaboration between the author and Prof. H.O. Meyer of IUCF. I am indebted to Hans for bringing the problem to my attention and for helping to clarify many of the complex issues.

REFERENCES

1. H. O. Meyer, et al., *Phys. Rev. Lett.* **81**, 3096 (1998).
2. S. K. Saha, et al., *Phys. Lett.* **B461**, 175 (1999).
3. H. O. Meyer, et al., *Phys. Rev. Lett.*, accepted for publication.
4. H. O. Meyer, in *Proceedings of the 8th International Conference on the Structure of Baryons*, D.W. Menze and B. Metsch, eds., (World Scientific, Singapore, 1999) p. 493.
5. L. D. Knutson, *Nucl. Phys.* **A198**, 439 (1972).
6. J. Zlomanczuk, et al., *Phys. Lett.* **B436**, 251 (1998).
7. J. Raynal, Ph.D. thesis, Saclay, 1965.
8. F. Mandl and T. Regge, *Phys. Rev.* **99**, 1478 (1955).
9. R. Handler, *Phys. Rev.* **138**, B1230 (1965).
10. C. L. Dolnick, *Nucl. Phys.* **B22**, 461 (1970).
11. R. W. Flammang, et al., *Phys. Rev. C* **58**, 916 (1998).

Review of Polarized Internal Gas Targets

Frank Rathmann

Physikalisches Institut II, Friedrich-Alexander Universität, 91058 Erlangen, Germany[1]

Abstract. Experiments utilizing polarized internal gas targets will be reviewed. Since a few years these targets are being used at electron and proton machines and the progress made in operating these targets on a routinely basis has been pronounced. This paper will focus on experimental aspects, such as the design and construction of storage cells and different methods employed for polarimetry.

I INTRODUCTION

Storage cells to enhance the target thickness provided by a source of polarized atoms have already been conceived some thirty years ago [1]. However, application of this technique to nuclear and particle physics became possible only through the advent of storage rings. The stored beam passes through the *internal* target about a million times per second and thus compensates for the low target thickness. In order to obtain higher target thicknesses compared to a free beam of polarized atoms the dwell time of an atom in the vicinity of the stored beam must be increased. For that purpose an open ended T-shaped storage cell is inserted into the ring through which the stored beam passes and into which the atomic beam is injected. Thereby the luminosity can be raised by about two orders of magnitude. Nowadays, internal targets with densities around 10^{14} atoms/cm^2 and nuclear vector or tensor polarization close to the theoretical maximum are operated routinely at a variety of storage rings.

This paper will focus on experimental aspects of polarized internal targets, such as the design of storage cells and currently employed techniques to determine the target polarization.

II A BIT OF HISTORY

In order to put the progress and developments in the field of internal polarized gas targets into perspective, a short account of major milestones is presented in

---

[1]) supported by the BMBF under contract no. 06 ER 831.

the following. Most of the experiments utilizing polarized internal targets are represented elsewhere in these proceedings, therefore the account given here is rather brief.

The suggestion by Haeberli [1] to use a storage cell to enhance the available target thickness produced by an atomic beam source in 1966 was motivated by the successful application of a storage container in hydrogen maser applications [2]. The proof of principle showing that this approach is actually feasible was carried out by the Wisconsin Group in 1980 using a 12 MeV beam of α-particles incident on a proton target [3]. The apparatus used is depicted in Fig. 1. The atomic

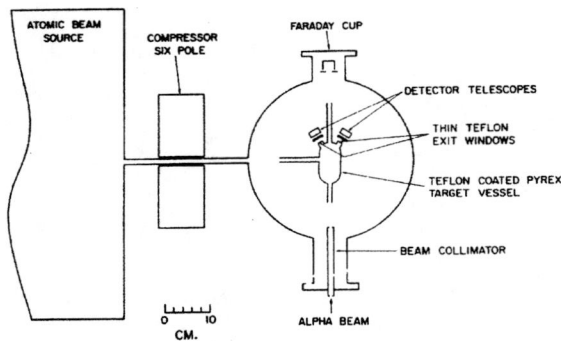

FIGURE 1. First Test of a storage cell with an external beam of 12 MeV α-particles incident on a hydrogen target produced by an atomic beam source (Figure from ref. [3]).

beam entered the storage vessel through a 10 cm long tube of 10 mm diameter. The storage volume (0.26 l) was coated with Teflon to avoid depolarization. The measured polarization after background subtraction was found to be compatible with the polarization of a free beam, thus essentially no depolarization occurred during an average of 900 wall collisions inside the vessel. The improvement in target thickness was a factor of six over that of a free beam.

In 1985 the FILTEX group proposed to use a polarized hydrogen target internal to LEAR at CERN [4]. The objective was to polarize a stored beam of antiprotons by spin-dependent attenuation in the polarized gas target (Spin-Filter-Method) and to subsequently carry out $\bar{p}p$-interaction studies. The goal was to reach a target density of 10^{14} atoms/cm^2 for one substate.

Since 1988 an internal polarized deuterium target is operated at the electron storage ring VEPP-3 in Novosibirsk to measure the tensor analyzing power T_{20} in $e\vec{d}$-scattering [5,6]. Substantial progress in atomic beam intensity could be achieved recently through the development of superconducting sextupole magnets [7].

In 1992, the FILTEX target was tested at the Heidelberg Test Storage Ring [8–11]. A target thickness of close to 10^{14} atoms/cm^2 in two substates was achieved, with a nuclear polarization of 0.8 for one substate. More details pertaining to the

FILTEX test experiment are presented in Sec. IV B 1.

In 1992, an optically pumped polarized $^3\vec{\mathrm{He}}$ target was installed at IUCF to study spin-dependent $\vec{p}\,^3\vec{\mathrm{He}}$-reactions [12–14]. This experiment represents the first measurement of a reaction where a polarized beam is incident on a polarized target internal to a storage ring. A target thickness of $1.5 \cdot 10^{14}$ atoms/cm^2 was reported [13]. Later in 1995, this target was installed at the HERMES internal target experiment at HERA/DESY to investigate the spin-structure of nucleons by means of deep inelastic scattering [15].

In 1996 the modified source of polarized hydrogen and deuterium atoms developed initially for the FILTEX experiment was installed at the HERMES experiment at HERA/DESY [15]. The storage cell setup of the HERMES experiment is discussed in Sec. IV B 3.

Since 1994 the PINTEX collaboration at IUCF [16] is operating a polarized hydrogen and deuterium gas target [17–19]. Details about specific aspects of this facility will be presented in Sec. (IV B 4).

Studies in $e\vec{d}$-reactions similar to the ones performed at VEPP-3 have been carried out at NIKHEF employing an electron beam incident on a tensor polarized internal deuterium gas target [20–22]. The ion-extraction method used for the determination of the target polarization will be discussed in more detail in Sec. V B.

Among the current polarized internal target experiments, the EDDA experiment at COSY is the only experimental installation utilizing a free atomic jet [23–25]. Recently, the group has performed first tests with a storage cell at COSY, which will be dealt with in Sec. IV B 5.

A new type of a polarized hydrogen and deuterium source based on spin-exchange optical pumping has recently been installed in the IUCF cooler [26]. First results in $\vec{p}\vec{d}$ elastic scattering have been obtained [27]. Details on the storage cell used in this experiment are given in Sec. IV B 6.

There are two new facilities presently being prepared for nuclear physics experiments employing polarized internal gas targets. The Bates Large Acceptance Spectrometer Toroid (BLAST) experiment [28] will utilize the slightly modified polarized target used at NIKHEF for studies in polarized ed reactions. At COSY a polarized internal target is being developed to study the $\vec{p}\vec{d}$-breakup reaction in a collinear kinematic with the magnetic spectrometer ANKE [29,30].

III POLARIZED INTERNAL TARGETS FOR STORAGE RINGS

The principal layout of the setup of a polarized internal target experiment is depicted in Fig. 2. The interaction zone produced by the target source if no storage cell is used is given by the volume of overlap between stored and atomic beam, which is typically of the order of a cm^3. The target sources for this type of setup are optimized for maximum volume density inside the interaction zone, which increases

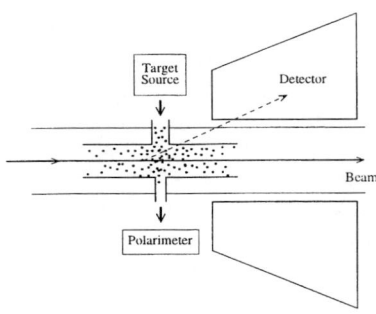

FIGURE 2. Polarized internal Target setup.

with decreasing atomic velocity. When a storage cell is used the target source should be optimized for maximum intensity into the acceptance of the feed tube of the cell. In this case the interaction region is extended with cell dimensions typically ranging from 25 to 60 cm in length with diameters of 10 to 30 mm and cylindrical, elliptical, squared or rectangular cross sections.

The main advantages of polarized gas targets over solid targets are: *i)* high isotopical purity, *ii)* possibility of rapid spin reversal up to 100 Hz in $\vec{H}$ or $\vec{D}$ targets, *iii)* low background, since no container walls intercept the particle beam, *iv)* no radiation damage, since the target gas is replenished every few ms. Polarized gas targets combined with beams internal to storage rings are therefore ideally suited for high precision experiments.

IV STORAGE CELLS FOR POLARIZED INTERNAL TARGETS

One important objective in the development of storage cells for polarized internal targets was the identification of suitable wall coating materials that inhibit depolarization of atoms stored inside the cell that are compatible with the ultra-high vacuum requirements of a storage ring.

A Cell Coating

Early studies of wall depolarization in storage cells were performed by Price and Haeberli [31]. These measurements were carried out with the apparatus shown in Fig. 3. A 50 keV beam of D^+ ions from a duoplasmatron source passes through a storage cell. Polarized $\vec{H}^0$ atoms from an atomic beam source are injected into the cell through a feed tube. A weak magnetic guide field of 5 mT is applied over the cell. Inside the cell the electron pick-up reaction $D^+(50\,\text{keV}) + \vec{H}^0 \to \vec{D}^0(50\,\text{keV}) + H^+$ takes place. In a weak magnetic guide field

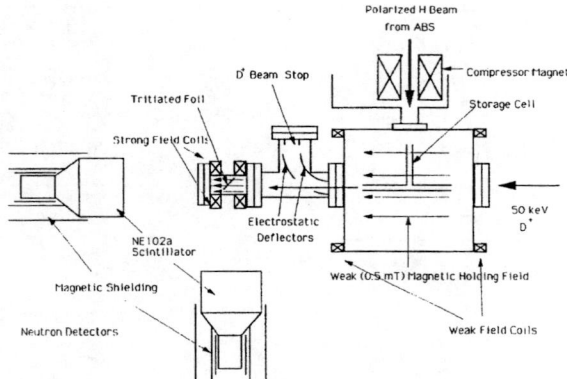

FIGURE 3. Experimental apparatus used for weak field depolarization measurements. (Reprinted from [31], © 1994, with permission from Elsevier Science.)

electrons and protons of hydrogen atoms in a state $|m_j = 1/2, m_I = 1/2>$ carry the same polarization. The electron pickup reaction in the cell transfers the initial polarization of the electrons through hyperfine interaction to the deuterium atom, which acquires a nuclear tensor polarization. The neutral $\vec{D}^0$ atoms impinge on a tritiated foil located in a strong field of 30 mT. From the neutron asymmetry of the $^3\mathrm{H}(\vec{d}, n)^4\mathrm{He}$ reaction the tensor polarization of the deuterons is deduced. Using this method as a polarimeter the depolarization behaviour of various wall coating materials was investigated. Teflon stands out for its low depolarization and good vacuum compatibility. In Fig. 4 a selection of results are shown for cells coated with Teflon FEP 120, TPFE 3170 and Fomblin oil.

B Design and Construction of Storage Cells

The design of a storage cell must be well adapted to the experimental requirements of an internal target facility, taking into account properties of the stored beam, e.g. beam dimension, detector acceptance, possible background reactions from the cell walls. Recently these design constraints for polarized internal targets have been reviewed by Meyer [32]. In this section the storage cells employed by several different experimental installations will be discussed.

1 FILTEX

A boundary condition initially imposed during the measurements by the TSR machine group was to allow for injection of the beam without the storage cell acting

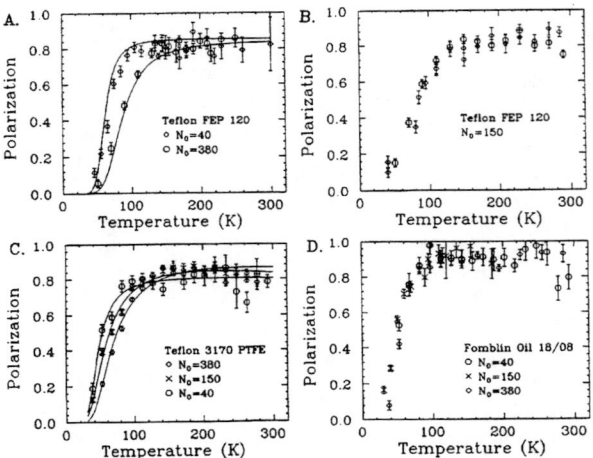

FIGURE 4. Polarization of hydrogen atoms in coated storage cells for different numbers of wall collisions N_0 as a function of the cell temperature. (Reprinted from [31], © 1994, with permission from Elsevier Science.)

as acceptance limiter. Therefore a storage cell that could be opened and closed like a clam-shell, shown in Fig. 5, was developed [8]. Only after the beam had been injected and cooled was the cell closed. The upstream end of the cell was attached to a cold head, such that the cell center could be kept at temperatures between 50-300 K. The longitudinal fins left and right of the cell were used to increase the heat conductivity. The rather low polarization observed for this type of cell ($\approx 50\%$ of the theoretical maximum) was caused by exposure of atoms to the badly-coated fins of the cell. Spot-welding the two half cells together [8] (see Fig. 5) substantially increased the polarization, once the TSR machine group was confident enough to give this approach a try. Eventually, for the FILTEX test experiment, closed cells manufactured from cylindrical aluminum tubes like shown on the right in Fig. 5 were used. All cells were of the same dimensions, 250 mm length with a diameter of 11 mm and a wall thickness of 0.2 mm. The first two cell types (clam-shell, spot-welded clam-shell) were coated with Teflon FEP 120, while the final cells had Teflon TPFE 3170 coatings. Results of the polarization measurements will be presented in Sec. V A 1. All cells were equipped with a small capillary near the cell center to allow for unpolarized gas inlet to facilitate background studies.

2 ^{3}He-Target at IUCF

The effects of depolarization in wall collisions in polarized ^{3}He gas targets are greatly inhibited due to the absence of hyperfine interaction. In that sense polarized ^{3}He atoms can be called "robust". The storage cells used for the experiments with

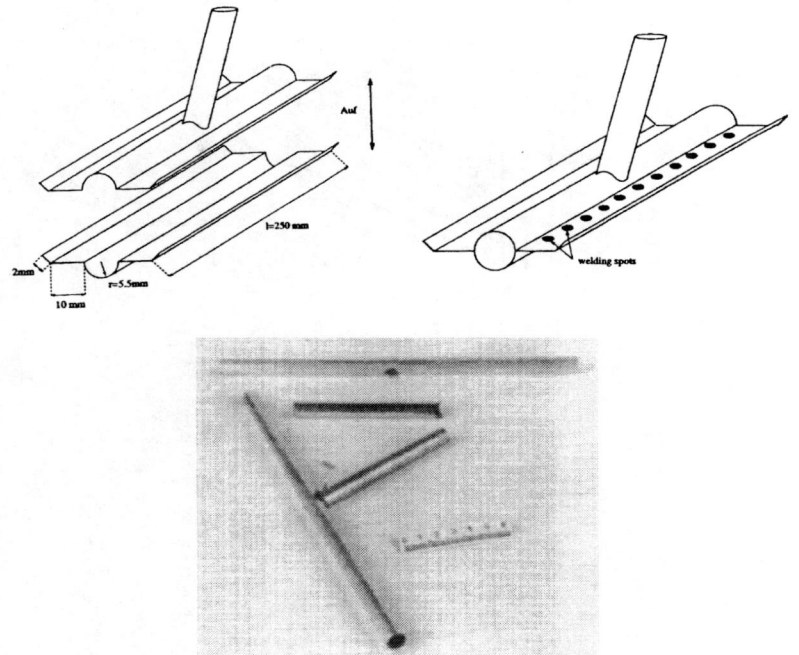

FIGURE 5. Storage cells used during the FILTEX test experiment at TSR. Top left: Clam-shell cell, top right: spot-welded clam-shell, and bottom: cell made from cylindrical tubing.

$^3\vec{\text{He}}$ at IUCF [12,13] therefore remained uncoated. Aluminized mylar sheets of 1.7 μm thickness were attached to a support frame consisting of 0.2 mm thick aluminum pieces to form a 400 mm long cell of rectangular cross section (h = 16.6 mm, w = 13.1 mm), as shown in Fig. 6. The experimental setup did not include any provision to cool the cell. The authors report a target thickness of $1.5 \cdot 10^{14}$ atoms/cm^2. If restrictions due to deadtime had not been encountered, an order of magnitude higher target thickness would have been possible.

3 HERMES

The storage cell for HERMES at HERA/DESY, shown in Fig. 6, is adapted to the beam distribution by an elliptical cross section (w = 30 mm, h = 10 mm). The cell walls consist of 75 μm thick aluminum, coated by drifilm to prevent depolarization [34]. The target cell is cooled to temperatures between 35-260 K by cold helium gas passing through cooling rails alongside of the target cell. A thin layer

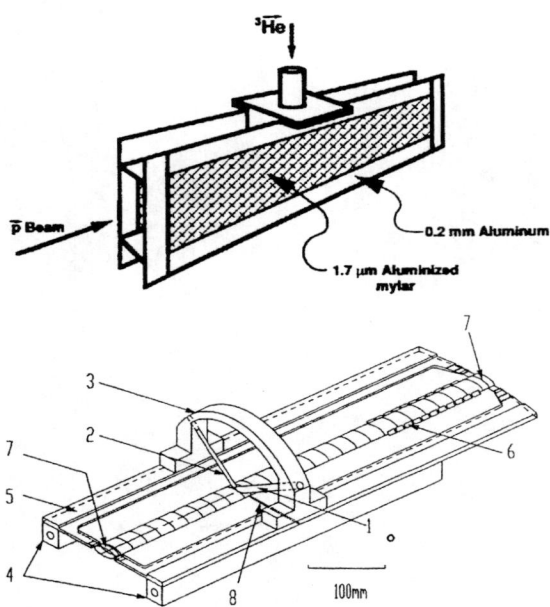

FIGURE 6. Storage cells used for the $\vec{p}\,{}^3\vec{\mathrm{He}}$ studies at IUCF (reprinted from [12], © 1995, with permission from Elsevier Science) (top) and the storage cell for HERMES [33] (bottom: 1) H/D feed tube, 2) sample tube, 3) support arch, 4) cooling rails, 5) support plates, 6) cell extension, 7) end support 8) unpolarized feed tube.)

of ice is formed inside the cell, which was found to prevent effectively both depolarization and recombination. This surprising result is in agreement with earlier measurements carried out at Heidelberg during the development of the HERMES target polarimeter by Braun [35]. The strong magnetic holding field of the HERMES target must be rather uniform in order to prevent depolarization of target atoms by induced hyperfine transitions that are caused by the periodic time structure of the HERA positron beam [36,37]. Some aspects of the HERMES target polarimeter are discussed in Sec. V C.

4 PINTEX

For a measurement of low energy recoil particles thin cell walls are required. A storage cell design that complies with this condition is the PINTEX cell, shown in Fig. 7, that was used for measurements of polarization observables in pp-elastic scattering at IUCF [38]. Thin sheets of Teflon foil of 450 μg/cm^2 thickness stretched over fins form a quadratic channel of about 1 cm^2 cross section [39,40].

The storage cells used for the measurements in pion production [18] are made from aluminum foil, coated by Teflon TPFE 3170 and similar in design to the cells used by the FILTEX group (see right side of Fig. 5), but with significantly thinner cell walls of only 25 μm.

Yet another cell design employed by the PINTEX group is used for measurements of the nuclear polarization of molecular hydrogen formed by recombination inside the cell, where a valve and a Cu-recombiner box are attached to the cell center. Thereby the atomic-to-molecular fraction inside the cell can be varied. Details about this ongoing experiment that also requires a strong longitudinal holding field can be found elsewhere in these proceedings [19].

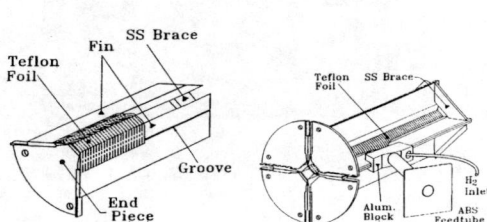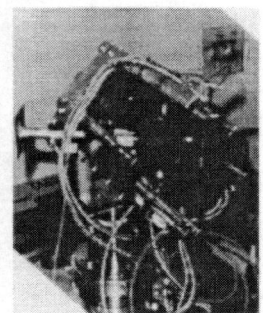

FIGURE 7. Storage cell used by PINTEX for measurements of polarization observables in pp-elastic scattering [38]. One quadrant is depicted on the left, while the fully assembled cell is shown in the middle. A photograph of the complete setup including mounting of target near silicon microstrip detectors is shown on the right. (Left and middle figure reprinted from [39], © 1994, with permission from Elsevier Science.)

5   EDDA

The first tests with a storage cell inside the COSY machine at Jülich have been performed by the EDDA collaboration [23,24]. Although the EDDA detector has not been designed to accommodate an extended target, employing a storage cell would still lead to an enhancement in target thickness by about a factor ten [41]. In this factor the significant reduction in beam intensity because of acceptance limitations due to the cell during injection ($\approx$50%) have not been taken into account. It should be noted that for the time-reversal-invariance experiment [24] the full target thickness contributes and an enhancement factor of about 60 applies. The test cell of EDDA is shown in Fig. 8. The 300 mm long cell ($h = 12$ mm, $w = 29$ mm) consists of a 0.2 mm thick aluminum body coated by Teflon PTFE 3170 that can be cooled down to 80-100K. Polarized atoms are injected into the cell through a conical 10 cm long feed tube (entrance/exit diameter 20/12 mm, shown on the right

in Fig. 8) that is not rigidly attached to the cell, in order to be able to remove the
cell from the beam position without breaking of vacuum.

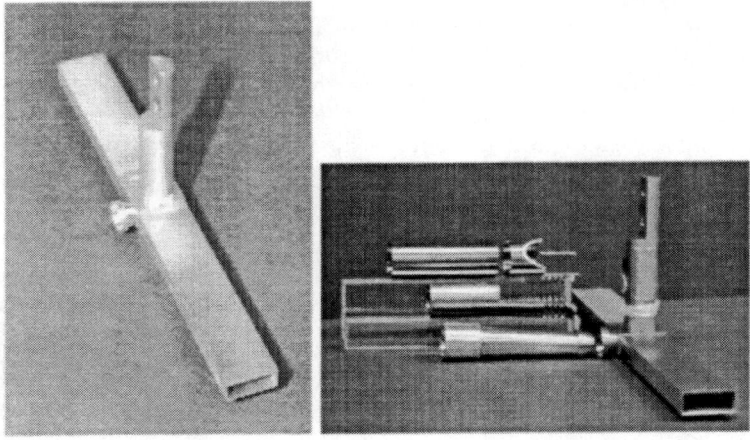

FIGURE 8. Storage cell used by EDDA at COSY. Cell body (left) and mechanical fixture to
hold the conical entrance tube in place (right) [41].

6 Laser-driven Spin-Exchange Target

The storage cell of the first internal target experiment utilizing a laser driven spin-exchange source presently installed at IUCF [26,27] is shown in Fig. 9. The spin exchange cell is directly attached to the storage cell. Depolarization is inhibited by drifilm coating. In order to prevent condensation of potassium vapour on the cell walls, the temperature of the cell is raised to about 180 °C. The 400 mm long cell of wall thickness 3.2 mm ($h = 19.1$ mm, $w = 38.1$ mm) has thin windows of 0.3 mm thickness on the left and right to reduce multiple scattering. With a flow of about 10^{18} polarized atoms/s a target thickness around $4 \cdot 10^{14}$ cm^{-2} with vector polarization $P_z = 0.25$ for both hydrogen and deuterium targets is reported [26].

V POLARIMETRY OF INTERNAL GAS TARGETS

Various different methods have been applied to the problem of measuring the nuclear polarization of an internal gas target: *i)* Polarimetry through a known reaction, *ii)* extraction of ions formed inside the storage cell and subsequent polarization analysis, and *iii)* polarization analysis of an extracted sample of neutral

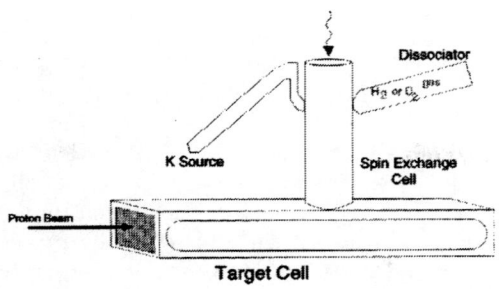

FIGURE 9. Storage cell for the optically pumped spin-exchange target at IUCF [26].

atoms from the cell. These three methods will be discussed in more detail in this section.

There have also been two optical methods applied to determine the polarization of an internal gas target, both of which are suitable as a relative polarization monitor only, and not well adapted for a precise absolute determination of the target polarization. The Balmer-Polarimeter measures the circular polarization of light from hydrogen atoms excited to $n = 3$ states by electron impact [42]. The initial electron polarization of the ground state atoms is partially transferred into angular momentum of the excited state. The emitted Balmer-light is therefore circularly polarized and a measurement of the initial electron polarization is possible. This method is restricted to polarized targets operated in a weak guide field, where to a good approximation electron and nuclear polarization of hydrogen atoms are equal. For the $^3\vec{\text{He}}$ target at HERMES an optical monitor was developed that measures the circular polarization of photons emitted in the $4^1D \rightarrow 2^1P$ transition [43].

It should be noted that a known (calibrated) reaction does not distinguish atoms from molecules inside the cell, nor from any other contaminant present in the cell. This method is therefore the first choice wherever feasible, because the target is sampled in the same way as in the experiment under consideration. The other methods measure the nuclear polarization of atoms, and with additional instrumentation also the atomic to molecular fraction in the cell. The magnitude of target polarization carried by the molecules is currently being investigated. A first measurement carried out at NIKHEF suggests a non-vanishing nuclear polarization of recombined deuterium molecules inside a storage cell $P_{\text{molecule}}(D_2^{\text{rec}})/P_{\text{atom}}(D) = 0.81 \pm 0.32$ [44]. A new experiment underway at IUCF is studying nuclear polarization in recombined hydrogen and deuterium in more detail [19].

A Known Reactions

Suitable for the determination of the nuclear polarization is any calibrated reaction that enables one to deduce the polarization of the target. Two examples of this kind are discussed below.

1 Example 1: 27 MeV αp scattering

Analyzing power and cross section for 27 MeV α particles incident on a proton target are well known [45]. Employing the detector system shown in Fig. 10, polarization and density of the FILTEX polarized hydrogen target could be measured efficiently at the Heidelberg Test Storage Ring [8]. The target was operated in a weak vertical magnetic guide field. Two measurements separated in time by about four weeks during which high proton currents were stored revealed that the polarization of the target is very stable. It was concluded that no damage to the wall coating occurred. The target was operated at a temperature of 125 K where the figure of merit is highest.

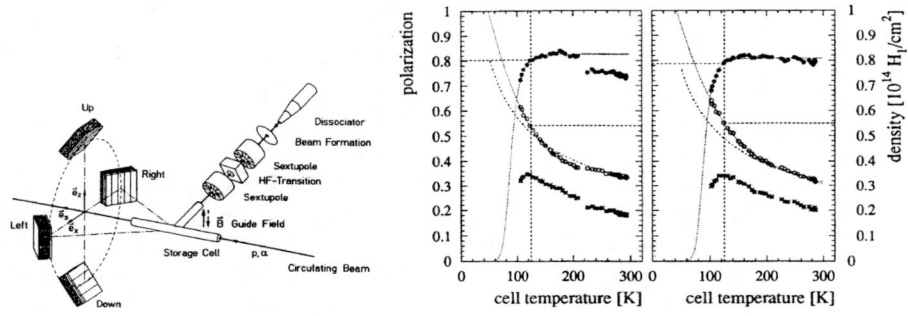

FIGURE 10. Polarized internal target setup of the FILTEX test experiment [8] (left). Four sets of scintillation counters are used to detect scattered α-particles. Polarization P ($\bullet$), density d_t ($\circ$) and figure of merit $P^2 \cdot d_t$ ($\star$) of the target as a function of the central cell temperature (right) for two measurements separated by four weeks based on the known αp-reaction [45].

2 Example 2: pp elastic scattering

The PINTEX detector system at IUCF, described in detail in ref. [38], is designed specifically for experiments involving both polarized beams and polarized targets. It enables a very precise determination of target and beam polarization through the analysis of pp elastic scattering. A precise analyzing power measurement in pp elastic scattering is available [46] that provides the basis of a polarization calibration. It is of great importance that this calibrated polarization can be exported to serve as a standard at other energies [47]. Through the analysis of elastic pp events it becomes possible to measure the target polarization along the cell axis, as shown in Fig. 11. It was also possible to deduce the effective reversal time (7 ± 1 ms), e.g. the time it takes to completely reverse or reorient the target polarization. High stability and high polarization characterize the longterm behaviour of the target as illustrated in Fig. 11. Typical target thicknesses are $2 \cdot 10^{13}$ atoms/cm^2, with atomic beam intensities into the feed tube of the storage cell of about $3.6 \cdot 10^{16}$ atoms/s, which translate into luminosities of about $5 \cdot 10^{28}$ cm^{-2}s^{-1}.

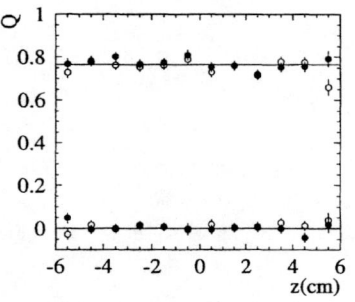 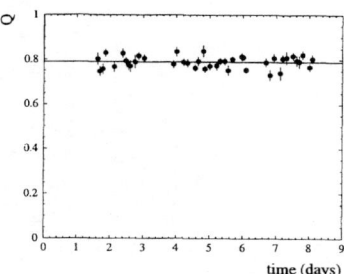

FIGURE 11. Target polarization as a function of the longitudinal position in the storage cell (left) and as a function of time during a run (right). (Figures from ref. [17].)

B Ion-Extraction

A first experiment of this kind made use of a low-energy electron beam to ionize the target atoms, which were then extracted and accelerated to 70 keV and incident on a tritiated foil for polarization determination by the ^{3}H($\vec{\text{d}}$, n)^{4}He reaction [48], as discussed in Sec. IV A. The method can be directly applied to cell targets internal to storage rings, because enough atoms are ionized by the orbiting particles. This method is particularly useful in applications, where a strong longitudinal holding field is applied over the target through which the ions are focussed. At the internal target at NIKHEF [20,49,50] this method was experimentally realized [51]. The ion-extraction system of the NIKHEF storage cell target, employing a spherical deflector is shown in Fig. 12.

One drawback of the experimental setup described above is that it is not directly applicable to polarized hydrogen targets, because of a lack of suitable low energy reactions to analyze the polarization of protons. The ^{6}Li($\vec{\text{p}}$,^{3}He)^{4}He reaction, mentioned in ref. [48], requires energies of at least 260 keV, where the analyzing power becomes sufficiently large. A more promising method for both hydrogen and deuterium gas targets might be the combination of an ion-extraction system and a Lamb-shift polarimeter [52]. Such a polarimeter system is presently being developed by Schieck et al. at the University of Cologne [53].

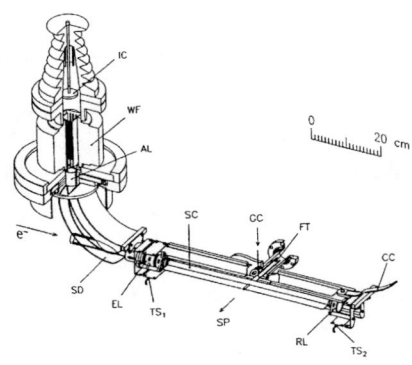

FIGURE 12. Storage cell and ion-extraction system used by the NIKEF group. Fig. reprinted from [51], © 1996, with permission from Elsevier Science. SC: storage cell, FT: feed tube, SP: sample port, RL: repeller lens, EL: extraction lens, SD: spherical deflector, WF: Wien-filter.

C Extraction of a Sample of Polarized Atoms

The polarimeter of the HERMES target determines the nuclear polarization of a small fraction of the target gas ($\approx 3\%$) that is extracted by a sample tube (see Fig. 6). A polarization analysis is performed through a measurement of hyperfine state occupation numbers in a so-called Breit-Rabi Polarimeter [54,35]. A schematical drawing of the polarimeter is shown in Fig. 13. The polarimeter consists of a set of strong and weak-field transition units and a sextupole magnet system. A chopper is used to suppress background. A target gas analyzer measures the atomic to molecular fraction of the target gas. Small systematic errors of 0.01 for the nuclear

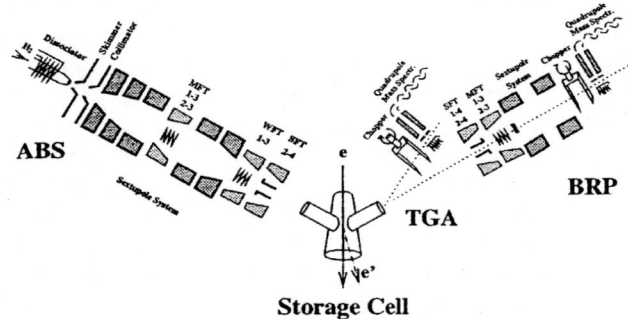

FIGURE 13. The HERMES target. The storage cell is fed by an atomic beam source (left). The composition of the target gas is measured via a target gas analyzer (TGA). The polarization of the target gas is measured by the Breit-Rabi Polarimeter (BRP). (Figure from ref. [54])

polarization of the extracted beam sample are reported. A more detailed account of the HERMES polarimeter is given elsewhere in these proceedings [36].

VI   CONCLUSIONS AND OUTLOOK

Much experience has been gained in operating polarized internal targets in the environment of electron and ion storage rings. In particular, an impressive body of information on systematic effects arising from interactions of beam and target, such as unwanted reactions with the walls of the storage cell, radiation damage of the wall coating, long term behaviour etc. has been obtained. The physics results cover a wide range of topics from deep inelastic polarized electron-proton scattering to measurements of spin correlation parameters and analyzing powers in elastic and inelastic proton-proton collisions. Even atomic physics experiments are underway to study polarization effects in recombined hydrogen and deuterium molecules. The experiments greatly benefit from the high purity of the target material and the high degree of polarization. Fast polarization reversal and the alignment along different axes eliminates systematic asymmetries to a high degree. Furthermore, the low target thickness turns out to be actually a benefit, because the lifetime and the high momentum resolution of the stored beam are hardly altered by the target. Polarized targets internal to a storage ring can thus be regarded as an ideal tool for high precision experiments. They have just begun to open a new door to hadronic interaction studies and one is inclined to predict they will continue to play a major role in years to come.

However, the need for higher target thicknesses prevails, particularly for electro-nuclear experiments. Most experiments could digest a factor ten higher target densities from sources of polarized H and D atoms used to feed storage cell targets. Future developments in this field address a new type of dissociator based on a surface wave plasma discharge [55], studies of a carrier jet system [56] and the development of high field sextupole magnet systems yielding higher fluxes of polarized atoms.

REFERENCES

1. W. Haeberli, Proc. 2nd Int. Symp. on polarization Phenomena, eds. P. Huber and H. Schopper, Experientia Suppl. **12** (Birkhäuser, Basel, 1966), p.64.
2. D. Kleppner *et al.*, Phys. Rev. A **138**, 972 (1965).
3. M.D. Barker *et al.*, AIP Conf. Proc. on Polarization Phenomena in Nuclear Physics, ed. G.G. Ohlsen, New York (1981), p. 931.
4. H. Döbbeling *et al.*, PROPOSAL CERN/PSSC/85-80 (1985) and Addendum (1986).
5. R. Gilman *et al.*, Phys. Rev. Lett. **65**, 1733 (1990).
6. I.A. Rachek, contribution to these proceedings.
7. L.G. Isaeva *et al.*, Nucl. Instr. Meth. **A411**, 201 (1998).
8. K. Zapfe, *et al.* Nucl. Instr. Meth. **A 368**, 293 (1996).
9. K. Zapfe *et al.*, Rev. Sci. Instrum. **66**, 28 (1995).

10. F. Stock et al., Nucl. Instr. Meth. **A343**, 334 (1994).
11. F. Rathmann et al., Phys. Rev. Lett. **71**, 1379 (1993).
12. C. Bloch et al., Nucl. Instr. Meth. **A354**, 437 (1995).
13. K. Lee et al., Phys. Rev. Lett. **70**, 738 (1993).
14. R.G. Milner, R.D. McKeown and C.E. Woodward, Nucl. Instr. Meth. **A274**, 56 (1989).
15. K. Ackerstaff et al., Nucl. Instr. Meth. **A417**, 230 (1998).
16. http://www.iucf.indiana.edu/Experiments/PINTEX/pintex.html
17. W. Haeberli et al., Phys. Rev. C **55**, 597, 1997.
18. J. Balewski, contribution to these proceedings and references therein.
19. T. Wise, contribution to these proceedings.
20. L.D. van Buuren, contribution to these proceedings.
21. Z.-L. Zhou et al., Phys. Rev. Lett. **82**, 687 (1999).
22. M. Bouwhuis et al., Phys. Rev. Lett. **82**, 3755 (1999).
23. W. Scobel, contribution to these proceedings.
24. P.D. Eversheim, contribution to these proceedings.
25. D. Albers et al., Phys. Rev. Lett. **78**, 1652 (1997).
26. M. Miller, private communication.
27. M. A. Miller et al., *Investigation of Three-Nucleon Force Effects in pd Elastic Scattering*, proposal to the Indiana University Cyclotron Facility; and R. V. Cadman et al., *Observation of three-nucleon force effects in proton-deuteron elastic scattering*, in preparation for Phys. Rev. Lett.
28. J.L. Matthews, contribution to these proceedings.
29. H. Ströher, contribution to these proceedings.
30. V.I. Komarov et al., *Exclusive deuteron break-up study with polarized protons and deuterons at COSY*, COSY Proposal #20 (1992); R. Baldauf et al., *Polarized Atomic Beam source for the ANKE-Spectrometer*, Annual Report IKP/COSY, p. 23 (1998).
31. J.S. Price and W. Haeberli, Nucl. Instr. Meth. **A349**, 321 (1994).
32. H.O. Meyer, Proc. Int. Workshop on Polarized Beams and Polarized Gas Targets, Cologne 1995, eds. H. Paetz gen Schieck and L. Sydow, World Scientific 1996, p. 355.
33. J. Stewart, Proc. 7th Int. Workshop on Polarized Gas Targets and Polarized Beams, eds. R.J. Holt and M.A. Miller, AIP Conf. Proc. **421**, p. 69.
34. G.E. Thomas et al., Nucl. Instr. Meth. **A257**, 32 (1987).
35. B. Braun, *Spin Relaxation of Hydrogen and Deuterium in Storage Cells*, PhD Thesis, Fakultät für Physik, Ludwig-Maximilians-Universität, München, http://www-library.desy.de/preprints.html#diss
36. H. Kolster, contributions to these proceedings.
37. K. Ackerstaff et al., Phys. Rev. Lett. **82**, 1164 (1999).
38. Measurements of pp-elastic polarization observables: See ref. [17], and F. Rathmann et al. Phys. Rev. C **58**, 658 (1998); B. von Przewoski et al., Phys. Rev. C **58**, 1897 (1998); B. Lorentz et al., submitted to Phys. Rev. C.
39. M.A. Ross et al., Nucl. Instr. Meth. **A344**, 307 (1994).
40. M.A. Ross et al., Nucl. Instr. Meth. **A326**, 424 (1993).
41. M. Glende, Die Speicherzelle für das EDDA Experiment am COSY in Jülich, PhD

Thesis in preparation, Universität Bonn.
42. J. Stenger, E. Steffens, and K. Zapfe, Nucl. Instr. Meth. **A330**, 21 (1993).
43. M.L. Pitt et al., Proc. Int. Workshop on Polarized Beams and Polarized Gas Targets, Cologne 1995, eds. H. Paetz gen Schieck and L. Sydow, World Scientific 1996, p. 413.
44. J.F.J. van den Brand et al., Phys. Rev. Lett. **78**, 1235 (1997).
45. P. Schwandt, T.B. Clegg and W. Haeberli, Nucl. Phys. **A163**, 432 (1971).
46. B. von Przewoski et al., Phys. Rev. C **44**, 44 (1991).
47. R.E. Pollock et al., Phys. Rev. E **55**, 7606 (1997).
48. J.S. Price and W. Haeberli, Nucl. Instr. Meth. **A326**, 416 (1993).
49. I. Paschier, contribution to these proceedings.
50. Z.L. Zhou et al., Nucl. Instr. Meth. **A379**, 212 (1996).
51. Z.L. Zhou et al., Nucl. Instr. Meth. **A378**, 40 (1996).
52. J.E. Brolley, G.P. Lawrence and G.G. Ohlsen, Proc. 3rd Int. Symposium on Polarization Phenomena in Nucl. Reactions, Madison 1970, eds. H.H. Barschall and W. Haeberli, Univ. of Wisconsin Press Madison, p. 846 (1971).
53. R. Engels, Verhandlungen der Deutschen Physikalischen Gesellschaft, 3/99, Frühjahrstagung Freiburg 1999, HK 36.14, p.140.
54. B. Braun, Proc. 7th Int. Workshop on Polarized Gas Targets and Polarized Beams, eds. R.J. Holt and M.A. Miller, AIP Conf. Proc. **421**, p. 156.
55. N. Koch and E. Steffens, Rev. Sci. Instrum. **70**, 1 (1999).
56. V.L. Varentsov et al., Proc. 7th Int. Workshop on Polarized Gas Targets and Polarized Beams, eds. R.J. Holt and M.A. Miller, AIP Conf. Proc. **421**, p. 381.

Nuclear Polarization of H_2 Molecules formed from Polarized Atoms

T. Wise[1], J.T. Balewski[2], W.W. Daehnick[3], J. Doskow[2], D. Friesel[2], W. Haeberli[1], H. Kolster[8], B. Lorentz[6], H.O. Meyer[2,4], P.V. Pancella[5], R.E. Pollock[2,4], B. v. Przewoski[2], P.A. Quin[1], F. Rathmann[6], T. Rinckel[2], Swapan K. Saha[3,7], B. Schwartz[1], A. Wellinghausen[2]

1) Dept. of Physics, University of Wisconsin-Madison, Madison, WI 53706, USA
2) IUCF, Milo B. Samson Lane, Bloomington, IN 47405, USA
3) Dept. of Physics and Astronomy, University of Pittsburgh, Pittsburgh, PA 15260, USA
4) Physics Department, Indiana University, Bloomington, IN 47405, USA
5) Western Michigan University, Kalamazoo, MI 49008, USA
6) Forschungs Zentrum Jülich GmbH, 52425 Jülich Germany
7) Bose Institute, Calcutta 700009, India
8) VU Amsterdam, De Boelelaan 1105, 10081 HV Amsterdam, Netherlands

Abstract. A planned experiment to measure the nuclear polarization of H_2 molecules formed by recombination of polarized H atoms is described. Polarization will be measured with a longitudinally polarized 200 MeV proton beam and a longitudinally polarized storage cell gas target at the IUCF cooler ring.

We are preparing to measure the nuclear polarization of hydrogen molecules formed by recombination of polarized atomic hydrogen gas. This measurement is motivated, in part, by the experience at HERMES where the degree of dissociation of the target gas is seen to vary with time and with the circulating beam current. The presence of molecules can contribute significantly to the systematic error of the target polarization if only the atom polarization is known. The experiment is scheduled to run at the IUCF proton cooler ring late in 1999.

A polarized atomic hydrogen beam is incident upon a recombination zone and subsequently drifts into an internal target cell located in a straight section of the IUCF cooler ring. The recombination zone is a bare copper box with a copper screen liner to increase the number of wall bounces. The internal target is constructed from Teflon-coated aluminum. The target assembly contains an internal valve that allows us to isolate the recombination zone. In that case, the atoms enter the target cell directly and see only Teflon coated surfaces. In this way we are able to rapidly alternate between a mostly atomic and a mostly molecular target. By a comparison of the target polarization for these two states we will measure the fraction of the atom polarization that survives the recombination process. The entire target, including the recombination

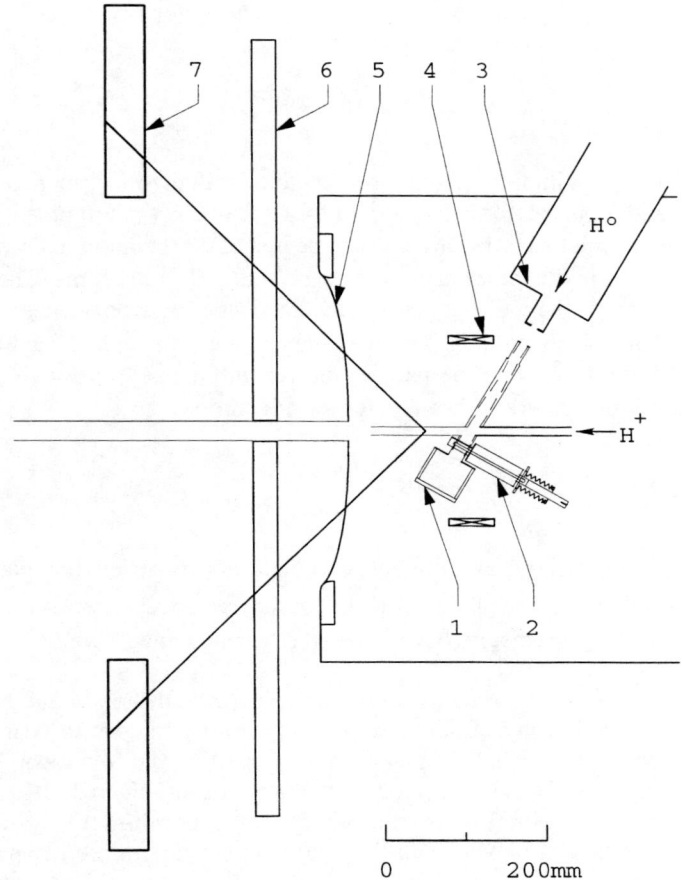

FIGURE 1. Schematic top view of target and detector assembly. A typical p-p elastic event is illustrated. The numbers point to: 1) copper recombiner, 2) internal valve, 3) atomic beam collimator, 4) super-conducting magnet coil, 5) exit window, 6) wire chambers, 7) detector scintillator. Items 1-4 are immersed in vacuum.

attachment, cell, and valve, is immersed in a strong (up to 0.6T) guide field oriented parallel to the cooler beam. A schematic of the target and detector arrangement is shown in Fig. 1.

The target polarization will be measured with a 200 MeV longitudinally polarized proton beam using the known (1) and large p-p elastic spin correlation coefficient, A_{zz}. At 200 MeV beam energy and at lab angles 30° and larger A_{zz} is above 0.85. Protons from p-p elastic scattering exit the vacuum chamber through a thin window and, after passing through wire chambers to establish azimuthal angle, enter a plastic scintillator. A timing coincidence and co-planarity is required to accept an event. A magnetic DC current transducer measures the beam current. The detector yields divided by the accumulated beam current can be written as

$$Y = Y_0[1 - \alpha(1 - \tfrac{1}{\sqrt{2}})] \cdot [1 + P_z Q \, A_{zz}], \tag{1}$$

Here Q is the (longitudinal) target polarization, P_z is the z-component of the beam polarization, and Y_0 is the normalized yield for a completely recombined target ($\alpha = 0$) if either beam or target are unpolarized. The degree of dissociation, α, is defined as the fraction of protons in the target that are present in the form of atoms. The first bracket in Eq. 1 accounts for the reduced gas conductance when H atoms (mass 1) recombine to form H2 molecules (mass 2). A measurement with an unpolarized beam ($P_z=0$) yields the change in α when the path to the recombiner is opened. With α known, measurements with a polarized beam determine the ratio

$$R = \frac{Q_{open}}{Q_{closed}}, \tag{2}$$

where Q_{open} and Q_{closed} are the target polarizations when the path to the recombination volume is opened or closed. It is not necessary to know the beam polarization or A_{zz} precisely because these terms cancel when measuring the polarization ratio.

The temperatures of the recombination attachment and the storage cell target are independently varied. The Teflon-coated storage cell will be set to near 150K where Teflon is known to have excellent polarization properties and low recombination. The recombiner temperature can be varied over the range from 30K to 320K.

To study the recombining properties of the recombination zone, we have constructed a test bench in Madison. We find strong temperature dependence with a minimum in the recombination near 80K. Additionally, in agreement with the effects seen at HERMES, we measure a strong decrease in recombination when the recombiner surface is coated with a water layer.

At the IUCF ring we will measure the fraction of the atomic polarization that survives the recombination process as a function of the temperature of the recombining surface and of the amount of water coverage. The effect of variation of the guide field strength will also be measured. If time allows additional recombining surfaces will be studied.

REFERENCES

1. B.Lorentz et al. "Measurement of Azz in p-p elastic scattering with a longitudinally polarized proton beam in the Indiana cooler", 12th International Symposium on High-Energy Spin Physics, (Amsterdam 1996), in Spin 96 Conference Proceedings, C.W. de Jager et al., eds., (World Scientific, Singapore 1997), p. 426, and Phys. Rev. C (submitted for publication).

Acceleration and Storage of Polarized Proton Beam at RHIC *

Thomas Roser

AGS Department, Brookhaven National Laboratory
Upton, NY 11973-5000

Abstract. High energy polarized beam collisions will open up the unique physics opportunities of studying spin effects in hard processes. However, the acceleration of polarized beams in circular accelerators is complicated by the numerous depolarizing spin resonances. Using a partial Siberian Snake and a rf dipole that ensure stable adiabatic spin motion during acceleration has made it possible to accelerate polarized protons to 25 GeV at the Brookhaven AGS. Full Siberian Snakes and polarimeters are being developed for RHIC to make the acceleration of polarized protons to 250 GeV possible. Spin rotators around two RHIC interaction regions will allow for the collision of longitudinally polarized protons at the STAR and PHENIX detectors.

INTRODUCTION

Polarized proton colliders will open up the completely unique physics opportunities of studying spin effects in hard processes at high luminosity, high energy proton-proton collisions. It will allow to study the spin structure of the proton, in particular the degree of polarization of the gluons and antiquarks, and also to verify the many well documented expectations of spin effects in perturbative QCD and parity violation in W and Z production.

Proton-proton collisions at high energies involve hard scattering of gluons and quarks as is shown schematically in Fig. 1. In this kinematic region factorization should hold and any asymmetry A measured for a high p_T reaction is a sum of corresponding asymmetries $\hat{a}$ at the parton level weighted by the actual degree of polarization of the initial partons given by the spin structure functions:

$$A = \sum_{subprocesses} \frac{\Delta a}{a} \times \frac{\Delta b}{b} \times \hat{a}\left(a + b \rightarrow c + d\right)$$

The subprocess asymmetries are predicted by the standard model and are often large. For example, $\hat{a}_{LL}$ in QCD is 50% or larger for most subprocesses and the

---

*) Work performed under the auspices of the U.S. Department of Energy

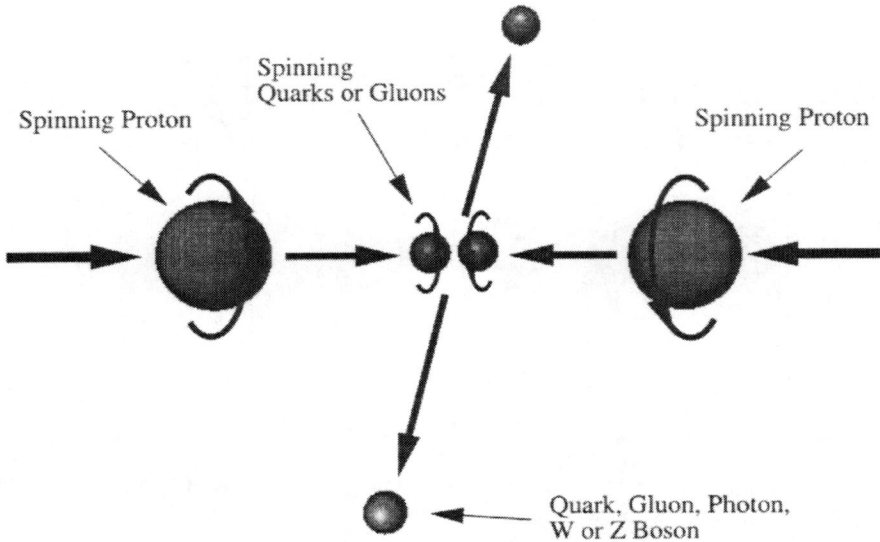

FIGURE 1. Schematic of parton collisions in high energy polarized proton collisions

parity violating $\hat{a}_L$ is unity in weak processes. By measuring different reactions and different types of asymmetries we can determine subprocess asymmetries, the spin structure functions of all partons, and also perform self consistency checks. This very ambitious program is greatly simplified by the fact that the spin structure functions of valence quarks are known from deep inelastic scattering measurements and that we can select reactions that are dominated by just one subprocess. A center-of-mass energy range of 200 to 500 GeV, as achievable in the Brookhaven Relativistic Heavy Ion Collider (RHIC) [1], is ideal in the sense that it is high enough for perturbative QCD to be applicable and low enough so that the average x value is about 0.1 or larger which guarantees significant levels of polarization for the valence quarks.

SPIN DYNAMICS, RESONANCES AND SIBERIAN SNAKES

Accelerating polarized beams requires an understanding of both the orbital motion and spin motion. Whereas the effect of the spin on the orbit is negligible the effect of the orbit on the spin is usually very strong. The evolution of the spin direction of a beam of polarized protons in external magnetic fields such as exist in a circular accelerator is governed by the Thomas-BMT equation [2],

$$\frac{d\vec{P}}{dt} = -\left(\frac{e}{\gamma m}\right)\left[G\gamma\vec{B_\perp} + (1+G)\vec{B_\parallel}\right] \times \vec{P}$$

where the polarization vector P is expressed in the frame that moves with the particle. This simple precession equation is very similar to the Lorentz force equation which governs the evolution of the orbital motion in an external magnetic field:

$$\frac{d\vec{v}}{dt} = -\left(\frac{e}{\gamma m}\right)\left[\vec{B_\perp}\right] \times \vec{v}.$$

From comparing these two equations it can readily be seen that, in a pure vertical field, the spin rotates $G\gamma$ times faster than the orbital motion. Here $G = 1.7928$ is the anomalous magnetic moment of the proton and $\gamma = E/m$. In this case the factor $G\gamma$ then gives the number of full spin precessions for every full revolution, a number which also called the spin tune ν_{sp}. At top RHIC energies this number reaches about 400. The Thomas-BMT equation also shows that at low energies ($\gamma \approx 1$) longitudinal fields $\vec{B_\parallel}$ can be quite effective in manipulating the spin motion, but at high energies transverse fields $\vec{B_\perp}$ need to be used to have any effect beyond the always present vertical holding field.

The acceleration of polarized beams in circular accelerators is complicated by the presence of numerous depolarizing spin resonances. During acceleration, a spin resonance is crossed whenever the spin precession frequency equals the frequency with which spin-perturbing magnetic fields are encountered. There are two main types of spin resonances corresponding to the possible sources of such fields: imperfection resonances, which are driven by magnet errors and misalignments, and intrinsic resonances, driven by the focusing fields.

The resonance conditions are usually expressed in terms of the spin tune ν_{sp}. For an ideal planar accelerator, where orbiting particles experience only the vertical guide field, the spin tune is equal to $G\gamma$, as stated earlier. The resonance condition for imperfection depolarizing resonances arise when $\nu_{sp} = G\gamma = n$, where n is an integer. Imperfection resonances are therefore separated by only 523 MeV energy steps. The condition for intrinsic resonances is $\nu_{sp} = G\gamma = kP \pm \nu_y$, where k is an integer, ν_y is the vertical betatron tune and P is the superperiodicity. For example at the AGS, $P = 12$ and $\nu_y \approx 8.8$.

Close to a spin resonance the spin tune deviates away from its value of $G\gamma$ of the ideal flat machine. For a resonance with strength ϵ, which is the total spin rotation due to the resonance driving fields, the new spin tune is given by the equation

$$\cos(\pi\nu_{sp}) = \cos(\pi G\gamma)\cos(\pi\epsilon).$$

Fig. 2 shows the solutions of this equation with and without a resonance. A similar calculation can be done for the effective precession direction or, as it is now often called, the stable spin direction. The stable spin direction describes

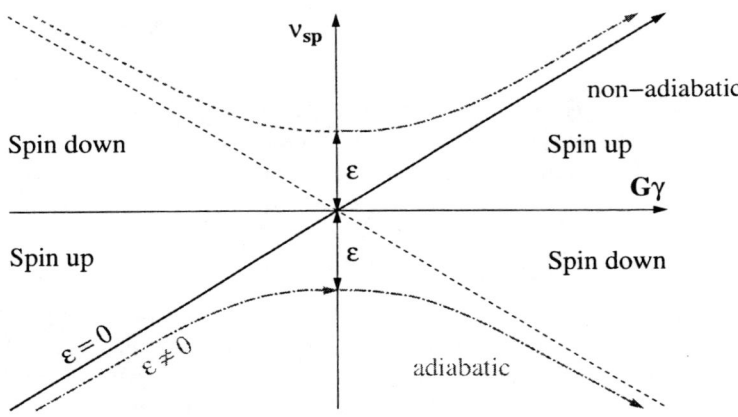

FIGURE 2. The evolution of the spin tune during the crossing of a resonance with strength ϵ

those polarization components that are repeated every turn. Note that both the stable spin direction and the spin tune are completely determined by the magnetic structure of the accelerator and the beam energy. The magnitude and sign of the beam polarization, however, depends on the beam polarization at injection and the history of the acceleration process.

The spin tune and stable spin direction calculations apply only to a time-independent static situation or if parameters are changed adiabatically. Far from the resonance the stable spin direction coincides with the main vertical magnetic field. Close to the resonance, the stable spin direction is perturbed away from the vertical direction by the resonance driving fields. When a polarized beam is accelerated through an isolated resonance at arbitrary speed, the final polarization can be calculated analytically [3] and is given by

$$P_f/P_i = 2e^{-\frac{\pi|\epsilon|^2}{2\alpha}} - 1,$$

where P_i and P_f are the polarizations before and after the resonance crossing, respectively, and α is the change of the spin tune per radian of the orbit angle. When the beam is slowly ($\alpha \ll |\epsilon|^2$) accelerated through the resonance, the spin vector will adiabatically follow the stable spin direction resulting in spin flip as is indicated in Fig. 2. However, for a faster acceleration rate partial depolarization or partial spin flip will occur.

Traditionally, the intrinsic resonances are overcome by using a betatron tune jump, which effectively makes α large, and the imperfection resonances are overcome with the harmonic corrections of the vertical orbit to reduce the resonance strength ϵ [4]. Both of these methods aim at making the resonance crossing non-adiabatic. They require very accurate adjustments at every resonance crossing which can become very difficult and time consuming.

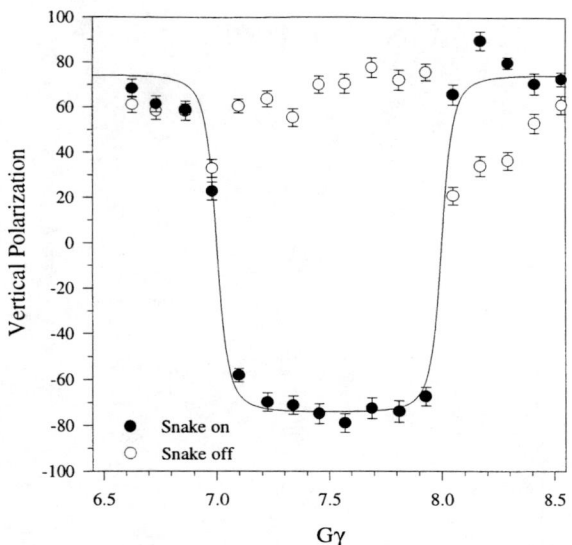

FIGURE 3. The measured vertical polarization as a function of $G\gamma$ for a 10% snake is shown with and without a snake. The solid line is the predicted energy dependence of the polarization.

Over the last ten years new techniques to cross both imperfection and intrinsic resonances adiabatically have been developed. The correction dipoles used to correct the imperfection resonance strength to zero were replaced by a localized spin rotator or 'partial Siberian snake' which makes all the imperfection resonance strengths large and causes complete adiabatic spin flip at every imperfection resonance [5]. The tune jump quadrupoles were recently replaced at the AGS by a single rf dipole magnet which creates a strong spin resonance by driving large coherent betatron oscillations, overpowering the effect of the intrinsic resonances.

At higher energies a 'full Siberian snake' [6], which is a 180° spin rotator of the spin about a horizontal axis, will keep the stable spin direction unperturbed at all times as long as the spin rotation from the Siberian snake is much larger than the spin rotation due to the resonance driving fields. Therefore the beam polarization is preserved during acceleration. An alternative way to describe the effect of the Siberian snake comes from the observation that the spin tune with the snake is a half-integer and energy independent. Therefore, neither imperfection nor intrinsic resonance conditions can ever be met as long as the betatron tune is different from a half-integer.

A local spin rotator can be constructed by using either a solenoid at lower energies or at high energy by a sequence of interleaved horizontal and vertical dipole magnets producing only a local orbit distortion. Since the orbit distortion is inversely proportional to the momentum of the particle, such a dipole snake is particularly effective for high-energy accelerators, e.g. energies above about $30\,GeV$.

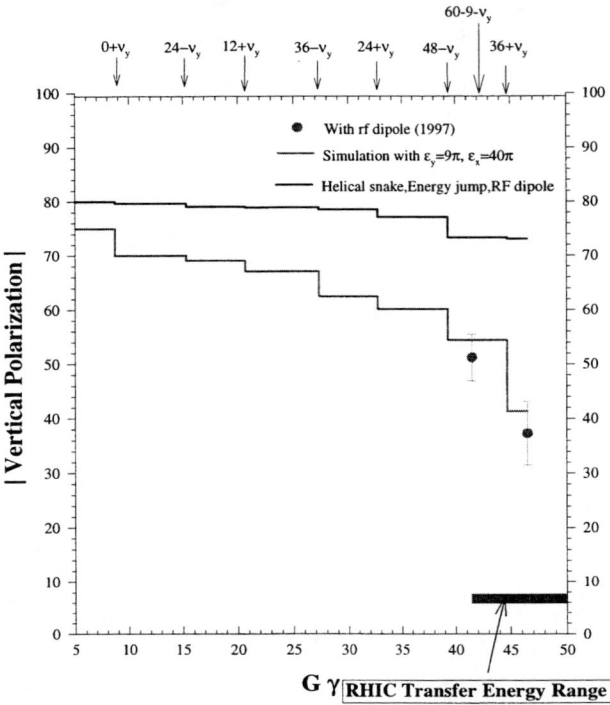

FIGURE 4. Vertical polarization versus $G\gamma$ measured in the AGS. The lower curve is the result of a spin tracking calculation for the experimental conditions. The upper curve simulates the use of a helical partial snake in the AGS.

ACCELERATING POLARIZED PROTONS IN AGS AND RHIC

Polarized proton beam experiments at the AGS have demonstrated the feasibility of polarized proton acceleration using a 5% partial Siberian snake. It was shown that a 5% snake is sufficient to avoid depolarization from imperfection resonances without using the harmonic correction method up to the required RHIC transfer energy of about 25 GeV. Fig. 3 shows the evolution of the beam polarization as the beam energy and therefore $G\gamma$ is increased [7]. As predicted the polarization reverses its sign whenever $G\gamma$ is equal to an integer. At this relatively low energy polarization is preserved even without snake but is partially lost at energies close to integer values of $G\gamma$.

More recently a novel scheme of overcoming strong intrinsic resonances using a rf dipole magnet was successfully tested [9]. Full spin flip can be achieved with a strong artificial rf spin resonance excited coherently for the whole beam by driving large coherent vertical betatron oscillations. If the rf spin resonance location is chosen near the intrinsic spin resonance, the spin motion will be dominated by the rf resonance and the spin near the intrinsic resonance will adiabatically follow the spin closed orbit of the rf spin resonance. With the rf dipole, a new dominant resonance near the intrinsic resonance is introduced to flip the spin, instead of enhancing the intrinsic resonance, as has been proposed earlier [8], which would also enhance the strength of the nearby coupling resonance. Fig. 4 shows the new record proton beam polarization achieved during the last AGS polarized beam experiment. The rf dipole was used to completely flip the spin at the four strong intrinsic resonances $0+\nu_y$, $12+\nu_y$, $36-\nu_y$, and $36+\nu_y$. The lower curve shown going through the data points was obtained from a spin tracking calculation simulating the experimental conditions. Most of the remaining polarization loss is caused by the coupling resonances. A new AGS partial snake using a helical dipole magnet would eliminate all coupling resonances. Spin tracking simulations of this condition are depicted by the upper curve in Fig. 4.

With full snakes in RHIC all depolarizing resonances should be avoided since the spin tune is a half-integer independent of energy. However, if the spin disturbance from small horizontal fields is adding up sufficiently between the snakes depolarization can still occur. This is most pronounced when the spin rotation from all the focusing fields add up coherently which is the case at the strongest intrinsic resonances. A simplistic rule of thumb would then suggest that as long as the total spin rotation of all the Siberian snakes is much larger than the total spin rotation per turn caused by the strongest spin resonance the polarization should be preserved during acceleration. This rule holds for the AGS partial Siberian snake with regard to the imperfection resonances. It would also predict that for a beam with a normalized 95% emittance of $20\,\pi\,mm\,mrad$ at least two snakes are needed for RHIC.

Polarized protons from the AGS are injected into the two RHIC rings to allow for up to $\sqrt{s} = 500\,GeV$ collisions with both beams polarized [1]. Fig. 5 shows the layout of the Brookhaven accelerator complex highlighting the components required for polarized beam acceleration. A new polarized source using the 'Optically Pumped Polarized Ion Source' (OPPIS) technique is being assembled at TRIUMF using components of the KEK OPPIS [10]. The new source will be able to deliver 10^{12} polarized protons per pulse which will be injected into RHIC as a single bunch with nominally 2×10^{11} polarized protons.

Of particular interest is the design of the Siberian snakes (two for each ring) and the spin rotators (four for each collider experiment) for RHIC. Each snake or spin rotator consists of four $2.4\,m$ long, $4\,T$ helical dipole magnet modules each having a full 360 degree helical twist [11]. Using helical magnets minimizes orbit excursions within the extend of the snake or spin rotator which is most important at injection energy. Nevertheless the bore of the helical magnets has to be 10 cm in diameter

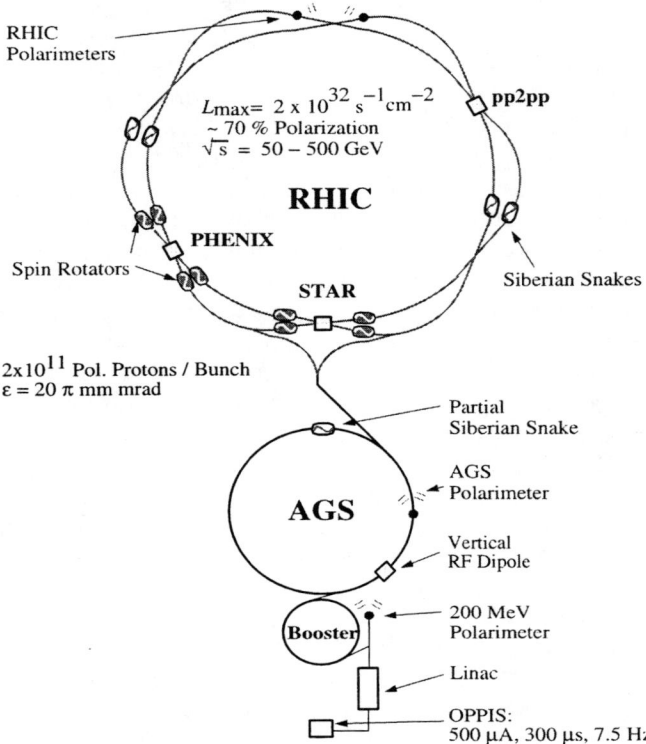

FIGURE 5. The Brookhaven hadron facility complex, which includes the AGS Booster, the AGS, and RHIC. The RHIC spin project will install two snakes per ring with four spin rotators per detector for achieving helicity-spin experiments.

to accommodate the 3 cm orbit excursions. Superconducting helical dipoles have been successfully tested at BNL using thin cable placed into helical grooves that have been milled into a thick-walled aluminum cylinder. A schematic picture of the helical dipole magnet is shown in Fig. 6. The first four full length magnets, enough for the first complete Siberian snake, have recently been successfully tested [12].

To verify that full polarization is preserved during acceleration in RHIC an elaborate spin tracking program was developed [13]. The acceleration through the energy region of the strongest resonance was simulated in great detail including a 1 mm rms misalignment of the quadrupoles, and sextupoles as well as the corrector dipoles used to correct the closed orbit. The result is shown in Fig. 7 for a beam with a normalized 95% emittance of $20\,\pi\,mm\,mrad$. The upper and lower curve

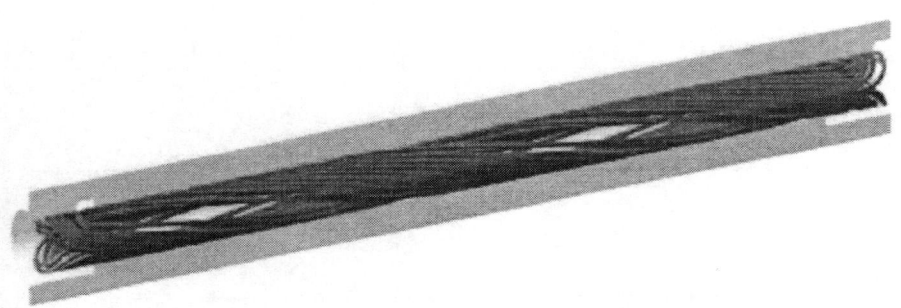

FIGURE 6. Schematic picture of the superconducting helical dipole shows the 16 helical current blocks and half of the iron yoke

show the result for the full beam and the particles at the edge of the beam, respectively. Although there is a significant decrease of the polarization at the energy of the resonance at $G\gamma = 5 \times 81 + (\nu_y - 12) = 422.18$, the polarization of the full beam is restored after accelerating completely through the resonance region. The simulation also shows that there is significant polarization loss at the edge of the beam. This fact highlights the need for a polarimeter that can measure polarization profiles.

Measuring proton polarization at high energies has been the focus of several workshops and experiments. The analyzing power of only very few reactions has been

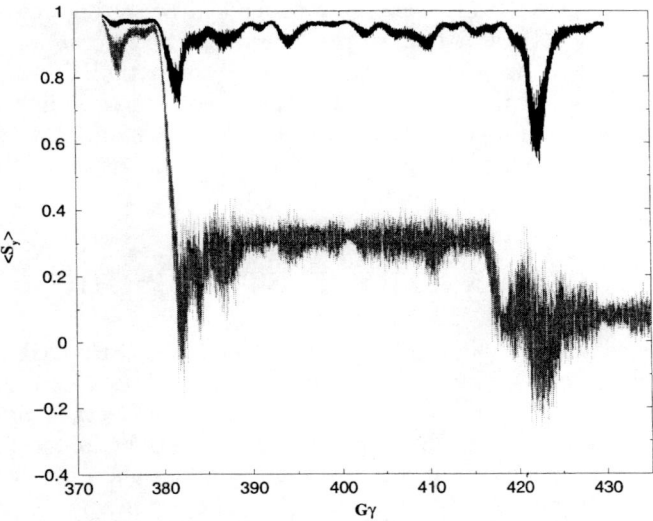

FIGURE 7. Result of spin tracking through the region of the strongest resonance in RHIC. The closed orbit was corrected with residuals of less than 0.2 mm.

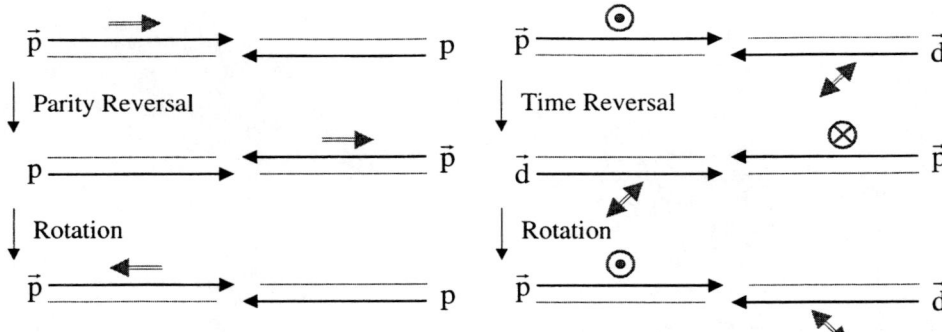

FIGURE 8. Schematic illustration of the implications of parity violation for the spin dependent proton-proton total cross section.

FIGURE 9. Schematic illustration of the implications of time reversal invariance for the spin dependent proton-deuteron total cross section.

measured at high energies and the magnitudes are typically rather small. Polarization sensitive interaction with an external electromagnetic field is also much smaller than for the much lighter electron for which Compton back scattering is typically used for high energy polarization measurement. Two methods are presently being considered for RHIC. Inclusive pion production from both hydrogen and carbon targets has been shown to have large analyzing power. Magnetic spectrometers are being designed for RHIC to allow for high rate data acquisition. The second method is based on the relatively large, energy independent analyzing power predicted for very small angle elastic scattering where the magnetic interaction interferes with the spin independent nuclear interaction [14]. Tests at the IUCF Cooler ring and with polarized beam at the AGS have demonstrated the possibility of detecting the recoil carbon nucleus from a ultra thin carbon fiber. A significant non-zero analyzing power at 22 GeV was also measured. For both methods the carbon fiber target could be scanned through the circulating beam to measure polarization profiles.

SYMMETRY TESTS AT RHIC

The polarized proton beams in RHIC can also be used with fixed internal targets. This offers the unique opportunity to study parity violation and violation of time reversal invariance in total cross section measurements over a very large energy range. As is shown schematically in Fig. 8 measuring a difference of the total cross section for the two helicity states of the beam would violate parity. This could be accomplished with high statistical accuracy by simply measuring the difference in the beam life time between the two helicity states [15]. A similar scheme can be used to measure the violation of time reversal invariance by using transverse polarized proton beam and a tensor-polarized deuterium gas target [16]. Such measurements pose a tremendous demand on the detailed understanding of systematic errors but the unique and novel environment of a storage ring may make such high precession experiments possible.

COMMISSIONING PLANS AND SCHEDULE

It is planned that, for the expected turn-on of RHIC in 2000, one ring will be equipped with two snakes and a polarimeter which will allow for initial commissioning of polarized beam acceleration. Starting after October 2000 all snakes and spin rotators will be in place for the first RHIC spin physics run with longitudinal polarization at the two detectors STAR and PHENIX.

REFERENCES

1. Design Manual - Polarized Proton Collider at RHIC, Brookhaven National Laboratory, July 1998, http://www.ags.bnl.gov/ rhicspin .
2. L.H. Thomas, Phil. Mag. **3**, 1 (1927); V. Bargmann, L. Michel, V.L. Telegdi, Phys. Rev. Lett. **2**, 435 (1959).
3. M. Froissart and R. Stora, Nucl. Instr. Meth., **1**, 297 (1960).
4. T. Khoe et al., Part. Accel. **6**, 213 (1975); J.L. Laclare et al., J. Phys. (Paris), Colloq. **46**, C2-499 (1985); H. Sato et al., Nucl. Inst. Meth., Phys. Res. Sec **A272**, 617 (1988); F.Z. Khiari, et al., Phys. Rev. D**39**, 45 (1989).
5. T. Roser, AIP Conf. Proc. No. 187, ed. K.J. Heller p.1442 (AIP, New York, 1988).
6. Ya.S. Derbenev et al., Part. Accel. **8**, 115 (1978).
7. H. Huang et al., Phys. Rev. Lett. **73**, 2982 (1994)
8. T. Roser, in Proc. of the 10th Int. Symp. on High Energy Spin Physics, Nagoya, Japan, p. 429 (1992).
9. M. Bai et al., Phys. Rev. Lett. **80**, 4673 (1998)
10. A.N. Zelinski et al., 'Optically-Pumped Polarized H- ION Sources for RHIC and HERA Colliders', Proceedings of the 1999 Particle Accelerator Conference, to be published.
11. V.I.Ptitsin and Yu.M.Shatunov, Helical Spin Rotators and Snakes, Proc. 3. Workshop on Siberian Snakes and Spin Rotators (A.Luccio and T.Roser Eds.) Upton, NY, Sept. 12-13,1994, BNL-52453, p.15;
12. E. Willen et al., 'Construction of helical magnets for RHIC', Proceedings of the 1999 Particle Accelerator Conference, to be published.
13. A.U. Luccio et al., 'Development of the spin tracking program SPINK', Proceedings of the 1999 Particle Accelerator Conference, to be published.
14. J. Schwinger, Phys. Rev. **73**, 407 (1948)
15. S.E. Vigdor, in Proc. Workshop on Future Directions in Particle and Nucl. Physics (D.F. Geesaman, Ed.), Brookhaven 1993, Rep. BNL-52389, 1993, p.173
16. H.E. Conzett, Phys. Rev. **C48**, 423 (1993)

Preparing the COSY-Ring for a Test of Time-Reversal-Invariance[1]

P.D. Eversheim* for the TRI Collaboration[2]

Institut für Strahlen- und Kernphysik, Universität Bonn, Germany

Abstract.
At the cooler synchrotron COSY at Jülich a novel (P-even, T-odd) true null test of time-reversal invariance (TRI) was proposed and accepted, that is supposed to measure TRI to an accuracy of 10^{-4} (Phase 1) or 10^{-6} (Phase 2). The parity conserving time-reversal violating observable is the total cross-section asymmetry Ay,xz. The measurement is planned as an internal target transmission experiment at the cooler synchrotron COSY. Ay,xz is measured using a polarized beam with an energy of about 1 GeV and a tensor polarized deuteron atomic-beam target.

INTRODUCTION

At present, the only link to a violation of time-reversal symmetry is given via the CPT-theorem and the observation of CP- and T-violation in the neutral kaon system. Although the CP- or T-violation could be accomodated by a complex phase of the Kobayashi- Maskawa-matrix or the Θ-term [1] allowed by QCD, other explanations go beyond the standard model, like for instance the extension of the Higgs sector, the superweak interaction [2], or the left-right symmetric models [3]. These extensions of the standard model may lead to interactions that are not related to the observed CP- or T- violation. Since the origin of the CP- or T-violation is not clear, further experimental tests of CP- or T-invariance are necessary. In this context clean theoretical interpretation is expected rather from tests involving

---

[1] Work supported by the BMBF und the Forschungszentrum Jülich
[2] D. Eversheim* (Spokesperson], M. Beyer†, J. Bisplinghoff*, H. Conzett‡, J. Dietrich§, J. Ernst*, O. Felden§, R. Gebel§, M. Glende*, F. Hinterberger*, R. Jahn*, W. Kretschmer¶ and H. Paetz gen. Schieck∥
* Institut für Strahlen- und Kernphysik, University Bonn, Germany
† FB Physik, University Rostock, Germany
‡ Lawrence Berkeley Laboratory, Berkeley, USA
§ IKP, Forschungsanstalt Jülich, Jülich, Germany
¶ Physikalisches Institut, University Erlangen, Germany
∥ Institut für Kernphysik, University Köln, Germany

elementary particles or few nucleon systems than from tests involving complex nuclei, where a possible time-reversal violating (TRV) effect may be diluted [4].

GENERAL CONSIDERATIONS

Usually P-even TRI tests compare two observables (cf. tests of detailed balance or P-A tests). Thus, the experimental accuracy was limited to 10^{-3}–10^{-2}. The accuracy can be increased by orders of magnitude if a true null experiment is performed i.e. a non-vanishing value of one single observable proves that the symmetry involved is violated. The term "true" stresses the concept that the intended test has to be completely independent from for instance: Final state interactions, special tensorial interactions or, Hamiltonians of a certain form. True null tests are based only on the structure of the scattering matrix as determined by general conservation laws [7].

Conzett [8] could show that a transmission experiment can be devised, which constitutes a true P-even, T-odd TRI null test. He suggested to measure the total cross-section asymmetry $A_{y,xz}$ of vector polarized spin 1/2 particles interacting with tensor polarized spin 1 particles.

The quantity $A_{y,xz}$ will be measured in an transmission experiment of an circulating vector polarized proton beam through an internal tensor polarized atomic deuteron target in the cooler synchrotron COSY. The deuteron target will be available from the EDDA-experiment (phase 1). In phase 2 a target cell will be added to the experimental set up and the control of systematic error contributions is improved. The transmission losses of the circulating polarized proton beam are measured with high precision. Thus, this experiment uses the COSY facility in three respects: As an accelerator, as an ideal forward spectrometer and as an detector.

An analysis of possible systematic error sources [9] shows that the most probable dominant contribution will arise from the observable $A_{y,y}$. For this to happen, two conditions have to be fulfilled: i) the deuteron beam from the atomic-beam target is not completely tensor polarized; and ii) a misalignment of the atomic-beam exists from the x-z direction into the y-direction. Therefore, in order to control the polarization of the atomic-beam, the beam-dump of the atomic-beam target has to provide the option for a polarimeter.

Under these conditions it seems to be possible to measure in phase 2 of the TRI test the analyzing power $A_{y,xz}$ with an accuracy of 10^{-6}. First estimates [5] indicate an order of magnitude improvement of the experimental bound of the TRV-part of the meson-nucleon ρ-coupling constant over the presently known best value [4]. Thus, this experiment has the potential to probe for this class of P-even, TRV tests the theoretical bounds as derived from the neutron EDM-experiments [6].

THE ATOMIC-BEAM TARGET

A polarized atomic beam target (ABT) has been built for the EDDA experiment. After the proton-proton elastic scattering program of EDDA with unpolarized beam and target has been finished, the next phase of this experiment has started to study elastic scattering excitation functions of spin observables. Meanwhile the analyzing power data have been taken with the ABT. In a next step, spin correlation observables, using a polarized beam and target, will be measured. Once this program has been finished, a test of time-reversal invariance will take place at the EDDA target station. The EDDA detector will serve as an efficient polarimeter for the tensor polarized deuteron beam of the ABT and the internal polarized proton beam of COSY.

Besides providing free polarized proton and deuteron beams the ABT was designed -in view of an increased luminosity- to feed also a target cell. As a consequence, the ABT should be able to deliver a density optimized beam for a free atomic beam, whereas for a target cell the polarized beam should be intensity optimized.

In June 1997 the ABS was tested for the first time at COSY with respect to polarization and atomic beam cross-section at the interaction region of the EDDA detector. Results obtained so far with the EDDA detector agree in all respects with the expectations derived from computer simulations and/or off-line measurements. At present the ABT can be characterized by the following data:

- Efficiency of the dissociator: $> 80\%$
- Efficiency of the RF-transitions: $> 97\%$

- Atomic-beam cross-section: 11 ± 1mm
- Intensity in the EDDA interaction regime: $5.7 \cdot 10^{16}$ atoms/s in two states
- Most probable velocity: 1200m/s at 30K nozzle temperature
- Polarization of the atomic-beam: $> 90\%$ in the center
  $> 85\%$ averaged over 15mm diameter
- Efficiency of the beam-dump: $> 90\%$

REFERENCES

1. R.J. Crewther et al., *Phys. Lett.* **88B** (1979) 123
2. L. Wolfenstein, *Ann. Rev. Nucl. Part. Sci.* **36** (1986) 137
3. J.C. Pati, A. Salam, *Phys. Rev.* **D10** (1974) 275
4. P.R. Huffman et al., *Phys. Rev. Lett.* **76** (1996) 4681
5. M. Beyer, *Nucl. Phys.* **A560** (1993) 895
6. W.C. Haxton et al., *Phys. Rev.* **D50** (1994) 3422
7. F. Arash, M.J. Moravcsik and G.R. Goldstein, *Phys. Rev. Lett.* **54** (1985) 2649
8. H.E. Conzett, *7th Intl. Conf. on "Pol. Phen. Nucl. Phys."*, Paris (1990) 2D
9. P.D. Eversheim, *Pol. Phen. Nucl. Part. Phys.*, 7-10. Jan. 1992, Trieste, Italy, Proc. 2nd Adriatico Research Conf., World Scientific, (1993) 142

NEW FACILITIES AND TECHNIQUES

WASA Detector: Towards Rare Pion and Eta Decays

H. Calén[a]

for the CELSIUS/WASA collaboration:
R. Bilger[c], M. Blom[b], D. Bogoslawsky[d], A. Bondar[f],
W. Brodowski[c], I. Chuvilo[j], H. Clement[c], V. Dunin[d], J. Dyring[b],
C. Ekström[a], K. Fransson[b], C-J. Friden[a], J. Greiff[b], L. Gustafsson[b],
S. Häggström[b], B. Höistad[b], M. Jacewicz[b], J. Johanson[b],
A. Johansson[b], T. Johansson[b], A. Khoukaz[b], K. Kilian[g],
N. Kimura[k], I. Koch[b], G. Kolachev[f], M. Komogorov[d], S. Kullander[b],
A. Kupsc[e], L. Kurdadze[f], A. Kuzmin[f], A. Kuznetsov[d],
P. Marciniewski[e], A. Martemyanov[j], B. Martemyanov[j],
B. Morosov[d], A. Mörtsell[b], A. Nawrot[e], W. Oelert[g],
S. Oreshkin[f], Y. Petukhov[d], A. Povtorejko[d], K. Przestrzelska[e],
J. Pätzold[c], D. Reistad[a], R.J.M.Y. Ruber[b], V. Sandukovsky[d],
U. Schuberth[b], T. Sefzick[g], V. Sidorov[f], B. Shwartz[f],
V. Sopov[j], J. Stepaniak[e], A. Sukhanov[f], A. Sukhanov[a],
P. Sundberg[b], V. Tchernychev[j], P. Thörngren[b], V. Tikhomirov[d],
A. Turowiecki[i], G. Wagner[c], Z. Wilhelmi[i], A. Yamamoto[k],
H. Yamaoka[k], J. Zabierowski[h], A. Zernov[d], J. Zlomanczuk[b]

[a] *TSL, Uppsala, Sweden;* [b] *ISV, Uppsala Univ., Sweden;* [c] *Tübingen University, Germany* [d] *JINR, Dubna, Russia;* [e] *INS, Warsaw, Poland;* [f] *BINP, Novosibirsk, Russia;* [g] *IKP, KFA, Jülich, Germany;* [h] *INS, Lodz, Poland;* [i] *Inst. of Exp. Phys., Warsaw Univ., Poland;* [j] *ITEP, Moscow, Russia;* [k] *KEK, Tsukuba, Japan;*

Abstract. The WASA 4π detector at the The Svedberg Laboratory in Uppsala is now being commissioned. This detector will make possible new detailed studies of many interesting rare processes in intermediate energy light-ion physics. WASA is built around a new target system, providing well defined internal hydrogen (and deuterium) targets of high density. It has a detection coverage of close to 4π sr for high energy photons and charged particles and it includes a strong magnetic field provided by an extremely thin-walled superconducting solenoid.

INTRODUCTION

The CELSIUS/WASA project [1] aims at high precision studies of rare processes in intermediate energy light-ion reactions. One of the main goals is to reach a sensitivity of down to 10^{-9} in the branching ratio for some particularly interesting eta decays (e.g. $\eta \to e^+e^-$ and $\eta \to \pi^0 e^+e^-$). The η-s (and other light mesons to be studied) are produced in proton-proton and proton-deuteron collisions. The necessary high luminosity of around 10^{32} cm^{-2}s^{-1} is provided by using small frozen hydrogen pellets as internal targets in the CELSIUS accelerator and storage ring. The maximum energy for protons in CELSIUS is 1360 MeV.

I THE PELLET TARGET

The newly developed target system (Fig. 1) provides a stream of pellets with 30 μm diameter, a velocity of 60 m/s and a flow of about 70 000 pellets per second. In the droplet chamber of the pellet generator, a jet of liquid hydrogen is broken up into uniformly sized and spaced droplets by an acoustical excitation of the injection nozzle. The droplets are then injected into vacuum through a second nozzle and due to the low pressure the droplets will freeze. A skimmer with a 1 mm diameter opening is introduced 0.8 m downstream of the pellet generator to reduce the pellet stream and give the desired target thickness. After 2.5 m, at the interaction region, the pellet beam has a diameter of about 3 mm.

Results from tests at CELSIUS have shown that a target thickness of 5×10^{15} atoms/cm^2 gives acceptable half-lives of the circulating ion beam, a few minutes

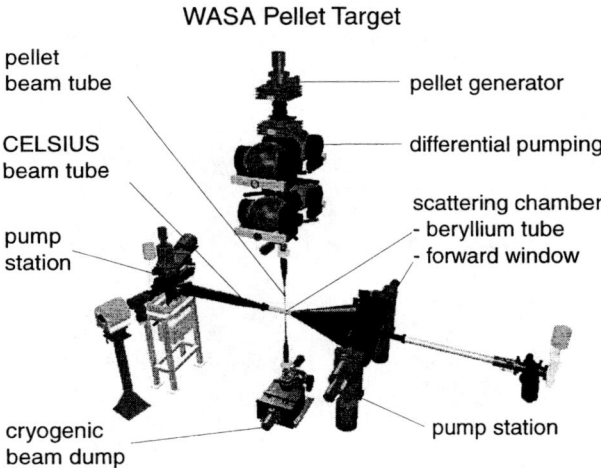

FIGURE 1. The pellet system and the CELSIUS beam vacuum chambers at WASA.

at energies around 1 GeV, as well as acceptable vacuum conditions, about 10^{-6} mbar, in the scattering chamber. With 10^{10} protons stored in CELSIUS, a target thickness of 3×10^{15} atoms/cm^2 is sufficient to get a luminosity of 10^{32} cm^{-2}s^{-1}.

II THE WASA DETECTOR

The WASA setup consists of a forward and a central part (Fig. 2). The forward detector is designed to measure the scattered projectiles and target recoil particles. The central part of the WASA detector is optimized for measuring electrons and

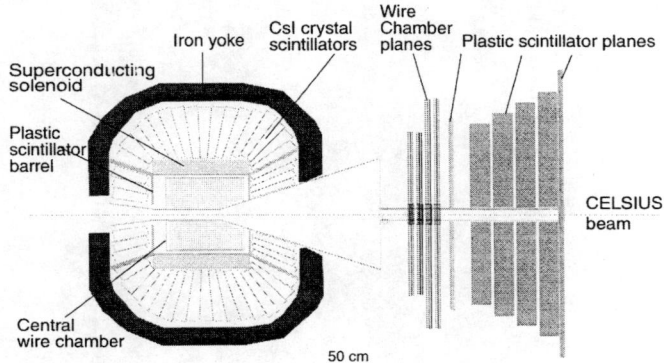

FIGURE 2. The WASA detector being cabled. The bottom picture gives a cross section in the horizontal plane

photons from the meson decays. It consists of a cylindrical wire chamber sitting inside of a superconducting solenoid that is surrounded by a CsI calorimeter. To avoid disturbance of the particles to be measured, the amount of structural material is kept low; the scattering chamber wall is made of 1.2 mm thin beryllium and the solenoid is made extremely thin walled (0.18 radiation lengths).

The solenoid can provide an axial field of up to 1.3 tesla. The coil is mechanically supported in the back end only, where also the cooling pipe with liquid helium is applied. In Fig. 3 the solenoid is seen with the cryostat outer wall made of 1 mm thick corrugated aluminium. The cryostat has one radiation shield only and is cooled by a small helium refrigerator. The magnet coil is split in two parts with a central gap of 40 mm to allow for the openings to let through the 8 mm diameter pellet pipe.

The cylindrical drift chamber consists of thin walled (25 μm) aluminized mylar straw tubes with diameters of 4, 6 and 8 mm and a maximum length of 424 mm. There are 1738 straw tubes grouped in 17 cylindrical layers. The radius of the inner layer is 41 mm and the radius of the outer layer is 203 mm. Nine layers are

FIGURE 3. The central detector with the solenoid, wire chamber and one half of the CsI calorimeter.

parallel to the beam direction and the other eight have small skew angles (6-9°) with respect to the beam axis. The straws of each layer are packed into an Al-Be semi-ring support. The semi-rings are assembled around the beryllium beam pipe. A complication in the design is that the thin tapering tube (diameter 5-10 mm) for the pellets has to go through the chamber. The coverage in scattering angle is from 24° to 159°. At a magnetic field of 1 tesla, the momenta of electrons will be measured with an accuracy of 1-3 % in the range of 20-600 MeV/c.

The wire chamber is enclosed in a plastic scintillator barrel, which provides fast signals for the trigger and information on deposited energy to be used for particle identification. The 146 elements of the barrel are read out by photomultiplier (PM) tubes via 50 cm long Plexiglas light guides.

The central calorimeter covers scattering angles from 20° to 169° (96% of 4π steradians) and its primary purpose is to detect photons, electrons and positrons and measure energies in the range 1 to 600 MeV. The calorimeter consists of 1012 sodium-doped CsI scintillating crystals placed in the volume between the superconducting solenoid and its field return yoke (iron yoke). The crystals, shaped as truncated pyramids, are placed in 24 layers along the beam. The crystals are wrapped with teflon and aluminized polyester and some are seen in Fig. 3. A layer in the central part consists of 48 identical crystals. The length of the crystals vary from 30 cm (16.2 radiation lengths) in the central part to 20 cm in the backward and 25 cm in the forward part.

The calorimeter will work in a relatively high intensity proton (and neutron) radiation environment and therefore radiation hardness of the crystals has been checked using a proton beam. The CsI(Na) emission spectrum is peaked at 420 nm which matches well to standard PM tubes with bi-alkali photocathodes. The crystal surfaces were treated in order to obtain uniform light output along the crystal axis, which is one of the crucial factors determining the energy resolution of the calorimeter. The read-out is done by PMs placed outside of the iron yoke and optically coupled to the crystals by 20–30 cm long Plexiglas light guides.

Each CsI and plastic scintillator of the detector setup is equipped with a light fiber attached to the crystal window providing light pulser calibration signals.

The forward detector of WASA (Fig. 2) covers scattering angles below 18° and measures charged particles. It consists of sixteen drift-chamber planes and a 50 cm thick stack of plastic scintillators. Particle identification is done by ΔE-E technique. Positively charged pions can be distinguished from negatively charged pion by the observation of delayed pulses from decay positrons. At scattering angles below 1° a spectrometer system using the accelerator magnets allows accurate measurements of deuterons and He nuclei.

Most of the detector components in the forward part and 112 CsI crystals from the central calorimeter have been used successfully for meson production studies at the CELSIUS cluster-jet target setup during many years [2].

The trigger system has to cope with a total event rate of 10 MHz at the design luminosity for proton-proton reactions. The rate of interesting events selected by the first level trigger for eta decay studies is estimated to be around 20 kHz. The data acquisition system, based on an ATM switch is designed for an event capacity of 40 kHz.

In the initial commissioning phase work has been concentrated on optimizing the pellet target operation and on tuning of the detector to be able to work in the new experimental environment. Preparations for high rate operation will continue and the design performance should be reached in the next year.

III CONCLUSION

WASA, a general facility for light-ion physics at TSL, has now been taken into operation. It is designed for studies of rare decays of π^0 and η. For decays including e^+e^--pairs (plus γs) the goal is to reach branching ratios at the 10^{-9} level. WASA will also be very well suited for a wide range of meson production reactions in the threshold region and it will give access to other rare processes e g the isospin breaking reaction $dd \to {}^4\mathrm{He}\ \pi^0$ and prompt production of e^+e^--pairs that can be studied with high accuracy. In the beginning the emphasis will be put on continued production studies of 2π and η in proton-nucleon reactions.

REFERENCES

1. The CELSIUS/WASA Collaboration, http://www.tsl.uu.se/wasa/
2. H. Calén et al., Nucl. Instr. and Meth. A379 (1996) 57.

Tagged Neutron Production with a Storage Ring[1]

Todd Peterson* for the TNT collaboration

*Indiana University Cyclotron Facility
Bloomington, Indiana 47408*

Abstract. We describe the ongoing development of TNT, the T-region Neutron Tagger. As a way of overcoming the problem of normalization in neutron scattering experiments, we are developing a facility to tag the production of neutrons on an event-by-event basis. The neutrons are produced using the reaction $p + d \to n + 2p$ with a 200 MeV circulating proton beam incident on a deuterium gas jet target in the Indiana Cooler. The tagging of a neutron is accomplished via the detection of the two low energy recoil protons in an array of double-sided silicon strip detectors. A tagged neutron beam makes possible absolute neutron cross section measurements, and the first experiment that will be done using this tagged neutron facility is a measurement of the np backscattering cross section. Some other possible experiments using tagged neutrons are also presented.

INTRODUCTION

A difficulty which has plagued neutron scattering experiments in the past is that of accurately determining the flux of neutrons incident upon the target. The uncertainty in neutron flux has often been the dominant source of systematic uncertainty, and the data from many neutron induced charged particle differential cross section measurements have been presented as relative only.

One specific area where an absolute neutron scattering cross section measurement would be of great impact is in np backscattering. Not only does the existing data suffer from the problem with normalization, but there are large datasets which are in apparent disagreement with one another. In spite of this, the Nijmegen group claims that their PWA [1] gives ±0.7% accuracy for the normalization, although none of the cross section data included in their fits has a normalization uncertainty that approaches this level of precision. A measurement with an absolute uncertainty near this level would provide an important test of the Nijmegen claim. Furthermore, the normalization of np cross sections is an issue in the ongoing controversy over the proper value of the charged pion-nucleon coupling constant, g_c^2. The largest source

---

[1] This work supported in part by the National Science Foundation.

of systematic error in recent determinations of g_c^2 using np scattering data, both from phase-shift analyses [2] and pole extrapolation [3], is due to the normalization uncertainty in the np differential cross section. In order to try to address these issues, we set about developing a tagged neutron facility with the goal of measuring the absolute np differential cross section at back angles to $\sim \pm 1\%$.

TAGGING NEUTRONS

The TNT (T-region Neutron Tagger) facility is located in the T-region of the Indiana Cooler. The neutrons are produced via the reaction $p + d \rightarrow n + 2p$ with a circulating proton beam of bombarding energy 200 MeV incident on a deuterium gas jet target (GJT). The presence of a 6° bending magnet in the ring allows for the centering of the secondary target at 14° exiting neutron angle. At small neutron angles the predominance of the $^1S_0$ final state interaction for the two protons results in a neutron beam of narrow energy spread ($\sim$10 MeV).

The "tagging" of the neutrons is accomplished by detection of the two recoil protons in a silicon detector array (the "tagger") located in vacuum. The detection of two protons in coincidence signals the production of a neutron, while energy and position measurements on the recoil protons allow for complete reconstruction of the four-momentum of the neutron. Knowledge of the neutron's four-momentum then makes it possible not only to determine whether the neutron is incident on the secondary target, but also where it is incident on the target and with what bombarding energy.

It is important to point out some aspects of this "beam" of tagged neutrons. First, not all neutrons that are incident on the secondary target are tagged. Simulations indicate that the tagging efficiency for neutrons incident on the secondary target will be $\sim$40%. However, by requiring that a tag be associated with all neutron scattering events, the untagged neutrons do not enter into the analysis. Second, not all neutrons that are tagged are incident on the secondary target. Good energy and position measurements on the recoil protons from the production reaction are thus required in order to accurately identify those neutrons that do impinge on the secondary target. We hope to achieve $\lesssim$100 keV neutron energy resolution (FWHM) and $\lesssim$1 mm position uncertainty for location on target. While the uncertainty in position has a lower limit arising from the granularity of the recoil detectors, the resolution achieved for both neutron impact position and energy ultimately depend on the energy resolution attained in the detection of the recoil protons. For a time-averaged primary luminosity of 10^{31} cm^{-2}s^{-1}, the expected tagged neutron rate is $\sim$2 kHz. The actual operating luminosity used will depend on the rate of false neutron tags arising from accidental coincidences.

The tagger consists of an array of four silicon double-sided strip detectors (DSSDs), each backed by a large area silicon pad detector. Each detector has an active area of approximately 6.4×6.4 cm^2 and thickness of 500 μm. The strips on the two sides of the DSSDs are orthogonal, and the readout pitch is $\simeq$0.5 mm,

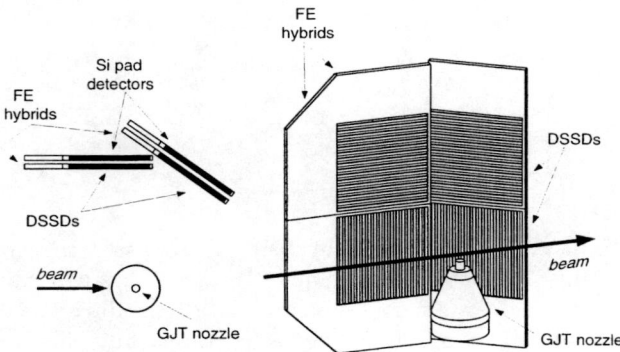

FIGURE 1. Top and side views showing the arrangement of the neutron tagging detectors with respect to the gas jet target (GJT).

yielding a total of 1024 channels (excluding the silicon pad detectors). The front-end electronics for the DSSDs are comprised of pairs of 32-channel application specific integrated circuits (ASICs) located in vacuum. The ASICs, the VA32_hdr2 and TA32C [4], provide pulse-height information for each channel along with a single fast logic output from each chip set for triggering and timing purposes. This allowance for self-triggering of the DSSD readout is a novel feature of the front-end electronics, and one essential to measuring the absolute tagged neutron flux on the secondary target. The trigger electronics are set up such that, in most cases, a neutron can be tagged even when both recoil protons impinge upon the same silicon detector. The silicon pad detectors located behind the DSSDs are needed in order to relieve neutron reconstruction ambiguities caused by recoil protons that have sufficient energy to penetrate through a DSSD. A schematic view of the arrangement of the tagging detectors is shown in Figure 1.

One issue with which we must contend is the fact that the gas jet target is an extended target. Because only one (x,y) position measurement is made on each recoil proton, we need to know the event origin in order to determine the proton angles and, therefore, the neutron angles. The event vertex is determined by calculating the magnitude of the outgoing neutron's momentum using both conservation of energy and conservation of momentum. The neutron's momentum using energy conservation is

$$p_{EC} = \sqrt{(E_i - E_{p1} - E_{p2})^2 - m_n^2} \qquad (1)$$

where E_i is the initial energy of the system (beam proton plus target deuteron), E_{p1} and E_{p2} are the energies of the two recoil protons, and the quantity is independent of the vertex position. In calculating the neutron's momentum using momentum conservation, we take the event origin to be along the central beam axis. The momentum is then only a function of z, the location along the beam direction,

$$\vec{p}_{MC}(z) = \vec{p}_i - \vec{p}_{p1}(z) - \vec{p}_{p2}(z). \qquad (2)$$

By forming the quantity,

$$\Delta p(z) \equiv p_{EC} - p_{MC}(z), \qquad (3)$$

we can determine the event origin by using a bisection method to find where $\Delta p(z) = 0$. Simulations show that the quantity $\Delta p(z)$ is single-valued, and the distribution of event origins from experimental data reproduce the expected product of the longitudinal density profile of the gas jet target and the tagging acceptance.

NEUTRON SCATTERING

For the np differential cross section measurement, the tagged neutrons will impinge upon a liquid hydrogen target of 4 cm thickness located approximately 1 m from the gas jet target and subtending a solid angle of $\approx$15 msr. The detection of neutron backscattering events will be achieved via the detection of the forward scattered proton in a detector array of large acceptance. This forward detector array, shown in Figure 2, consists of 9 planes of wire chambers for proton tracking, as well as a large area ΔE/start scintillator and 20 element scintillator hodoscope for triggering and energy measurements. Nearly complete azimuthal coverage is provided for neutron scattering with $\theta_{cm} \geq 90°$. The angular resolution for scattered protons will be about 0.5°.

As previously mentioned, the four-momentum reconstruction of the neutron will allow us to predict where it passes through the target. We can then do a traceback to the target volume for the forward scattered proton and determine the distance of closest approach of the two tracks. This comparison of the interaction position in the target from scattered proton tracking to that predicted from the neutron tag will provide a check on the accuracy of the four-momentum reconstruction of the tagged neutrons. Good agreement here is necessary in order to attain confidence in the neutron flux determination, since this will depend critically on our ability to determine whether the tagged neutrons which do not undergo scatterings in the secondary target actually passed through it.

Because the detectors comprising the scintillator hodoscope have reasonable neutron detection efficiency ($\approx$20%) in this energy range, we will also be able to collect some neutron scattering data via the detection of neutrons in coincidence with the scattered protons. This coincidence data can be obtained simultaneously with the primary scattering data and extends the angular coverage to $\theta_{cm} \gtrsim 60°$. Although this mid-angle data will be collected with poorer statistics, it will facilitate comparison of the measured angular distribution to the total np cross section.

The experiment has been designed with many built-in crosschecks aimed at controlling possible sources of systematic error. In addition to the primary event

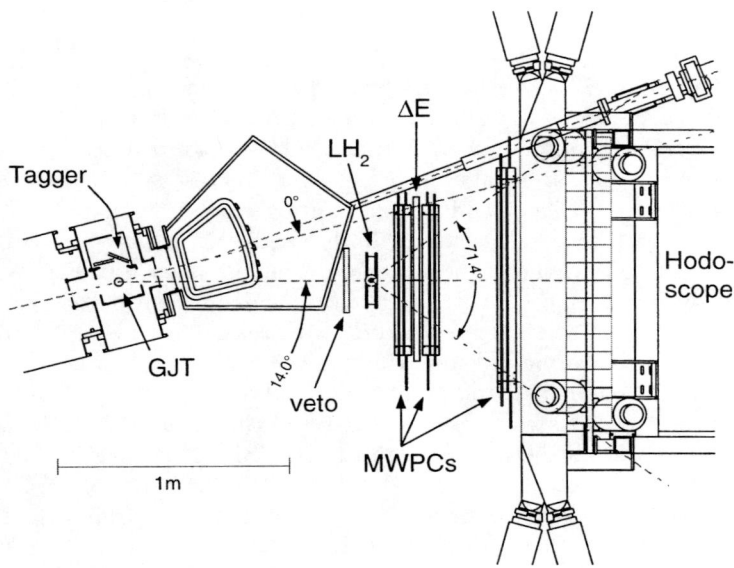

FIGURE 2. A top view of the experimental setup for the np scattering experiment. The circulating proton beam is bent by 6° (indicated by the upper right track) by a dipole magnet, allowing the secondary target and forward detector stack to be centered at 14° neutron emission angle.

streams that measure the neutron flux and np backscattering, data from pd elastic scattering are acquired simultaneously. The pd elastic scattering data will be used as a luminosity monitor to aid in normalizing empty target background runs to full-target runs. The forward going protons from pd scattering also provide a tagged proton beam with which to measure pp scattering in the LH_2 for the purpose of monitoring the product of the target thickness and solid angle of the forward detector array. Because tagged neutrons will also be incident on the scintillator hodoscope, the setup allows for a direct measurement of the hodoscope neutron detection efficiency. This efficiency data provides calibration information for the mid-angle coincidence events, while at the same time giving us an additional means for testing the accuracy of the tagged neutron reconstruction. A particularly powerful crosscheck will be made possible by the combination of the large area target together with a prediction for where each neutron passes through it. By subdividing the target into smaller target bins in the analysis, a comparison can be made of the cross section determinations for each bin. The requirement of agreement for these logically separate measurements will impose a stringent test on the accuracy of the neutron flux determination. Taken as a whole, these many crosschecks should make it possible to approach the goal of $\pm 1\%$ absolute accuracy.

STATUS

We have conducted a test of the neutron tagging method, where a two-dimensional position-sensitive silicon detector (2D-PSD) used in previous Cooler experiments took the place of the full tagger. The forward detector stack for the neutron scattering measurements was in place for this test. While the 2D-PSD had much lower tagging efficiency and poorer position resolution than the final tagger will have, this beam test was useful for both commissioning the forward detectors and demonstrating that the general tagging principle is sound. Figure 3 demonstrates that we were able both to predict the neutron position on target and to detect and trace back the scattered proton with reasonable precision in that first test. This test run also yielded important information for the design of the final tagger system. In particular, the points falling above the diagonal in Figure 3(a) were determined to come from cases where one of the recoil protons possessed sufficient energy to pass all the way through the 2D-PSD. This effect is clearly seen in the simulated events shown in Figure 3(b). It was the neutron reconstruction errors from these cases that led us to include silicon pad detectors behind the DSSDs, as described previously in the section on tagging neutrons.

After extensive development work on the interface electronics for the tagger readout, the tagger underwent its first beam test in July of 1999. The primary focus of this run was on setting up the trigger logic and timing for the various event streams. Data was collected for *pd* elastic scattering, where the deuteron is detected in the tagger array. Figure 4 shows the energy deposited in the tagger versus the position of the forward-going charged particle at the first wire chamber. The kinematic

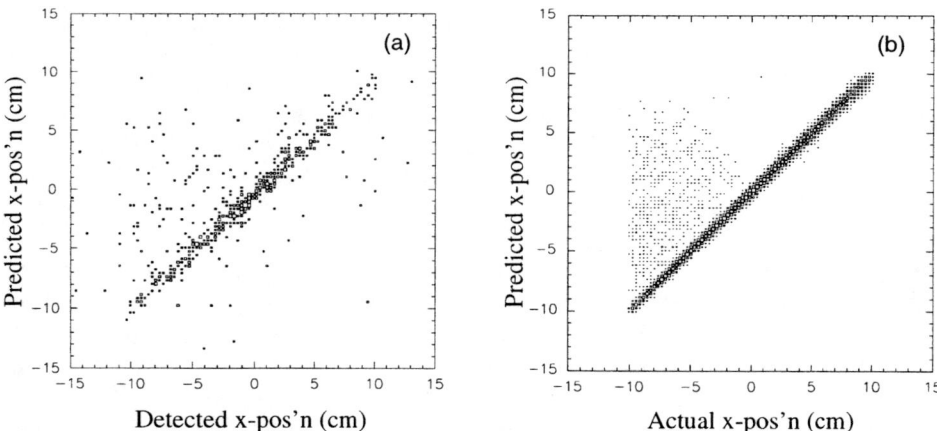

FIGURE 3. (a) Plot of predicted neutron position on the secondary target vs. the target position determined from the scattered proton traceback. (b) Results of a simulation showing the predicted neutron position on the secondary target vs. the actual position on target.

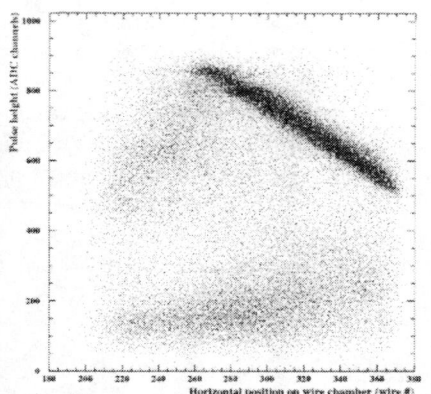 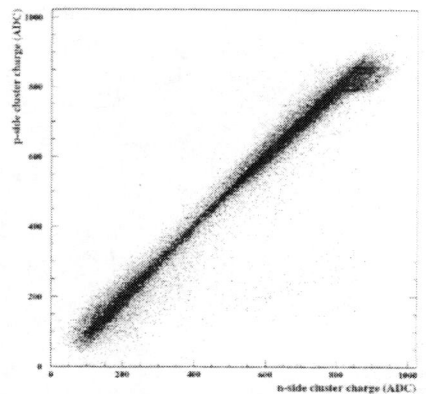

FIGURE 4. Plot of the pulse height in a silicon detector versus the position of the forward-going particle in the first wire chamber. The dark band corresponds to pd elastic scattering events where the deuteron is incident on the silicon detector array.

FIGURE 5. Plot of the charge correlation for the two sides of a silicon detector for the events shown in Figure 4.

correlation from the pd elastic events can be clearly distinguished. An important issue in the tagged neutron reconstruction will be the ability to correctly assign the (x, y) coordinates to the recoil protons in cases where they impinge on the same silicon detector. This task depends on the energy resolution and gain matching of the readout of the two sides of the detector. Figure 5 shows the charge correlation between the two sides of a detector for the events appearing in Figure 4. Given the improvement in noise reduction subsequent to the first beam test, we are confident that we should be able reliably reconstruct the vast majority of events with two particles in a single tagging detector.

The liquid hydrogen target for use in the np scattering experiment is currently under construction. Additional tagging tests and initial np scattering measurements will be conducted using a plastic scintillator target. The primary np scattering data taking with the LH$_2$ target is scheduled to take place in 2000.

FUTURE POSSIBILITIES

Other possible experiments using the tagged neutron facility are under consideration. The method should be extendable to other neutron energies in a straightforward manner, although further study is necessary to determine how the tagged neutron flux varies with incident proton beam energy. Depending on the results of the initial np scattering measurement, further np measurements at other energies may be performed. By simply replacing the hydrogen in the LH$_2$ target with deuterium, an absolute measurement of the nd elastic scattering cross section could be made. Such an experiment might shed light on the nature of the three-nucleon force, which has been predicted [5] to play an important role in the region of the angular distribution minimum.

It should also be possible to create a beam of polarized tagged neutrons, taking advantage of the sizable spin transfer coefficient, D_{SS}, in the neutron production reaction. One could then, for example, measure the cross section and analyzing power in the reaction $\vec{n} + p \to 2p + \pi^-$ near threshold. In this case it is the event-by-event knowledge of the neutron four-momentum that facilitates a near threshold measurement not possible with traditional neutron sources due to the rapid cross section variation with energy.

CONCLUSION

The TNT facility currently under development will be a unique facility for medium energy neutron scattering experiments. The method used to tag neutrons capitalizes on the low energy recoil detection made possible by the storage ring environment. The ability to produce a beam of tagged neutrons opens up a number of interesting experimental possibilities. In particular, knowledge of the neutron flux will make possible an absolute np differential cross section measurement.

ACKNOWLEDGMENTS

I wish to thank the members of the TNT collaboration for their contributions to the work presented here. The TNT collaboration, in its current guise, consists of L.C. Bland, J. Doskow, W.W. Jacobs, T. Kinashi, A. Klyachko, T. Rinckel, E.J. Stephenson, S.E. Vigdor, S.W. Wissink, and Y. Zhou at IUCF; J. Rapaport from Ohio University; and B. Bergenwall, J. Blomgren, C. Johansson, J. Klug, P. Nadel-Turonski, L. Nilsson, and N. Olsson from Uppsala University. I am also indebted to many members of the IUCF technical staff, especially D. Bilodeau and W. Hunt, for their help in building TNT.

REFERENCES

1. V.G.J. Stoks, R.A.M. Klomp, M.C.M. Rentmeester, and J.J. de Swart, Phys. Rev. C **48**, 792 (1993).
2. D.V. Bugg and R. Machleidt, Phys. Rev. C **52**, 1203 (1995).
3. J. Rahm *et al.*, Phys. Rev. C **57**, 1077 (1998).
4. Manufactured by Integrated Detectors & Electronics AS, Veritasveien 9, N-1322 Hovik, Norway.
5. H. Witala *et al.*, Phys. Rev. Lett. **81** (1998) 1183.

Stochastic Cooling and Extraction at COSY

R. Maier, U. Bechstedt, J. Dietrich, K. Henn, A. Lehrach[1], B. Lorentz, D. Prasuhn, A. Schnase, H. Schneider, R. Stassen, H. Stockhorst, R. Tölle

Forschungszentrum Jülich GmbH, Institut für Kernphysik, D-52425 Jülich, Germany
[1] *Brookhaven National Laboratory, P.O. Box 5000, Upton, NY 11973, USA*

Abstract. COSY is a cooler synchrotron and storage ring delivering high precision proton beams with momenta between 300 MeV/c and 3400 MeV/c for experiments in medium energy physics. Two cooling systems are used. An electron cooler that reaches up to a momentum of 600 MeV/c and a stochastic cooling system that covers the upper momentum range from 1500 to 3400 MeV/c. Horizontal and vertical as well as longitudinal stochastic cooling by the filter method are successfully applied for internal experiments with cluster targets. Third order resonant extraction and stochastic extraction are used to serve external experiments. A nearly ideal rectangular spill of 5 to more than 50 s has been achieved with the stochastic feeding system. In addition, a polarized beam was successfully extracted.

INTRODUCTION

The cooler synchrotron COSY Jülich [1] delivers up to $3 \cdot 10^{10}$ protons in the momentum range from 300 MeV/c to 3.4 GeV/c. The machine is equipped with an electron cooler operating up to 600 MeV/c. Stochastic cooling [2] enhances the beam quality in the range from 1.5 to 3.4 GeV/c. Stochastic extraction [2] serves the external users at different momenta with the desired spill lengths in the order of minutes. Polarized protons are available up to 3.3 GeV/c. A special tune jump system has been incorporated to overcome depolarizing resonances. At present four internal experiments are in operation and the extracted beam is fed to three external experiment devices. The scheduled beam time for COSY in the year 1998 has amounted 7172 hours. With an up-time of 6766 hours the accelerator complex has proven an extremely good reliability of nearly 94%.

STOCHASTIC COOLING

Stochastic cooling in COSY [2] is used routinely for internal target experiments with thin cluster targets and long flat top times, allowing equilibrium experiments between target heating and stochastic cooling. The profit of the stochastic cooling for the COSY-11 experiment is evident in figure 1. When the stochastic cooling in the longitudinal and transverse planes is switched off the counting rate decreases rather rapidly due to the emittance growth of the beam and the smaller overlap between beam and target. After switching on the cooling amplifiers the emittance decreases again and the counting rate increases and stays constant for one hour flat top time. The flat top time could be even increased to two hours with constant luminosity in the experiment.

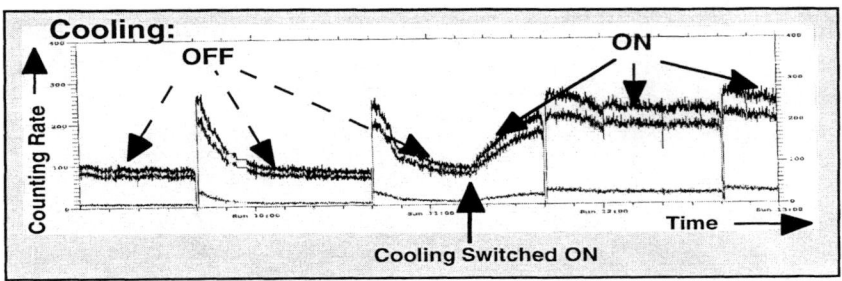

FIGURE 1. Experiment counting rates with stochastic cooling at 3.3 GeV/c OFF and ON.

In the first longitudinal cooling experiments the notch filter was built out of solid air-filled coaxial lines leading to frequency dispersion and small notch depths. In a further approach the solid notch filter was replaced by an optical fibre notch filter. Due to the large operating frequency of the optical system the dispersion can be neglected. As a result, the notch depth could be increased by 10 dB. In a next step a variable delayed notch filter was installed and tested. The optical fibre is supplemented by two prisms whose distance can be altered by a linear motor. First promising tests were already carried out. The system is intended for future experiments where cooling within a supercycle with different flat top momenta is necessary.

STOCHASTIC EXTRACTION AND POLARIZED BEAMS

Stochastic extraction [2] is now used in the wide momentum range from 800 MeV/c to 3.4 GeV/c. A digital noise generator developed at COSY is used for stochastic extraction and beam shaping. An extraction efficiency of about 80% is realized with a spill duration up to several minutes. In a recent experiment a polarized proton beam at momentum 0.8 GeV/c was extracted over a period of 20 s with $2 \cdot 10^5$ protons per sec for the first time at COSY (figure 2). An extraction efficiency of 20% was received. A possible explanation of the decreasing polarization is the fact that during extraction depolarizing resonances are crossed which are close to the extraction resonance. Further studies will follow. Preceding machine studies were carried out in developing accelerator schemes to provide a stored polarized proton beam up to the maximum momentum 3.4 GeV/c. Figure 3 shows the successful acceleration of a polarized beam to 3.3 GeV/c. Correction dipoles have been used to excite total spin flips. The solenoids of the electron cooler acting as partial snake have been used to conserve polarization at the $\gamma G = 2$ resonance. With superperiodicity $P = 6$ only one resonance ($\gamma G = 8 - Q_y$) is excited which flips the spin with polarization losses. This diminution could be avoided by applying a tune jumping system.

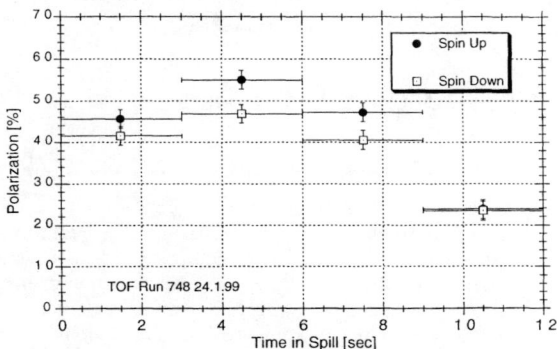

FIGURE 2. Polarization of extracted protons at 800 MeV/c within the first 12 sec of the 20 sec-spill.

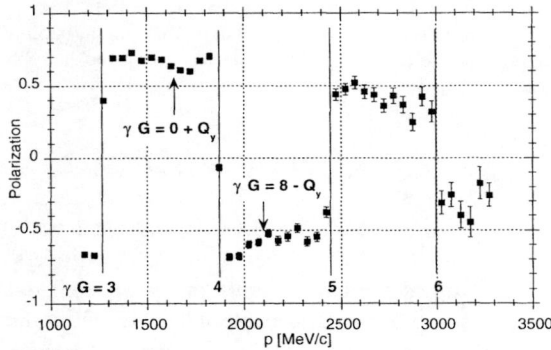

FIGURE 3. Acceleration of a polarized proton beam up to 3.3 GeV/c. The vertical lines indicate the imperfection resonances $\gamma G = 3, 4, 5, 6$. Intrinsic resonances are marked by arrows.

During acceleration to maximum energy the transition energy is shifted upwards with the horizontally focusing quadrupoles in the arcs leading to a reduced superperiodicity. Additional intrinsic resonances are excited (e.g. $\gamma G = 0 + Q_y$) which are suppressed by tuning the vertically focusing quadrupoles in the arcs.

REFERENCES

1. Maier, R., *Nucl. Instr. and Meth.* **A 390**, 1-8 (1997)

2. Stockhorst, H., et al., 'The Performance of COSY', Proc. 1998 Particle Accelerator Conf., 1998, Stockholm, 22.-26.6.98, Sweden, pp. 1595-1597, and references therein.

Status of the ESR and Prospects for Radioactive Ion Beams

B. Franzke, K. Beckert, L. Groening, F. Nolden, M. Steck

Gesellschaft für Schwerionenforschung mbH, Darmstadt, Germany

Abstract. This contribution informs about the status of the Experimental Storage Ring ESR at GSI after nearly 10 years of experiments with electron-cooled primary and secondary heavy ion beams and discusses possible upgrades for the near future. The major fraction of the ESR-beam time was dedicated to atomic spectroscopy experiments with fully stripped or few electron heavy ions up to uranium. Remarkable nuclear physics results came, so far, mainly from the precise Schottky mass spectrometry with cooled radioactive fragments. The experimental program of a future – very likely modified – ESR aims at nuclear physics experiments with secondary, neutron rich or neutron deficient nuclides with life-times in the order of 1 s. Luminosities of up to $1 \times 10^{29}\,\mathrm{cm}^{-2}\,\mathrm{s}^{-1}$ for collisions with protons of an internal H_2-target and up to $1 \times 10^{28}\,\mathrm{cm}^{-2}\,\mathrm{s}^{-1}$ for electron-nucleus interactions in the collider mode will be required. These values might be in reach by improving both the collection efficiency and the cooling speed for the secondary nuclei and by further increasing the primary beam intensity at the heavy ion accelerator complex UNILAC/SIS [1].

STATUS OF THE ESR

In nearly ten years of operation the ESR [4] was used mainly for atomic physics experiments with electron cooled, stored beams of fully stripped or few electron heavy ions up to uranium. Examples are experiments on inner shell excitation in coincidence with radiative and non-radiative electron capture [3], di-electronic recombination of Li-like Au- and U-ions with cooler electrons [4] and hyperfine transitions in H-like $^{209}Bi^{82+}$-ions by means of resonance fluorescence from collinear laser light [5].

There are also a few remarkable nuclear physics experiments making use of specific possibilities offered by the synchrotron/storage ring facility SIS/ESR in combination with the production and separation of projectile fragments at the FRS-facility [6]. The first experimental proof of the so-called bound-β-decay of fully stripped ^{163}Dy- and ^{187}Re-ions to H-like ^{163}Ho- and ^{187}Os-ions [7,8], respectively, may be mentioned in this context. More than 150 experimental mass values for radioactive nuclides could be measured with essentially improved accuracy by means of the novel method of Schottky mass spectrometry [9,10]. In the isochronous mode of operation the ring is also being used as a multi-turn time-of-flight (TOF) mass spectrometer. Masses of nuclides with half-lives as low as a few microseconds can be determined with rather good accuracy this way [11,12].

Mainly because of low luminosity, elastic and inelastic nuclear scattering experiments with radioactive nuclei in inverse kinematics, i.e. using target recoil analysis, were not feasible in the past. Meanwhile, both the primary beam intensity from SIS and the thickness of the internal H_2-jet [13] have been increased essentially. Internal target experiments with abundant, projectile near nuclear fragments, for instance $^{56}Ni^{28+}$ produced from primary $^{56}Ni^{28+}$, could presently be done at luminosities $\geq 1 \times 10^{27} \, cm^{-2} \, s^{-1}$. Before discussing further developments considered for the next decade we give a short overview on the present status of ESR-operation.

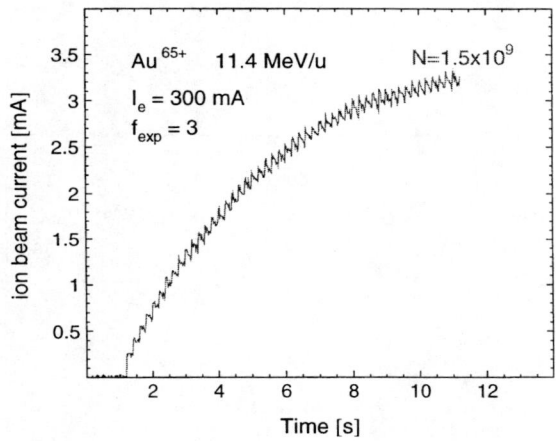

FIGURE 1. Fast accumulation of an Au^{65+}-beam at 11.4 MeV/u in the SIS by means of multiple multi-turn injection supported by electron cooling. The electron current was 500 mA. The quantity f_{exp} is the electron beam expansion factor, given by the magnetic field strength at the electron gun divided by that in the cooler solenoid. The ion beam current is measured with a DC current transformer.

Primary Beam Intensities

During the past two years the number of stored high-Z ions (Au^{79+} to U^{92+}) per ESR-injection has been increased by nearly a factor of 100 to a few times 10^8 by means of several improvements

- fast beam accumulation at the heavy ion synchrotron SIS by multiple multiturn injection supported by electron cooling [14] (see fig. 1),
- strongly reduced transverse emittances (also because of electron cooling in SIS) of the fast extracted beam for the ESR (see fig. 2),
- slightly improved transverse acceptance of the injection channel and better ion optical matching to the injection orbit of the ESR,

- single bunch transfer to the ESR after rebunching of 4 SIS-bunches to a single bunch (see fig. 3).

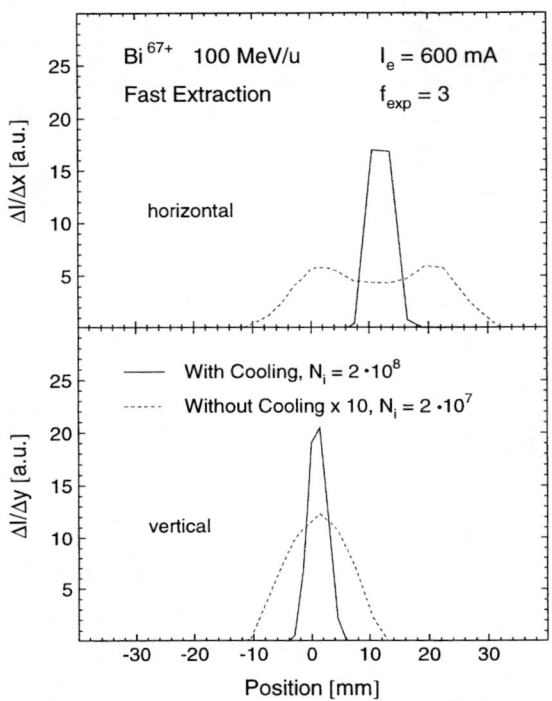

FIGURE 2. Transverse intensity distributions of a Bi^{67+}-beam – measured by means of beam profile grids ("harps") – in the transfer line to the ESR. The wide (dashed) profiles are measured without electron cooling in SIS, the narrow (solid) ones with cooling at injection energy.

Beam currents up to the space charge limit for all ion species will be available at the SIS after the installation of a new high current pre-stripper section HSI [15] at the UNILAC, which serves as SIS-injector (see table 1). The new linac section is being commissioned and shall be in full operation by the end of 1999.

The number of ions, N_i, stored in the ESR after single bunch transfer depends on the stripping yield for the desired charge state at the corresponding ion energy and on the transverse emittances of the SIS beam (see fig. 2). If electron cooling had be applied in the SIS more than 80% of the ions – multiplied by the stripping efficiency – can be transferred to injection orbit of the storage ring.

TABLE 1. Number of primary nuclei per SIS cycle for a few representative ion species. For specific energies up to 400 MeV/u – as used for experiments at the ESR – the SIS cycling period in the case of fast beam ejection is approximately 3 s.

Nucleus	without EC in SIS (1997)	with EC in SIS (1998)	with new HCI* (2000)
$^{20}Ne^{10+}$	5×10^{10}	1×10^{11}	1×10^{11}
$^{40}Ar^{18+}$	1×10^{10}	5×10^{10}	5×10^{10}
$^{58}Ni^{26+}$	1×10^{9}	2×10^{10}	5×10^{10}
$^{84}Kr^{34+}$	1×10^{9}	1×10^{10}	4×10^{10}
$^{132}Xe^{48+}$	5×10^{8}	5×10^{9}	3×10^{10}
$^{197}Au^{65+}$	5×10^{7}	5×10^{8}	2×10^{10}
$^{238}U^{73+}$	5×10^{7}	5×10^{8}	2×10^{10}

* incoherent space charge limit at SIS

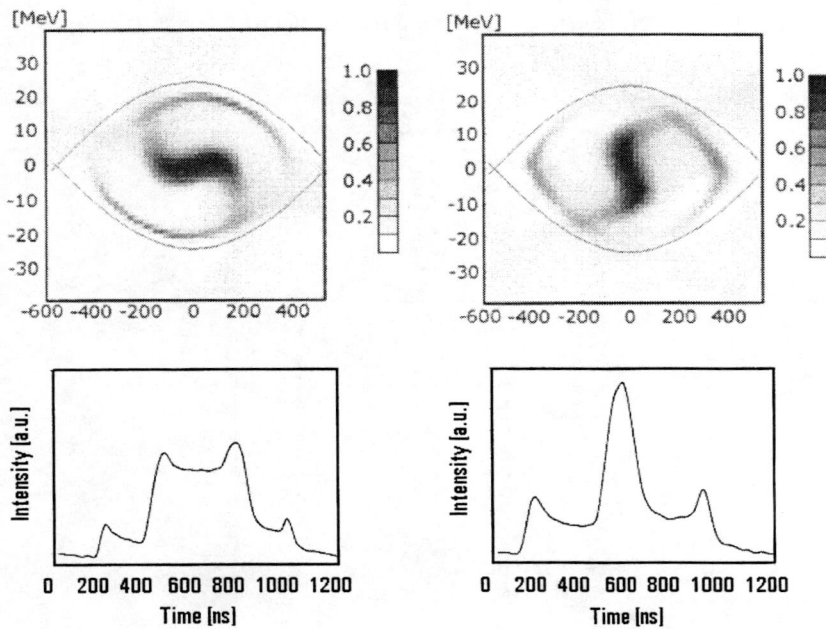

FIGURE 3. Distribution of beam particles in the longitudinal phase plane after bunching of a coasting beam into a 1st harmonic rf-bucket on the flat top of the SIS-cycle. The histograms (upper diagrams) are calculated from consecutive bunch probe signals (lower diagrams) recorded at times differing by a quarter of a synchrotron period T_s [16]. The beam bunch is transferred to the ESR, when the bunch signal is peaked optimally (right). With an improved bunch compression scheme approximately 90% of the beam particles shall be compressed into a bunch length of 50 ns [17].

Electron Cooling Results

The practical upper limit for N_i is determined by the beam quality required for internal experiments in compared to the beam parameters in the equilibrium between electron cooling (EC) on one hand, and intra-beam scattering (IBS) and angular scattering in the internal gas jet target on the other hand. The results of systematic measurements for many ion species in the equilibrium between EC and IBS [18] are plotted in fig 4 versus N_i. The relative spread in the longitudinal momentum, $\delta p/p$, is derived from the frequency spread in the Schottky spectrum of the cooled beam. The transverse emittances, ε_x and ε_y, either by evaluating beam size determination by beam scraping or beam profiles from position sensitive residual gas ion detectors. The general tendency of the results is that ε_x and ε_y grow approximately $\propto N_i^{0.5}$ and $\delta p/p \propto N_i^{0.3}$ Reasonably small emittances of $1\,\pi$ mm mrad and $\delta p/p$ of $\leq 10^{-3}$ are achieved with N_i not grater than a few times 10^9. The dependence of the equilibrium beam parameters on the ion species or charge state is relatively weak as well as that on the electron density, n_e, in the 2 m long electron cooler (see fig. 4).

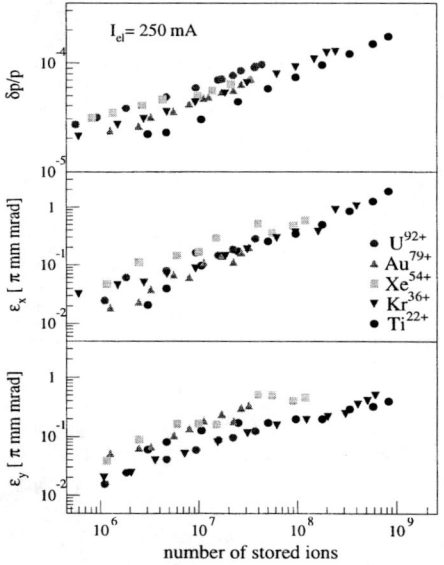

FIGURE 4. Equilibrium emittances and momentum spread of electron cooled beams vs. number of stored ions [18]. The electron current of 250 mA corresponds to an electron density of 5×10^6 cm^{-3}. The effective length of the cooler section (1.8 m) covers about 1.7% of the ESR orbit circumference.

For ion species with nuclear charge $Z \geq 18$ there is obviously a critical N_i, where beam emittances and momentum spread drop immediately by approximately an order of magnitude to their ultimate minimum, e.g. $\delta p/p \leq 10^{-6}$. At $N_i < 2\text{-}4 \times 10^3$ the average distance between beam ions is so large (a few cm) that the IBS rates become extremely small. On the other hand, the electron cooling rate is practically independent of N_i. The observed behavior is similar to the transition from the gaseous to the liquid phase [19]. The true momentum spread might be even much smaller than deduced from the measured frequency spread of the Schottky band. The limited stability of power supplies for the ring magnets and high voltage supplies of the electron cooler might finally cause the measured spread in $\delta f/f$ of a few times 10^{-7}.

Status of Operation

The ring operation had to be adapted to the requirements from different experiments. In general injection and accumulation of beams take place at energies between 250 MeV/u and 400 MeV/u, where the fraction of fully stripped or H-like ions for all species is high enough. At present, the maximum acceleration voltage of 250°kV at the electron cooler determines the upper limit of the injection energy, which is 20% below the design limit. A few examples of recent achievements in ring operation are given below.

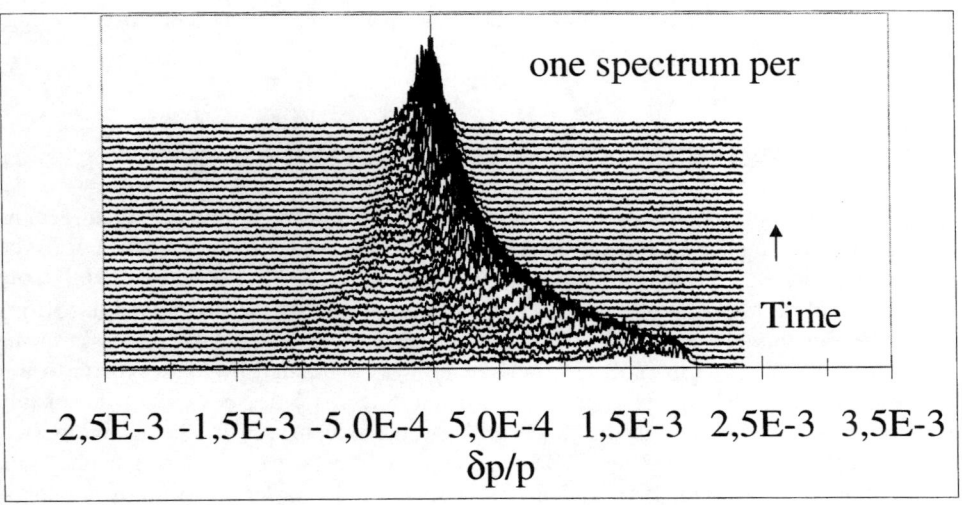

FIGURE 5. First test of momentum cooling of an Ar^{18+}-beam at 390 MeV/u with the stochastic pre-cooling device at the ESR.

- Deceleration: Several atomic spectroscopy experiments require deceleration to energies far below 100°MeV/u. For that purpose we apply simultaneously rf-amplitudes and electron cooling to achieve small longitudinal emittance and, at the same time, minimum transverse emittances, i.e. optimal conditions for the deceleration process. Presently beams can be decelerated down to 30 MeV/u with only 20% beam loss and to 11.5 MeV/u with approximately 80% loss. The loss occurs mainly in the very last phase of the deceleration ramp. It is assumed that better compensation for eddy current fields might help to reduce the effect.

- Internal target operation: The internal target, a supersonic gas jet in a 4-stage differential pumping and skimming system, can be operated with all gases from O_2 to Xe in the normal gas jet mode with thickness values $1-5 \times 10^{13}$ atoms/cm^2. A similar thickness was achieved also with H_2 in the so-called cluster mode, where the gas under 20 bar pressure is cooled to liquid nitrogen temperature before entering the Laval nozzle. With 1×10^9 stored and electron cooled ions circulating at 1.6 MHz (300 MeV/u) we can offer a luminosity of nearly 1×10^{29} cm^{-2} s^{-1} for internal target experiments.

- Stochastic pre-cooling: A stochastic pre-cooling device for all phase planes has been recently commissioned. The result of a first test of momentum cooling is illustrated in fig. 5. The operation of the cooling system has to be optimized still. It is foreseen as the first cooling stage for "hot" radioactive beams, before final electron cooling is applied.

PROSPECTS FOR RADIOACTIVE BEAMS

At this time, there is only poor experience at the ESR with the injection and accumulation of nuclear fragment beams from the magnetic fragment separator FRS. A few years ago, a secondary ^{56}Ni^{28}-beam was injected to the ESR at 250 MeV/u and electron cooled. After accumulation of approximately 1×10^5 secondary nuclei the beam was used for an internal H_2-jet target experiment investigating inelastic scattering of protons in inverse kinematics. At that time the collection efficiency was estimated by FRS-experts to be in the order of 1%. A few years later suitable beam diagnostic devices have been installed between FRS and ESR and in the ring itself, and the structure of the beam line between FRS and ESR was improved in 1998 for somewhat larger acceptance and for better ion optical matching between FRS and ESR. Despite of the described improvements, the collection efficiency for projectile near fragments is still estimated to be only about 5%. This low value is mainly determined by the limited acceptance of the ESR injection channel, which is only a relatively small part of the total acceptance of the multi-purpose storage ring. Essentially improved efficiency figures for production, collection, cooling and accumulation of fragment beams -- as considered for a future upgrade for the nuclear fragment beam facility at GSI -- can be attained only by a complete re-design of the existing facility. This includes all relevant subsystems between the ion optics for the production target and the experimental storage ring. A possible scenario is discussed in the following.

Objectives of an Upgrade for Fragment Beams

Main objectives of future improvements of the facility for experiments with cooled beams of nuclear fragments are

- high luminosity of up to 1×10^{30} cm^{-2} s^{-1} for internal target experiments with approximately 1×10^8 stored exotic nuclei and protons or heavier target nuclei in internal jet or fiber targets,
- high luminosity of up to 1×10^{28} cm^{-2} s^{-1} in collisions between stored electrons and 1×10^8 stored exotic nuclei for the determination of electrical charge distributions in or charge radii of exotic nuclei,
- high luminosity in collinear laser experiments aiming at measurements of nuclear magnetic moments of exotic nuclei,
- experimental access to short-lived nuclei (≥ 0.1 s), i.e. to nuclei as far as possible away from stability,
- improved conditions for other high resolution experiments, e.g. measurements of nuclear masses and life times, as already done at ESR by applying the Schottky or TOF mass spectrometry.

Prior to any consideration about a further increase of the primary beam intensity the envisaged upgrade program aims at highest efficiency in the production, accumulation and cooling of secondary nuclei. For that we have to meet the following requirements

High production energy: The first (basic) requirement to achieve optimal efficiency in the (useful) production, accumulation and cooling of the secondary nuclear fragment beams is a high energy for primary ions. It is necessary to get sufficiently high yields of fully stripped nuclei for all ion species and to make use of the kinematic focusing of the secondary beams after the production target. The maximum specific energies of SIS-beams ≥ 1000 MeV/u for the heaviest ions are obviously not far from being optimal.

Thick production target: In order to achieve a high conversion efficiency is necessary to apply thick production targets. At the FRS the typical thickness of the target is a few g/cm^2. The mean energy loss of the ions is between 10% and 20%.

Optimal target optics: In order to minimize the emittance growth in the transverse planes due to the production dynamics the beam optics at the target has to provide small transverse (sub-mm) beam size and vanishing dispersion function at the target.

Short beam bunch: In order to minimize the emittance in the longitudinal phase plane the beam bunch at the target has to be as short as possible. A bunch length of approximately 50 ns that is envisaged by the SIS bunch compression project would be quite sufficient.

Dedicated collector/cooler ring CR

The highest efficiency for accumulation and cooling of secondary nuclear fragment beams can be attained by separating the major tasks, namely the efficient collection and fast stochastic pre-cooling of the fragments in the first step, and the accumulation, electron cooling and subsequent experiments in the second step. A dedicated collector ring CR collector and pre-cooler ring would have the following characteristics:

- large transverse and momentum acceptance,

- beam injection using the full ring acceptance, i.e. applying full aperture kickers,

- single bunch injection followed by bunch rotation and adiabatic de-bunching in order to reduce the beam momentum spread before starting with fast stochastic cooling as practiced in the former CERN AC-ring [18],

- flexible ring optics, optimized for injection and fast stochastic cooling, but also applicable for mass spectrometry in the isochronous mode,

- fast stochastic cooling in all phase planes with an overall cooling time of about 100 ms for a beam of 1×10^8 reference nuclei ($^{238}U^{92+}$) 750 MeV/u,

- final beam bunching to h=1 for the transfer to the experimental storage ring.

TABLE 2. Some preliminary basic parameters of the CR (see also [21]).

circumference of mean orbit	≈160 m
maximum magnetic bending power, $B\times\rho$	113 T m
reference nucleus	$^{238}U^{92+}$
maximum specific energy for reference nucleus	750 MeV/u
useful momentum spread acceptance, $\Delta p/p_{max}$	±2.5%
useful transverse acceptances, A_x, A_y	250 π mm mrad
injection, ejection	full aperture kickers
rf harmonic, number of bunches	1-2
bunch rotation rf, voltage	1.5-3 MHz, 1-3 MV
beam cooling, cooling time	stochastic, 0.1 s
final momentum spread	±0.1%
final transverse beam emittances	5 π mm mrad
ultra-high vacuum (UHV) pressure	$\leq 10^{-10}$ mbar

Additional support to the stochastic cooling might be achieved by a dynamic variation of the frequency dispersion $\eta=(df/f)/(dp/p)$ during cooling process. But this has to be investigated carefully. Preliminary major parameters for the CR are listed in table 3. Design considerations related mainly to fast stochastic cooling are discussed in a separate contribution [21].

Modification of the ESR

The properties of an optimal storage ring for beam accumulation and experiments with electron cooled beams are very similar to that of the ESR in its present shape. However, the injection and accumulation procedure could be simplified, if the nuclear fragment beams are pre-cooled in a separate ring to qualities of primary beams. Beam loss during injection and accumulation procedures are avoided this way. If only a single beam bunch is injected, one could consider alternative accumulation methods besides the rf-stacking being practiced at present. For the so-called off-phase injection at the first harmonic (h = 1) the bunch is injected at 180^0 with respect to the center of the rf-bucket and then moved into the bucket by electron cooling forces. The big advantage of this kind of accumulation is that the beam orbit stays constant all the time. Hence, there is a good chance to achieve an optimal duty factor for experiments with cooled circulating fragment beams. A few examples of the available intensity – after all the improvements described above – are given in table 3.

TABLE 3. Particles per synchrotron cycle for several secondary reference nuclei in front of a new fragment separator. The numbers in the last column ("long term plan") could be attained with a new 100 Tm synchrotron using the SIS as booster.

Nucleus	Life time	without EC in SIS, 1997	with EC in SIS 1998	with new HCI 2000	long term plan 200X
$^{56}Ni^{28+}$	6 d	1×10^6	2×10^6	1×10^7	1×10^9
$^{66}Ni^{28+}$	55 h	5×10^3	5×10^4	4×10^6	4×10^8
$^{73}Ni^{28+}$	0.8 s	5×10^2	5×10^3	4×10^5	4×10^7
$^{104}Sn^{28+}$	20 s	5×10^2	5×10^3	7×10^4	7×10^6
$^{128}Sn^{28+}$	1 h s	3×10^4	3×10^5	2×10^7	2×10^9
$^{132}Sn^{28+}$	40 s	2×10^3	2×10^4	2×10^6	2×10^8

transmission of existing FRS:30% -40% transmission of new FRS: 50% - 80% (included fission products)
present transmission FRS-ESR ≈ 30% transmission new FRS - new CR: 100%

The re-design of the ESR lattice would aim at more space for experiments without giving up important quality figures as, for instance, comfortable transverse momentum acceptances, flexible ion optics (isochronous mode), energy variation (deceleration), and optimal electron cooling. Besides the necessary increase of the ring circumference from 110 m to approximately 160 m the following changes are under consideration:

- increase of the maximum magnetic bending power from 10 Tm to 13 Tm (same as for CR) by replacing the bending magnets,
- modification of the lattice structure with a special straight section with suitable insertions for electron-nucleus-collision experiments being under discussion,
- more space for experimental equipment around the internal gas jet target (e.g. for spectrometers, projectile recoil detectors, magnetic orbit chicanes etc.) and additional shorter straight sections for other experimental setups as, for instance, a separate electron target and special equipment for collinear laser experiments.

CONCLUSIONS

Prior to a further increase of primary beam intensity at the production target to values between 10^{11} and 10^{12} ions per beam bunch one should enhance the efficiency of production, storage, fast cooling, accumulation of exotic beams.

Access to short-lived nuclei and efficient collection of nuclear fragment beams require a dedicated large acceptance collector/cooler ring, where cooling times of 100 ms might be feasible. Mass measurements with short-lived nuclei could be done also in the CR by operating it at the transition point, i.e. in the isochronous mode.

In order to establish much better conditions for future nuclear physics experiments with cooled fragment beams [22], the bending power of the ESR should be raised moderately to about 13 Tm. At the same time the ring size could be increased for several reasons. Space for additional experimental equipment around the internal target, e.g. spectrometers, magnetic orbit chicanes and projectile recoil detectors will be certainly claimed for. A special long straight section with all necessary devices would be needed if electron-nucleus-collision experiments should be planned. More shorter straight sections, e.g. for a new electron target and for additional laser spectroscopy devices, would be desirable in addition.

An important step to reach much higher primary and secondary beam intensities has been done recently by the construction of the new high current injector HSI at the UNILAC. Up to 4×10^{10} uranium ions and 2×10^{11} neon ions per cycle shall be attained at the SIS within a few years. This powerful injector linac will play also an important role for future extensions of the GSI accelerator and fragment beam facilities.

REFERENCES

1. N. Angert, "Status and Development of the GSI Accelerator Facilities", *Proc. of 5th European Particle Accelerator Conference, Sitges 1996*, edited by S. Myers et al., IOP Publishing, Bristol and Philadelphia, 1996, pp. 125-129

2. B. Franzke, *Nucl. Instr. and Meth. in Phys. Res.* **B24/25**, 18 (1987)

3. H. F. Beyer et al., *Phys. Lett. A* **184**, 435 (1994)
   and Th. Stöhlker et al., *Phys. Rev Lett.* **71**, 2184 (1993)

4. C. Brandau, F. Bosch, G. Dunn, B. Franzke, A. Hofknecht, C. Kozhuharov, P. Mokler, A. Müller, F. Nolden, S. Schippers, Z. Stachura, M. Steck, T. Stöhlker, T. Winkler, A. Wolf, *Hyperfine Interactions* **114**, 45 (1998)

5. I. Klaft, S. Borneis, T. Engel, B. Fricke, R. Grieser, G. Huber, T. Kühl, D. Marx, R. Neumann, S. Schröder, P. Seelig, L. Völker, *Phys. Rev Lett.* **73**, 2425 (1994)

6. H. Geissel et al., *Nucl. Instr. and Meth. in Phys. Res.* **B70**, 286 (1992)

7. M. Jung, F. Bosch, K. Beckert, H. Eickhoff, H. Folger, B. Franzke, A. Gruber, P. Kienle, O. Klepper, W. Koenig, R. Mann, R. Moshammer, F. Nolden, U. Schaaf, G. Soff, P. Spädtke, M. Steck, T. Stöhlker, K. Sümmerer, *Phys. Rev Lett.* **69**, 2164 (1992)

8. F. Bosch, T. Faestermann, J. Friese, F. Heine, P. Kienle, E. Wefers, K. Zeitelhack, K. Beckert, B. Franzke, O. Klepper, C. Kozhuharov, G. Menzel, R. Moshammer, F. Nolden, H. Reich, B. Schlitt, M. Steck, T. Stöhlker, T. Winkler, T. Takahashi, *Phys. Rev Lett.* **77**, 5190 (1996)

9. B. Franzke K. Beckert, H. Eickhoff, F. Nolden, H. Reich, U. Schaaf, B. Schlitt, A. Schwinn, M. Steck, T. Winkler *Physica Scripta* **T59**, 176 (1995)

10. T. Radon, T. Kerscher, B. Schlitt, K. Beckert, T. Beha, F. Bosch, H. Eickhoff, B. Franzke, Y. Fujita, H. Geissel, M. Hausmann, H. Irnich, H.C. Jung, O. Klepper, H.-J. Kluge, C. Kozhuharov, G. Kraus, K.E.G. Löbner, G. Münzenberg, Yu. Novikov, F. Nickel, F. Nolden, Z. Patyk, H. Reich, C. Scheidenberger, W. Schwab, M. Steck, K. Sümmerer, H. Wollnik, *Phys. Rev. Letters* **78**, 4701 (1997), and C. Scheidenberger et al., invited talk at this conference

11. M. Hausmann et al., accepted for publication in Nucl. Instr. and Methods A (1999)

12. T. Radon, invited talk at this conference

13. A. Kraemer, Thesis, University Frankfurt/M., in preparation, to be published in *Nucl. Instr. and Meth. A*, (1999)

14. M. Steck, K. Blasche, H. Eickhoff, B. Franczak, B. Franzke, L. Groening, T. Winkler, "Commissioning of the Electron Cooling Device in the Heavy Ion Synchrotron SIS", *Proc. of 6th European Particle Accelerator Conference, Stockholm, 22-26 June 1998*, edited by S. Myers et al., IOP Publishing, Bristol and Philadelphia, 1998, pp.

15. U. Ratzinger, "The New GSI Pre-Stripper Linac for High Current Heavy Ion Beams", *Proc. of the 1996 LINAC Conf., Geneva 1996*, edited by C. Hill et al., CERN 96-07, pp. 288-292

16. Hancock, S., Knaus, P., and Lindroos, M., "Tomographic Measurements of Longitudinal Phase Space Density", *Proc. of 6th European Particle Accelerator Conference, Stockholm, 22-26 June 1998*, edited by S. Myers et al., IOP Publishing, Bristol and Philadelphia, 1998, pp. 1520-1522

17. K. Blasche, O. Boine-Frankenheim, H. Eickhoff, M. Emmerling, B. Franczak, I. Hofmann, K. Kaspar, U. Ratzinger, P. Spiller, "Bunch Compression in the Heavy Ion Synchrotron SIS at GSI", *Proc. of 6th European Particle Accelerator Conference, Stockholm, 22-26 June 1998*, edited by S. Myers et al., IOP Publishing, Bristol and Philadelphia, 1998, pp. 1347-1349

18. M. Steck, K. Beckert, H. Eickhoff, B. Franzke, F. Nolden, H. Reich, B. Schlitt, T. Winkler, "Cooling of Radioactive Isotopes for Schottky Mass Spectrometry" in *Trapped Charged Particles and Fundamental Physics*, edited by D.H.E. Dubin and D. Schneider, AIP 1999, pp. 87-94

19. M. Steck, K. Beckert, H. Eickhoff, B. Franzke, F. Nolden, H. Reich, B. Schlitt, T. Winkler , *Phys. Rev. Lett.* **77**, 3803 (1996)

20. H. Koziol, S. Maury, "Parameter List for the Antiproton Accumulator Complex", CERN 95-15 (AR/BD)

21. F. Nolden, A. Dolinski, B. Franzke, "Design Considerations for a Collector Ring of Nuclear Fragments at GSI", contribution to this conference

22. G. Muenzenberg, invited talk at this conference

Calorimetric Low Temperature Detectors for High Resolution X-ray Spectroscopy on Stored Highly Stripped Heavy Ions

A. Bleile[1,2], P. Egelhof[1,2], H.-J. Kluge[2], U. Liebisch[1], D. Mc Cammon[3], H.J. Meier[1,2], O. Sebastián[1], C.K. Stahle[4], T. Stöhlker[2] and M. Weber[1]

(1) Inst. für Physik, Univ. Mainz, Germany; (2) GSI, Darmstadt, Germany; (3) Dep. of Physics, Univ. of Wisconsin, Madison, USA; (4) NASA/Goddard Space Flight Center, Greenbelt, USA

Abstract. The precise determination of the Lamb shift in heavy hydrogen-like ions provides a sensitive test of QED in very strong Coulomb fields, not accessible otherwise, and has also the potential to deduce nuclear charge radii. A brief overview on the present status of such experiments, performed at the storage ring ESR at GSI Darmstadt, is given. For the investigation of the Lyman-α transitions in Au^{78+}- or U^{91+}-ions with improved accuracy a high resolving calorimetric low temperature detector for hard x-rays ($E \leq 100$ keV) is presently developed. The detector modules consist of arrays of silicon thermistors and of x-ray absorbers made of high Z material to optimize the absorption efficiency. The detectors are housed in a specially designed $^3He/^4He$ dilution refrigerator which fits to the geometry of the ESR target. The detector performance presently achieved is already close to fulfill the demands of the Lamb shift experiment. For a prototype detector an energy resolution of $\Delta E_{FWHM} = 75$ eV is obtained for 60 keV x-rays.

1. MOTIVATION

The precise experimental test of the theoretical predictions of quantum electrodynamics (QED) on corrections to the classical Coulomb interaction potential is still - at least for high Z systems - one of the outstanding and most challenging problems of atomic physics. In the hydrogen atom, or in hydrogen-like ions, the QED corrections give rise to the so called Lamb shift, which is a small deviation of the binding energies from those predicted by the relativistic Dirac-Coulomb theory (see fig. 1). Whereas in light systems, where QED predictions were confirmed to a high precision (1), the higher order contributions are almost negligible, they increase strongly with higher Z. On the other hand, the theoretical predictions of QED, which are usually performed in series expansions in Zα (α being the fine structure constant), become most critical for the heaviest systems, where Zα approaches values close to unity. Therefore a precise determination of the Lamb shift in hydrogen-like very heavy ions represents one of the most sensitive tests of QED in strong electromagnetic fields, not accessible otherwise.

The level scheme of the hydrogen-like U^{91+} ion is displayed in fig. 1. The binding energy of the 1s-energy level is about -132 keV, thus yielding transition energies for the Lyman-α lines of about 100 keV. The 1s Lamb shift is predicted to be $L_{1s} = 463$ eV. Besides the QED contributions self energy ($\approx 80\%$) and vacuum polarization ($\approx -20\%$) the effect of finite nuclear size ($\approx 40\%$) also contributes considerably to this

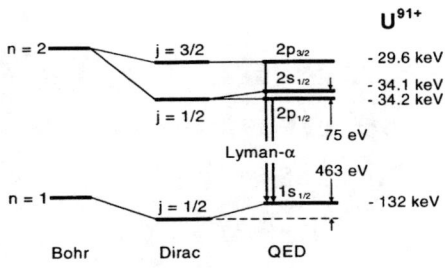
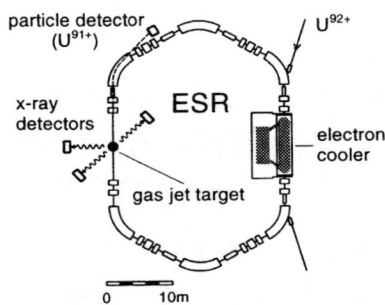

Figure 1. Level scheme for hydrogen-like U^{91+} according to various atomic models. The numbers on the right indicate the electronic binding energies.

Figure 2. Experimental setup for Lamb shift measurements on hydrogen-like very heavy ions at the internal gas jet target of the storage ring ESR.

value. Therefore, apart from the experimental uncertainties, the accuracy on the QED test will be finally limited by the uncertainties on the finite nuclear size effects, estimated for ^{238}U to be about ≈ 0.4 eV. This provides the Lamb shift experiments also with the potential to deduce nuclear charge radii by testing QED in one isotope with a well known nuclear structure, and by investigating the Lamb shift for the corresponding chain of isotopes, which may in future be also extended to exotic isotopes off stability.

2. LAMB SHIFT MEASUREMENTS ON HYDROGEN-LIKE HEAVY IONS

To determine the 1s Lamb shift of heavy ions the transition energies of the Lyman-α transitions are to be measured with high accuracy and compared with theoretical predictions from the Dirac theory. Such experiments are performed at the experimental storage ring ESR of GSI Darmstadt (2) (see also fig. 2). A beam of bare U^{92+} ions is injected, stored and cooled in the ESR and interacts with an internal gas target. This may lead to the capture of one electron and to the population of a 2p state, which promptly decays to the 1s state. The emitted Lyman-α x-rays are detected by x-ray detectors surrounding the internal target in coincidence with the charge exchanged U^{91+} ions. The latest experimental results on the Z = 79 and Z = 92 systems are compared with theoretical predictions in (2). The experimental results agree well with the theoretical predictions and provide already a test of QED for the high-Z domain on the level of 3%. However, the experimental errors ($\pm$ 13 eV) are about one order of magnitude larger than the theoretical ones ($\pm$ 1 eV). Thus the experimental accuracy has to be improved considerably for a more stringent test of QED and/or for the determination of precise nuclear charge radii. One major contribution to the experimental error is the poor energy resolution of $\Delta E \geq 500$ eV obtained with Ge-detectors, which must be improved to at least $\Delta E \leq 50 - 100$ eV in order to reach an absolute accuracy of about $\delta E = \pm 1$ eV in the determination of the center of gravity of the transition energy. This was the motivation for the design of a calorimetric low temperature detector for the present application.

3. DEVELOPMENT OF CALORIMETRIC LOW TEMPERATURE DETECTORS FOR HARD X-RAYS

The detection principle of a calorimetric low temperature detector is schematically displayed in fig. 3 (for more details see ref. (3)). An incident photon with energy E enters an absorber, leading primarily to ionization and production of high energy phonons. After the decay of electronic excitations the main part of the incident energy will be converted into heat (thermal phonons), leading to a temperature rise $\Delta T = E/C$ of the absorber, where C denotes the heat capacity of the absorber. The observed thermal signal is read out by a thermometer coupled to the absorber. Finally the deposited heat is transferred via thermal coupling to the heat sink, described by the coupling constant k. The dynamic behavior of the detector is determined by the thermal time constant $\tau = C/k$. The amplitude of the thermal signal being inversely proportional to the heat capacity C, it is obvious that a high sensitivity of the detector is, due to the T^3-dependence of the Debye law, achieved at low temperatures. As thermometers for reading out the temperature rise ΔT of the calorimetric detector thermistors with a strong temperature dependence of their resistance at low temperatures, for example semiconductor thermistors or superconducting phase transition thermometers may be used.

The potential advantages of calorimetric detectors over conventional ionization detectors are: the smaller energy gap for the creation of an elementary excitation, leading to a better counting statistics of the detected quanta (phonons); the more complete energy detection because both, the energy deposited in phonons and in ionization contribute to the signal; the flexibility in the choice of the absorber material (to be optimized with respect to the detection efficiency); the small noise power at the low operating temperatures. Therefore calorimetric low temperature detectors promise a considerable improvement of the energy resolution in combination with a still reasonable detection efficiency. The fundamental limit on the energy resolution of a calorimetric detector is given by thermodynamical fluctuations of the energy content in the absorber (phonon noise) and the Johnson noise of the thermistor (3). For the detectors discussed in the present contribution this limit is as low as $\Delta E = 6$ eV.

To meet the experimental conditions required by the Lamb shift experiment the calorimetric detector should have a relative energy resolution of $\Delta E/E \leq 1 \cdot 10^{-3}$ for $E_\gamma = 50 - 100$ keV and a total detection efficiency (including detector solid angle)

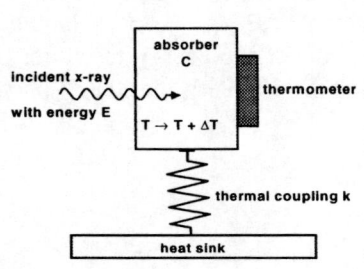

Figure 3. Operation principle of a calorimetric low temperature detector.

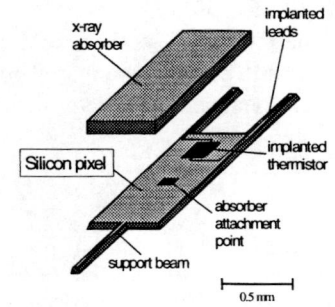

Figure 4. Layout of a pixel of the present calorimetric x-ray detector.

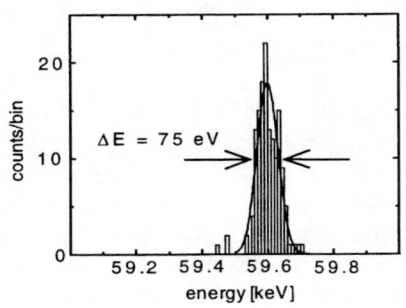

Figure 5. Energy spectrum observed with a calorimetric low temperature detector for 59.6 keV photons. For the photopeak an energy resolution of $\Delta E_{FWHM} = 75$ eV is obtained.

of $\geq 10^{-6} - 10^{-5}$, which may be reached with a photopeak efficiency of $\geq 30\%$ and an active detector area of ≥ 100 mm^2. The detector modules designed for the present experiment consist of silicon thermistors, which are used as thermometers, and of x-ray absorbers glued on the top of the thermistors by means of an epoxy varnish. A schematic view of the layout of a detector pixel is displayed in fig. 4. The pixel is made from a wafer of silicon containing an implanted thermistor and the implanted leads. Thermistor arrays (4), consisting of 36 pixels each, are provided from the collaborating groups from Madison and Goddard. The final detector concept foresees three calorimeter arrays, the active area of 1 pixel being about 1 mm^2.

For the Lamb shift measurement the experimental setup was optimized with respect to energy resolution and detection efficiency for hard x-rays at the experimental area of the ESR. In order to reach sufficient photopeak efficiency, the absorber should be a high Z material and should have a volume of $V \geq 1$ mm$^2 \cdot 40$ µm. Systematical tests of suitable absorber materials, such as Sn, HgTe, Re, fulfilling also the conditions of low heat capacity and rapid and complete thermalization, were recently performed. To obtain a reasonable detection solid angle the detector arrays have to be located as close as possible to the interaction zone at the internal target of the ESR. To realize this concept a special ^{3}He/^{4}He-dilution refrigerator with a side arm which fits to the internal target geometry was designed. The detector arrays are mounted on the cold finger at the end of the side arm and can be irradiated through a system of aluminum coated mylar windows. In order to suppress low-frequency microphonics the first amplifier stage is positioned close to the detectors inside the side arm of the cryostat. The cryostat reaches a base temperature of 11.5 mK, and a cooling power of 400 µW. The operating temperature of the detectors may be chosen between 50 mK and 100 mK.

The detector performance presently achieved is already close to fulfill the demands of the Lamb shift experiment. The best results were obtained with Sn as absorber material. The energy spectrum obtained for a detector with a 0.3 mm$^2 \cdot 66$ µm Sn absorber for 59.6 keV photons, provided by an ^{241}Am source, is displayed in fig. 5. For the photopeak at 59.6 keV an energy resolution of $\Delta E_{FWHM} = 75$ eV is obtained. This result may be compared to the theoretical limit of the energy resolution for a conventional semiconductor detector which is about $\Delta E \approx 380$ eV for 60 keV photons.

REFERENCES

1. M. Weitz et al., Phys. Rev. Lett. 68 (1992) 1120.
2. H.F. Beyer and T. Stöhlker, in Frontier Tests of QED and Physics of the Vacuum, E. Zavattini B. Akalov, C. Rizzo (eds.), Heron Press, Sofia 1998, p. 354, and references therein.
3. P. Egelhof, Adv. in Solid State Phys. 39 (1999) 61, and references therein.
4. C.K. Stahle et al., Nucl. Instr. Meth. A 370 (1996) 173.

Schottky Mass Spectrometry at the ESR with a new data acquisition system

M. Falch[1], K.E.G. Löbner[1], Th. Kerscher[1], F. Attallah[2], F. Bosch[2], B. Franzke[2], H. Geissel[2], M. Hausmann[2], O. Klepper[2], H. Kluge[2], C. Kozhuharov[2], G. Münzenberg[2], F. Nolden[2], Y. Novikov[4], Z. Patyk[2], T. Radon[2], C. Scheidenberger[2], M. Steck[2], M. Winkler[2], H. Wollnik[3]

[1] Sektion Physik, LMU München, Am Coulombwall 1, 85748 Garching, Germany[1]
[2] Gesellschaft für Schwerionenforschung mbH, Planckstraße 1, D-64291 Darmstadt, Germany
[3] II. Physikalisches Institut, JLU Gießen, Heinrich-Buff-Ring 16, 35392 Gießen, Germany
[4] St. Petersburg Nuclear Physics Institut, Gatchina 188350, Russia

Abstract.
Schottky Mass Spectrometry turned out to be an excellent tool for systematic nuclear mass measurements. The newly developed data acquisition system [1] which yields several advantages for the data analysis, was used for the first time in Nov. 1997. Instead of online Fourier transformation as in former measurements [2,3] the signal is now digitized with a sampling rate of 640kHz (16 bits of resolution) and stored as time-captures. With a continuous data rate of 1.2MB/s nearly 1TB of raw data were stored. The Fast Fourier Transformation offline analysis allows the study of the dynamical behaviour of the stored ions in the ESR including cooling forces and decay. The new spectra contain enough reference nuclei to determine the deviations from the constant momentum compaction factor α_p.

Equation (1) shows the relation between the revolution frequency of the ions and their mass to charge ratio, with the momentum compaction factor α_p. The second term on the right side contains the relativistic Lorentz factor γ and the γ-value at the transition energy γ_t. This term gets very low by reducing the spread of the ion velocity dv by applying electron cooling or by operating the storage ring in the isochronous mode ($\gamma = \gamma_t$) [4]. For the Schottky Mass Measurements the electron cooling is used to reduce the velocity spread $\Delta v/v$ of the ions to values below 10^{-6}.

$$\frac{df}{f} = -\alpha_p \frac{d(m/q)}{(m/q)} + (\gamma^{-2} - \gamma_t^{-2})\gamma^2 \frac{dv}{v} \quad (1)$$

---

[1] This work is supported by the BMBF under contract number 06LM363, the GSI Darmstadt and the Beschleunigerlaboratorium München.

Fig.1 shows a schematic view of the Schottky diagnosis setup. The revolution frequency of the ions is about 2 MHz. The pickup electrodes detect the influenced pulses from each ion. To increase the frequency resolution an external resonant circuit running on the 30th harmonic close to 60MHz is used. The signals are sampled with a fast ADC of the new TCAP-system and stored as time captures.

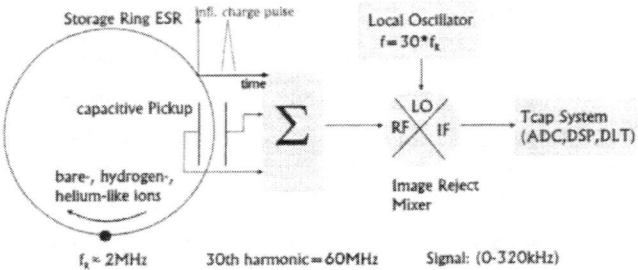

FIGURE 1. Schematic view of the Schottky setup with the storage ring ESR.

Advantages of continuously data writing over real-time FFT: [2,3]
i) more efficient use of the beam time by two orders of magnitude, **ii)** possibility of determining the function of the momentum compaction factor, see Fig.3, **iii)** offline corrections of frequency variations, **iv)** offline optimization of the FFT parameters.

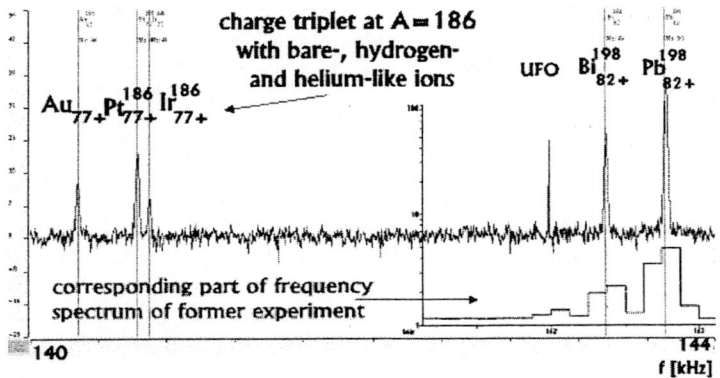

FIGURE 2. Part of high resolution Schottky spectrum. FFT with $70*2^{19}$ data points, frequency resolution 1.19Hz/channel, bandwidth 320kHz covering the whole momentum acceptance of the ESR, see text. Sketch of former spectra to demonstrate increase of resolution.

The offline FFT calculation has no limit in the number of input points. Using $70*2^{19}$ data points a frequency resolution of 1.19Hz/channel is achieved. The bandwidth of 320kHz covers the whole momentum acceptance of the ring. Fig.2 shows a small part of a broadband frequency spectrum. It shows a charge-triplet and -doublet with a so-called unidentified frequency object which can now be detected because of its smaller FWHM compared to the real peaks. Below, part of a spectrum with 6.25 Hz/ch (bdw 10kHz) of the former measurements is shown. The decrease in measuring time of broadband spectra allows the determination of the momentum-compaction-factor α_p as a function of the orbit length. (Fig.3). This is very important for a correct isotope-identification of the frequency peaks.

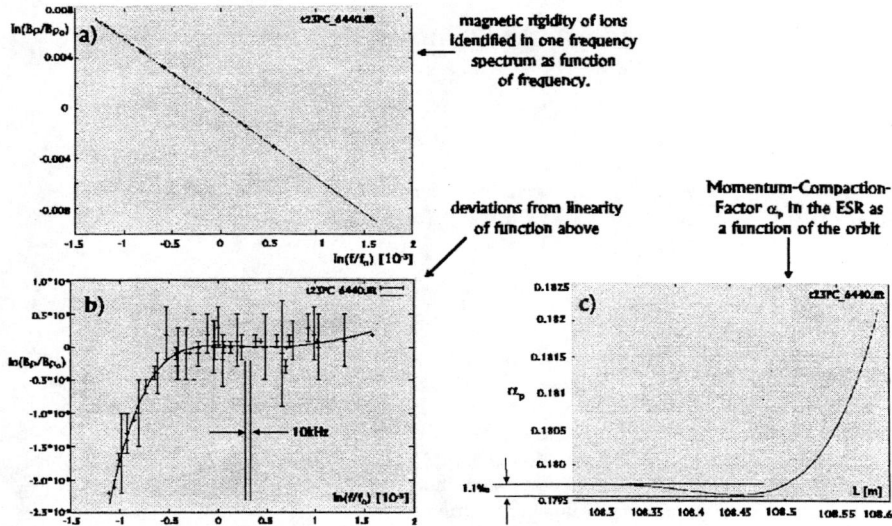

FIGURE 3. a) magnetic rigidity $B\varrho = p/q$ of identified ions in one frequency spectrum, b) without linear part of the $B\varrho$-function, c) α_p as function of the orbit length.

A new feature is the offline drift correction. Due to some jitter in the ESR settings, the width of the frequency peaks increases. As the data are stored as time captures it is possible to detect offline these frequency variations and correct the spectra. A decrease of the peak width up to 30% in some cases and of 5% as an average value over 6600 spectra could be achieved. The data analysis is still in progress. Up to now 259 nuclei from Ru to Po with $T_{1/2} \geq 6sec$ could be identified in the spectra. Several isomeric states could be resolved. The lowest resolved excitation energy of only 166keV was ^{161}Lu. This shows the limit of the frequency resolution of $\Delta f/f = 2.36 * 10^{-7}$ which yields a mass resolution of $\Delta m/m = 1.3 * 10^{-6}$ which is an improvement by a factor 2 compared to former measurements.

REFERENCES

1. M. Falch et al., *Annual Report LMU-München*, 108 (1996)
2. Th. Kerscher, *Ph.D. thesis LMU-München*, (1996)
3. T. Radon et al., *Phys. Rev. Lett.*, **78**,4071 (1997)
4. J. Stadlmann et al., *these proceedings*

Considerations for a Collector Ring of Nuclear Fragments at GSI

F. Nolden*, A. Dolinsky†, B. Franzke*

Gesellschaft für Schwerionenforschung
Planckstraße 1
D-64291 Darmstadt, Germany
†*Institute for Nuclear Research*
Prospekt Nauki 47
Kiev, 252650, Ukraine

Abstract. A possible scenario for an extension of the present GSI accelerator complex would include a dedicated ring for the collection and cooling of secondary beams. Both radioactive ions from the upgraded Fragment Separator FRS and optionally antiprotons from a source yet to be built are possible beams to be treated in the ring. After injection and subsequent active debunching, fast stochastic cooling in all three phase space dimensions will be applied. Very short overall cooling times of $\leq$ 100ms are envisaged for highly charged nuclear fragments.

PURPOSE OF THE RING

The Collector Ring (CR) could be an important part of an upgrade of the GSI accelerator and storage ring facilities [1]. It would be placed between the Fragment Separator (FRS) and the existing experimental storage ring ESR. Its main purpose is the fast reduction of the phase space volume occupied by the secondary beams emerging from the FRS.

The required phase space reduction is achieved by the operation of a fast stochastic cooling system with cooling times of the order of 100 ms. A sufficiently large acceptance for the secondary beams is reached by large vacuum chamber apertures and a fast full aperture injection kicker. Optional cooling of antiprotons emerging from a dedicated source yet to be built is a part of the present design study, too.

The ring is preliminarily designed for a magnetic rigidity of 13 Tm which leads to qualitatively different parameters for the Lorentz β and γ of secondary fragments with a mass over charge ratio $A/q \leq 2.7$ and high charge states on the one hand and, on the other hand, antiprotons with $A/q \approx 1$ and $q = 1$. The different energies, in turn, lead to a different layout of the ring optics, as will be discussed below.

A fast full aperture kicker will inflect the beam to the closed orbit. The required transverse acceptances $A_{x,y}$ of the injected beams are demanding. The energy and

TABLE 1. Energy and acceptance of the CR at 13 Tm

species	A/q	γ	β	$\delta p/p_{\text{inj}}$	$\delta p/p_{\text{debunched}}$	$A_{x,y}$
nuclear fragments	2.7	1.8443	0.8402	±2.5%	±0.4%	50 π mm mrad
$\bar{\text{p}}$	1.0	4.3018	0.9726	±3.0%	±0.75%	250 π mm mrad

phase space parameters are listed in table 1 for the different beams.

Single bunches will be injected with a short pulse length of $\approx$ 50 ns. Therefore the injected momentum width can be reduced by a factor of five using active debunching in an rf system similar to the former CERN AC complex [2].

REQUIREMENTS FOR FAST STOCHASTIC COOLING

The experiments following the CR require a reduction to 5 π mm mrad in the transverse phase planes and $\delta p/p = \pm 0.1\%$. The desired cooling times are given by lifetime limits in the case of nuclear fragments. Very short cooling times τ of the order of 100 ms are envisaged. For antiprotons, 4 s would correspond to the injector synchrotron repetition rate and are presumed to be sufficient.

The maximum number N of particles (radioactive fragments or antiprotons) to be cooled will be limited to 10^8. An effective cooling bandwidth $W \approx 1$ GHz is required because of the given τ and N. The frequency band 0.9-1.7 GHz is favourable because hardware presently installed at the ESR could be recycled. This choice together with the momentum width after debunching leads to important constraints for the ring optics. The pick-up signal must arrive at the kicker at the same time as the particle that induced the signal. The relative time of flight difference $\delta T/T_{p-k}$ between pick-up and kicker of particles with momentum difference $\delta p/p_0$ is

$$\frac{\delta T}{T_{p-k}} = -\eta \frac{\delta p}{p_0}$$

Here, η is the frequency slip factor $\eta = \gamma^{-2} - \alpha_p$. α_p is the momentum compaction factor. It is proportional to the mean dispersion function inside the dipole magnets. If the signal transmission time is T_{p-k} for all signals, a phase error $\delta\phi = \Omega \delta T$ will arise (we calculate T_{p-k} for a path length of 75 m.) This effect is also known as undesired mixing between pick-up and kicker. For efficient cooling, $|\delta\phi|$ must be bounded at the the upper limit $\Omega_{\max}$ of the cooling band, or $|\eta| < |\delta\phi|/(\Omega_{\max} T_{p-k} |\delta p/p_0|)$.

For fragment beams one expects a rather comfortable signal-to noise ratio of the Schottky signal at the pick-ups. However, because of the low value of γ, the α_p limit

TABLE 2. Frequency slip and momentum compaction

species	T_{p-k} [ns]	$|\delta\phi|$	$|\eta|_{\max}$	α_p	γ_T
nuclear fragments	298	$\pi/3$	0.0823	$\geq$ 0.2117	$\leq$ 2.1734
$\bar{\text{p}}$ above γ_T	257	$\pi/15$	0.0190	$\leq$ 0.0730	$\geq$ 3.7012

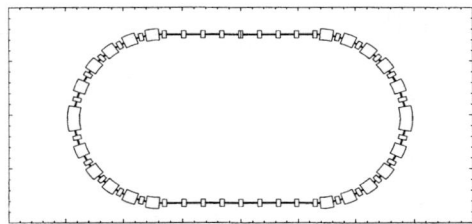

FIGURE 1. Preliminary CR ring layout. The distance between major ticks is 10 m.

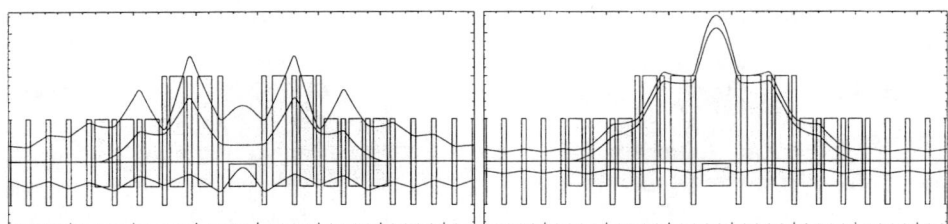

FIGURE 2. Preliminary layout of the CR ion optics for antiprotons (left) and fragment beams (right) in one half (i.e. 75 m) of the ring. Shown are the dispersion function and the horizontal beam envelope (above the horizontal straight line) as well as the vertical envelope (below). The narrow shaded structures indicate quadrupole magnets, the broader ones dipoles. The vertical axis extends from -150 mm to 350 mm. The central dipole has a horizontally focussing gradient.

due to the synchronicity problem becomes severe. Even if only a limit $|\delta\phi| \leq \pi/3$ is required, the ring must be operated at a large minimum value for α_p, and the dispersion function inside some of the dipole magnets becomes rather large. That enforces wide horizontal apertures and may affect the total cost of the machine. Palmer cooling will be used for longitudinal cooling.

For antiprotons the signal-to-noise problem is crucial and concentrated Schottky bands are desirable. By using notch filter techniques instead of Palmer cooling, a large fraction of the noise in the frequency gaps between the bands can be suppressed. Then a sharper limit $|\delta\phi| \leq \pi/15$ is required. Operation of the CR *above transition* reduces the α_p difference for the two beam options. This facilitates the optical design and reduces the number of required optical elements. However, as γ_T is roughly equal to the horiontal tune, a reduction of α_p tends to decrease the horizontal acceptance, which is very demanding for antiprotons.

The resulting parameters are listed in table 1. A very preliminary design is shown in figures 1 and 2.

REFERENCES

1. B. Franzke, this conference
2. H. Koziol, S. Maury, Parameter List for the Antiproton Accumulator Complex, CERN 95-15 (AR/BD)

Particle Detectors for Beam Diagnosis and for Experiments with Stable and Radioactive Ions in the Storage-Cooler Ring ESR

Otto Klepper and Christophor Kozhuharov

GSI Darmstadt, Planckstr.1, 64291 Darmstadt, Germany

Abstract. A survey of the 12 available detector positions and of their applications is given. Emphasis is on the new positions for beam diagnosis.

INTRODUCTION AND CONCEPT

Based on the pocket concept, 12 permanent positions for particle detectors are provided in the Experimental Storage Ring (ESR) at GSI or in the adjacent beam lines. This means, detectors of various kinds (multiwire chambers, scintillators etc.) are separated from the ultra-high vacuum (UHV) of the storage ring and, therefore, can easily be exchanged or temporarily be removed during the baking out of the vacuum chamber. The first 5 positions [1] were originally intended for detecting ions that change their mass-over-charge ratio in the electron cooler or in the gas-jet target on the straight sections of the ring and then leave the primary beam in the next bending magnet.(See the "Experimental positions" at the ring chamber in Fig.1.) As described in Ref [1], some of these detectors turned out to be indispensable for the injection and the storage procedure of low-intensity radioactive beams. In order to disentangle the problems with frequently using the same pocket in one experiment for two very different purposes, special detector positions 'beam diagnosis' were installed. Similar detector stations have also been added to the injection and to the (slow-)extraction beam-line (Fig.1).

BEAM DIAGNOSIS

The 4 pockets 'beam diagnosis' (Fig.1), newly installed in the ring chamber and in the injection beam-line, and the corresponding detectors are identical (except their length): The size of the entrance windows (50-μm thick stainless steel) and of the sensitive areas of the gas-filled multiwire chambers are 100x60 mm^2 and

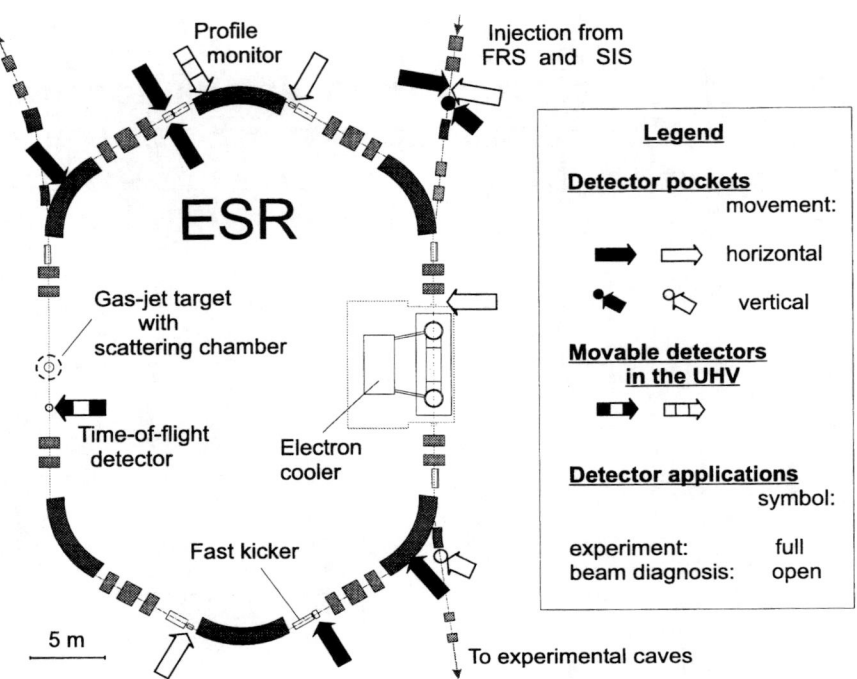

FIGURE 1. Layout of the storage and cooler ring ESR. The arrows indicate the position (and the direction of the movement) of the pockets with particle detectors in the ring chamber or in the UHV region of adjacent beam lines. For completeness also the two bakable detectors are shown that are mounted directly in the UHV: the time-of-flight detector for mass measurements and the beam-profile monitor based on the ionization of the residual gas. In the target chamber around the gas jet no particle detectors are presently mounted.

100x56 mm^2, respectively. (Gas mixture: argon + CO_2 + heptane; 80 : 20 : 1.5) The x and y cathode planes (wire distance: 1 mm; readout spacing: 2 mm) serve to measure current profiles. Variation of the anode bias (100 to 1500 Volt) and the integration time (0.1 ms to 6 s) enable us to adjust the profiles to the GSI standard beam-profile display.

Low-intensity radioactive ion beams. These detectors provide the beam diagnosis for the 'hot' fragments with their large emittances between the Fragment Separator FRS and the ESR and in the ring during the first turn. These current grids are the only single-pass diagnosis for this kind of injected bunches prior to their storage and to the availability of the standard Schottky beam-diagnosis.

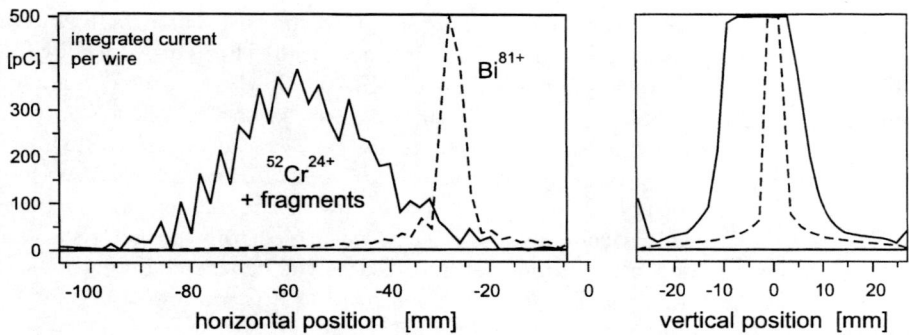

FIGURE 2. Current profiles taken at the detector position following the fast kicker. Both ring positions have roughly the same dispersion of ≈ 6 m. The distances on the axes are relative to the central orbit. <u>Full curve:</u> 100-ns long bunch of ^{52}Cr ions and of their fragments injected from the FRS at 344·A MeV. The horizontal beam profile has been optimized to match the acceptance (about -30 to -100 mm) of the kicker. (Integration time 0.1 ms). <u>Dashed curve:</u> DC-beam of bismuth ions down-charged in the electron cooler from a coasting, cooled 310·A MeV Bi^{82+} beam circulating at $x \approx +50$ mm on the inner side of the ring.(Integration time 0.2 s)

Operation control of the storage ring. The x-y beam profiles and the rate of ions down-charged in the electron cooler allow us to optimize and to monitor the cooling of a stored beam of ions heavier than about xenon. This also holds for the slow charge-exchange-extraction of ions out of the ESR to the experimental caves via electron capture in the cooler or in the gas jet.The corresponding detector is similar to the four detectors mentioned above.- The thickness of the entrance window corresponds to a range of heavy ions of about 12·A MeV. So far beam profiles of lead ions with energies down to 25·A MeV have been observed.

Fig.2 presents an example for the two different applications given above. The detectors [2] were built in the GSI detector laboratory. New ones are in preparation to overcome the zig-zag pattern of the horizontal profile that is due to the wiring of the chamber.

EXPERIMENTS

The detector pockets 'experiment' (Fig.1) are used for various detectors, standard or user supplied ones, according to the experimental demands. The positions at the ring and the position-sensitive multi-wire proportional chambers used so far are described in ref. [1]. Meanwhile the height of the entrance windows of the pockets has been increased from 20 to 44 mm for better adaption to the experiments with short-lived, uncooled fragments. The new detector stations in the injection beam line are positioned at a place that is the 4th FRS focus during the coupling of FRS und ESR.The two 'experimental' positions serve to optimize the separation in the

FRS of the isotope(s) of interest for the injection into the ring. The fragments are identified and their position and emittances are measured. (For this procedure the primary beam from the heavy-ion synchrotron SIS is slowly extracted for a few seconds.) The current-profile monitor ('beam diagnosis') can be used for both the long 'spills' and the 100-ns long injection bunches (fast extraction from SIS).

Atomic physics investigations of electron-capture and ionization processes. Such reactions are studied in both the electron cooler and in the gas-jet target. Single counting of ions as well as coincidences with x rays, electrons, target recoil-ions and laser pulses have been performed. In most cases the charge-changing ions are detected by fast scintillators. Details are given in ref. [3]: a review of the physics with highly charged ions in storage-cooler rings.

Nuclear half-life measurements. The nuclear physics experiments with particle detectors have essentially been beta half-life measurements. A review of the performed experiments and of the future potential is given in ref. [4]. For half-lives down to the sub-ms region, the daughter nuclei may be counted as a function of the storage time by a detector intercepting the new trajectory after the decay. With the presently available detector positions, such application is somewhat restricted depending on the atomic number and on the decay mode. In two cases of bound-state β-decay of (very long-lived) bare nuclei, a few hundred hydrogenlike daughter nuclei have been detected after their selective ionization and due to the resulting trajectory change.

OUTLOOK

So far the particle detectors have been more often used for beam diagnosis and for atomic physics than for nuclear physics investigations. With the completion of the present intensity upgrade of the SIS, higher intensities of (mono-isotopic and mixed) radioactive ion beams will be available from the FRS. An improved transfer efficiency from the FRS into the ESR, that is currently worked on and that the extended beam diagnosis described here will contribute to, will result in higher intensities of stored fragments and facilitate new experiments. So nuclear physics experiments and the corresponding use of particle detectors, as e.g. in half-life measurements of short-lived nuclei or in reaction studies [5], will play a stronger role in the future ESR activities than before.

REFERENCES

1. Klepper O., et al., *Nucl. Instr. and Meth.* **B70**, 427 (1992).
2. Stelzer H., *Nucl. Instr. and Meth.* **130**, 103 (1991).
3. Mokler, P. H., et al., *Advances in Atomic and Optical Phys.*, Vol.**37**, 297 (1996).
4. Klepper O., *Nucl. Phys.* **A626**,199c (1997).
5. Peter M., *Nuclear Spectroscopy by Proton Scattering at* ^{56}Ni, Ph. D. thesis, Techn. Univ. München, 1996; Peter M., et al., GSI Scientific Report 1995, p. 29; unpublished.

PHYSICS WITH STORED HEAVY IONS

Gross properties of exotic nuclei investigated at storage rings and ion traps

C. Scheidenberger[1,2], G. Bollen[3,2], F. Bosch[1], A. Casares[4],
H. Geissel[1,4], A. Kholomeev[4], G. Münzenberg[1,5],
H. Weick[4,1], H. Wollnik[4],

[1] *GSI, D-64291 Darmstadt*
[2] *CERN, CH-1211 Geneva 23*
[3] *Ludwig-Maximilians-Universität, D-85748 München*
[4] *Justus-Liebig-Universität Gießen, D-35392 Gießen*
[5] *Johannes Gutenberg-Universität Mainz, D-55099 Mainz*

Abstract. Properties of exotic nuclei like atomic masses, decay modes, and half-lives can be ideally investigated in storage rings and ion traps. Some experiments can be carried out under conditions which prevail in hot stellar plasmas. The experimental potential of storage and cooling of exotic nuclei is illustrated with recent experimental results and an outlook to future experiments is presented.

INTRODUCTION

Beams of exotic heavy nuclei and of antiprotons are used for a large variety of investigations in physics ranging from the test of fundamental symmetries and physical constants [1,2] to structure studies of complex nuclei [3–5]. The common feature of these experiments is that the particles under investigation are not available from conventional ion sources but have to be produced in dedicated nuclear reactions [1,4,6]. Therefore, it was necessary to develop techniques and devices for production, separation, accumulation, bunching, storage, and cooling of the reaction products: electromagnetic separators, storage rings, and ion traps [7].

In the first part of this paper the concepts and experimental techniques for production and handling of exotic nuclear beams in ion traps and storage rings are outlined. The characteristic features of the production mechanisms fusion, fragmentation, spallation, and fission are briefly described together with the two basic separation concepts ISOL (Isotope Separation On-Line) and in-flight separation. The second part concentrates on precision mass measurements of exotic nuclei at ISOLDE (CERN) and SIS-FRS-ESR (GSI). The third part highlights a new class of experiments which became possible in the experimental storage ring ESR with ions which can be called 'exotic' because of their high charge state: manipulation of the nuclear lifetimes in dependence of their ionic charge state of bare and few-electron heavy ions. Finally new ideas for sub-millisecond direct mass measurements of fusion-reaction products at SHIP and of projectile or fission fragments at the FRS are presented and an outlook to new applications is given.

THE EXPERIMENTAL TECHNIQUES
Sources of exotic nuclei

Antiprotons are produced in collisions of intense proton beams with heavy target materials like copper or iridium. Due to energy and momentum conservation the laboratory threshold energy amounts to six times the proton rest mass and the optimum yield lies around 26GeV/c. For the production of heavy exotic nuclei there are basically three methods available: 1) proton- or neutron-induced target fragmentation (spallation) or fission, 2) projectile fragmentation or fission, and 3) heavy ion fusion. While the first method is usually related to ISOL devices [6] the latter two yield energetic reaction products which can be separated in-flight with recoil separators and fragment separators [8]. The statistical nature of projectile fragmentation leads to the formation of a large variety of ion species, in particular exotic nuclei with unusual proton-to-neutron ratios, which predominantely populate the area between the stable isotopes and the proton dripline ranging from the projectile atomic number down to the lightest elements. Target spallation, where the fragmentation reactions are induced by a light ion beam (protons, neutrons or light ions), can be described in a similar way [4]. Electromagnetic dissociation (ED) mainly takes place in collisions with high-Z target nuclei and leads to the evaporation of neutrons and/or (in the case of heavy projectiles like lead or uranium) to projectile fission. At an ISOL system the role of projectile and target is reversed and the heavy target nuclei fission after proton bombardment. In both cases predominantly neutron-rich fission fragments are produced because the fission event mostly takes place after excitation of the giant dipole resonance (~10MeV) thus leading to cold fission. In-flight separated and short-lived heavy ion fusion reactions take place at projectile energies just sufficient to cross the Coulomb barrier, which is of the order of 5MeV/u. The projectile and target nucleus merge to a compound nucleus, which de-excites evaporating nucleons, the number of which depends on the excitation energy.

Separation, accumulation, storage, and cooling of exotic nuclei with storage rings and ion traps

Heavy ion fusion products can be investigated at the GSI velocity filter SHIP [9] and at the gas-filled separator NASE [10]. In this low-energy domain the heavy recoils carry many electrons and are distributed over many different ionic charge states with typical abundances of the order of 10% each [11]. To obtain a high collection efficiency and transmission the many different charge states are 'focused' in gas-filled separators on an average charge state $\bar{q} \simeq (v/v_0) \cdot Z^{1/3}$ [12].
At the high-energy heavy-ion facilities SIS-FRS-ESR relativistic exotic nuclei are generated in a production target placed at the entrance of the fragment separator FRS [13]. At relativistic velocities the ions emerge from the production target mainly as bare or few-electron projectiles [14]. Two separation schemes are applicable at the FRS: pure magnetic $B\rho$-analysis and a combined $B\rho$–ΔE–$B\rho$-analysis. In the first case all fragments with the same magnetic rigidity $B\rho$ are transmitted within an acceptance of $\Delta B\rho/(B\rho) \simeq \pm 1$ % and a multi-component 'cocktail beam' can be injected into the storage ring ESR [15]. In the second case the FRS is operated as momentum-loss achromat and an energy degrader is placed in the dispersive midplane. This mode is used either when it is necessary to reduce the number of different species in cocktail beams and/or to suppress high-intensity beam components or when an experiment requires monoisotopic secondary beams. Figure 1 illustrates both scenarios showing the spectra of stored cooled Ni fragments using pure $B\rho$ analysis (left spectrum) and using $B\rho$–ΔE–$B\rho$-separation of bare ^{52}Mn (right spectrum), respectively [16].
The shown spectra have been obtained with Schottky noise diagnosis and FFT frequency analysis [17,18]. The area of each peak, the noise power, is proportional to the number of stored ions. The sensitivity of this method is so high, that even single ions [19] can be detected and their revolution frequency can be determined. Therefore, Schottky diagnosis is well suited for experiments with stored exotic secondary beams, where only low intensities are available due to the small production cross sections. The width of the frequency peaks is proportional to the relative longitudinal velocity spread $\delta v/v$ and therefore efficient cooling is required for high frequency and mass resolution of multi-component beams (see below). For precooling of the hot fragments stochastic Palmer cooling is available [20], which reduces the phase space down to $\delta p/p = \pm 0.1$ %. With electron cooling [21,22] the longitudinal momentum spread can be further reduced down

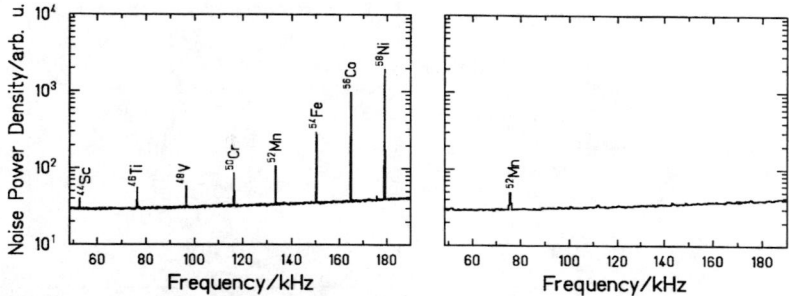

FIGURE 1. Schottky noise power spectra of cooled bare Ni fragments stored in the ESR [16]. **Left panel:** pure $B\rho$ analysis was applied and thus many different bare fragments with similar m/q are transmitted through the FRS. **Right panel:** by the use of an intermediate degrader the combined $B\rho$–ΔE–$B\rho$-analysis was applied leading to a monoisotopic secondary beam; the FRS was tuned for separation of ^{52}Mn^{25+}.

to $\delta p/p \leq 1 \cdot 10^{-6}$ for stored beams containing less than a few thousand ions [21].

At ISOLDE the exotic nuclei produced and stopped in the thick production target [23] are released from the target, ionized, and are then separated with electromagnetic separators, the general purpose separator GPS and the high-resolution separator HRS, respectively. The secondary ion yields at ISOL-type facilities are not only determined by the product of driver beam intensity, target thickness and production cross section but also, and sometimes even more crucial are diffusion and release processes from the target, transfer efficiencies to the ion source, ionization probabilities and finally radioactive decay during the entire process [6]. With respect to the individual physical and chemical properties and pecularities a large variety of target–transfer-line–ion-source–combinations has been developed in more than 30 years [6] to meet the specific experimental requirements like high secondary beam intensities, few or even no isobaric contaminants, element selectivity, and short half-lives. Presently beams of 70% of the chemical elements can be provided and the nuclei with the shortest half-lives are ^{14}Be ($T_{1/2} = 4.4$ms) and ^{11}Li ($T_{1/2} = 8.5$ms). The ions are usually singly ionized and typical ion beam emittances are of the order of 30π mm mrad at 60keV beam energy.

Several schemes for ion-beam accumulation, cooling, and bunching in ion traps have been developed [24]. At the ISOLTRAP experiment [25,26] a Penning trap was used as an accumulator, cooler, and buncher for the low-energy (typically 30keV or 60keV) secondary beams at ISOLDE [27,28]. This cooler trap is a segmented linear Penning trap with cylindrical electrodes, filled with He buffer gas with a pressure $\sim 10^{-5} \dots 10^{-4}$mbar. To the various segments different voltages are applied for the formation of a longitudinal potential well for the ions as shown in figure 2 [28]. The height of the potential step at the entrance of the trap is attuned to the kinetic energy of the injected ions so that they can pass this barrier once at their arrival. Then, the ions loose kinetic (and thus potential) energy in collisions with the buffer gas atoms and are slowed down until they are thermalized in the central potential well of the trap: the longitudinal ion motion has been cooled and an ion bunch was formed. For the transverse motion the dissipation of energy leads to an increase of the magnetron motion amplitude. This instability can be avoided for ions with mass-to-charge ratio m/q when an azimuthal rf-field with a frequency equal to the cyclotron frequency $\omega_c = q/m \cdot B$ is applied [29]. Ions with the proper m/q-ratio are centered in the trap whereas other ions are lost. Hence, this process is mass selective and can be used for mass separation. This scheme for capture, cooling, bunching, and separation of secondary beams has been pioneered at ISOLTRAP and will be used at the radioactive beam experiment REX-ISOLDE [30] and at the SHIPTRAP [31] project.

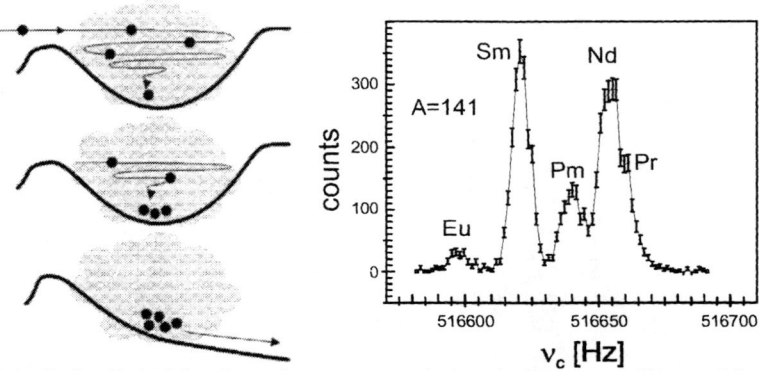

FIGURE 2. Left: Principle of continuous accumulation, buffer gas cooling, and bunching of ions in a cooler trap [28]. **Right:** This mass scan for $A = 141$-isobar shows the number of ejected ions as a function of the rf-frequency after mass-selective cooling.
Reprinted from [28], © 1997, with permission from Elsevier Science.

Facilities combining storage rings and ion traps

New precision experiments will become possible with ion traps capturing decelerated beams extracted from storage rings, for instance at the antiproton decelerator AD [1] at CERN or at the heavy ion trap experiment HITRAP [32] downstream of the ESR. In the latter experiment the ESR is used for accumulation and subsequent deceleration of highly or even fully stripped heavy ions from primary or secondary beams down to specific kinetic energies of the order of few MeV/u. After extraction and further deceleration the highly-charged ions will be captured in a Penning trap for experiments. Among other experiments like accurate mass measurements of stable und unstable isotopes, the main interest will be devoted to g-factor measurements of the bound electron in heavy one-electron ions, which is a sensitive test for QED [33]. Systematic hyperfine-structure investigations in isotopic chains will yield new insights on details of the magnetic moments and structure of nuclei.

The ATHENA [34] and the ATRAP [35] experiments at the AD are aiming at the capture of decelerated antiprotons and of positrons for the production and study of cold antihydrogen. In view of testing CPT-invariance these experiments focus on high-precision comparisons of matter and antimatter properties like inertial masses and atomic energy levels and transitions. So far the proton and antiproton masses agree within $9 \cdot 10^{-11}$ [36].

EXPERIMENTS AND RESULTS
Direct mass measurements

In this chapter some experiments on direct mass measurements of short-lived nuclei are presented. For recent reviews covering also other techniques the reader is refered to [37,38]. The motivation for accurate mass measurements is manyfold and the field has attracted lively interest over many decades. There are whole regions which are subject of general interest in the study of nuclear structure and effects as indicated in figure 3 and there are single isotopes whose masses are of particular interest, like for instance ^{32}Ar (see below).

Accurate mass measurements yield important information on nuclear structure and stability. The fundamental question: 'where are the limits of stability of nuclei' can be addressed in terms of the experimental determination of the drip line for neutrons and protons. Experiments exploring new regions of the chart of nuclei will possibly allow the discovery of new phenomena like new deformations, new regions of collectivity, and new interactions [39]. Some examples of present interest are briefly mentioned:

- nuclear binding energies and proton- and neutron-separation energies are key parameters for the study of skin and halo nuclei
- the existence of Thomas-Ehrman [40–42] shift in heavy nuclei

- proton-neutron pairing [39] and Wigner's SU(4) symmetry [43] can be studied in heavy $N = Z$ nuclei [44]
- study of deformation and shape-coexistence phenomena in the neutron-deficient region around the $Z = 82$ shell
- nuclear properties in the area around ^{208}Pb are important for the adjustment of nuclear-model parameters because it serves as a 'test ground' for the superheavy elements.
- Shell effects far off stability and shell quenching occuring in extremely neutron-rich nuclei have large impact on astrophysical theories modeling the paths of nucleosynthesis in stellar environments and the abundances of heavy nuclei

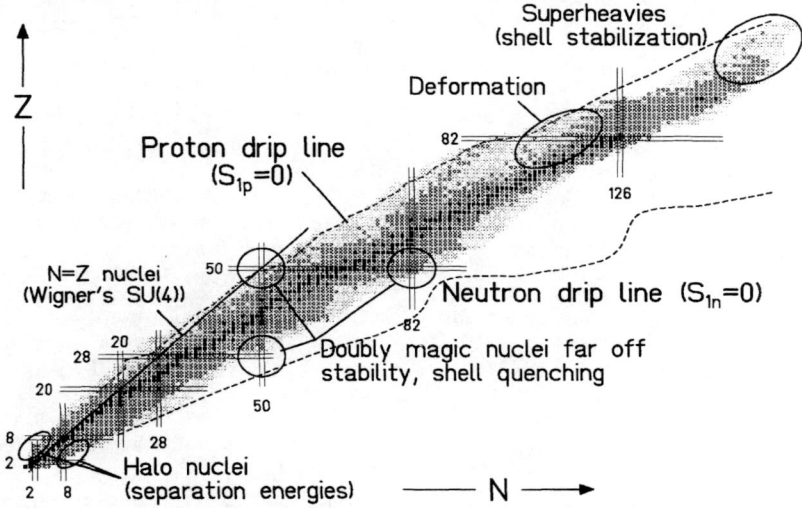

FIGURE 3. Schematic view of the chart of nuclei. Stable isotopes are indicated in black, known and unknown masses [45] in dark and light gray, respectively. Those candidates and areas which are of particular interest for mass measurements are indicated.

High-accuracy mass determinations with ISOLTRAP

The tandem Penning trap mass spectrometer ISOLTRAP [25,26] allows mass determinations of exotic nuclei with high accuracy. The setup consists of three main parts, a gas-filled rf-quadrupole for capture, cooling, and bunching of the ISOLDE beam, a linear cooler Penning trap for further cooling and for separation of isobaric contaminations (see above), and a hyperbolic high-precision Penning trap for mass measurements. The mass determination in the trap is based on the determination of the cyclotron frequency ω_c of the ion in a magnetic field of strength B according to $\omega_c = \omega_+ + \omega_- = B \cdot q/m$. The eigenmotions ω_+ and ω_- can be excited with dipole fields and the sum or difference of eigenfrequencies can be excited with a quadrupole field. At ISOLTRAP the sum $\omega_+ + \omega_-$ is excited by an azimuthal quadrupole field for the determination of the true cyclotron frequency ω_c. Due to the excitation the ions gain radial energy which is converted into axial energy when they pass through the inhomogeneous fringe field of the magnetic field at ejection. The measurement of the drift time to reach the detector as a function of the applied radio frequency yields a resonance curve with a characteristic minimum at ω_c. The magnetic field strength B is calibrated with ions of accurately known masses or, in other words, the unknown mass is obtained from the ratio of cyclotron frequency of the reference ion and that of the ion under investigation.

Under ideal conditions the achievable mass resolving power depends only on the excitation time T_{rf} and the Fourier-limited line width is $\Delta \nu_c (\text{FWHM}) = 0.9/T_{rf}$. A mass-resolving

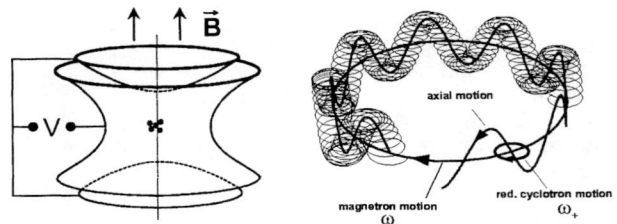

FIGURE 4. Left: Schematic drawing of the hyperbolic high-precision Penning trap for mass measurements. The magnetic field and the applied rf-voltage are indicated. **Right:** Schematic view of the three eigenmotions of an ion in a Penning trap [26].
Reprinted from [26], © 1996, with permission from Elsevier Science.

power $m/\Delta m$(FWHM) $= 8 \cdot 10^6$ has been reached for $A = 133$ ions and for barium and caesium nuclei a mass accuracy of $\delta m/m \simeq 1 \cdot 10^{-7}$ is typically achieved for ions in the same mass range, corresponding to $\delta m \simeq 14$keV [46].
In a recent experiment by Adelberger and co-workers [47] the (possible) weak scalar interaction is probed by studying positron-neutrino correlations in the pure Fermi transition of ^{32}Ar (for further reading see [48]). In this $0^+ \to 0^+$ β-decay the positron-neutrino correlation is derived from the energy spectrum of the delayed protons, which reflects the recoil momentum of the daughter nucleus of the superallowed decay. With the results one can set constraints on the possible existence of scalar weak interactions. The result of the experiment is consistent with the Standard Model [47] but the analysis contains a systematic uncertainty which is due to the mass uncertainty of ± 50keV in ^{32}Ar. The authors used the isospin-multiplett mass equation [49] to obtain a more precise estimate. This result can be put on much safer ground with direct mass measurements which are accuracte to $\simeq 5$keV. It is now planned to determine the atomic mass of ^{32}Ar, which has a half-life of only 98ms, with this accuracy with the ISOLTRAP setup.

Mass determinations of very short-lived ions with MISTRAL and with Isochronous Mass Spectrometry (IMS)

The challenge in direct mass measurements of exotic nuclei is to develop experimental techniques which on the one hand allow accurate mass determination and on the other hand are fast in order to access nuclei far from stability which mostly have lifetimes much shorter than one second. These requirements are met by in-flight techniques as employed at MISTRAL (Mass measurements at ISolde using a Transmission RAdiofrequency spectrometer on-Line) [50] or with isochronous mass spectrometry IMS at the ESR [51].
In the MISTRAL experiment it is again the cyclotron frequency of the ion under investigation which is to be determined, here by means of two successive modulations of the kinetic energy while two revolutions in a Smith-type radiofrequency spectrometer [52,53]. Ions for which the net effect of the radiofrequency modulation and demodulation vanishes are transmitted through the slit system (see figure 5) according to the condition

$$\omega_{\text{rf}} = (n + 1/2) \cdot \omega_c \tag{1}$$

where n is an integer harmonic. The mass-resolving power of the instrument is proportional to n and inversely proportional to the width of the slit. Therefore, beam cooling is essential for a high transmission which is particularly important for experiments with low-intensity beams. The radiofrequency is scanned resulting in spectra as shown in the upper left part of figure 5 and the unknown masses are determined from a reference mass from the ratio of their measured cyclotron frequencies. An relative mass accuracy on the $8 \cdot 10^{-7}$-level is reached for nuclei around $A = 30$ [54,55].
In recent experiments the masses of $^{23...31}$Na isotopes have been determined, ^{31}Na having a half-life as short as 17ms [54,55]. The achieved mass accuracy was typically 20keV and

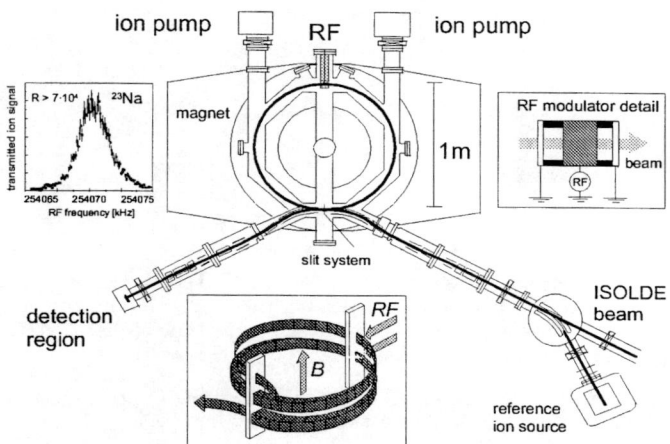

FIGURE 5. Schematic view of MISTRAL [50] showing the injection beam line, where ions delivered from ISOLDE are alternately injected with reference ions from an internal source, the spectrometer, and the detection region. The insets show the signal of a ^{23}Na beam transmitted through the spectrometer as a function of the modulation frequency (top left), the detail of the rf-modulator (top right), and the view of the trajectory envelope (center).

the resolving power (FWHM) was in the vicinity of $1 \cdot 10^5$. In addition atomic masses of the elements ranging from Ne to Al have been performed in order to investigate the region of the $N = 20$-shell closure in more detail [5]. This shell closure is located far from stability and for Ne the neutron dripline comes close (see figure 3). Irregularities of the trends of two-neutron separation energies might be an indication for vanishing shells far from stability but there are also experimental hints for deformation effects [5].

Another approach for fast direct mass measurements are spectrometers with an isochronous ion optical characteristics like the TOFI spectrometer [56]. The time-of-flight of an ion through such a spectrometer is to first order independent of it's velocity and depends only on it's mass-to-charge ratio. The ESR when operated in an isochronous mode can be used as a multi-turn time-of-flight mass spectrometer [51,57], if the ions are injected with velocities v such that their Lorentz factor γ matches the ion-optical parameter γ_{tr} which characterizes the transition point of the storage ring. In this case ions with identical m/q have identical revolution times T because their velocity differences are compensated by appropriate flightpath differences [51]. From the equation for the mass-resolving power

$$\frac{\Delta(m/q)}{m/q} = \gamma_{tr}^2 \cdot \frac{\Delta T}{T} + (\gamma_{tr}^2 - \gamma^2) \cdot \frac{\Delta v}{v} \qquad (2)$$

one can easily see that $\Delta T/T$ becomes a direct measure for $\Delta(m/q)/(m/q)$ if the isochronicity condition $\gamma = \gamma_{tr}$ is fulfilled. An appropriate ion optical setting has been developed for $\gamma_{tr} = 1.37$ [58] and its performance has been investigated with ion beams, whose mean momenta were modified with the electron cooler. As can be seen from figure 6, there is only a very weak dependence of the revolution time on the ion velocity: within the storage momentum acceptance of approximately $\Delta p/p \simeq 0.55\%$ the revolution time is affected by $\Delta T/T \simeq 0.01\%$ only.

For direct mass measurements of short-lived nuclei the revolution time T is obtained from a time-of-flight measurement of each turn for each individual ion. The measurements start with the injection into the ESR and the ions can be observed for typically 10 to 50 μs corresponding to approximately 20 to 100 turns of the ions before they are lost. The unknown masses are obtained from isotopes with known masses by linear interpolation according to equation (2). Therefore the FRS must provide several different isotopes simultaneously, but a preselection of injected ions with the $B\rho$–ΔE–$B\rho$-separation method

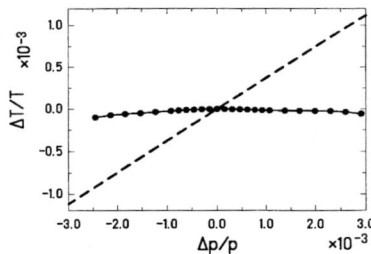

FIGURE 6. Relative change of revolution time $\Delta T/T$ versus relative momentum deviation $\Delta p/p$. The data points have been obtained with an isochronous ion-optical setting for $\gamma_t = 1.37$ with bare ^{58}Ni ions [58]. The dashed line shows the time-of-flight variation for the ESR standard ion optics ($\gamma_t \simeq 2.5$) for comparison.

must be applied aiming at a balance of isotopes with known and unknown masses in order to avoid deadtime effects of the detection system due to those fragments with known masses which appear with high rates. In a recent experiment with fragments of a ^{52}Cr beam the masses of the neutron-deficient isotopes ^{41}Ti, ^{43,44}V, ^{45}Cr, and ^{48}Mn could be determined for the first time. A mass resolution of $m/\Delta m(\text{FWHM}) = 1 \cdot 10^5$ and a mass accuracy of the order of 100 keV was achieved, for details and most recent results see ref. [59]. Although these nuclei have half-lives of the order of $T_{1/2} \sim 100$ms, the measurement time was of the order of 50μs or even less. Therefore, the same performance and accuracy is to be expected for nuclei with half-lives of the order of $T_{1/2} \sim 10\mu$s!

Those extremely short-lived isotopes can only be accessed with in-flight techniques and therefore unprecedented mass measurements are now possible with IMS. Moreover, new experiments aiming at the discovery and investigation of isomers are feasible now. Direct mass measurements with longer-lived nuclei are possible in the ESR with Schottky Mass Spectrometry.

Large-area mass determinations with Schottky Mass Spectrometry (SMS)

Another approach to cancel the second term in equation (2) is to minimize the relative velocity spread $\Delta v/v$ by means of electron cooling [22]. This technique is known as Schottky Mass Spectrometry [60,18] because the frequency determination is performed with Schottky diagnosis. So far, mass-resolving power $m/\Delta m(\text{FWHM}) \simeq 6.5 \cdot 10^5$ has been reached. Further improvements are expected from a still better stabilization of the ESR magnet power supplies, narrower temperature distributions of the cooling electrons, the new time-capture data aquisition [61], and the use of an ultra low noise resonant Schottky pickup cavity [62]. Schottky spectroscopy has no deadtime effects and hence it is desirable to inject many different isotopes simultaneously in order to get many correlations between isotopes with known and unknown masses at a time, therefore the FRS is operated as a pure magnetic-rigidity analyzer. SMS is presently the only experimental technique where a large mass surface can be mapped.

In experiments carried out so far with Bi primary beams more than 200 masses of neutron-deficient isotopes in the region from caesium to uranium, whose mass was unknown before [45], have been directly measured [16,63,64] with an accuracy of typically 100 keV. The lower limit of accessible half-lives lies around 5s due to the time necessary for cooling of the hot fragment beams and the time for measurement, which depends on the desired frequency bandwidth and frequency resolution and is of the order of few seconds to few minutes.

At the end of this section table 1 summarizes the main features and the performance of the experimental setups for direct mass measurements described in this paper. The experiments behind SHIP are described in the outlook.

TABLE 1. Characteristic features and parameters of the described experiments and techniques for direct mass measurements

	ISOLDE		SIS-FRS-ESR		SHIP	
production and separation	ISOL		fragmentation, ED, in-flight		heavy ion fusion, in-flight	
spectrometer	Penning trap	Smith type	storage ring		Penning trap	MR-TOF
mass range	$1 \leq A \leq 238$				$A \geq 238$	
accessible half-lives	$\geq$ 100ms	$\geq$ 1ms	$\geq$ 1s	$\geq$ 1μs	$\geq$ 1s	$\geq$ 100μs
cooling	buffer gas	no	electrons	no	buffer gas	no
mass resolving power $m/\Delta m$ (FWHM)	$6.0 \cdot 10^{5a}$	$1.0 \cdot 10^{5b}$	$6.5 \cdot 10^{5c}$	$9.0 \cdot 10^{4d}$	$10^6 \ldots 10^{8\ e}$	$\geq 10^{5\ e}$
mass accuracy	15keV^a	10...20keV^b	100keV^c	100keV^d	15keV^e	50keV^e

^{a} for masses around $A = 130$ and an excitation time of 0.9s, the mass resolving power can become as high as $8.0 \cdot 10^6$ in this mass range using an excitation time of 12s [46]
^{b} for masses around $A = 30$ [55]
^{c} independent of mass [16,61]
^{d} for masses around $A = 45$ [59]
^{e} design goal

Exemplary results

In this section results obtained at the ISOLTRAP experiment and with SMS at the ESR will be presented and discussed. As some representative results the mass values obtained for the even-even barium and polonium isotopes are displayed in figure 7 in comparison with the predictions of different mass models. The Finite-Range Droplet Model (FRDM) [65] is a macroscopic-microscopic nuclear model where the macroscopic part accounts for the bulk contribution to the binding energy and the microscopic part accounts for shell and pairing effects. The Extended Thomas-Fermi model with Strutinski Integral (ETFSI) [66] combines the self-consistent Hartree-Fock approach with a shell correction term obtained using the Strutinski-integral method. Therefore this model is in principle similar to the macroscopic-microscopic approach, but in ETFSI the two parts are connected much tighter because the same Skyrme-type force governs both parts. A fully microscopic model is the Relativistic Mean Field approach (RMF) [67], which yields ground-state properties of even-even nuclei such as binding energies, nuclear radii and deformation parameters. The effective force NL3, which has been fitted to ground-state properties of 10 spherical nuclei, is used to describe the effective nucleon-nucleon interaction. The best agreement between calculated and experimental masses is found

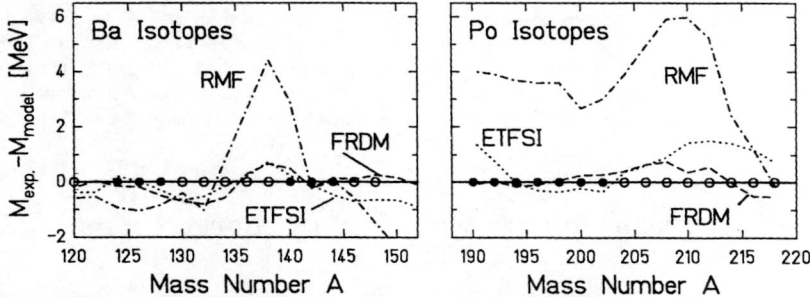

FIGURE 7. Comparison of experimental masses of even-A barium [46,45] (left panel) and polonium isotopes [64,45] (right panel) with the predictions of various nuclear models as indicated. The mass values obtained from the present experiments are shown with full dots while the previously known masses from ref. [45] are shwon with open symbols.

for the macroscopic-microscopic FRDM, whereas the fully microscopic RMF calculation shows the largest discrepancies. However, it is important to note that the total binding energy of polonium isotopes is of the order of 1500 MeV and thus the RMF results are accurate to 0.3%.

In all mentioned calculations deformation is included. The phenomenon of deformation and shape coexistence is well established in the Pb region [68–70]. Recent studies within the framework of the Particle-Core Model (PCM) show that the interplay between spherical structures and an intruder band in the light polonium isotopes leads to strong deformations [71]. The ground state of ^{192}Po is expected to have oblate shape, whereas for still lighter Po isotopes prolate ground state shapes are predicted.

The experimental mass values can be used for the analysis of deformation and of the nuclear shell structure. A macroscopic-microscopic model has been fitted to the experimental mass values obtained for the barium isotopes and the deformation parameter β_2 revealing the quadrupole deformation observed for the nuclear potential has been extracted [46]. In the left panel of figure 8 the theoretical values are compared with data obtained from optical isotope shift measurements. Both sets of data clearly reveal the shell closure at the 'magic' neutron number $N = 82$ in terms of the vanishing deformation and show increasing prolate deformation for the lighter and for the heavier barium isotopes, respectively. The absolute values of both sets of data show fair agreement (for a more detailed discussion see [46]).

In the right panel of figure 8 the shell closure at ^{210}Po corresponding to the magic neutron number $N = 126$ is clearly seen from the microscopic part of the nuclear binding energy $E_{\text{mic.}}$, which is defined as the experimental ground-state binding energy minus the macroscopic energy according to the FRDM [65]. Within the macroscopic-microscopic model the macroscopic liquid-drop term varies only smoothly with neutron number and the two major contributions to the microscopic energy are shell-correction energy and pairing energy. It has been pointed out earlier [72,73] that deformations set in at $A \simeq 199$ and the calculated potential-energy surface of polonium nuclei becomes shallow for $A \leq 194$.

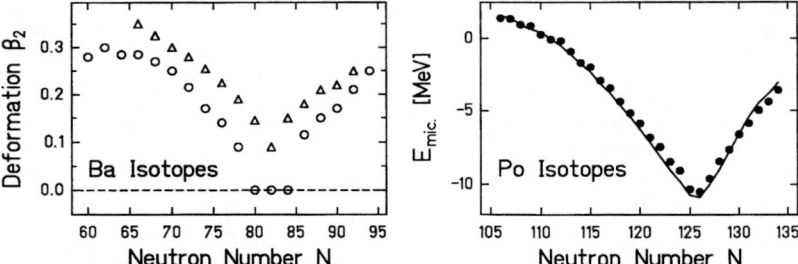

FIGURE 8. Shell structure effects explored with a macroscopic-microscopic model. The vanishing quadrupole deformation observed for barium isotopes [46] at $N = 82$ (left panel) and the pronounced minimum of microscopic energy [65] for polonium isotopes at $N = 126$ (right panel) clearly indicate the corresponding shell closure. The triangles in the left panel represent the result of isotope-shift measurements and the circles are calculation results (for details see [46]).

Half-life measurements with highly-charged ions in the ESR

Pioneering experiments

With the combination of the fragment separator FRS and the experimental storage ring ESR it is for the first time possible to store exotic nuclei in a storage ring [74] and to investigate nuclear β-decay properties in selected ionic charge states. This offers the unique possibility to manipulate and to study nuclear half-lives in dependence of the number of attached electrons [75] and to study few-electron ions or bare nuclei under conditions which usually occur only in hot plasmas of a star where nucleosynthesis takes place [76,77]. With bare ions, where orbital electron capture (EC) is impossible, one can study or search for weak β^+-decay channels [78,79]. For instance it is a very important question for supernovae scenarios, whether there is a weak β^+-branch in the doubly-magic ^{56}Ni [80].

By inverting the time-arrow of orbital electron capture one comes back to a β^--decay, but whith the created electron bound in an inner shell of the daughter atom, the 'bound-state β^--decay', β_b. Whereas this process represents only a scarce decay branch in neutral atoms because there are no inner-shell vacancies, it becomes the more important the higher the mother atom is ionized as for instance in hot stellar plasmas. For those highly ionized atoms β_b-decay enhances the Q-value of the β-decay into the continuum, β_c, by approximately the binding energy of the created electron. To observe β_b-decay in a terrestrial laboratory one has to preserve a high ionic charge state for an extended period of time (of the order of hours) which is nowadays only possible in the ultra-high vacuum conditions in storage rings or ion traps. Hence, bound-state β-decay was experimentally observed for the very first time at the heavy ion storage ESR [81]. For this pilot experiment a striking example has been chosen: ^{163}Dy, which is stable as a neutral atom because the Q-value for continuum β-decay to ^{163}Ho is negative, $Q_{\beta_c} = -2.56$keV, might decay as a bare ion by β_b-decay to the ground state of ^{163}Ho with a positive Q-value of roughly 50keV for the electron being emitted into the K-shell of the daughter atom. This decay has indeed been observed and the measured half-life of (48^{+5}_{-4})d agrees nicely with the theoretically expected half-life of 50d [81].

In a second experiment β_b-decay of bare ^{187}Re has been investigated at the ESR [82]. The couple ^{187}Re/^{187}Os serves, together with the couple ^{238}U/^{232}Th, as a 'clock' for the age of our Milky Way galaxy, due to the half-life of $42 \cdot 10^9$yr for the neutral ^{187}Re atom. However, for bare ^{187}Re^{86+} the experiment yielded a half-life dramatically reduced by nine orders of magnitude of only 33 ± 2yr, due to an open β_b-decay channel to the first excited state of ^{187}Os at 10keV excitation energy, which is not accesible for neutral ^{187}Re [82]. In this case, the nuclear matrix element, the ft-value, for the transition to the excited osmium state was experimentally not known before, in contrast to the dysprosium experiment where the corresponding ft-value could be determined from the measured half-life of the orbital EC-decay of the ground state of neutral ^{163}Ho into the ground state of ^{163}Dy. This crucial dependence of the β-half-life on the ionic charge state shows that the Re/Os-couple and also other eon clocks based on long-lived radioactive nuclides can provide constraints for the age of our galaxy only within a detailed and reliable model of the history of nucleosynthesis and chemical evolution of stars [83]. That means that those 'nuclear' clocks are, basically, on the same footing as the well-known 'astronomical' clocks like globular clusters, white dwarfs etc., which also have to be 'calibrated' by the time scale of stellar evolution. The hope that radioactive nuclear couples could serve as clocks, independent of the widely unknown galactic history, has to be withdrawn.

Future experiments exploring weak-interaction decays

A forthcoming experiment concerning bound-state beta decay will address the couple ^{205}Tl/^{205}Pb [84]. Neutral ^{205}Pb decays to the ground state of ^{205}Tl by orbital electron capture with a half-life of $1.5 \cdot 10^7$yr and a Q-value of about 50keV. The ^{205}Tl nucleus, on the other hand, can be transformed to the first excited state of ^{205}Pb at 2.3keV excitation energy by capturing solar neutrinos from the pp-cycle ($E_{\max} = 420$keV). Therefore, the relative ^{205}Tl/^{205}Pb abundance in deep-lying, thallium-rich ore bodies provides, when corrected for background, the product of the solar pp-neutrino flux, averadged over the lifetime of the ore bodies, and of the pp-neutrino capture probability into the 2.3keV state of ^{205}Pb, which is the dominant capture channel for pp-neutrinos. In this respect, ^{205}Tl serves as an integrating solar pp-neutrino detector like ^{71}Ga does in the GALLEX and

SAGE experiments, but with two significant differences: the threshold for the pp-neutrino energy is much lower (50keV as compared to 235keV for ^{71}Ga) and the integration time of the solar neutrino flux is quite different (typical lifetime of the ore bodies of some million years as compared to a few days in the ^{71}Ga experiments). Now, for bare or hydrogen-like ^{205}Tl bound-state β-decay into ^{205}Pb becomes energetically allowed with a positive Q-value of about 40keV for the bound electron being in the K-shell of ^{205}Pb. Since the ground state of ^{205}Tl, ground state of ^{205}Pb and 2.3keV state of ^{205}Pb have the spins $I = 1/2^+, 5/2^-$, and $1/2^-$, respectively, and since the second excited state of ^{205}Pb is as high as 263keV ($I = 5/2^-$), β_b-decay from the ^{205}Tl ground state feeds practically exclusively the 2.3keV level of ^{205}Pb, as it is the case for the capture of pp-neutrinos. The only way to determine precisely and reliably the unknown nuclear matrix element for this transition and, hence, the solar pp-neutrino capture probability into this state, is a measurement of the half-life of bare or hydrogen-like ^{205}Tl (as proposed by P. Kienle [85]), from which the ft-value can be simply extracted. Systematics of near-lying nuclear levels with similar shell structure gives as a very rough half-life estimate of about 1 year for bare ^{205}Tl. Hence, an experimental determination of the half-life of bare ^{205}Tl in the ESR should be even more simple than in the case of ^{187}Re with a half-life of 33 years.

In a next step we plan to measure for interesting cases β-lifetimes for well-defined ionic charge states, by adding or removing a certain number of electrons in the atomic cloud. Within some constraints, the charge state of heavy ions can be chosen arbitrarily and in this way the weak interaction decays can be influenced. The general behavior of the EC/β^+-branching ratio for bare, H-like, and He-like nuclei with atomic number $Z = 82$ is shown in figure 9. For a given Z and a fixed charge state the EC/β^+-branching ratio depends only on the Q_{EC}-value, which is connected with the β^+-endpoint energy $E_{\max}$ by $Q_{EC} = E_{\max} + 2 \cdot m_e c^2$, where $m_e c^2$ is the electron rest mass.

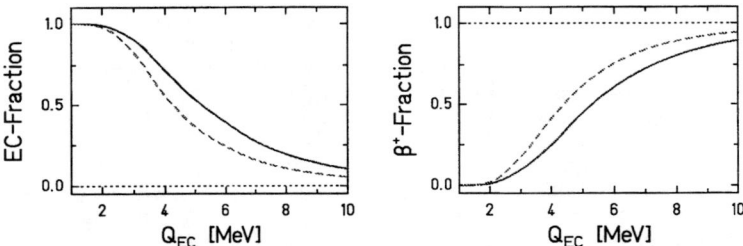

FIGURE 9. Branching ratio for EC and β^+-decay as a function of the Q_{EC}-value. The figure shows the EC probability (left panel) and the β^+-decay probability (right panel) calculated according to ref. [86] for bare (dottel lines), hydrogen-like (dashed lines), and helium-like nuclei (solid lines) with atomic number $Z = 82$.

An other effect, which has been known theoretically for a long time [87], but which could not be confirmed experimentally, can now be investigated at the ESR: the influence of screening of the atomic electrons on the β-decay functions. Not only the low-energy part of β-ray spectra is influenced by the screening of the nuclear charge by atomic electrons but also the β-decay constant λ_β. The origin of this modification is a rather simple one, namely that λ_β is proportional to the integrated Fermi-function, that function which has been introduced by Fermi to account for the interaction between the Coulomb field and the emitted β-particles. The presence or absence of shell electrons modifies the Coulomb field of the nucleus and for a calculation of the screening effect it is therefore necessary to correct the wave function of the β-particle at the origin for the modified Coulomb field [86–89]. The screening effect increases with atomic number Z and with decreasing Q-value. It is very weak for β^--decay and much stronger for β^+-decay. In figure 10 the ratio $B = \lambda_\beta^*/\lambda_\beta$, is shown as a function of the β-endpoint energy $E_{\max}$ for $Z = 82$ for electrons (left) and for positrons (right) for $Z = 36$ (dashed line) and for $Z = 82$ (solid line). Here, λ_β^* denotes the β-decay-constant with screening and λ_β denotes the same quantity without screening. Thus, the ratio B is equal to the ratio of the integrated Fermi-function with screening and the integrated Fermi-function without screening, which can be found in ref. [86].

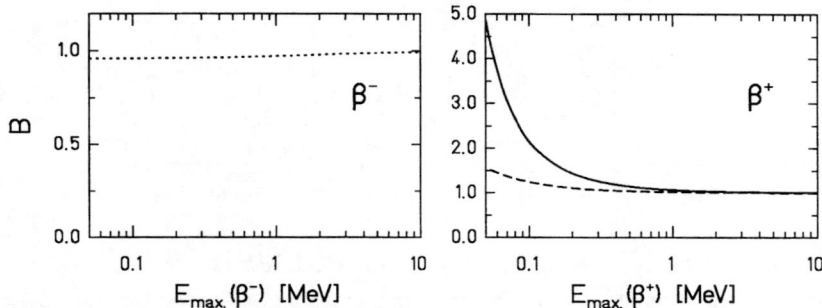

FIGURE 10. Screening effect depicted in terms of the ratio B (see text) as a function of the β-endpoint energy $E_{\max}(\beta)$. **Left:** β^--decay of a $Z = 82$-nucleus. **Right:** β^+-decay for $Z = 36$ (dashed line) and for $Z = 82$ (solid line). Note the different scales of the ordinate!

Schottky spectroscopy at the ESR is taylor-made for systematic and detailed investigations of the Fermi function in β-decay, for several reasons:
1. Beta-unstable ions in a well-defined ionic charge state carrying $0, 1, 2, \ldots$ electrons can be produced and separated with the FRS and injected, cooled and stored in the ESR for a long time.
2. Both, the mother- and daughter atom appear separated in the Schottky spectrum and can be unambiguously identified by their specific fingerprint. This is obvious for the common β^- and β^+-decay, where the ionic charge state will be changed by the decay. Moreover, it is even possible for EC- and β_b-decay where the ionic charge state does not change, if the corresponding Schottky lines are resolved, i. e., if the Q-value is sufficiently high. That supposes a Q-value exceeding a few hundred keV only (in the cases of ^{163}Dy and ^{187}Re discussed above the Q-values were only about 50keV and, thus, a more complicated detection method had to be applied, cf. refs. [81,82].
3. From the Schottky signals the number of the corresponding atoms can be determined as a function of time which provides the β-lifetime. The unique feature, that both, mother- and daughter atoms, can be counted simultaneously but independently at all times improves significantly the accuracy in the lifetime evaluation.

From the ratio of the β_b-decay probability $\lambda_{\beta_b}(n)$ (resp. EC-decay probability $\lambda_{\rm EC}(n)$) into (resp. from) a well-defined state n of the bound electron and of the corresponding continuum decay probability λ_{β_c} of the neutral atom (resp. β^+-continuum decay probability λ_{β^+}) according to Bahcall [90] one gets immediately and independent of the nuclear matrix element, which drops out in this comparison, the ratio of the electron wave function in the final (resp. initial) state $\Psi_n(R)$ at the nuclear surface R and the Fermi function $f(Z, W_0)$

$$\lambda_{\beta_b}(n)/\lambda_{\beta_c} = \lambda_{\rm EC}(n)/\lambda_{\beta^+} = n_f \cdot Q_n^2 \cdot |\Psi_n(R)|^2/f(Z, W_0) \tag{3}$$

where n_f is the number of free (resp. occupied) places in the electron shell, Q_n the corresponding Q-value, Z the nuclear charge of the daughter (resp. mother) atom and

W_0 the nuclear energy release in units of the electron rest energy for the corresponding continuum decay. Two regimes seem to be most interesting for this comparison:
1. concerning EC/β^+: the region just below and above the β^+-threshold, where the Q_{EC}-value is near $2m_ec^2$. Here, the Fermi function has by far the strongest dependence on Q_{EC}. A comparison of precisely measured decay probabilities of bare (pure β^+), hydrogen- and helium-like atoms could provide subtle and completely new information on both, the Fermi function and the density of the 1s electron wave function at the origin.
2. concerning β_b/β_c: bound-β decay of bare nuclei mainly into the K-shell. Most promising for this purpose are the nuclides ^{206,207}Tl and ^{205,206}Hg because of several reasons: the Q-values are in all cases about 1.5MeV, a separation of the Schottky lines for mother- and bound-beta daughter atoms, respectively, seems to be feasible. For those rather small Q-values and high atomic numbers $Z = 81, 80$ one expects a very strong β_b-branch to the K-shell, of the order of 20% to 30% of the total β-decay probability. Finally, those high Z and small Q-values are most sensitive probes for the 1s electron wave function as well as for the Fermi function.

OUTLOOK AND CONCLUSIONS

After the first generation of experiments with exotic nuclei at storage rings and ion traps new experimental facilities are coming up like the already mentioned AD experiments [1] and the HITRAP [32] project. At both facilities the ions are decelerated in storage rings and then captured, bunched, cooled and finally trapped. However, when one wants to investigate nuclei far off stability with very short half-lives, say below milliseconds, which can be investigated at in-flight separators exclusively, it is necessary to develop new schemes for deceleration and new investigation methods, which are fast and highly efficient. The fastest way of deceleration is the slowing down in matter, which brings even relativistic heavy ions to rest in time spans less than one nanosecond [91]. Schemes, where a fragment separator or a recoil separator is coupled to a stopping and bunching unit (which is similar to an IGISOL-type gas cell), are presently under discussion in many laboratories and are a new way to experiments with isotopically clean low-energy secondary beams with sub-millisecond half-lives. These fast deceleration schemes are completed by new fast and sensitive methods for measurements.
Here, we want to propose a new system for deceleration of heavy exotic nuclear beams combined with a new device for sub-millisecond direct mass measurements. Figure 11 shows a schematic view of the system, which consists of three main parts: a dispersive ion-optical system for fast slowing down and range bunching, a gas-filled stopper cell with an extraction, cooling, and bunching unit, and a multi-reflection time-of-flight mass spectrometer. The whole system can be coupled to an in-flight separator. In the first unit, the separated swift exotic nuclear beams are slowed down and range bunched taking advantage of the combination of a dispersive dipole stage (3QD3Q2Q) with the slowing down in homogeneous and wedge-shaped degraders [91]. The beam is then stopped in a gas cell, extracted and cooled and bunched in a gas-filled quadrupole [92]. Then the pulse of low-energy monoisotopic exotic nuclei is injected into an energy-isochronous multi-reflection time-of-flight mass analyzers [93] for direct mass measurements. These 'table-top' devices, in which ions are reflected back and forth between two or more electrostatic ion mirrors are presently under development for spectroscopic investigations at extraterrestrial sites, for instance for the exploration of the surface composition of comets [94]. The mass determination with such a spectrometer is based on the time-of-flight measurement of ions with keV energies in a given pathlength. In order to reach a high mass resolving power, which is proportional to the pathlength, the ions are multiply reflected between electrostatic grid-free ion mirrors before their arrival time is detected with a micro-channel-plate detector. The ion motion is achromatic and to second order energy isochronous and therefore the time-of-flight is directly proportional to the ions mass. Figure 12 shows the separated mass doublet CO$^+$-N$_2^+$ after 31 reflections in a one meter long spectrometer. A mass resolving power $m/\Delta m(\text{FWHM}) = 51000$ is achieved

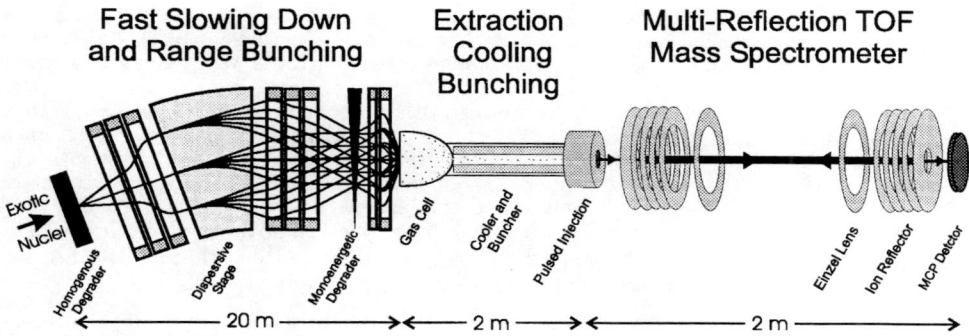

FIGURE 11. Schematic view (not to scale!) of a setup for sub-millisecond direct mass measurements with monoisotopic exotic nuclei showing the three main components for 1) fast slowing down and range bunching [91], 2) extraction, cooling, and bunching [92], and 3) direct mass measurements [93].

after a measurement time as short as 724μs. As can be seen from the right part of the figure, the mass resolving power is directly proportional to the number of reflections or the measurement time. Without the data aquisition and analysis system being opti-

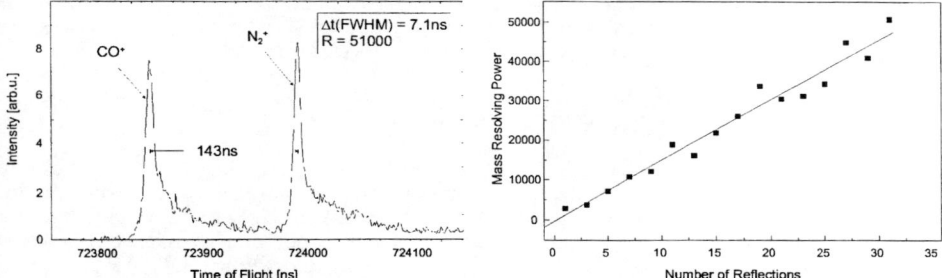

FIGURE 12. **Left:** part of a time-of-flight spectrum showing the mass doublet of N_2–CO after 31 passages through the one meter long multi-reflection time-of-flight mass spectrometer. **Right:** The mass resolving power increases quasi linearly with the number of reflections [94].

mized for this purpose, a mass accuracy of 1ppm (corresponding to a mass uncertainty of 50keV for an $A = 50$-ion) has been reached with such a spectrometer [95]. Present limitations are due to power-supply instabilities ($\Delta U/U \simeq 5 \cdot 10^{-5}$) and the size of the initial phase space. For longer-lived species the mass resolving power can be increased by a longer flight time, either by increasing the number of reflections or by increasing the pathlength between the mirrors. Also higher kinetic energies (presently $1.5 \cdot qe$ kV) and higher ionic charge states can lead to an increased mass resolving power. From this point of view multi-reflection time-of-flight mass spectrometers have a good potential for sub-millisecond mass measurements and will ideally complement the Penning trap mass measurements at the planned SHIPTRAP facility [31], which are better suited for longer-lived isotopes. The SHIPTRAP facility, which is designed also for other experiments with transuranium elements such as chemistry, optical spectroscopy, nuclear reaction studies, and nuclear spectrometry, will yield important experimental mass data which have been awaited for long time in the field of superheavy element research. The key parameter is the amount of microscopic energy which stabilizes superheavy elements (SHE). Experimental masses will allow a decisive determination of the location and strength of nuclear shells in SHE and are stringent tests of existing mass models. The data are also important for the prediction of reaction Q-values, the precise knowledge of which is crucial for SHE production, because systematic studies of cold-fusion reactions at SHIP have shown that the window of excitation energies becomes steadily narrower when going to heavier and heavier elements.

Finally it should be noted that there are also other possible applications for a multi-reflection time-of-flight spectrometer: 1) It can be used as a fast isobar separator: when the extraction potential is applied for a short time interval, only those ions with the proper flight time, i. e., with the proper mass or mass range, respectively, are extracted. 2) if the loss rate of the ions can be minimized one can use these devices as storage devices and, e. g., for the transport of antiprotons [96]. The practical advantages are low weight of few kilograms only, small size, and little power consumption (note that multi-reflection time-of-flight mass spectrometers are adapted to the needs of space missions).

In conclusion, storage rings and ion traps are facilities which are well suited for the investigation of gross properties of exotic nuclei. Presently several different and complementary experimental setups are in operation, which provide important nuclear structure data of short-lived nuclei far-off stability, a field living vividly from the interplay of experiment and theory. Schottky Mass Spectrometry with stored cooled ions is capable of mapping large sectors of the mass surface, Penning trap measurements yield the most accurate mass values and time-of-flight techniques give access to the shortest-lived nuclei. Besides the mass values themselves the experimental results allow the study of shell effects, deformation, and decay properties at the limits of stability. The coming experiments at GSI will focus on mass measurements with fission products stored in the ESR and with fusion-reaction products at SHIP. Unprecedented half-life measurements with bare and few-electron ions will lead to new insights into β-decay. The next experiments have the potential to obtain new information on the Fermi function, the density of the 1s electron wave function at the origin, and the role of electron-electron correlations.

ACKNOWLEDGEMENTS

For fruitful and stimulating discussions we are full of gratitude to G. Audi, M. Hausmann, F. Herfurth, O. Klepper, H.-J. Kluge, C. Kozhuharov, M. D. Lunney, C. Monsanglant, Yu. N. Novikov, T. Radon, M. de Saint Simon, J. Stadlmann, H. Schatz, S. Schwarz, and C. Thibault. One of us (C. S.) would like to thank CERN and the ISOLDE collaboration for a one-year Scientific Associateship and the MISTRAL and the ISOLTRAP groups for their hospitality during the stay.

REFERENCES

1. J. Eades, F. J. Hartmann, *Rev. Mod. Phys.* **71** (1999) 373.
2. S. Peil, G. Gabrielse, *Phys. Rev. Lett.* **83** (1999) 1287.
3. B. Jonson, A. Richter, accepted for publication in *Hyp. Int.*
4. H. Geissel *et al.*, *Annu. Rev. Nucl. Part. Sci.* **45** (1995) 163.
5. C. Détraz, D. J. Vieira, *Annu. Rev. Nucl. Part. Sci.* **39** (1989) 407.
6. H. Ravn, *Phil. Trans. R. Soc. Lond.* **A356** (1998) 1955.
7. For an overview see: *Proceedings of Stori 1996*, eds. F. Bosch and P. Egelhof, *Nucl. Phys.* **A626** (1997).
8. G. Münzenberg, *Nucl. Instr. Meth.* **B70** (1992) 265.
9. G. Münzenberg *et al.*, *Nucl. Instr. Meth.* **161** (1979) 65.
10. V. Ninov *et al.*, *Nucl. Instr. Meth.* **A357** (1995) 486.
11. T. S. Nikolaev, T. S. Dimitriev, *Sov. Phys. Tech. Phys.* **15** (1971) 1383.
12. N. Bohr, *K. Dan. Vidensk. Selsk. Mat. Fys. Medd.* **18(8)** (1948).
13. H. Geissel *et al.*, *Nucl. Instr. Meth.* **B70** (1992) 286.
14. C. Scheidenberger *et al.*, *Nucl. Instr. Meth.* **B142** (1998) 441.
15. B. Franzke *et al.*, *Nucl. Instr. Meth.* **B24/25** (1987) 18.

16. H. Geissel et al. in *Proceedings of the 2nd International Conference on Exotic Nuclei and Atomic Masses (ENAM98)*, Bellaire, Michigan, USA.
17. J. Borer et al., *Proceedings IXth Conf. High Energy Accelerators*, Stanford (1974) 53.
18. B. Schlitt et al., *Hyp. Int.* **99** (1996) 117.
19. B. Franzke et al., *GSI Scientific Report 1995* **GSI96-1** (1996) 159.
20. F. Noldenet al., *Nucl. Phys.* **A626** (1997) 491c.
21. M. Steck et al., *Phys. Rev. Lett.* **77** (1996) 3803.
22. M. Steck et al., *Nucl. Phys.* **A626** (1997) 495c.
23. H. L. Ravn et al., *Nucl. Instr. Meth.* **B126** (1997) 176.
24. R. B. Moore et al., *Physica Scripta* **T59** (1995) 93, and
    W. M. Itano et al., *ibid.*, 106.
25. H.-J. Kluge et al., in *Proceedings of the International Conference on Nuclear Shapes and Nuclear Structure at Low Excitation Energies*, eds. M. Vergnes, D. Goutte, P. H. Heenen, J. Sauvage, Antibes (1994) 83.
26. G. Bollen et al., *Nucl. Instr. Meth.* **A368** (1996) 675.
27. M. König et al., *Int. J. Mass Spectrom. Ion Processes* **142** (1995) 95.
28. H. Raimbault-Hartmann et al., *Nucl. Instr. Meth.* **B126** (1997) 378.
29. G. Savard et al., *Phys. Lett.* **A158** (1991) 247.
30. D. Habs et al., *Nucl. Instr. Meth.* **B139** (1998) 128.
31. Latest news on SHIPTRAP can be found in the World Wide Web at the following site: http://www-aix.gsi.de/~shiptrap/
32. M. Diederich et al., *Phys. Scripta* **T80** (1999) 437. More information about the HITRAP project is available from http://www-aix.gsi.de/~eurotrap/hitrap.htm
33. W. Quint, *Physica Scripta* **T59** (1995) 203, and
    G. Werth, *ibid.*, 206.
34. M. H. Holzscheiter et al., Proposal SPSLC96-47/P302, presented to the CERN SPSLC on 20. October 1996, available from http://www.cern.ch/~athena/
35. G. Gabrielse et al., Proposal SPSLC97-8/P306, presented to the CERN SPSLC on 25. March 1997, available from http://hussle.harvard.edu/~atrap/
36. G. Gabrielse et al., *Phys. Rev. Lett.* **82** (1999) 3198.
37. G. Bollen, *Nucl. Phys.* **A626** (1997) 297c.
38. W. Mittig et al., *Annu. Rev. Nucl. Sci.* **47** (1997) 27.
39. W. Nazarewicz et al., *Phys. Scripta* **T56** (1995) 9.
40. J. B. Ehrman, *Phys. Rev.* **81** (1951) 412.
41. R. G. Thomas, *Phys. Rev.* **88** (1952) 1109.
42. S. Aoyama et al., *Phys. Rev.* **C57** (1998) 975.
43. E. P. Wigner, *Phys. Rev.* **51** (1937) 106.
44. P. Van Isacker et al., *Phys. Rev. Lett.* **74** (1995) 4607.
45. G. Audi, et al., *Nucl. Phys.* **A624** (1997) 1.
46. F. Ames et al., *Nucl. Phys.* **A651** (1999) 3.
47. E. G. Adelberger et al., *Phys. Rev. Lett.* **83** (1999) 1299, and *Phys. Rev. Lett.* **83** (1999) 3101.
48. J. Deutsch, P. Quin, in *Precision Tests of the Standard Electroweak Model* (ed. Paul Langacker, World Scientific, Singapure, 1993).
49. M. S. Antony et al., *At. Data Nucl. Data Tables* **33** (1985) 447.
50. M. D. Lunney et al., *Hyp. Int.* **99** (1996) 105, and
    M. D. Lunney et al. in *Proceedings of the 2nd International Conference on Exotic Nuclei and Atomic Masses (ENAM98)*, Bellaire, Michigan, USA.
51. J. Trötscher et al., *Nucl. Instr. Meth.* **B70** (1992) 455.
52. L. G. Smith, *Proc. 3rd Int. Conf. on Atomic Masses*, University of Manitoba Press, Winnipeg (1967) 811.
53. M. de Saint Simon et al., *Nucl. Instr. Meth.* **B70** (1992) 459.
54. C. Toader et al., in *Proceedings of the International Conference on Trapped Charged Particles and Fundamental Physics*, Asilomar, AIP Conf. Proc. 457, eds. Daniel H. E.

Dubin, Dieter Schneider (1999) 95.
55. C. Monsanglant et al., accepted for publication in *Proceedings of the International Conference on Experimental Nuclear Physics Facing the Next Millenium*, Seville (1999).
56. J. M. Wouters et al., *Nucl. Instr. Meth.* **B26** (1987) 286.
57. H. Wollnik et al., *Nucl. Phys.* **A626** (1997) 327c.
58. M. Hausmann, doctoral thesis, Justus-Liebig-Universität Gießen (1999), and M. Hausmann et al., accepted for publication in *Nucl. Instr. Meth.* **A**.
59. J. Stadlmann et al., contribution to this conference.
60. B. Franzke et al., *Physica Scripta* **T59** (1995) 176.
61. M. Falch et al., contribution to this conference.
62. C. Gonzalez, F. Pedersen, *Proceedings Particle Accelerator Conference 1999*, New York (1999), 474.
63. T. Radon et al., *Phys. Rev. Lett.* **78** (1997) 4701.
64. T. Radon et al., submitted for publication in *Nucl. Phys.* **A**.
65. P. Möller et al., *At. Data Nucl. Data Tab.* **59** (1995) 185.
66. Y. Aboussir, J. M. Pearson, *At. Data Nucl. Data Tab.* **61** (1995) 127.
67. G. A. Lalazissis et al., *At. Data Nucl. Data Tab.* **71** (1999) 1.
68. G. Ulm et al., *Z. Phys.* **A325** (1986) 247.
69. M. P. Carpenter et al., *Phys. Rev. Lett.* **78** (1997) 3650.
70. W. Nazarewicz, *Phys. Lett.* **B305** (1993) 195.
71. A. M. Oros et al., *Nucl. Phys.* **A645** (1999) 107.
72. I. Ragnarsson et al., *Physica Scripta* **29** (1984) 385.
73. R. Bengtsson et al., *Physica Scripta* **29** (1984) 402.
74. H. Geissel et al., *Phys. Rev. Lett.* **68** (1992) 3412.
75. F. Bosch, *Physica Scripta* **T59** (1995) 221.
76. D. D. Clayton, *Astrophys. J.* **139** (1964) 637.
77. K. Takahashi, K. Yokoi, *Nucl. Phys.* **A404** (1983) 578.
78. H. Irnich et al., *Phys. Rev. Lett.* **75** (1995) 4182.
79. O. Klepper, *Nucl. Phys.* **A626** (1997) 119c.
80. D. D. Clayton, *Principles of stellar evolution and nucleosynthesis*, McGraw-Hill Book Company, New York, 1968.
81. M. Jung et al. *Phys. Rev. Lett.* **69** (1992) 2164.
82. F. Bosch et al., *Phys. Rev. Lett.* **77** (1996) 5190.
83. F. Bosch, *Phys. Scripta* **T80A** (1999) 28.
84. M. S. Freedman et al., *Science* **193** (1976) 1117, and M. S. Freedman, *Nucl. Instr. Meth.* **A271** (1988) 267.
85. P. Kienle, *Nucl. Instr. Meth.* **A271** (1988) 277.
86. H. Behrens, J. Jänecke, *Numerische Tabellen für beta-Zerfall und Elektronen-Einfang* (ed. H. Schopper, Springer Verlag, Berlin, 1969).
87. M. E. Rose, *Phys. Rev.* **49** (1936) 727.
88. W. Bühring, *Nucl. Phys.* **61** (1965) 110.
89. B. Crasemann, *Nucl. Instr. Meth.* **112** (1973) 33.
90. J. N. Bahcall, *Phys. Rev.* **124** (1961) 495.
91. H. Weick et al., accepted for publication in *Nucl. Instr. Meth.* **B**.
92. G. Bollen et al. in *Proceedings of the 2nd International Conference on Exotic Nuclei and Atomic Masses (ENAM98)*, Bellaire, Michigan, USA.
93. H. Wollnik in *AIP Conf. Nuclear Structure, Gatlinburg*, Proc. **184**, ed. C. Baktash (1999).
94. A. Casares et al., in *Proceeding of the 47th ASMS Conference on Mass Spectrometry and Allied Topics, Dallas* (1999).
95. A. Casares, H. Wollnik, private communication (1999).
96. C. H. Tseng, G. Gabrielse, *Hyp. Int.* **76** (1993) 381.

Reaction Studies with Exotic Nuclei in Storage Rings

Gottfried Münzenberg [1] and Gerhard Schrieder [2]

[1] *Gesellschaft für Schwerionenforschung mbH (GSI) Darmstadt, D 64291 Darmstadt and Johannes Gutenberg-Universität Mainz, Germany*
[2] *Institut für Kernphysik, Technische Universität Darmstadt, D 64289 Darmstadt, Germany*

Abstract.
Already the first experiments to explore nuclear ground-state properties of exotic nuclei with heavy-ion storage rings proved the research potential of precision experiments with the new experimental technique. In this contribution the perspectives for reaction studies in storage rings with energetic exotic nuclei at internal targets and in a small electron – heavy ion collider will be addressed. The feasibility of such experiments will be discussed.

I INTRODUCTION

The investigation of unstable exotic nuclei in heavy-ion storage rings opens up a new field of precision experiments. Heavy-ion cooler rings have been used at near Coulomb-barrier energies (TSR, Heidelberg, ASTRID, Aarhus) as well as at relativistic energies (Celsius, Stockholm, COSY, Jülich, ESR, Darmstadt). Storage rings are complementary to ion traps, however, with the capability to accept large phase-space and in-flight separated ions at full energy. The tremendous research potential of storage-ring experiments with exotic nuclear beams has been outlined in a number of publications [1–3].

So far the GSI Experimental Storage Ring (ESR) is the only facility for research with heavy exotic nuclei. New activities have been started in this field: the MUSES project [4] in RIKEN, a new project at GSI to extend the research program successfully started at the ESR, and recently a discussion in the US on the research possibilities with exotic nuclei in storage rings.

A highlight in nuclear structure research was the systematic mapping of the mass surface along the neutron deficient isotopes from tin to bismuth with a precision of 100 keV [5–7]. In these experiments not only the body of known masses could be extended by 10% but also new information on the extension of the shell region near ^{208}Pb and the onset of deformation was obtained.

The new phenomena expected and partly observed for exotic nuclei far-off stability are

- spatial separation of neutron and proton matter, the formation of nuclear skins
- extended wave-functions for weakly bound systems at the limits of stability, the existence of nuclear halos
- strong coupling to the continuum for weakly or almost unbound valence nucleons
- many-body correlations

These phenomena, not sufficiently well described in the conventional nuclear models, will have tremendous effects on the structure of exotic nuclei far-off stability.

The first-generation experiments at GSI already reveal the specific problems of the storage and investigation of short-lived exotic nuclei. Energetic secondary beams for nuclear reactions are produced by projectile fragmentation in peripheral collisions of relativistic heavy ions or by fission of uranium in-flight. Fission fragments are of specific interest as they will cover the hitherto inaccessible area of neutron rich nuclides. Subsequent in-flight separation allows to provide mono-isotopic beams or well defined isotope cocktails of short-lived nuclei far-off stability [8]. A disadvantage of this method for reaction studies is that the separated projectile fragments conserve the large phase-space from the production process and therefore have low phase-space density. To cope with the poor beam quality event-by-event particle tracking or energy-loss spectrometers are used to allow for precision experiments.

The beam quality of in-flight separated fragments is considerably improved by beam cooling in a heavy-ion storage-and-cooler ring [9]. To fully profit from the short separation time of in-flight separation, cooling times of the order of 0.1 s for beams are required. Fast stacking could enhance the intensity of stored beams even for short-lived exotic species. Next generation fragment separators will transmit beams with angular spreads of ± 30 mrad and momentum spreads of ± 2.5 % to transport fission fragments with sufficient efficiency. The large-emittance beams with rigidities of 14 Tm to 18 Tm (e. g. Uranium of 1 AGeV) need the development of new injection and cooling schemes, stochastic pre-cooling seems most appropriate [9].

An efficient use of the exotic beam for nuclear studies is only possible with a ring system comprising a collector- and cooler-ring and a separate second storage and Cooler ring dedicated to the experiments. The ring system should include internal targets of light nuclei such as hydrogen and helium for scattering experiments. The use of heavy targets as xenon for Coulomb excitation of the interacting fragments is under discussion. The implementation of a small electron ring, operated in the collider mode would open up new perspectives for structure research, not possible with existing equipment.

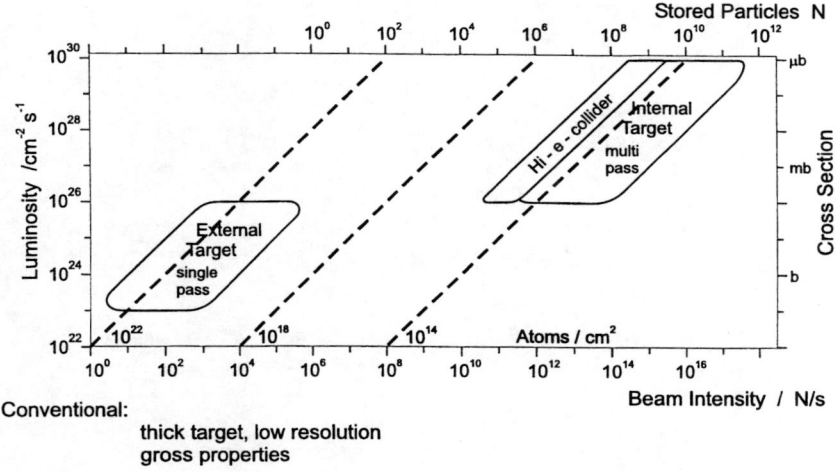

FIGURE 1. Luminosities for single-pass experiments and experiments in a storage ring. The circulation frequency of the stored ion is 10^6 s^{-1}. The domains of conventional single-pass experiments and of internal targets inside storage rings are indicated. For comparison the working range of a small electron collider is also displayed.

II REACTION STUDIES WITH EXOTIC NUCLEAR BEAMS – THE PRESENT SITUATION

The principal problem for nuclear reaction studies in storage rings is the luminosity. Assuming a reaction rate of 1 s^{-1}, luminosities of 10^{23} cm^{-2}s^{-1} are sufficient for elastic scattering experiments where the reaction cross sections are close to the geometrical cross section, structure investigations need 10^{25} cm^{-2}s^{-1} to 10^{28} cm^{-2}s^{-1}.

For reaction studies with unstable nuclear beams the technique of reversed kinematics has been developed: The energetic radioactive nucleus under investigation is directed onto a stable target [10]. The first generation of experiments with radioactive beams concentrated on the measurement of matter radii by measuring the total reaction cross section in a thick target. With cross sections of the order of barns and a target thickness of 10^{23} atoms cm^{-2} such experiments can be performed with extremely low intensities of down to one projectile per second.

The momentum distributions from the break-up of halo nuclei give insight into the momentum distributions of the halo nucleons and are therefore another important information about the halo structure. To cope with the broad momentum distributions of the projectiles high resolution experiments are performed in energy-loss spectrometers, placing the break-up target in the dispersive plane of the analyzing system [11]. For the new generation of large-acceptance spectrom-

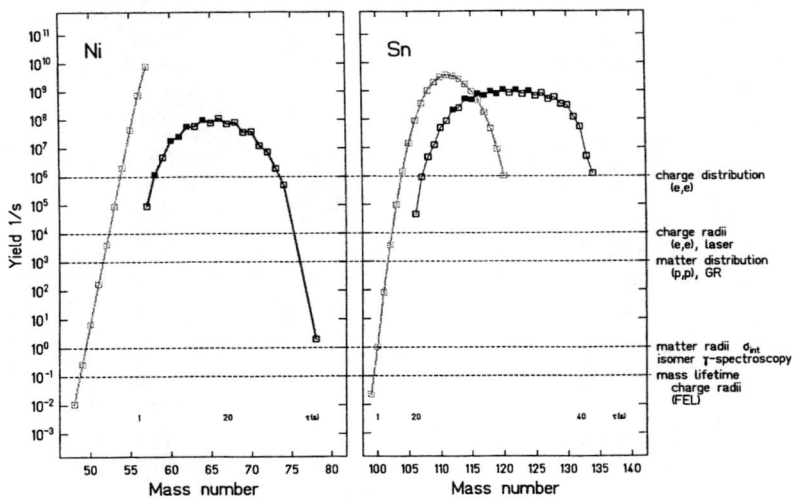

FIGURE 2. Yields for nickel and tin isotopes for driver beams of 10^{12} s^{-1}.

eters which accept momentum widths in excess of two per cent, dispersion will create extended focal planes which causes detector problems e. g. in the use of efficient target calorimeters. Cooled beams would be the ideal tool to overcome these problems.

For structure investigations in reverse kinematics with moderate resolution, e. g. in proton scattering, the targets must be sufficiently thin to allow e. g. for the undisturbed detection of the recoiling (target-) proton. The typical target thickness is 10^{18} cm^{-2} to 10^{20} cm^{-2}. To obtain sufficient resolution, the momentum vector of the projectile must be measured event-by-event which limits the beam energy. Experiments of this type are ideally performed in a storage ring with cooled beams and internal targets. Fig. 1 shows the thin-target limit for single-pass and high-resolution experiments, respectively. The domain of the in-ring experiments starts when thin targets and small cross-sections require projectile intensities in excess of 10^9 s^{-1} for single-pass experiments to achieve sufficient luminosity. These high intensities are only achievable for nuclei close to stability. In a storage ring the ions circulate with 10^6 s^{-1}. The equivalent of 10^9 s^{-1} is created by 10^3 stored ions.

III INTENSITIES OF EXOTIC BEAMS

Future heavy-ion driver accelerators [14,15] will deliver beams with intensities of $2 \cdot 10^{12}$ s^{-1} for all elements throughout the periodic table with energies from 100 AMeV to 1 AGeV. Systematic precision studies of fragmentation and fission yields provide a solid basis of experimental data and reliable predictions of production

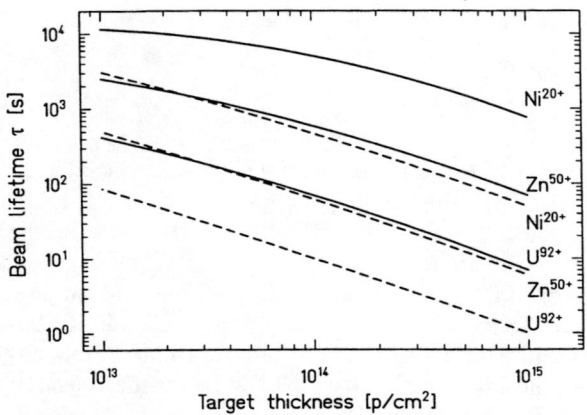

FIGURE 3. Lifetimes of stored beams in dependence of the target thickness for hydrogen targets and energies of 500 AMeV (solid line) and 100 AMeV (dashed line). The lifetime has been calculated for vacuum in the ESR of 10^{-10} mbar, and a cooler current of 200 mA at 500 AMeV and 100 mA at 100 AMeV (courtesy T. Stoelker [19]).

rates [16,17]. Fig. 2 displays the rates for the nickel and tin isotopic chains. They are of specific interest as they cover broad isospin ranges and contain magic nuclei close-to and far-off stability. The nickel chain ranges from the proton- to the neutron dripline. The yield curves exhibit a flat maximum with intensities in excess of 10^8 s^{-1}. Such intensities are well suited for reaction studies in storage rings (fig. 2). Along the steep slopes of the yield, techniques of increasing sensitivity and with less detailed structure information must be used. Precise and detailed structure information close to stability will be combined with basic information such as mass and interaction radii at the limits of stability. With this pyramid of information a conclusive picture of the development of nuclear structure, such as the distribution of neutron and proton matter, the diffuseness of the nuclear matter in presence of neutron- or proton skins, and the development of single-particle structure – in particular of nuclear shells towards the limits of stability – can be explored. This strategy is in close analogy to the strategy used in the investigation of the heaviest elements at the top of the nuclear chart [18].

IV STORAGE OF EXOTIC HEAVY-ION BEAMS: LIFETIMES AND ATOMIC INTERACTIONS

Limiting factors of the beam lifetimes of exotic nuclei are:

- atomic interactions with the internal reaction target, with the residual gas in the storage ring

- electron capture in the cooler
- nuclear reaction rates
- radioactive decay

As pointed out earlier [2] atomic interactions play a crucial role in the beam loss when running with an internal target, specifically at "low" energies below 100 AMeV. Nuclear reaction rates are negligible for the beam loss at all energies considered for structure research.

In principle the figure of merit for experiments with internal targets is determined by the ratio of the nuclear to the atomic cross section. This holds however only for an ideal ring where atomic interactions occur only in the reaction target and nuclei have indefinitely long lifetime. Fig. 3 displays the beam lifetimes [19] as predicted for the Experimental Storage Ring (ESR) at GSI on the basis of experimental data. The calculations are for hydrogen targets of 10^{13}, 10^{14}, and 10^{15} atoms cm^{-2} thickness and include interactions with the residual gas in the ring and with the cooler electrons. For nickel and tin of 500 AMeV beam lifetimes are of the order of an hour to several hours. The most unfavourable case is uranium where the lifetime at 500 AMeV drops to 7 s and at 100 AMeV even to 1s for the thickest target, with 10^{15} cm^{-2}. Consequently in the energy regime discussed here the most severe restriction for beam lifetimes is the radioactive decay of the stored nuclei. If we assume a cycling rate of the driver accelerator of 1 s^{-1} and cooling times of the order of 0.1 s, an increase of the number of stored ions by stacking should be feasible already for nuclides with half-lives of at least 10 seconds. For the example of nickel this covers the isotopes from ^{56}Ni to ^{69}Ni, and for tin the isotopes from ^{104}Sn to ^{132}Sn which corresponds to a broad isospin range.

For optimum performance, a storage ring system must include a dedicated collector ring for beam preparation such as stacking and cooling and a separate experimental storage ring [9]. To efficiently inject the projectile- and fission fragments which cover a large phase-space an optimum matching of the separator emittance and the collector acceptance is required. In addition phase-space compression is necessary. Modern fragment separators will be equipped with condenser systems to compress the transversal phase-space. A severe problem is however the compression of the longitudinal phase-space. Passive methods such as mono energetic degraders [20] or active methods like bunch rotation (de-bunching) in an accelerator section need be investigated. These problems not solved yet need intense ion optical calculations and test experiments at existing machines such as the ESR.

The large-acceptance collector ring includes the fast cooling and the capability of deceleration or acceleration to inject the exotic nuclei into the experimental ring with the energy appropriate for the experiments. A combination of stochastic and electron cooling is most promising for projectile fragments because of their large momentum spread. Fast coolers need still to be developed.

Low energy reaction studies such as transfer- and fusion reactions with cooled and stored beams cannot be carried out with internal targets because of the fast destruction of the circulating beam by atomic interactions with the electrons of the target atoms. New experimental methods such as low energy proton- heavy ion colliders or the use of traps with bare nuclei as reaction targets need being developed.

V HADRONIC SCATTERING

Reaction experiments with cooled beams of exotic nuclei and internal gas targets are a new field for experimental technique and new physics information. Besides radioactive species the study of sufficiently long-lived isomers, separated in the cooler ring seems within reach. Reaction experiments with cooled beams allow for highest precision in beam energy and momentum definition of the scattered nuclei. The atomic energy loss and scattering in targets with densities of 10^{13} atoms/cm^2 to 10^{15} atoms/cm^2 are negligible. The luminosity for 10^6 stored fragments circulating in the ring reacting with an internal target of 10^{15} cm^2 corresponds to 10^{27}/s cm^2 including the circulation frequency of 10^6 /s. This corresponds to rates of 1/s at 1 mb. Thin targets would help to provide precise data down to low proton energies.

Elastic scattering (p,p) at low energies gives access to the optical potential and at medium energies to nuclear matter distributions. The physics background and first examples of structure studies by inelastic proton scattering have been discussed in the previous conference [2]. At high energies inelastic scattering (p,p'), (α,α') and charge-exchange allow to study the nuclear electric multipole response. Isovector excitations in particular are sensitive to the thickness of the neutron skin. The (p,n) and (^{3}He,t) reactions near zero degree and at energies of 0.1 to 0.5 AGeV are excellent tools to investigate spin and isospin response of nuclei. Knockout reactions explore the single particle structure and nucleon momentum distributions. Fission of unstable nuclei, e. g. of neutron deficient actinides at well defined excitation energy by measuring the recoil proton would give more insight into fission dynamics and the persistence of nuclear shells with increasing temperature, a crucial question e. g. for the synthesis of superheavy elements. Coulomb excitation in heavy gas targets could be used to investigate the inverse capture, already discussed as relevant for cosmic nucleosynthesis.

Experiments in storage rings in reversed kinematics, beside high resolution will have a number of technical advantages compared to conventional techniques as already demonstrated in current experiments with exotic beams [12]. Coincidences between the large-angle scattered recoil proton and the scattered fragment allow for extremely clean and background-free experimental conditions not achievable with conventional methods: Furthermore the high energy of the scattered fragments allows for unambiguous mass and charge indentification. Nucleons emitted from the energetic fragments are kinematically focussed in forward direction. The energetic breakup neutrons are detected with high efficiency so that all reaction products and their kinematic properties are measured completely. The experimental progress with experiments in storage rings will be the use of thin targets, which allows e. g.

TABLE 1. Relevant parameters for an e-A collider [22].

Circumference, m	16
Maximum electron energy, E_{max}, MeV	400
Injection energy, E_{inj}, MeV	100
Maximum number of stored electrons, N_e	$3 \cdot 10^{11}$
Maximum magnetic field, B, T	1.6
Synchrotron radiation loss at 400 MeV, keV/turn	3

high resolution experiments and the identification and momentum determination of all reaction products in the exit channel.

VI ELECTRON SCATTERING

A small heavy-ion - electron (HI-e) collider for nuclear structure research has been first discussed at JINR Dubna in the frame of the K4-K10 project [13], which was never realised. At present HI-e - colliders are planned at RIKEN, Tokyo in the frame of the MUSES project [4]. Such a system has also been discussed at GSI since several years [1,21,22]. Recently an international workshop was held in the frame of the NuPECC workinggroup [14] on the "Next Generation Radioactive Beam Facility in Europe". A small HI - e collider for structure research has been strongly recommended as part of a next generation fragmentation facility. Very recently also in the US the discussion has been started.

Such a small collider would not only give completely new access to the nuclear structure of exotic nuclei, but also allow for new quality of experiments free of background and with a complete determination of all reaction products and their kinematic parameters, similar as discussed above for inverse hadronic scattering. The bare nuclei can be stored. Scattering is free from interactions with the electron cloud as observed in conventional experiments.

The small electron-synchrotron with a maximum electron energy of about 400 MeV could be implemented e. g in a 10 Tm experimental heavy-ion storage ring. The most important data are summarised in Table 1. Operating both rings in the co-propagating or counter-propagating mode, respectively, center-of-mass energies ranging from 40 MeV to 800 MeV would be covered [22].

Elastic electron scattering provides unambiguous information about charge radii and diffuseness. Fig. 4 shows as an example the matter and charge radii for the carbon isotopes. The development of the neutron skin towards the neutron dripline is clearly observed [24]. Towards the proton rich isotopes an indication for a proton skin is visible. The theoretical prediction of the interaction cross-section (dashed line) fits the stable nuclides, but fails immediately beyond. The core radii have been extracted form the charge-changing cross sections. They remain constant along the isotopic chain [23]. The still open question is to which extent these reactions where at least one proton is removed, are characteristic for the proton radius. It would

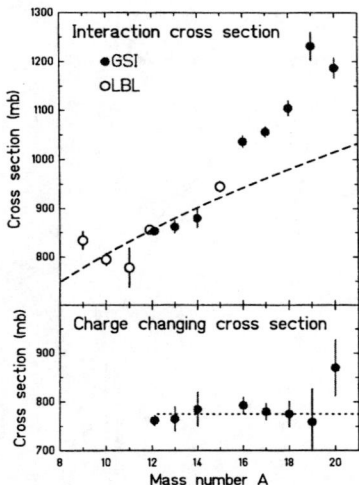

FIGURE 4. Matter distributions of the carbon isotopes, compared to a conventional mode developed from stable nuclei (upper part) and the corresponding charge-changing cross sections (lower part).

be of extreme importance to see whether the halo compresses the nuclear core, how the diffuseness of the proton distribution is influenced by the skin thickness, in other words, whether the neutron-proton interaction causes the protons to follow the neutrons. The knowledge of the diffuseness is a prerequisite to learn about the presently not well understood spin-orbit force. Formation and strength of shells at the limits of stability are not safely predictable at present.

The luminosities required for charge radii measurements are 10^{25} cm^{-2}s^{-1}, the determination of the diffuseness needs 10^{27} cm^{-2}s^{-1}. A luminosity of 10^{25} cm^{-2}s^{-1} for instance can be achieved with 10^4 ions stored in the ring [22]. The calculated formfactors for the tin isotopes [26] are displayed in Fig. 5. The slope at the maximum or the position of the first minimum determines the nuclear charge radius, the diffuseness in determined by the slope of the line connecting the first end the second maximum.

Inelastic electron scattering is an extremely clean probe for electromagnetic excitation. Contrary to conventional experiments colliding HI-e beams allow a 4π detection of the heavy recoils. With such a reaction trigger background originating e. g. from the radiation tail of elastic scattered electrons is absent.

Coulomb excitation, as explored in current experiments, interacts with nuclei by γ-cascades leads to multi-step excitation and not well defined excitation energy. Electrons permit single-step excitation to well defined states and well defined ex-

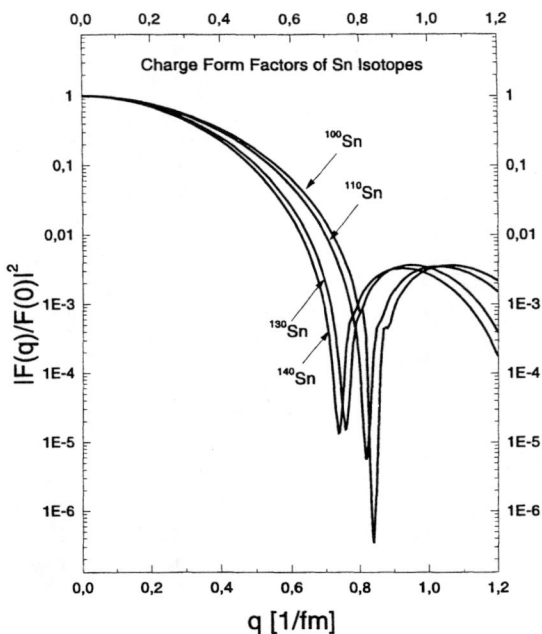

FIGURE 5. Formfactors for the tin isotopes [26] (courtesy of H. Lenske).

citation energies. Moreover, since the electromagnetic interaction is well known, form factors and transition densities can be determined accurately. Because the form factors $F_l(Q)$ are strongly l-dependent, by variation of the momentum transfer specific multipole components can be excited and thus a high spin selectivity is achieved. Inelastic electron scattering is an excellent probe to study bound and unbound states, the excitation of electric and magnetic giant resonances to investigate the equation for state at zero temperature, the fission process at defined temperature of the fissioning system, electromagnetic dissociation as the reversed process of capture, and to extract cross sections for astrophysical application. Here it should be recalled that isomeric beams will be available. Electron scattering also gives access to nucleon-momentum distributions, which is of specific interest for weakly bound systems at the limits of stability.

As an experimental example to compare conventional electron scattering with scattering in the collider we will discuss the reaction ^{48}Ca(e,e'n)^{47}Ca. The experiment at the DALINAC at TU Darmstadt [25] has been performed with a luminosity of the order 10^{31} s^{-1} to obtain $2 \cdot 10^5$ counts in one week. The same experiment in a collider would need a luminosity of only 10^{28} cm^{-2} s^{-1}, because of higher solid-angle efficiency due to forward focussing of the reaction products, and the enhanced detection efficiency.

I is noteworthy to mention that the luminosities achievable with a HI-e collider are comparable to those achievable with internal gas targets, provided the same number of nuclei is circulating in the heavy-ion ring (Fig. 1). The cross sections for electromagnetic interactions as compared to hadronic however are reduced by a factor of $\alpha = 1/137$, the amount of the finestructure constant. To cope with the long beamtimes expected, electron and hadronic scattering could be measured simultaneously.

An experimental problem to be solved is the large solid-angle electron spectrometer which must resolve 100 KeV with a high angular resolution. To limit the resolution to a reasonable value, the electron energy in the laboratory system should not be too high, e. g. for an electron momentum of 100 MeV/c a resolution of $0.5 \cdot 10^{-4}$ 10 000 would be sufficient.

An option under investigation is to which extent the electron ring could be optionally used as a low-energy storage ring for protons or heavy ions e. g. for operation in a collider mode with the heavy-ion ring to provide proton targets or targets of bare nuclei or interdisciplinary research.

VII   INTERACTIONS WITH PHOTONS

High-resolution collinear laser spectroscopy is a well known tool to measure nuclear moments and charge radii. The implication of the new generation of power lasers to improve the sensitivity are discussed but still need investigation.

The electron synchrotron or the high-current electron injector could serve as an intense photon source. Photons in the 100 eV region, produced by ondulators, or of high energy around 100 keV, produced e. g. by channelling radiation, would allow for spectroscopy of hydrogen or lithium-like systems where the highest precision for the measurement of nuclear charge distributions could be achieved. A challenge would be single-atom spectroscopy which may become possible with the SASE free electron lasers.

VIII   OUTLOOK

The investigation of exotic nuclei in storage- and cooler rings is a new field of structure research at highest precision with a high research potential. This technique will allow experiments not possible with present methods. The high research potential of storage rings is convincingly demonstrated in the first generation of experiments: the direct mass measurements and the determination of nuclear β - half-lives of bare or few-electron atoms.

Reaction studies in storage rings are second-generation experiments, requiring increased beam intensities and new techniques to efficiently store and cool projectile fragments or fission products. These are the only way to experiments at highest precision with unstable nuclei. HI-e–colliders will open up a new quality of experiments to explore the electromagnetic response of exotic nuclei not possible with conventional methods.

Useful discussions with H. Geissel, H. Lenske, B. Franzke, A. Skrinski, C. Schlegel, Th. Stoehlker and J. Wambach are gratefully acknowledged

REFERENCES

1. G. Münzenberg, Int. Workshop on the Physics and Techniques of Secondary Nuclear Beams, Dourdan, France, 1992, Eds. J. F. Bruandet, B. Fernandez, and M. Bex, Editions Frontieres, Gif-sur Yvette, France, (1992) 253
2. W. Henning, Nuclear Physics at Storage Ringe, STORI96, Bernkastel-Kues 1996, Nucl. Phys. A626(1997)225c
3. I. Tanihata, Nuclear Physics at Storage Ringe, STORI96, Bernkastel-Kues 1996, Nucl. Phys. A626(1997)531c
4. T. Katayama, Nuclear Physics at Storage Ringe, STORI96, Bernkastel-Kues 1996, Nucl. Phys. A626(1997)545c and this conference
5. C. Scheidenberger et al., this conference
6. T. Radon, this conference
7. Th. Kerscher, this conference
8. H. Geissel, G. Münzenberg and K. Riisager, Ann. Rev. Nucl. Part. Sci. 45(1995)163
9. B. Franzke, this conference
10. T. Nilsson, F. Humbert, W. Schwab, H. Simon, M.H. Smedberg et al., Nucl. Phys. A598(1996)418
11. T. Baumann, M.G. Borge, H. Geissel, H. Lenske, K. Markenroth et al., Phys. Lett. B439(1998)256
12. T. Aumann, D. Aleksandrov, L.Axelsson, T. Baumannn et al., Phys. Rev C59(1999)1252
13. Yu. Ts. Oganessian et al. Int. Rep.E7-91-75,Dubna 1991,217
14. B. Jonson (chair), NuPECC Report on the Next Generation Radioactive Beam facilities in Europe, in prepration
15. The new in-flight facility at GSI, under discussion
16. K. Sümmerer and B. Blank, submitted to Phys. Rev. C
17. J. Benlliure, A. Grewe, M. de Jong, K.-H. Schmidt, and S. Zhdanov, Nucl. Phys. A628(1998)468
18. G. Münzenberg, Rep. Prog. Phys. 51(1988)5
19. T. Stoelker, Priv. comm. 1999
20. H. Geissel et al., Nucl. Instr. Meth. B70(1992)286
21. G. Münzenberg, J. Friese, H. Geissel, I. Meshkov, G. Schrieder, and E. Syresin, Nuclear Physics at Storage Ringe, STORI96, Bernkastel-Kues 1996, Nucl. Phys. A626(1997)249c
22. I. Meshkow, Nuclear Physics at Storage Ringe, STORI96, Bernkastel-Kues 1996, Nucl. Phys. A626(1997)459c
23. A. Ozawa, GSI Ann. Rep. 1997, p 21
24. A. Ozawa et al., submitted to Phys. Lett. B,
25. S. Strauch, Nucl. Phys. A649(1999)85c
26. H. Lenske, private communication

First Isochronous Time-of-Flight Mass Measurements of Short-Lived Projectile Fragments in the ESR

J. Stadlmann[1,2], H. Geissel[1,2], M. Hausmann[1,2], F. Nolden[2], T. Radon[2], H. Schatz[2], C. Scheidenberger[2], F. Attallah[2], K. Beckert[2], F. Bosch[2], M. Falch[3], B. Franczak[2], B. Franzke[2], Th. Kerscher[3], O. Klepper[2], H.-J. Kluge[2], C. Kozhuharov[2], K.E.G. Löbner[3], G. Münzenberg[2], Yu.N. Novikov[4], M. Steck[2], Z. Sun[2], K. Sümmerer[2], H. Weick[1,2], H. Wollnik[1]

[1] *II. Physikalisches Institut, JLU Gießen, Heinrich-Buff-Ring 16, D-35392 Gießen, Germany*
[2] *Gesellschaft für Schwerionenforschung mbH, Planckstraße 1, D-64291 Darmstadt, Germany*
[3] *Sektion für Physik, LMU München, Am Coulombwall 1, D-85748 Garching, Germany*
[4] *St.Petersburg Nuclear Physics Institute, Gatchina 188350, Russia*

Abstract. We present a new method for precise mass measurements of short-lived hot nuclei. These nuclei were produced via projectile fragmentation, separated with the FRS and injected into the storage ring ESR being operated in the isochronous mode. The revolution time of the ions is measured with a time-of-flight detector sensitive to single particles. This new method allows access to exotic nuclei with half-lives in the microsecond region. We report on first results from this novel method obtained with measurements on neutron-deficient fragments of a chromium primary beam with half-lives down to 50 ms. A precision of $\delta m/m \leq 5 \cdot 10^{-6}$ has been achieved.

EXPERIMENTAL METHOD

Accurate knowledge of masses for very unstable nuclei provides new insight in nuclear structure far from stability and is required for nuclear astrophysics studies. We present first results of a novel technique to measure previously unknown masses of nuclei with lifetimes down to μs at the storage ring ESR at GSI. Short-lived neutron deficient nuclei in the vicinity of the proton drip-line were produced by fragmentation of a $(415 - 440)$ MeV/u ^{52}Cr beam on a 2.5 g/cm^2 beryllium target. After separation by the fragment separator FRS via the $B\rho - \Delta E - B\rho$-method [1] the nuclei were injected and stored in the ESR [2]. The method of mass measurement can be understood by

$$\frac{\Delta(m/q)}{(m/q)} = -\gamma_t^2 \frac{\Delta f}{f} + \left(\gamma_t^2 - \gamma^2\right) \frac{\Delta v}{v}, \quad (1)$$

the first-order relation between the mass-to-charge ratio m/q, the revolution frequency f, and the velocity v of ions circulating in the ESR. γ is the relativistic Lorentz factor and γ_t is the transition point, a parameter for the ion-optical setting of the ring. The second term in Eq.(1) has to be small for precise mass measurements despite of the large velocity spread Δv of the exotic fragments. This condition is achieved immediately after injection by operation of the ESR at its transition point ($\gamma_t = \gamma$) [3,4]. The revolution times were measured with a time-of-flight detector placed in the ESR [4–6]. It detects single ions with an efficiency of about 60%. The circulating ions penetrating a thin carbon foil (a few 10 $\mu g/cm^2$) release electrons which are isochronously transported to multi channel plates (MCP) delivering fast signals (see Fig. 1). The signals were recorded with a storage oscilloscope capable of recording up to $8 \cdot 10^9$ samples per second. About five particles are injected simultaneously and are observed in up to 200 turns. The revolution time for one turn is about 530 ns. Therefore, this method can be applied down to nuclides with lifetimes in the μs range in principle. This method is complementary to Schottky Mass Spectrometry (SMS) [7,8], where the $\Delta v/v$-term in Eq.(1) is minimized by electron cooling requiring minimum lifetimes of a few seconds.

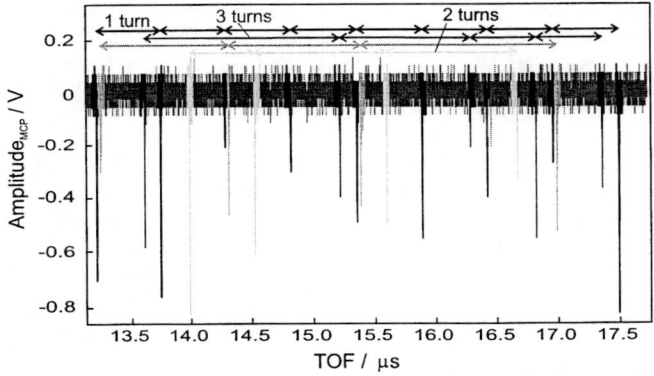

FIGURE 1. Signals from the TOF detector measured with a fast digital sampling oscilloscope. Signals belonging to different ions in the ring are marked in different shades of grey.

RESULTS

For each ion the corresponding time signals were extracted from the data and used to determine the revolution times. Fig. 2 shows a typical spectrum of the revolution times, containing the data of about 3500 injections. The different isotopes were identified unambiguously by their revolution times. We obtained data on about 70 neutron deficient nuclides between C and Mn with lifetimes down to 50 ms. The mass resolution is $\Delta m/m = 10^{-5}$. A comparison to data in the literature [9] shows nice agreement, see Fig.3. A precision of about $\delta m/m \leq 5 \cdot 10^{-6}$ has been achieved for nuclides with sufficient statistics. Five nuclides of unknown mass have been observed (see Fig. 2). The data is presently being analyzed.

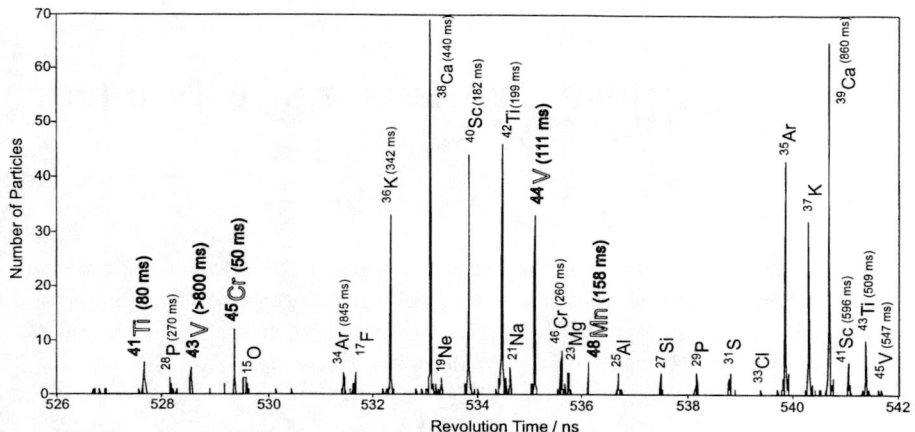

FIGURE 2. A time-of-flight spectrum of stored hot fragments created by chromium fragmentation in a beryllium target is shown. Only half-lives smaller than 1 s are given in the figure. Nuclides of unknown mass are marked by outlined letters.

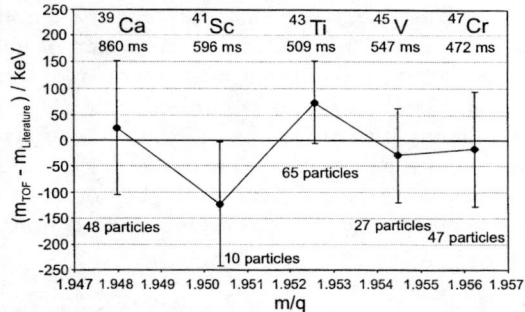

FIGURE 3. Preliminary masses of nuclei with short half-lives measured with our new method compared to literature values [9]. The numbers of detected particles are indicated.

REFERENCES

1. H. Geissel et al., Nucl. Instr. and Meth. B70 (1992), 286.
2. B. Franzke, Nucl. Instr. and Meth. B24/25 (1987), 18.
3. H. Wollnik et al., GSI-Report 86-1 (1986), 372.
4. M. Hausmann et al., accepted for publication in Nucl. Instr. and Meth. A.
5. H. Wollnik et al., Proceedings of Stori96,
   Ed. by F. Bosch and P. Egelhof, Elsevier Science B.N., ISBN 0375-9474.
6. J. Trötscher et al., Nucl. Instr. and Meth. B70, (1992), 455.
7. T. Radon et al., Phys. Rev. Lett. 78 (1997), 4701.
8. M. Falch et al., these proceedings.
9. G. Audi et al., Nucl. Phys. A624 (1997), 1.

Studying Phase Transitions in Nuclear Collisions

I.N. Mishustin

The Kurchatov Institute, Russian Research Center, 123182 Moscow, Russia;
The Niels Bohr Institute, Blegdamsvej 17, DK-2100 Copenhagen Ø, Denmark;
Institute for Theoretical Physics, J.-W. Goethe University, Robert-Meyer Str. 8-10,
D-60054 Frankfurt am Main, Germany

Abstract

In this talk I discuss three main topics concerning the theoretical description and observable signatures of possible phase transitions in nuclear collisions. The first one is related to the multifragmentation of equilibrated sources and its connection to a liquid-gas phase transition in finite systems. The second one is dealing with the Coulomb excitation of ultrarelativistic heavy ions resulting in their deep disintegration. The third topic is devoted to the description of a first order phase transition in rapidly expanding matter. The resulting picture is that a strong collective flow of matter will lead to the fragmentation of a metastable phase into droplets. If the transition from quark-gluon plasma to hadron gas is of the first order, it will manifest itself by strong nonstatistical fluctuations in observable hadron distributions.

INTRODUCTION

A general goal of present and future experiments with heavy-ion beams is to study the properties of strongly interacting matter away from the nuclear ground state. The main interest is focussed on searching for and studying possible phase transitions. Several phase transitions are predicted in different domains of temperature T and baryon density ρ_B. There is no doubt that there should be a first order phase transition of the liquid-gas type in normal nuclear matter. This follows simply from the existence of the nuclear bound state at the saturation density $\rho_0 \approx 0.15$ fm^{-3}. Therefore, at $\rho_B < \rho_0$ and low temperatures, $T < T_c \sim 10$ MeV, the matter will organize itself in the form of a mixed phase with droplets of nuclear liquid surrounded by the nucleon gas. The only problem is whether relatively small amounts of excited nuclear matter produced in nuclear collisions and its limited lifetime are sufficient to observe this phase transition. Based on recent data on the nuclear caloric curve [1] and temperature fluctuations [2] I am tempting to give a positive answer to this question. This topic will be discussed in the first part of the talk after a short description of the Statistical Multifragmentation Model (SMM) [3, 4] which provides a basis for theoretical analysis.

The situation at high T and nonzero baryon chemical potential μ ($\rho_B > 0$) is not so clear, although everybody is sure that the deconfinement and chiral transitions should occur somewhere. The phase structure of QCD is not yet fully understood. Reliable lattice calculations exist only for $\mu = 0$ ($\rho_B = 0$) where they predict a second order phase transition or crossover at $T \approx 160$ MeV. As model calculations show, the phase diagram in the (T, μ) plane may contain a first order transition line (below called the critical line) terminated at a (tri)critical point [5, 6]. Possible signatures of this point in heavy-ion collisions are discussed in ref. [7]. Under certain non-equilibrium conditions, a first order transition is also predicted for symmetric quark-antiquark matter [8].

A striking feature of central heavy-ion collisions at high energies, confirmed in many experiments (see e.g. [9, 10]), is a very strong collective expansion of matter. The applicability of equilibrium concepts for describing phase transitions under such conditions becomes questionable. In the last part of the talk I demonstrate that non-equilibrium phase transitions in rapidly expanding matter can lead to interesting phenomena which, in a certain sense, can be even easier to observe [11].

In the middle part of the talk I address a question which is closely related to the main topic of this conference. It illustrates how the knowledge accumulated in intermediate-energy heavy-ion physics can be used for ultrarelativistic heavy-ion colliders. Namely, I will discuss the excitation of nuclei by Lorentz-contracted and strongly-enhanced Coulomb fields of ultrarelativistic heavy ions. As well known, this process can be treated in terms of equivalent photons. Their flux grows linearly with the squared nuclear charge, and their characteristic energy is proportional to the relative Lorentz factor of colliding nuclei. This is why this Coulomb excitation of nuclei becomes especially important in high-energy heavy-ion colliders such as RHIC and LHC. The calculations [12, 13] show that in such colliders the equivalent photon spectrum extends far above the Giant Resonance region, into the GeV domain. The absorption of such a photon by a nucleus leads to its high excitation and subsequent disintegration. This might be an important factor determining a lifetime of ultrarelativistic heavy-ion beams.

STATISTICAL MULTIFRAGMENTATION AND LIQUID-GAS PHASE TRANSITION

When a nucleus is suddenly heated up to a temperature T it starts expanding to adjust a new equilibrium density $\rho_0(T)$ which is less than the equilibrium density at zero temperature ρ_0. If the initial temperature is high enough the expansion is unlimited. At some stage of expansion the system enters into the spinodal region, where the homogeneous distribution of matter becomes thermodynamically unstable. Therefore, the nucleons form smaller and bigger clusters or droplets with density close to ρ_0. This clusterization process resembles a liquid-gas phase transition in ordinary fluids. In the transition region the matter is very soft in a sense that the sound velocity is close to zero (soft point). This means also that the expansion is slow and the system has enough time to find a most favorable cluster-size distribution maximizing the entropy.

At a later stage of expansion the system reaches a so-called freeze-out state when clusters cease to interact with each other. This break-up state of the system can be described within a statistical approach. In 1985 we have constructed a Statistical Multifragmentation Model (SMM) [3, 4] which up to now is one of the most successful realizations of this approach for finite nuclear systems. The model and its numerous applications are described in detail in a recent review [14]. A similar model was also constructed by Gross [15]. In this talk I outline only some general features of the SMM and give a few examples of how it works.

It is assumed that at break-up the system consists of primary hot fragments and nucleons in thermal equilibrium. Each break-up channel or partition, f, is specified by the multiplicities of different species, N_{AZ}, constrained by the total baryon number A_0 and charge Z_0. The total fragment multiplicity is defined as $M = \sum_{AZ} N_{AZ}$. The probabilities of different break-up channels are calculated in an approximate microcanonical way according to their statistical weights,

$$W_f \propto \exp\left[S_f(E^*, V, A_0, Z_0)\right], \qquad (1)$$

where S_f is the entropy of a channel f at excitation energy E^* and volume V.

Translational degrees of freedom of fragments are described by the Boltzmann statistics while the internal excitations of individual fragments with $A > 4$ are calculated according to the quantum liquid-drop model. An ensemble of microscopic states corresponding to a break-up channel f is characterized by a temperature T_f which is determined from the energy balance equation

$$\frac{3}{2}T(M-1) + \sum_{(A,Z)} E_{AZ}(T) N_{AZ} + E_f^C(V) - Q_f = E^* . \qquad (2)$$

Here the first term comes from the translational motion, the second term includes internal excitation energies of individual fragments, the third term is the Coulomb interaction energy and the last one is the Q-value of the channel f. The excitation energy E^* is measured with respect to the ground state of the compound nucleus (A_0, Z_0). It is fixed for all fragmentation channels while the temperature T_f fluctuates from channel to channel.

The total break-up volume is parametrized as $V = (1 + \kappa)V_0$, where V_0 is the compound nucleus volume at normal density and the model parameter κ is the same for all channels. The entropy associated with the translational motion of fragments is determined by the "free" volume, V_f, which is only a fraction of the total break-up volume V. In the SMM $V_f(M)$ is parametrized in such a way that it grows almost linearly with the primary fragment multiplicity M or equivalently, with the excitation energy $\varepsilon^* = E^*/A_0$ of the system [14].

At given inputs A_0, Z_0 and ε^* the individual multifragment configurations are generated by the Monte Carlo method. After the break-up the hot primary fragments propagate in a common Coulomb field and loose their excitation. The most important de-excitation mechanisms included in the SMM [14] are the simultaneous Fermi break-up of lighter fragments ($A \leq 16$) and the evaporation from heavier fragments, including the compound-like residues. In refs. [17, 18, 19, 20, 21, 22] one can find fresh examples showing how well the SMM works in describing the multifragmentation of thermalized sources.

In ref. [16] an equation of state of a multifragment system was calculated for the grand canonical version of the SMM. As expected, it shows clear signs of a liquid-gas phase transition with a critical temperature of about 7 MeV. In the transition region the pressure isotherms are very flat indicating that the sound velocity is very small. In this region the compressibility and specific heat have nonmonotonic behaviour.

The most interesting prediction of the statistical model, a plateau in the caloric curve $T(\varepsilon^*)$, was formulated already in 1985 [23]. Since that time it was a challenge for experimentalists to measure the nuclear caloric curve. First measurements were performed at GSI by the ALADIN collaboration only in 1995 [24]. They have shown an impressive agreement with the theoretical prediction. These results have initiated an avalanche of other measurements and a lively discussion in the community (see latest ALADIN results in ref. [25]).

Most temperature measurements are based on the Albergo method [26] relating the temperature to the double ratio of isotope yields. The analysis shows (see for instance [1, 27]) that the temperatures extracted by this method are very sensitive to the side-feeding and nuclear structure effects. According to SMM the observed light isotopes are produced mainly by the secondary decays of hot primary fragments. This leads to a difference between the isotopic temperatures and true thermodynamical temperature at freeze-out (see detailed analysis in ref. [1]). In particular, isotopic temperatures have typically a less pronounced plateau than the true temperature, which can even have a backbending. In recent years several comparisons have been made (see examples in refs. [1, 18, 28] which generally show very good agreement between the theory and experiment.

One should bear in mind that the ALADIN caloric curves are measured for a wide ensemble of decaying sources associated with the projectile or target spectators produced in peripheral nuclear collisions. The thermodynamical significance of such observations would increase if the measurements were done for a fixed source size with a varying excitation energy. Also the temperature measurements on the event-by-event basis would make it possible to study its fluctuations and therefore the heat capacity of the nuclear system. Such an analysis was performed recently by the Bologna group [2] in the study of quasi-projectile (QP) fragmentation in peripheral Au+Au collisions at 35 A MeV. In this analysis only the events with reconstructed QP charges $70 < Z_{QP} < 88$ were included. The excitation energy, determined by a calorimetric method, varied for these events from 0.5 to about 8

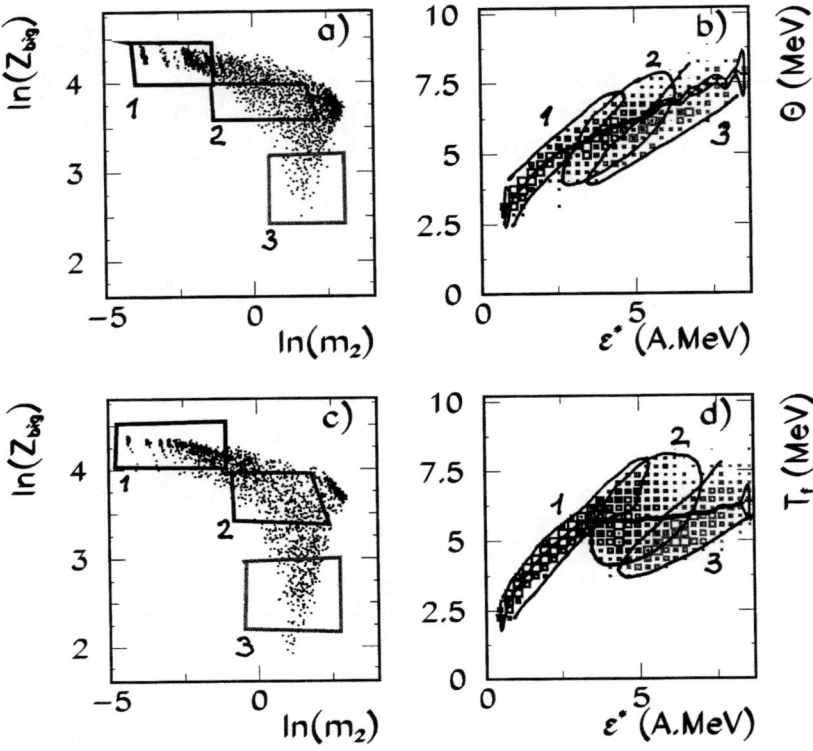

Figure 1: Experimental (a) and SMM generated (c) Campi scatter plots Three cuts are introduced to select liquid-like events (Cut 1), gas-like events (Cut 3) and critical events (Cut 2). Panels b) and d) show the correlation between temperatures and excitation energies for experimental and SMM generated events. Sizes of the squares are proportional to the yields. The regions populated by the events from the three cuts are encircled. The solid lines in panels b) and d) show the mean temperatures for all the events at given ε^*.

Mev/nucleon.

Measuring temperatures event by event is of course a nontrivial task. An attempt of estimating the event "temperature" (below denoted by θ) was made in ref. [2]. The idea is to apply the energy balance equation (2), which is used in the SMM, but now for the experimental events. Of course, this requires certain assumptions on how the observed partitions, involving cold reaction products, are related to the original partitions consisting of hot primary fragments. Therefore, it was assumed that the light particles detected in a partition were produced by de-excitation of hot primary fragments. For reconstructing a primary partition these light particles were shared among the detected fragments proportionally to their charges and assuming the charge-to-mass ratio as in the entrance channel. Applying this procedure for the asymptotic SMM events showed that the correlation between the microcanonical temperature and excitation energy was reproduced within 5%.

Fig. 1 shows the scatter plots in the (T, ε^*) plane for experimental (b) and SMM generated (d) events. The ensemble-averaged temperatures are indicated by the solid lines. Their behaviour is typical for caloric curves measured by other methods. In addition to the flattening of the average temperature at $T \approx 6$ MeV, one can clearly see the broadening of the distributions in the transition region at $\varepsilon^* = 4 \div 8$ MeV/nucleon. The quantity characterizing energy fluctuations is heat capacity. For a canonical ensemble at constant volume it can be expressed as

$$C_V = \frac{\sigma_E^2}{T^2} = \frac{\langle E^2 \rangle - \langle E \rangle^2}{T^2}. \qquad (3)$$

It is not clear whether the constant volume condition applies to actual freeze-out configurations but nevertheless studying the energy fluctuations provides an additional and important information compared to the average characteristics. Indeed, applying Eq. (3) for scatter plots of Fig. 1 reveals a peak in C_V at temperatures around 6 MeV. This behaviour was also predicted theoretically a long time ago [23, 15].

Another way of characterizing the critical behaviour is to analyze the conditional moments of fragment multiplicity distributions introduced by Campi [29]. Fig. 1 a) shows, for each event j, the experimental correlation between the logarithm of the charge of the largest fragment, $\ln(Z_{big}^{(j)})$, and the logarithm of the corresponding second moment of the multiplicity distribution, $\ln(m_2^{(j)})$ (Campi scatter plot). Fig. 1 c) shows the same for events generated by the SMM. As expected for a system experiencing a phase transition these plots exhibit two branches: an upper branch with an average negative slop, corresponding to under-critical events, and a lower branch with a positive slop that corresponds to super-critical events. The two branches meet in a central region signalling the approach to a critical point. This trend is nicely reproduced in Fig. 1 by both the experiment and the theory.

We have made three cuts in these scatter plots selecting the upper branch (Cut 1), the lower branch (Cut 3) and the central region (Cut 2) and analyzed the events falling in each of the three zones. The fragment charge distributions in these three zones exhibit shapes going from a U-shape in Cut 1, characteristic of the evaporation events at low excitation energies, to an exponential one in Cut 3, characteristic of the vaporization events at high excitations. In Cut 2 a power-low fragment charge distribution $Z^{-\tau}$ with $\tau \approx 2.2$ is observed as expected according to the Fisher's droplet model for fragment formation near the critical point of a liquid-gas phase transition [30] (see also an interesting analysis of ref. [22]).

The contributions of these three types of events to the caloric curves are shown in Fig. 1 for experiment (b) and for theory (d). It is clearly seen for both the data and the SMM, that in Cuts 1 and 3 besides normal events there are unusual events (although with low probability) which lie far from the average $T(\varepsilon^*)$ behaviour. These are compound-like states with very high temperatures and vaporization events with low temperatures. For these events one can make the analogy respectively with an overheated liquid and a super-cooled gas in the ordinary liquid-gas phase transition. Here we see the advantage of a finite system where not only the most probable states but also the metastable states can be produced with a finite probability. In my opinion, the observation of these metastable states is the best indication that we are dealing here with the first order phase transition of the liquid-gas type. These interesting questions were further studied in ref. [31].

ELECTROMAGNETIC EXCITATION OF ULTRARELATIVISTIC HEAVY IONS

It has become clear in recent years [12, 13] that high nuclear excitations can be induced by the Coulomb fields of ultrarelativistic heavy ions. Following the famous Waizsäcker-Williams method the Lorentz contracted Coulomb field of an ultrarelativistic projectile in the rest frame of a target nucleus (and vice versa) can be represented as a beam of equivalent or virtual photons. The flux of equivalent photons with energy E_γ in a collision of nuclei with charge Z at impact parameter b is given by the standard formula

$$N(E_\gamma, b) = \frac{\alpha Z^2}{\pi^2} \frac{x^2}{\beta^2 E_\gamma b^2} \left[K_1^2(x) + \frac{1}{\gamma^2} K_0^2(x) \right], \qquad (4)$$

where α is the fine structure constant, $\beta = v/c$ and $\gamma = \sqrt{1-\beta^2}$ is the relative Lorentz factor. The variable x in the modified Bessel functions $K_{0,1}(x)$ is defined as $x = E_\gamma b/(\beta\gamma\hbar c)$. Since $K_{0,1}$ drop exponentially at large arguments, the main contribution to the virtual photon flux comes from the region $x \sim 1$. Thus the characteristic energy of virtual photons grows linearly with γ. This explains why the relativistic Coulomb excitation is very important for ultrarelativistic heavy-ion beams where both γ and Z are large. For colliding beams $\gamma = 2\gamma_{beam}^2 - 1$ that gives $2 \cdot 10^4$ and 10^7 for RHIC and LHC respectively. This brings the spectrum of virtual photons into the GeV energy domain, i. e. much above the traditionally studied Giant Resonance (GR) and Delta-resonance regions. The absorption of

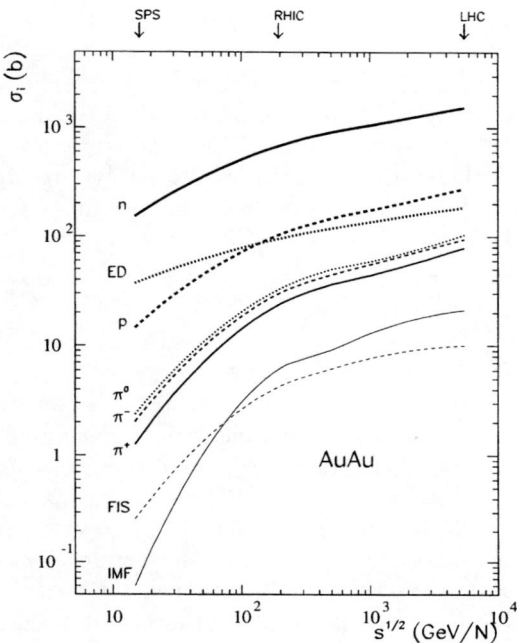

Figure 2: Theoretical predictions for inclusive cross sections for emitting nucleons, pions, intermediate mass fragments (IMF: $3 \leq Z \leq 30$) and fission fragments (FIS: $30 < Z \leq 50$) in the electromagnetic dissociation of Au nuclei as functions of the c.m. energy $\sqrt{s}$ (the SPS, RHIC and LHC energies are indicated by arrows). The thick dotted line shows the total ED cross section for Au beams.

such a high-energy photon will result in a very high nuclear excitation sufficient for its total disintegration.

In ref. [32] the description of nuclear photoabsorption was extended to the photon energies much above the GR region, where the excitation of individual nucleons and multiple pion production are the dominant reaction channels. A model of electromagnetic dissociation (ED) taking into account these high-energy photon absorption channels was constructed in ref. [12]. According to this model, the fast hadrons produced after the photon absorption initiate a cascade of subsequent collisions with the intranuclear nucleons leading to the fast particle emission and heating of a residual nucleus. This stage is described by the Intranuclear Cascade Model (INC). At a later stage the nucleus undergoes de-excitation by means of the evaporation of nucleons and lightest fragments, binary fission or multifragmentation. The latter process becomes important at ultrarelativistic beam energies, when the excitation energy of residual nuclei exceeds 3-4 MeV/nucleon. This stage of the reaction is described by the SMM.

To include all the processes described above, in ref. [13] we have developed a specialized computer code RELDIS aimed at the Monte Carlo simulation of the Relativistic ELectromagnetic DISsociation of nuclei. The simulation begins with generating the single- or double-photon absorption process. Then the INC model is used to calculate the fast particle emission and the characteristics of residual nuclei. Finally, de-excitation of thermalized residual nuclei is simulated by the SMM.

The cross section of the photo-nuclear (γA) reaction induced by a photon of energy E_γ is expressed as

$$\frac{d\sigma_{ED}}{dE_\gamma} = \sigma_{\gamma A}(E_\gamma) \int_{b_{min}}^{\infty} N(E_\gamma, b) 2\pi b \, db, \qquad (5)$$

where $b_{min} \approx (R_p + R_t)$ is a minimal impact parameter for heavy-ion collisions without nuclear overlap, $\sigma_{\gamma A}(E_\gamma)$ is an appropriate photo-absorption cross section, either measured for the A-nucleus with real photons or calculated within a model. The total ED cross section, σ_{ED}, is obtained by integrating Eq. (5) by dE_γ from 0 to ∞. The calculations [12] show that the total ED cross sections for RHIC and LHC are very large, 100 b and 200 b respectively. Accordingly, the ED reaction rates are much higher than those for nuclear interactions, although the ED events are much less violent. For instance, at expected RHIC luminosity $L \approx 10^{27}$ cm^{-2}s^{-1} the ED reaction rate will be 10^5 interactions per second. Together with the electron capture reactions the ED processes will be the important factors reducing the lifetime of ultrarelativistic heavy-ion beams compared with the proton ones.

We have applied the RELDIS code for calculating the ED characteristics for several heavy-ion beams. The model is in a reasonable agreement with experimental data, when available. We have also made predictions for the reactions: 160A GeV Pb+Pb (SPS), 100A+100A GeV Au+Au (RHIC) and 2.75A+2.75A TeV Pb+Pb (LHC). The inclusive (multiplicity weighted) cross sections for emitting nucleons, pions and nuclear fragments in the electromagnetic dissociation of one of the colliding Au nuclei are shown in Fig. 2 as functions of the incident c.m. energy. Nuclear fragments are divided in two groups: fission fragments (30<Z≤ 50) and Intermediate Mass Fragments (IMFs, 3≤Z≤30), which are associated with the multifragmentation. One can clearly see a steep rise in the yields of all species, especially IMFs, when the incident energy grows from the SPS to RHIC and LHC domain. The inclusive cross section for neutron emission is especially large, above 1000 b at RHIC and LHC. The average neutron multiplicities are predicted to be 4.1, 7.2 and 8.8 at SPS, RHIC and LHC respectively.

The predicted neutron multiplicity distributions are shown in Fig. 3. They have a nontrivial structure. There is a strong peak at 1n emission channel associated with the GR decay. On the other hand, there is a long tail of multiple neutron emission associated with more violent reaction channels, from the direct knock-out and evaporation from the compound nucleus to fission and multifragmentation. This is where our model including all these channels shows its strength. One can see, for example, that the probability to emit more than 20 neutrons is quite noticeable

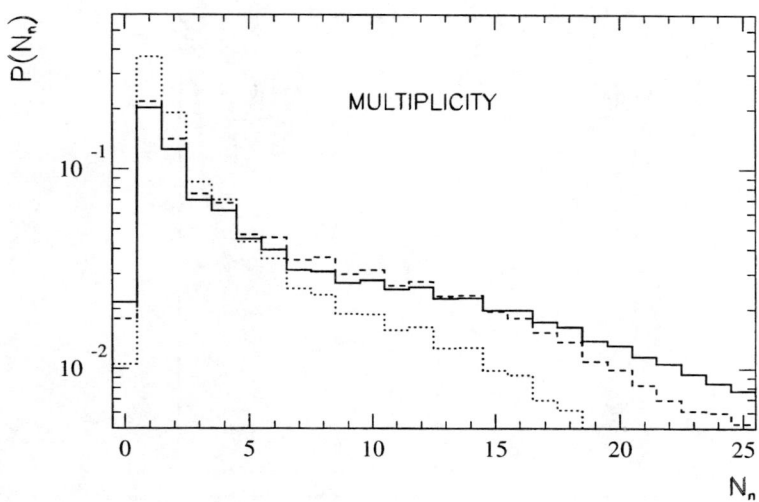

Figure 3: Normalized neutron multiplicity distributions for the electromagnetic dissociation of Pb nuclei at LHC and SPS (solid and dotted histograms, respectively) and Au nuclei at RHIC energies (dashed histogram). Calculations are made with RELDIS code.

($\approx 5\%$ at RHIC). These results might be important for designing neutron-sensitive zero-degree calorimeters at RHIC and LHC. One of such proposals was made recently in ref. [33] but only the 1n channel was considered there.

FIRST ORDER PHASE TRANSITION IN FAST DYNAMICS

The implications of a strong collective expansion on the liquid-gas phase transition were discussed in ref. [34]. Here I will focus on consequences of the strong collective flow of matter for a possible first order chiral transition. I will assume that the collective velocity field is described locally by the Hubble law, $v(r) = H \cdot r$, where the Hubble "constant" H may in general depends on time.

To make the discussion more concrete, I adopt a picture of the chiral phase transition predicted by the linear sigma-model with constituent quarks [35]. Then the mean chiral field $\Phi = (\sigma, \pi)$ serves as an order parameter. The model respects chiral symmetry, which is spontaneously broken in the vacuum where $\sigma = f_\pi$, $\pi = 0$. The effective thermodynamic potential $\Omega(T, \mu; \Phi)$ depends, besides Φ, on temperature T and baryon chemical potential μ. The schematic behaviour of $\Omega(T, \mu; \Phi)$ as a function of the order parameter field σ at $\pi = 0$ is shown in Fig. 4. The minima of Ω correspond to the stable or metastable states of matter under the

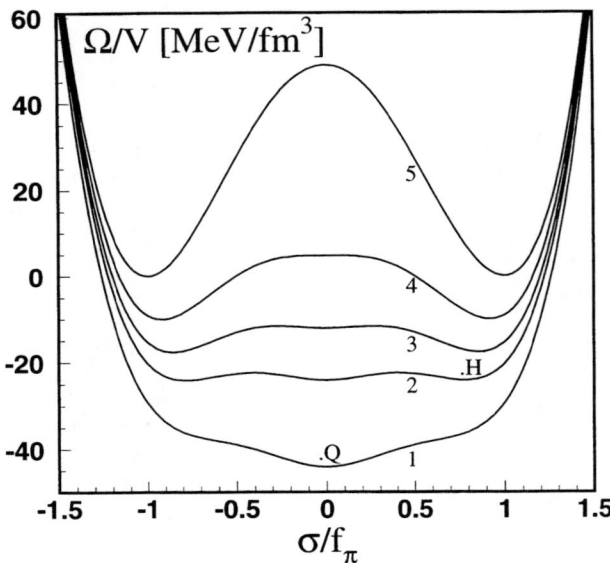

Figure 4: Schematic view of the effective thermodynamic potential per volume Ω/V as a function of the order parameter field σ at $\pi = 0$, as predicted by the linear σ-model in the chiral limit $m_\pi = 0$ [35]. The curves from bottom to top correspond to the different stages of the isentropic expansion of homogeneous matter starting from $T=100$ MeV and $\mu=750$ MeV (curve 1). The upper curve 5 is the vacuum potential. The other curves are discussed in the text.

condition of thermodynamical equilibrium, where the pressure is $P = -\Omega_{min}/V$. The curves from bottom to top correspond to different stages of the isentropic expansion of homogeneous matter. Each curve represents a certain point on the (T, μ) trajectory. As one can see from the figure, the model of ref. [35] reveals a rather weak first order phase transition, although some other models [5, 6] predict a stronger transition. The discussion below is quite general.

Assume that at some early stage of the reaction the thermal equilibrium is established, and partonic matter is in a "high energy density" phase Q. This state corresponds to the absolute minimum of Ω with the order parameter close to zero, $\sigma \approx 0$, $\pi \approx 0$, and chiral symmetry restored (curve 1). Due to a very high internal pressure, Q matter will expand and cool down. At some stage a metastable minimum appears in Ω at a finite value of σ corresponding to a "low energy density" phase H, in which chiral symmetry is spontaneously broken. At some later time, the critical line in the (T, μ) plane is crossed where the Q and H minima have equal depths, i.e. $P_H = P_Q$ (curve 2). At later times the H phase becomes more favorable (curve 3), but the two phases are still separated by a potential barrier. If the expansion of the Q phase continues until the barrier vanishes (curve 4), the system will find itself in an absolutely unstable state at a maximum of the thermodynamic potential. Therefore, it will freely roll down into the lower energy state corresponding to the H phase. This situation is known as a spinodal instability.

As well known, a first order phase transition proceeds through the nucleation process. According to the standard theory of homogeneous nucleation [36], supercritical bubbles of the H phase appear only below the critical line, when $P_H > P_Q$. In rapidly expanding matter the nucleation picture might be very different. As shown in ref. [11], the phase separation in this case can start as early as the metastable H state appears in the thermodynamic potential, and a stable interface between the two phases may exist. An appreciable amount of nucleation bubbles and even empty cavities may be created already above the critical line.

The bubble formation and growth will also continue below the critical line. Previously formed bubbles will now grow faster due to increasing pressure difference, $P_H - P_Q > 0$, between the two phases. It is most likely that the conversion of Q matter on the bubble boundary is not fast enough to saturate the H phase. Therefore, a fast expansion may lead to a deeper cooling of the H phase inside the bubbles compared to the surrounding Q matter. Strictly speaking, such a system cannot be characterized by the unique temperature. At some stage the H bubbles will percolate, and the topology of the system will change to isolated regions of the Q phase (Q droplets) surrounded by the undersaturated vapor of the H phase.

The characteristic droplet size can be estimated by applying the energy balance consideration, proposed by Grady [37, 38] in the study of dynamical fragmentation of fluids. The idea is that the fragmentation of expanding matter is a local process minimizing the sum of surface and kinetic (dilational) energies per fragment volume. As shown in ref. [34], this prescription works fairly well also for multifragmentation of expanding nuclei, where the standard statistical approach fails.

Let us imagine an expanding spherical Q droplet of radius R, embedded in the background of the dilute H phase. The change of the thermodynamic potential, $\Delta\Omega$, compared to the uniform H phase can be easily estimated within the thin-wall approximation [11]. According to the Grady's prescription, the quantity to be minimized is $\Delta\Omega$ per droplet volume, $V \propto R^3$, that is

$$\left(\frac{\Delta\Omega}{V}\right)_{droplet} = -(P_Q - P_H) + \frac{3\gamma}{R} + \frac{3}{10}\Delta\mathcal{E}H^2R^2 \ . \tag{6}$$

Here $\Delta\mathcal{E} = \mathcal{E}_Q - \mathcal{E}_H$ is the difference of the bulk energy densities of the two phases, γ is the interface energy per unit area. One should notice that the last term, i.e. the change in the collective kinetic energy, is positive because $\mathcal{E}_Q > \mathcal{E}_H$. This term acts here as an effective long-range potential, similar to the Coulomb potential in nuclei. Since the bulk term does not depend on R the minimization condition constitutes the balance between the collective kinetic energy and the interface energy. This leads to an optimum droplet radius

$$R^* = \left(\frac{5\gamma}{\Delta\mathcal{E}H^2}\right)^{1/3} . \tag{7}$$

One can say that the metastable Q phase is torn apart by a mechanical strain associated with the collective expansion. This phenomenon has a direct analogy with the fragmentation of pressurized fluids leaving nozzles [39, 40]. In a similar way, splashed water forms droplets which have little to do with the equilibrium liquid-gas phase transition.

In the lowest-order approximation the characteristic droplet mass can be calculated as $M^* \approx \Delta \mathcal{E} V$. It is natural to think that nucleons and heavy mesons are smallest droplets of the Q phase. For numerical estimates I take $\gamma = 10$ MeV/fm^2 and $\Delta \mathcal{E} = 0.5$ GeV/fm^3, i.e. the energy density inside the nucleon. For the Hubble constant I consider two possibilities: $H^{-1} = 20$ fm/, representing a slow expansion from a soft point, and $H^{-1} = 6$ fm/c typical for a fast expansion. Substituting these values in Eq. (7) one gets R^*=3.4 fm and 1.5 fm for the slow and fast expansion respectively. These two values of R^* give M^* of about 100 GeV and 10 GeV, respectively. At ultrarelativistic energies the collective expansion is very anisotropic, with the strongest component along the beam axes. For the predominantly 1-d expansion one should expect the formation of slab-like structures with intermittent layers of Q and H phases.

After separation the droplets recede from each other according to the global Hubble expansion, predominantly along the beam direction. Hence their center-of-mass rapidities are in one-to-one correspondence with their spatial positions. Presumably they will be distributed more or less evenly between the target and projectile rapidities. Since rescatterings in the dilute H phase are rare, most hadrons produced from individual droplets will go directly into detectors. One can guess that the number of produced hadrons is proportional to the droplet mass. Each droplet will give a bump in the hadron rapidity distribution around its center-of-mass rapidity. If emitted particles have a Boltzmann spectrum, the width of the bump will be $\delta y \sim 2\sqrt{T/m}$, where T is the droplet temperature and m is the particle mass. At $T \sim 100$ MeV this gives $\delta y \approx 2$ for pions and $\delta y \approx 1$ for nucleons. These spectra might be slightly modified by the residual expansion of droplets and their transverse motion. The resulting rapidity distribution in a single event will be a superposition of contributions from different droplets, and therefore it will exhibit strong non-statistical fluctuations. The fluctuations will be more pronounced if primordial droplets are big, as expected in the vicinity of the soft point. If droplets as heavy as 100 GeV are formed, each of them will emit up to ~ 300 pions within a narrow rapidity interval, $\delta y \sim 1$. Such bumps can be easily resolved and analyzed. The fluctuations will be less pronounced if many small droplets shine in the same rapidity interval. Critical fluctuations of similar nature were discussed in ref. [41].

Some unusual events produced by high-energy cosmic nuclei have been already seen by the JACEE collaboration [42]. Unfortunately, they are very few and it is difficult to draw definite conclusions by analyzing them. We should be prepared to see plenty of such events in the future RHIC and LHC experiments. It is clear that the nontrivial structure of the hadronic spectra will be washed out to a great extent when averaging over many events. Therefore, more sophisticated methods of the event sample analysis should be used. The simplest one is to search for non-statistical fluctuations in the hadron multiplicity distributions measured in a fixed rapidity bin [43]. One can also study the correlation of multiplicities in neighbouring rapidity bins, bump-bump correlations etc. Such standard methods as intermittency and commulant moments [41], wavelet transforms [44], HBT interferometry [45] can also be useful. All these studies should be done at different collision energies to identify the phase transition threshold. The predicted dependence on the Hubble constant and the reaction geometry can be checked in collisions with different ion masses and impact parameters.

CONCLUSIONS

- The statistical approach (SMM) works well in situations when thermalized sources are well defined and no significant collective flow is present.

- The quantitative agreement of SMM with recent data on the caloric curve and temperature fluctuations provides a strong indication on the nuclear liquid-gas phase transition. The nuclear heat capacity has a peak at $T \approx 6$ MeV.

- A first order phase transition in rapidly expanding matter should proceed through the nonequilibrium stage when a metastable phase splits into droplets. The primordial droplets should be biggest in the vicinity of a soft point when the expansion is slowest.

- Hadron emission from droplets of the quark-gluon plasma should lead to large nonstatistical fluctuations in their rapidity spectra and multiplicity distributions. The hadron abundances may reflect directly the chemical composition in the plasma phase.

- Electromagnetic excitation of nuclei in ultrarelativistic heavy-ion colliders is an important reaction mechanism leading to the deep nuclear disintegration. The multiple neutron emission associated with this process may be used for monitoring ultrarelativistic heavy-ion beams.

- And finally, we should use the lessons of the liquid-gas phase transition for future studies of the deconfinement-hadronization and chiral phase transitions in relativistic heavy-ion collisions.

ACKNOWLEDGMENTS

The author is grateful to J.P. Bondorf and A.D. Jackson for many fruitful discussions. I thank A.S. Botvina, M. D'Agostino, A. Mocsy, I.A. Pshenichnov, O. Scavenius for cooperation. Discussions with D. Diakonov, A. Dumitru, J.J. Gaardhoje, M.I. Gorenstein, W. Greiner, B. Jakobsson, L. McLerran, R. Mattiello, W.F.J. Müller, W. Reisdorf, L.M. Satarov, H. Stöcker, E.V. Shuryak, W. Trautmann and

V. Viola are greatly appreciated. I thank the Niels Bohr Institute, Copenhagen University, and the Institute for Theoretical Physics, Frankfurt University, for kind hospitality. This work was carried out partly within the framework of a Humboldt Award, Germany.

References

[1] J.P. Bondorf, A.S. Botvina and I.N. Mishustin, *Phys. Rev.* **C58**, R27 (1998).

[2] M. D'Agostino, A.S. Botvina, M. Bruno, A. Bonasera, J.P. Bondorf, I.N. Mishustin, F. Gulminelli, R. Bougault, N. Le Neindre, P, Desesquelles, E. Geraci, A. Pagano, I. Iori, A. Moroni, G.V. Margagliotti, G. Vannini, *Nucl. Phys.* **A650**, 329 (1999).

[3] J.P. Bondorf, R. Donangelo, I.N. Mishustin, C.J. Pethick, H. Schulz, K. Sneppen, *Nucl. Phys.* **A443**, 321 (1985).

[4] I.N. Mishustin, *Nucl. Phys.* **A 447** (1985) 67c.

[5] J. Berges and K. Rajagopal, *Nucl. Phys.* **B538**, 215 (1999).

[6] M.A. Halasz, A.D. Jackson, R.E. Shrock, M.A. Stephanov and J.J.M. Verbarshot, *Phys. Rev.* **D58**, 096007 (1998).

[7] M. Stephanov, K. Rajagopal and E. Shuryak, *Phys. Rev. Lett.* **81**, 4816 (1998).

[8] I.N. Mishustin, L.M. Satarov, H. Stoecker and W. Greiner, *Phys. Rev.* **C59**, 3243 (1999).

[9] W. Reisdorf and FOPI Collaboration, *Nucl. Phys.* **A612**, 493 (1997).

[10] P. Braun-Münzinger and J. Stachel, *Nucl. Phys.* **A638**, 3c (1998).

[11] I.N. Mishustin, *Phys. Rev. Lett.* **82**, 4779 (1999).

[12] I.A. Pshenichnov, I.N. Mishustin, J.P. Bondorf, A.S. Botvina, A.S. Iljinov, *Phys. Rev.* **C57**, 1920 (1998).

[13] I.A. Pshenichnov, I.N. Mishustin, J.P. Bondorf, A.S. Botvina, A.S. Iljinov, *Phys. Rev.* **C60**, 044901 (1999).

[14] J.P. Bondorf, A.S. Botvina, A.S. Iljinov, I.N. Mishustin and and K.Sneppen, *Phys. Rep.* **257** (1995) 133.

[15] D.H.E. Gross, *Rep. Progr. Phys.* **53** (1990) 605.

[16] S. Das Gupta, J. Pan, I. Kvasnikova, C. Gale, *Nucl. Phys.* **A621**, 897 (1997).

[17] A.S. Botvina et al., *Nucl. Phys.* **A584** (1995) 737.

[18] H. Xi and ALADIN Collaboration, *Z. Phys.* **A359**, 397 (1997).

[19] M. D'Agostino et al., *Phys. Lett.* **B371**,175 (1996).

[20] R. Bougault et al., *in Proceedings of the XXXV International Winter Meeting on Nuclear Physics (Bormio, February 3-7, 1997)*; Preprint LPCC 97-04, April 1997.

[21] V.E. Viola et al., *in Proceedings of the International Workshop on Gross Properties of Nuclei and Nuclear Excitations XXVII: Multifragmentation (Hirschegg, Austria, 17-23 January 1999)*, p. 93; Preprint INC-40007-136, Indiana University, 1999.

[22] R.P. Scharenberg, B.K. Srivastava and EOS Colaboration, *in the same Hirschegg Proceedings as above*, pp. 237, 247.

[23] J.P. Bondorf, R. Donangelo, I.N. Mishustin, H. Schulz, *Nucl. Phys.* **A444**, 460 (1985).

[24] J. Pochodzalla and ALADIN Collaboration, *Phys. Rev. Lett.* **75**, 1040 (1995).

[25] W.F.J. Müller, in the same Hirschegg Proceedings as above, p. 200.

[26] S. Albergo et al., *Nuovo Cimento* **89**, 1 (1985).

[27] M.B. Tsang, W.G. Lynch, H. Xi and W.A. Friedman, *Phys. Rev. Lett.* **78** 3836 (1997).

[28] Al. H. Raduta, Ad.R. Raduta, *Phys. Rev.* **C59**, 323 (1999).

[29] X. Campi, *J. Phys.* **A19**, L917 (1986); *Phys. Lett.* **B208**, 351 (1988).

[30] M.E. Fisher, *Rep. Prog. Phys.* **30**, 615 (1967).

[31] F. Gulminelli and Ph. Chomaz, *Phys. Rev. Lett.* **82**, 1402 (1999).

[32] A.S. Iljinov, I.A. Pshenichnov, N. Bianchi, E. De Sanctis, V. Muccifora, M. Mirazita and P. Rossi, *Nucl. Phys.* **A616**, 575 (1997).

[33] A.J. Baltz, C. Chasman, and S.N. White, nucl-ex/9801002.

[34] I.N. Mishustin, *in Proceedings of the 6th International Conference on Nucleus-Nucleus Collisions (Gatlinburg, June 2-6, 1997)*; *Nucl. Phys.* **A630**, 111c (1998).

[35] L.P. Csernai, I.N. Mishustin and A. Mocsy, *Heavy Ion Phys.*, **3**, 151 (1996); A. Mocsy, M.Sc. thesis, University of Bergen, 1996.

[36] L.P. Csernai, J.I. Kapusta, *Phys. Rev. Lett.* **69**, 737 (1992); *Phys. Rev.* **D46**, 1379 (1992).

[37] D.E. Grady, *J. Appl. Phys.* **53**(1), 322 (1981).

[38] B.L. Holian and D.E. Grady, *Phys. Rev. Lett.* **60**, 1355 (1988).

[39] J.A. Blink and W.G. Hoover, *Phys. Rev.* **A32**, 1027 (1985).

[40] H. Buchenau et al., *J. Chem. Phys.* **92**, 6875 (1990).

[41] N.G. Antoniou, *Nucl. Phys.* **B71**, 307 (1999).

[42] T.H. Barnett et al., *Phys. Rev. Lett.* **50**, 2062 (1983).

[43] M.J. Tannenbaum and E802 Collaboration, *Phys. Rev.* **C52**, 2663 (1995).

[44] N. Suzuki, M. Biyajima and A. Ohsawa, hep-ph/9503403.

[45] H. Heiselberg, A.D. Jackson, hep-ph/9809013.

Experiments with RHIC

Gary D. Westfall

*National Superconducting Cyclotron Laboratory and
Department of Physics and Astronomy
Michigan State University
East Lansing, Michigan 48824-1321*

Abstract. Experiments with the Relativistic Heavy Ion Collider (RHIC) will begin December 1999. RHIC consists of two superconducting rings capable of accelerating and storing Au beams of 100 GeV/nucleon and proton beams of 250 GeV. Four experiments are being prepared for RHIC; STAR, PHENIX, PHOBOS, and BRAHMS. These detector systems are designed to search for signals of the quark gluon plasma in Au-Au collisions. A spin physics program using polarized protons will also be carried out at RHIC.

RHIC PHYSICS

Search for the QGP

The Quark Gluon Plasma

According to the standard cosmological model, the temperature of the cosmic background radiation exceeded 200 MeV during the first 10 µs of the big bang. Thus before 10 µs, the universe was filled with quarks and gluons rather than hadrons. In this quark gluon plasma (QGP), quarks and gluons are not confined in hadrons but are free to move around and interact. This deconfinement is related to the restoration of chiral symmetry. Chiral symmetry is a symmetry of quantum chromodynamics (QCD) in the limit of vanishing quark masses. In normal matter, u and d quarks have masses and chiral symmetry is broken. At high temperatures and densities, chiral symmetry is restored, one has massless quarks and the σ and π mesons have the same mass. Lattice QCD calculations have shown that deconfinement and chiral restoration take place at the same time.

In Figure 1 lattice gauge calculations are shown in which the energy density, ε, and the pressure, p, divided by the temperature, T, to the fourth power are calculated as a function of T.[1] A clear phase transition is observed at a critical temperature of 150 MeV.

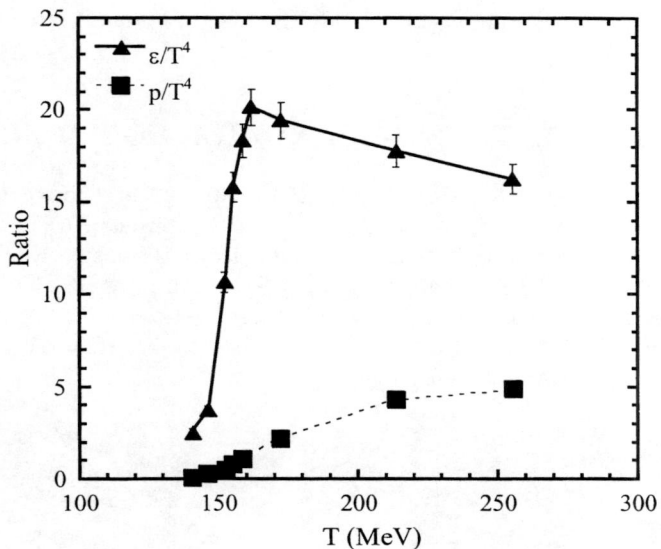

FIGURE 1. Lattice gauge calculations showing the phase transition confined to deconfined.[1]

Forming a QGP in the Lab

To form a QGP in the lab, one needs to collide heavy nuclei together at energies high enough to initiate hard QCD scattering between the quarks and gluons of the heavy nuclei. In these reactions, the mean free path is on the order of 1 fm, which is small compared with the diameter of gold nuclei, which is about 14 fm. The thermalization time is a few fm/c so that one may be able to describe these reactions a thermalized. The energy density needed to create the QGP is on the order of 3 GeV/fm^3, which is many times the energy density of normal nuclear matter (0.15 GeV/fm^3).

To produce the enough energy to create the QGP, a collider is necessary. For example, to create the same energy densities as those reached by RHIC with 100 GeV/nucleon Au + 100 GeV/nucleon Au, a fixed target accelerator with an energy of 21.7 TeV/nucleon would be required. This accelerator would be roughly comparable to one ring of the SSC. To match the 250 GeV p + 250 GeV p energies, a fixed target accelerator with an energy of 137 TeV would be required.

Spin Physics

Another major thrust of RHIC is the study of spin physics. These studies will be facilitated by polarized proton running at RHIC. The search for the origin of the spin of the proton has shown that only a fraction of the proton's spin is carried by the quarks. The measurement of spin dependent gluon distribution function, tests of QCD, and studies of fundamental symmetries will be carried out. A detailed contribution concerning spin physics at RHIC can be found elsewhere in this conference.

THE RELATIVISTIC HEAVY ION COLLIDER

The Relativistic Heavy Ion Collider (RHIC) is located at Brookhaven National Laboratory on Long Island, New York. RHIC consists of an ion source (keV/nucleon), a tandem Van de Graaff (MeV/nucleon), a booster synchrotron (500 MeV/nucleon for Au), the Alternating Gradient Synchrotron (AGS, 10 GeV/nucleon for Au), and two storage/acceleration rings (100 GeV/nucleon for Au). For proton running, a linac is used to inject the protons directly into the AGS. A photograph of the RHIC tunnel is shown in Figure 2[2].

FIGURE 2. Photograph of the two rings in the RHIC tunnel.

RHIC completed its engineering run August 16, 1999. During the engineering run, beam was captured in the blue ring and accelerated up to partial field. Beam was circulated in the yellow ring. Subsequent inspection after the engineering run revealed that damage to the beam chamber bellows between various parts of the magnet lattice had caused the observed beam loss and will be corrected before first production run, which is scheduled for May, 1999. The beams produced by RHIC are summarized in Table 1.

TABLE 1. RHIC Beams

Beam	Luminosity (cm^{-2}s^{-1})	Hold Time
100 GeV/n Au + 100 GeV/n Au	2×10^{26}	10 hours
250 GeV p + 250 GeV p	2×10^{32}	days

RHIC will run asymmetric beams such as a 100 GeV proton beam in one ring and a 100 GeV/nucleon Au beam in the other ring. Polarized protons will be accelerated up to 2 x 250 GeV with the polarization being controlled with a series of Siberian snakes.

RHIC has six intersection regions. Four of these intersections are instrumented. The RHIC layout including the detectors is shown in Figure 3[3].

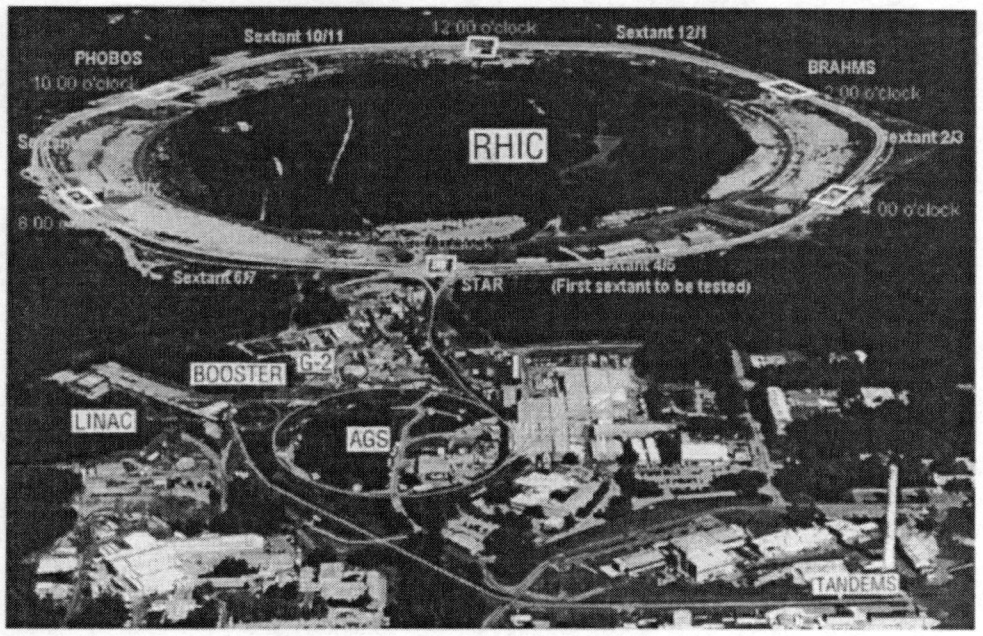

FIGURE 3. Overview of the RHIC complex showing the tandem Van de Graaff injector, the booster synchrotron, the AGS, and the two RHIC rings.

RHIC EXPERIMENTS

Four experiments are planned for RHIC. Two large experiments, STAR and PHENIX, will occupy the intersection regions at 6 o'clock and 8 o'clock respectively while two more focused experiments, PHOBOS and BRAHMS, will be located at 10 o'clock and 2 o'clock respectively.

STAR

STAR is the Solenoidal Tracker At RHIC. STAR consists of a large time projection chamber (TPC), a 0.5 T room temperature solenoidal magnet, a central trigger barrel/time-of-flight (CTB/TOF), a silicon vertex tracker (SVT), a barrel and endcap electromagnetic calorimeter (EMC), and two forward time projection chambers. A three dimensional drawing of STAR is shown in Figure 4[4].

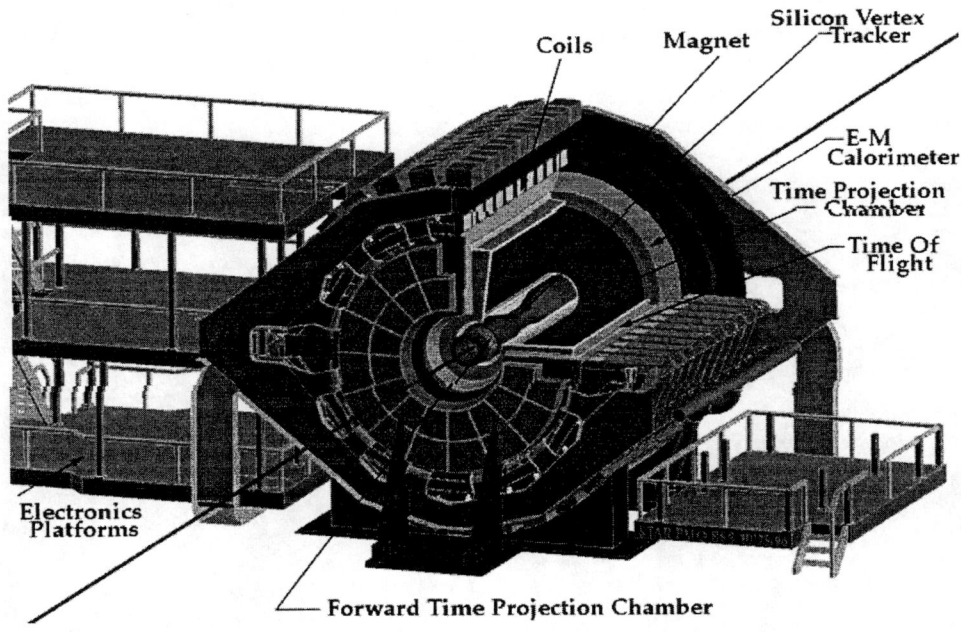

FIGURE 4. Drawing of the STAR Detector.

STAR is designed to have a very large acceptance and to be able to handle a large multiplicity of tracks. STAR will be able to study the hadronic signals of the quark gluon plasma such as excess strangeness production and disoriented chiral condensates (DCCs). STAR can study RHIC collisions event-by-event removing the ensemble averaging over events. The addition of the EMC allows STAR to also address hard QCD phenomena and to carry out spin physics measurements. In Figure 5 one can see the huge number of tracks the TPC can handle[5].

Figure 6 shows a representative method of attacking the observation of the QGP using STAR. The production rate of particles with strangeness, especially multi-strange anti-baryons, is higher for the QGP than for a hadronic gas[6].

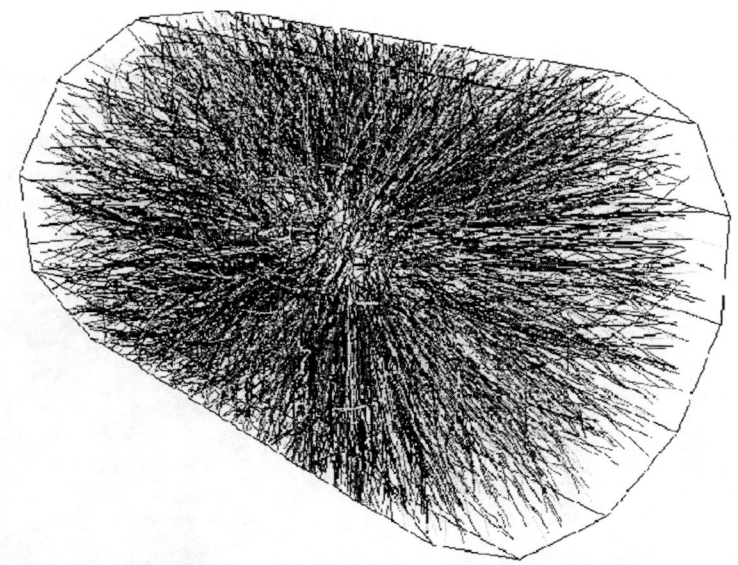

FIGURE 5. Depiction of tracks reconstructed in the STAR TPC.

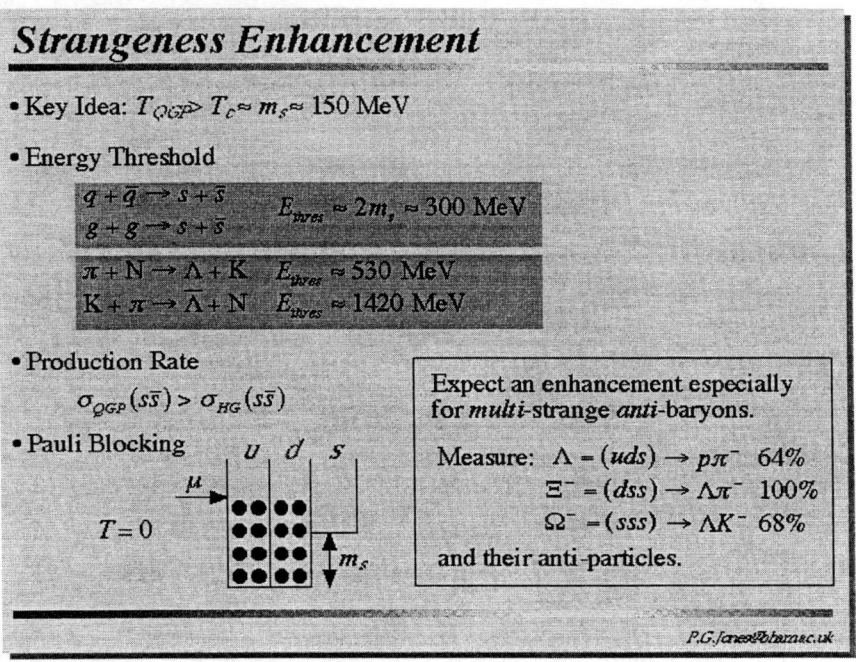

FIGURE 6. Example of studying the QGP using hadronic signals in STAR.

PHENIX

PHENIX is the Pioneering High ENergy Ion eXperiment. The east and west central arms (inner detectors, tracking system, ring imaging cerenkov detector, time-of-flight system, and electromagnetic calorimeters) are instrumented to detect electrons, photons, and charged hadrons. The north and south muon arms are instrumented with tracking chambers and particle identifiers to detect muons. A cut-away drawing of PHENIX is shown in Figure 7[7]. An illustration of the tracking abilities of PHENIX is shown in Figure 8[8].

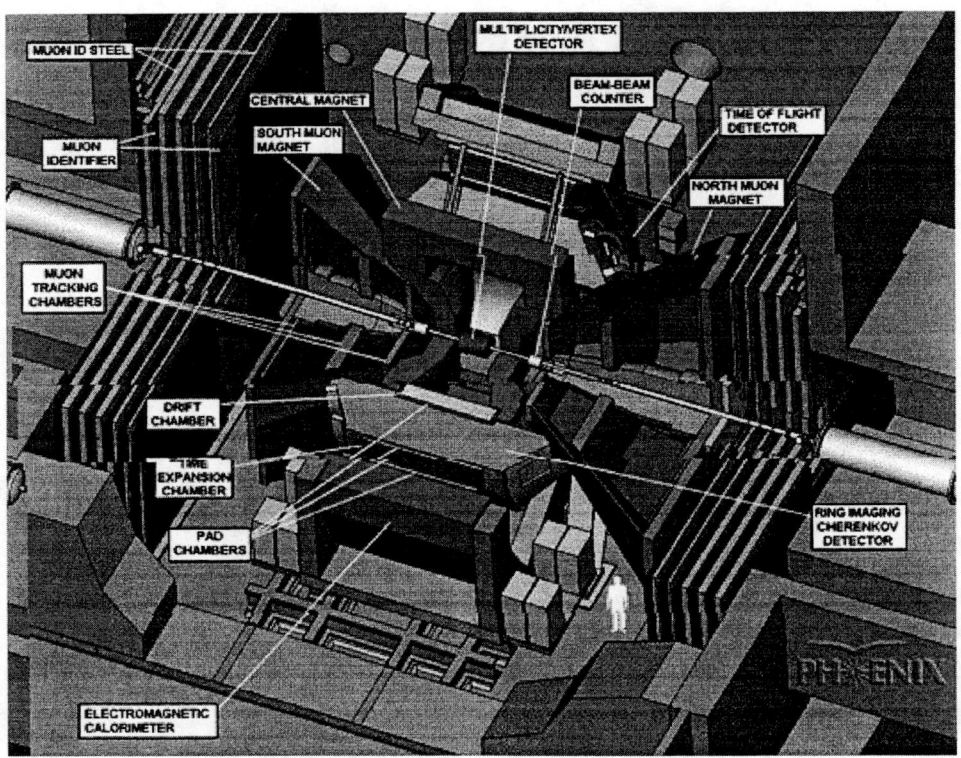

FIGURE 7. Cut-away drawing of the PHENIX detector showing multiplicity vertex detector (MVD), drift chamber, time expansion chamber (TEC), pad chambers, ring imaging cerenkov detector (RICH), electromagnetic calorimeter (EMC), and the two muon arms.

PHENIX will be able to probe deeply into the earliest times of Au-Au collisions at RHIC using its ability to reconstruct muon pairs and the electromagnetic calorimeter. The study of the suppression of J/ψ and the measurement of direct photons radiated from the quark-gluon plasma will give direct evidence for the QGP.

PHENIX will contribute strongly to the spin physics program by allowing the measurement of direct photons, jets, and high p_t hadrons.

The diverse tools PHENIX has allows a multi-pronged attack on the QGP as illustrated in Figure 9[9].

FIGURE 8. Drawing showing a Au-Au event being analyzed with the PHENIX tracking system (drift chambers, pad chambers, TEC).

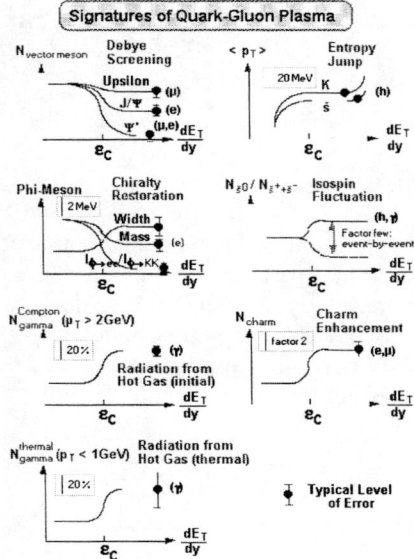

FIGURE 9. Examples of methods of extracting signals related to the quark-gluon plasma using PHENIX.

PHOBOS

PHOBOS is a "table-top" experiment in the sense that it is very compact. PHOBOS has large coverage for measuring charged particle multiplicity event-by-event from Au-Au collisions. PHOBOS has excellent coverage of low p_t particles at mid-rapidity which is essential to the study of phenomena such as DCCs. The silicon detectors combined with the magnetic field all momentum measurements and good particle identification over a large range of rapidity. PHOBOS has a high rate capability, which allows the writing of minimum bias events to tape as well as selected events. The scope of the experiment allows the possibility to reconfigure the experiment if data show unusual increase in multiplicity.

A perspective drawing of PHOBOS is shown in Figure 10[10].

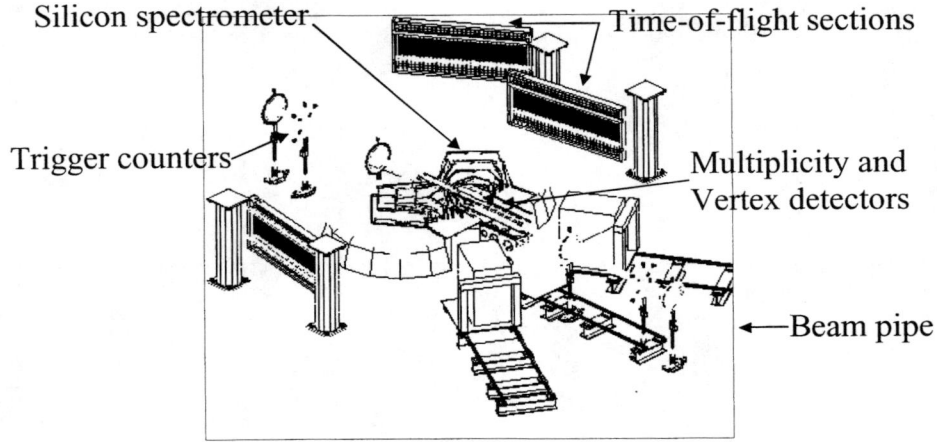

FIGURE 10. Perspective drawing the PHOBOS detector.

PHOBOS covers approximately 1% of the solid angle but offers excellent coverage in pseudorapidity and has excellent particle identification that will allow the stringent testing of the basic understanding of Au-Au collisions at RHIC. A simulation showing the acceptance of PHOBOS is shown in Figure 11[11]. In this figure, the charged particle multiplicity per unit rapidity is reconstructed using the multiplicity array and compared with the central detectors.

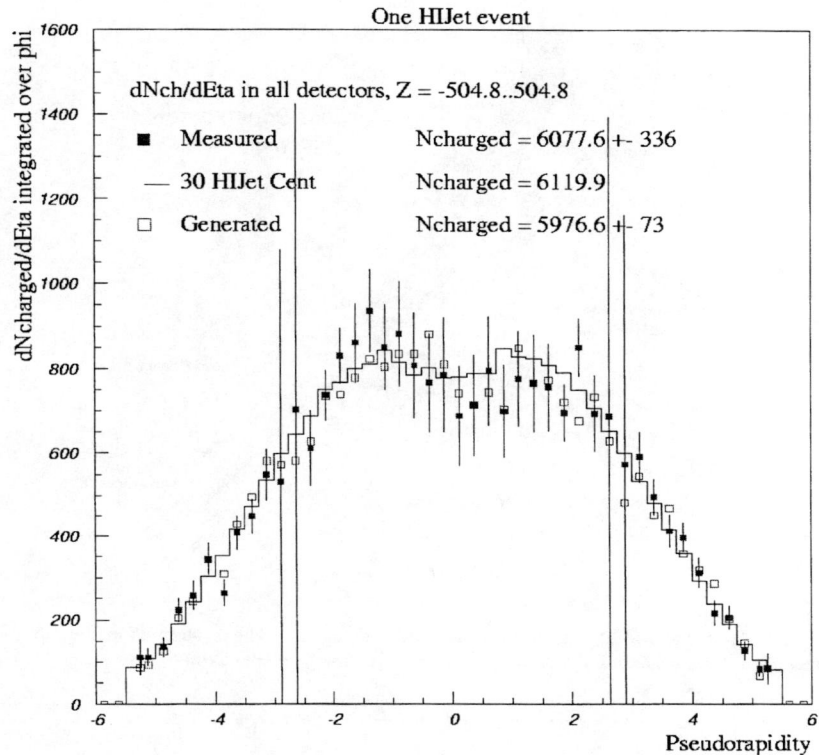

FIGURE 11. A simulation of the acceptance of PHOBOS in pseudorapidity. Note that the acceptance covers out to the pseudorapidity of the beams.

BRAHMS

BRAHMS is the Broad RAnge Hadron Magnetic Spectrometers experiment at RHIC. BRAHMS is designed to detect particles with a large range in p_t. Small solid angle, high resolution, and good particle identification characterize the BRAHMS experiment. BRAHMS uses two spectrometers, one at forward rapidities and one at mid-rapidity. BRAHMS also incorporates event characterization using a multiplicity counter and beam-beam counters.

A perspective drawing of BRAHMS is shown in Figure 12[12]. The forward and midrapidity spectrometers are visible along with the event characterization detectors. In Figure 13[13] a simulation is shown illustrating the acceptance of BRAHMS and comparing that acceptance with the predictions of various models. Clearly BRAHMS will be able to distinguish between the different predictions of the models.

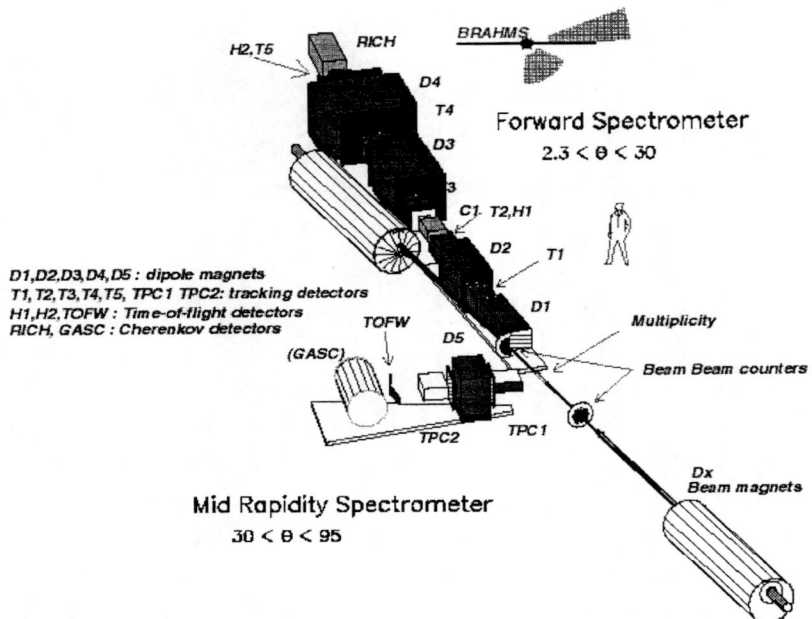

FIGURE 12. Perspective drawing of the BRAHMS detector. The system is made up of two spectrometers, one forward and one at midrapidity, combined with event characterization.

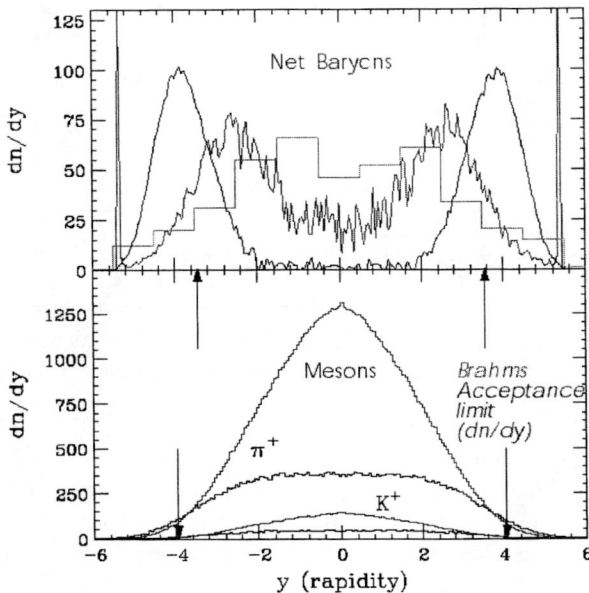

FIGURE 13. A simulation of the acceptance of BRAHMS showing the limits in y compared with the prediction of several models.

ACKNOWLEDGMENTS

This talk was prepared in part by using the excellent WWW facilities of RHIC, STAR, PHENIX, PHOBOS, and BRAHMS.

REFERENCES

1. T. Blum, Leo Kärkkäinen, and D. Toussaint, and Steven Gottlieb, Phys. Rev. D51, 5153 (1995).
2. http://www.collider.bnl.gov/images/ring1.jpg.
3. http://www.collider.bnl.gov/images/ring.jpg.
4. http://www.star.bnl.gov/STAR/img/images/star/star_detector.gif
5. http://www.star.bnl.gov/afs/rhic/star/doc/www/imagelib/event_images/STAR_GL/stargl_far.gif
6. STAR Strangeness Working Group, http://www.star.bnl.gov/afs/rhic/star/doc/www/viewgraphs/index.html
7. http://www.phenix.bnl.gov/phenix/WWW/figures/jpg/2_9703phnx1.jpg
8. From movie by J. Mitchell, http://www.phenix.bnl.gov/WWW/software/luxor/ani/birdpart.mpg
9. http://www.rhic.bnl.gov/phenix/WWW/figures/jpg/qgp_sig.gif
10. R. Pak, private communication.
11. Rudolf Ganz, <u>Advances in Nuclear Dynamics 3</u>, ed. A. Mignerey and W. Bauer, Plenum Press, 1997, p. 181.
12. http://www.rhic.bnl.gov/brahms/WWW/figures/brahms_pers.ps
13. http://www.rhic.bnl.gov/brahms/WWW/figures/Venus.gif

PHYSICS WITH STORED ELECTRONS

The HERMES Experiment

Richard G. Milner

MIT-Bates Linear Accelerator Center, P.O. Box 846, Middleton, MA 01949, USA

Abstract. The study of the spin structure of the nucleon is a fundamental problem in strong interaction physics. The HERMES experiment at DESY, Hamburg, Germany is carrying out measurements to probe the spin structure of the nucleon using a new technique. Polarized internal gas targets of hydrogen, deuterium, and ^{3}He are used with the 27.5 GeV longitudinally polarized positron (or electron) beam of the HERA collider to measure both inclusive and semi-inclusive spin-dependent deep-inelastic scattering from the nucleon. In addition, HERMES has observed a negative spin asymmetry in the photoproduction of hadron pairs with high transverse momenta. This is interpreted as the first direct experimental evidence for a positive gluon polarization in the nucleon. The azimuthal single-spin asymmetry measured in semi-inclusive pion production at HERMES is presented and interpreted as an effect of a new T-odd fragmentation function. HERMES also has carried out precision measurements of the ratios of unpolarized nuclear cross-sections. The data indicate a sizable nuclear dependence in the ratio of longitudinal to transverse cross-sections at low Q^2

INTRODUCTION

Deep-inelastic scattering (DIS) is well established as a powerful tool for the investigation of nucleon structure. The measured structure functions have been successfully interpreted in terms of parton distributions. From unpolarized scattering the momentum distributions of both quarks and gluons are to-day known over a wide kinematic range. Despite the much improved experimental precision the understanding of the spin structure of the nucleon in terms of quarks and gluons remains a challenge. Only a fraction of the nucleon spin can be attributed to the quark spins and the contributions from strange quarks and gluons is almost unknown. Semi-inclusive polarized deep-inelastic scattering experiments are one of the most promising means to resolve this problem. Using the correlation between the struck quark and the hadron observed in the final state (flavor tagging), the separate spin contributions of quarks and antiquarks can be determined. Additionally, the gluon polarization can be measured by isolating the photon-gluon fusion process, i.e. by measuring double-spin asymmetries in charm production or hadron pair production at high transverse momenta.

LONGITUDINAL QUARK POLARIZATION

Hadron production in DIS is described by the absorption of a virtual photon by a point-like quark and the subsequent fragmentation into a hadronic final state. The two processes can be characterized by two functions: the quark distribution function $q_f(x, Q^2)$, and the fragmentation function $D_f^h(z, Q^2)$. The semi-inclusive DIS cross section $\sigma^h(x, Q^2, z)$ to produce a hadron of type h with energy fraction $z = E_h/\nu$ is then given by

$$\sigma^h(x, Q^2, z) \propto \sum_f e_f^2 q_f(x, Q^2) D_f^h(z, Q^2). \tag{1}$$

Here the sum is over quark and antiquark types $f = (u, \bar{u}, d, \bar{d}, s, \bar{s})$. In the target rest frame, E_h is the energy of the hadron, $\nu = E - E'$ and $-Q^2$ are the energy and the squared four-momentum of the exchanged virtual photon, $E(E')$ is the energy of the incoming (scattered) lepton and e_f is the quark charge in units of the elementary charge. The Bjorken variable x is calculated from the kinematics of the scattered lepton according to $x = Q^2/2M\nu$ with M being the nucleon mass. It is assumed that the fragmentation process is spin independent, i.e. that the probability to produce a hadron of type h from a quark of flavor f is independent of the relative spin orientations of quark and nucleon. The spin asymmetry A_1^h in the semi-inclusive cross section for production of a hadron of type h by a polarized virtual photon is then given by

$$A_1^h(x, Q^2, z) = \frac{\sum_f e_f^2 \Delta q_f(x, Q^2) D_f^h(z, Q^2)}{\sum_f e_f^2 q_f(x, Q^2) D_f^h(z, Q^2)} \frac{(1 + R(x, Q^2))}{(1 + \gamma^2)} \tag{2}$$

where $\Delta q_f(x, Q^2) = q_f^{\uparrow\uparrow}(x, Q^2) - q_f^{\uparrow\downarrow}(x, Q^2)$ is the polarized quark distribution function and $q_f^{\uparrow\uparrow(\uparrow\downarrow)}(x, Q^2)$ is the distribution function of quarks with spin orientation parallel (anti-parallel) to the spin of the nucleon. The ratio $R = \sigma_L/\sigma_T$ of the longitudinal to transverse photon absorption cross sections appears in this formula to correct for the longitudinal component that is included in the experimentally determined parametrizations of $q_f(x, Q^2)$ but not in $\Delta q_f(x, Q^2)$. The term $\gamma = \sqrt{Q^2}/\nu$ is a kinematic factor.

At HERMES, polarized quark distributions have been extracted from a combination of inclusive and semi-inclusive asymmetry data on ^{3}He and hydrogen. As the wave function for ^{3}He is dominated by the configuration with the two protons paired to zero spin, most of the asymmetry from ^{3}He is due to the neutron. The inclusive (semi-inclusive) asymmetry $A_1^{(h)}$ was extracted from the measured asymmetry $A_\parallel^{(h)}$ using the relation

$$A_1^{(h)} = A_\parallel^{(h)} / [D(1 + \gamma\eta)], \tag{3}$$

where D is the depolarization factor for the virtual photon and η is a kinematic factor. In Eq. (3) the approximation is used that the contribution of the second spin structure function g_2 to $A_1^{(h)}$ can be neglected. In the kinematic region of the HERMES measurement, g_2 was previously measured to be consistent with zero for the proton and neutron [1,2]. In each kinematic bin the value of $A_{\parallel}^{(h)}$ was extracted from the measured counting rates using

$$A_{\parallel}^{(h)} = \frac{N_{(h)}^{\uparrow\downarrow} L^{\uparrow\uparrow} - N_{(h)}^{\uparrow\uparrow} L^{\uparrow\downarrow}}{N_{(h)}^{\uparrow\downarrow} L_P^{\uparrow\uparrow} + N_{(h)}^{\uparrow\uparrow} L_P^{\uparrow\downarrow}}, \qquad (4)$$

where $N^{\uparrow\uparrow}$ ($N^{\uparrow\downarrow}$) are the numbers of DIS events for target polarization parallel (anti-parallel) to the beam polarization, and $N_h^{\uparrow\uparrow}$ ($N_h^{\uparrow\downarrow}$) are the corresponding numbers of hadrons in coincidence with a DIS event. Here, $L^{\uparrow\uparrow(\uparrow\downarrow)}$ are the luminosities for each spin state corrected for dead time, and $L_P^{\uparrow\uparrow(\uparrow\downarrow)}$ are the luminosities corrected for dead time and weighted by the product of beam and target polarizations for each spin state.

The inclusive and semi-inclusive asymmetries for positively and negatively charged hadrons on both targets were extracted [5]. The measured spin asymmetries $A_1^h(x, Q^2, z)$ were integrated in each x bin over the corresponding Q^2-range and the z-range from 0.2 to 1 to yield $A_1^h(x)$. The inclusive results were measured at a similar energy to SLAC [3,4,1] and the hadron asymmetries on hydrogen were also measured by SMC [10]. The HERMES data are in good agreement within the quoted uncertainties. The agreement of the HERMES data with the SMC data, taken at 6-12 times higher average Q^2, shows that the semi-inclusive asymmetries are Q^2 independent within the present accuracy of the experiments.

Eq. (2) is used to extract polarized quark distribution functions from semi-inclusive asymmetries. It can be written as

$$A_1^h(x) = \sum_f P_f^h(x) \frac{\Delta q_f(x)}{q_f(x)} \frac{(1 + R(x))}{(1 + \gamma^2)} \qquad (5)$$

where $P_f^h(x)$ are the so-called integrated purities defined as

$$P_f^h(x) = \frac{e_f^2 q_f(x) \int_{0.2}^1 D_f^h(z)\, dz}{\sum_{f'} e_{f'}^2 q_{f'}(x) \int_{0.2}^1 D_{f'}^h(z')\, dz'}. \qquad (6)$$

The inclusive asymmetry A_1 is similarly expressed by replacing P_f^h by P_f where $P_f(x) = e_f^2 q_f(x) / \sum_{f'} e_{f'}^2 q_{f'}(x)$. After integrating over z, Eq. (5) together with the corresponding inclusive case can be written in matrix form

$$\vec{A}(x) = \mathcal{P}(x) \cdot \vec{Q}(x) \qquad (7)$$

where the vector $\vec{A} = (A_{1p}, A_{1p}^{h^+}, A_{1p}^{h^-}, A_{1He}, A_{1He}^{h^+}, A_{1He}^{h^-})$ contains as elements the measured asymmetries. The vector $\vec{Q}(x)$ contains the quark and antiquark polarizations. The matrix $\mathcal{P}$ contains the effective integrated purities for the proton and

^{3}He as well as the $(1+R)/(1+\gamma^2)$ factor. These purities describe the probability that the virtual photon hit a quark of flavor f when a hadron of type h is detected in the experiment.. They include the effects of the acceptance of the experiment and have been determined with a Monte Carlo simulation using the LUND string fragmentation model, a model of the detector, the CTEQ Low–Q^2 parametrizations [11] for the unpolarized parton distributions and values for R from Ref. [12]. The LUND fragmentation parameters were tuned to fit the measured hadron multiplicities. Eq. (7) can be solved for $\vec{Q}(x)$ by minimizing

$$\chi^2 = \left(\vec{A} - \mathcal{P}\cdot\vec{Q}\right)^{\mathrm{T}} \mathcal{V}_{\vec{A}}^{-1} \left(\vec{A} - \mathcal{P}\cdot\vec{Q}\right). \tag{8}$$

where $\mathcal{V}_{\vec{A}}$ is the covariance matrix of the asymmetry vector $\vec{A}$. In the fit procedure constraints were imposed on the sea polarization to improve statistical significance. In view of rather ambiguous theoretical model predictions [13,14], two alternatives were chosen for relating the spin distributions of the sea flavors. As a first possibility it was assumed that the polarization $\Delta q_s(x)/q_s(x)$ of sea quarks is independent of flavor $\frac{\Delta u_s(x)}{u_s(x)} = \frac{\Delta d_s(x)}{d_s(x)} = \frac{\Delta s(x)}{s(x)} = \frac{\Delta \bar{u}(x)}{\bar{u}(x)} = \frac{\Delta \bar{d}(x)}{\bar{d}(x)} = \frac{\Delta \bar{s}(x)}{\bar{s}(x)}$. As an alternative, a pure singlet spin distribution of the sea is considered: $\Delta u_s(x) = \Delta d_s(x) = \Delta s(x) = \Delta \bar{u}(x) = \Delta \bar{d}(x) = \Delta \bar{s}(x)$. Because of the assumption on the sea quark polarization, the polarizations of the strange quarks and of the total sea are equal: $(\Delta s(x) + \Delta \bar{s}(x))/(s(x) + \bar{s}(x)) = \Delta q_s(x)/q_s(x)$. For $x > 0.3$ the sea polarization is set to zero and the corresponding effect on the results for the non-sea polarizations is included in their systematic uncertainties. The up quark polarizations are clearly positive and the down quark polarizations are negative over the measured range of x. Their absolute values are largest at large x and remain different from zero in the sea region. The sea polarization is compatible with zero over the measured range of x.

The systematic uncertainties of this measurement were determined from the uncertainties on the measured asymmetries, the unpolarized parton distributions and the purities. The uncertainty on the unpolarized parton distributions was derived by comparing different parametrizations of the world data. The uncertainty coming from the symmetry assumption of the sea polarization was derived by comparing the results produced by the two different assumptions described above. The fitted quark polarizations change by typically less than 0.01. The uncertainty in the purities was determined by comparing different fragmentation models and varying the fragmentation parameters in the Monte Carlo code.

The polarized quark distributions $\Delta q_f(x)$ were determined by forming the products of the polarizations $\Delta q_f(x)/q_f(x)$ and the unpolarized parton distributions from Ref. [11] at $Q^2 = 2.5$ GeV^2. It was assumed that the polarization is independent of Q^2 within the Q^2 range of this measurement. This assumption is justified by the weak Q^2 dependence predicted by QCD and by the experimental result that there is no significant Q^2 dependence observed in the inclusive and in the semi-inclusive asymmetries. The results are shown in Figs. 1 and 2. In Fig. 2, the

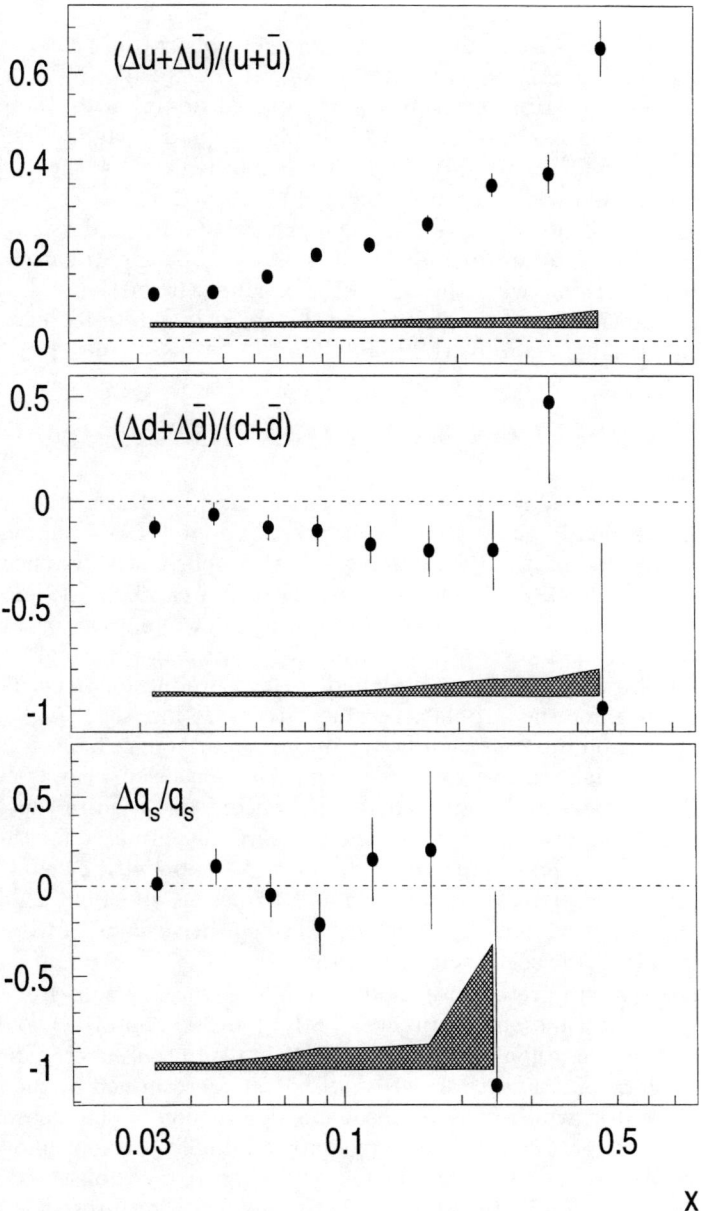

FIGURE 1. The flavor decomposition of the quark polarizations as a function of x derived from the HERMES inclusive and semi-inclusive asymmetries. The sea polarization is assumed to be flavor symmetric in this analysis. The error bars shown are statistical and the bands represent the systematic uncertainties.

upper plots show the polarized valence quark distributions $x\Delta u_v(x)$ and $x\Delta d_v(x)$, derived from the relation $\Delta q_v(x) = (\Delta q(x) + \Delta \bar{q}(x)) - 2\Delta \bar{q}(x)$. Since for scattering off sea quarks the contribution from u_s and $\bar{u}$ quarks dominates, the polarized $x\Delta \bar{u}(x)$ sea distribution is shown in the lower plot. Fig. 2 includes results from SMC [10] obtained at $Q^2 = 10$ GeV^2, which are shown here for the x-range explored by HERMES and which are extrapolated to $Q^2 = 2.5$ GeV^2 by assuming a Q^2-independent polarization $\Delta q(x)/q(x)$. The positivity limit and a parametrization of data from Ref. [15] are included in Fig. 2. The parametrization and the SMC results are consistent with the HERMES results within the statistical and systematic uncertainties. The uncertainties of the HERMES data for $x\Delta u_v(x)$ and $x\Delta \bar{u}(x)$ are much smaller than for the SMC data.

GLUON POLARIZATION

The precise inclusive and semi-inclusive polarized deep inelastic scattering data nowadays available clearly show that the quarks account for only about 25% of the nucleons spin. One possible explanation for this deficit is a significant gluon polarization in the nucleon. A direct measurement of ΔG can be performed via the isolation of the photon gluon fusion process. Experimental signatures of this process are charm production and the production of jets (hadrons) with high transverse momenta. Both charm production and high p_T jet production have been used successfully to determine the unpolarized gluon structure function $G(x)$. Here, we present the first data on the spin asymmetry in the production of hadron pairs with opposite charge and high transverse momentum. The measurement was performed using the HERMES polarized hydrogen target and did not require the detection of the scattered lepton [6]. Fig. 3 shows the measured asymmetry for the highest transverse momenta accessible at HERMES. For h^+h^- pairs with $p_T^{h_1} > 1.5$ GeV/c and $p_T^{h_2} > 1.0$ GeV/c, the asymmetry is found to be $A_{||} = -0.28 \pm 0.12$ (stat) ± 0.02 (sys). This negative value is in contrast to the positive asymmetries typically measured in deep inelastic scattering from protons.

The asymmetry is interpreted in a model including contributions from deep inelastic scattering, vector-meson dominance (VMD) processes and the two direct LO QCD processes: photon gluon fusion and the QCD Compton effect. The relative contributions of these processes to the cross section are determined by the PYTHIA Monte Carlo generator, which gives a reasonable description of the measured cross section of hadron pair production at intermediate and high transverse momenta. In the same region of phase space where the negative asymmetry is observed the cross section is found to be dominated by the photon gluon fusion process. Assuming a zero spin asymmetry for VMD processes, the gluon polarization is determined from the measured asymmetry and the known subprocess asymmetries for the QCD processes to be $\Delta G/G = 0.41 \pm 0.18$ (stat) ± 0.03 (exp syst) at $0.06 < x_G < 0.28$ and $\hat{p}_T^2$ of 2.1 (GeV/c)^2. This value of $\Delta G/G$ is compared in Fig. 4 with several phenomenological LO QCD fits to the data on $g_1(x, Q^2)$. Note that the error bar

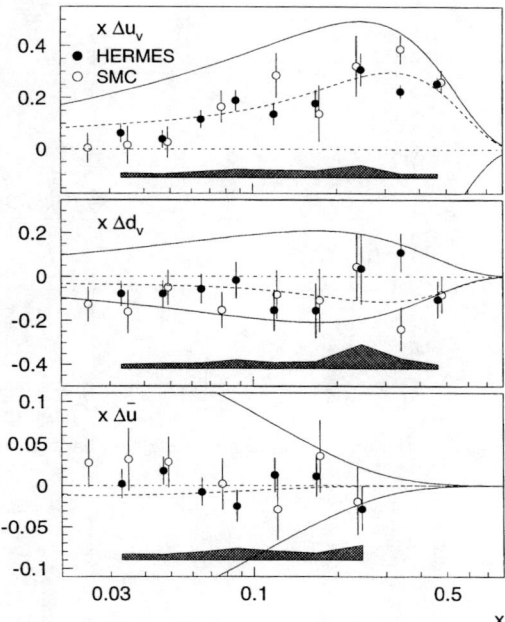

FIGURE 2. The spin distributions at $Q^2 = 2.5$ GeV^2 separately for the valence quarks $x\Delta u_v(x)$, $x\Delta d_v(x)$ and the sea quarks $x\Delta \bar{u}(x)$ as a function of x. The error bars shown are the statistical and the bands the systematic uncertainties. The distributions are compared to results from SMC, extrapolated to $Q^2 = 2.5$ GeV^2. The error bars of the SMC result correspond to its total uncertainty. The solid lines indicate the positivity limit and the dashed lines are the parametrization from Gehrmann and Stirling ('Gluon A', LO) [15].

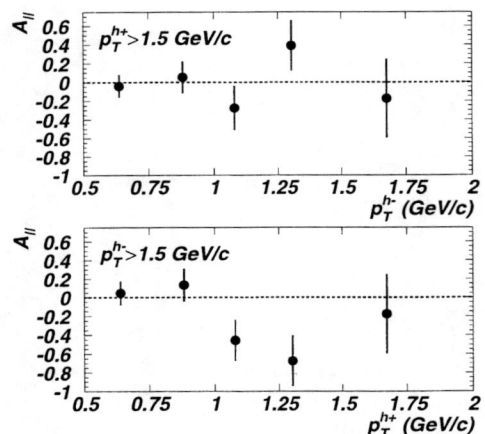

FIGURE 3. Asymmetry in the photoproduction of high p_T hadron pairs as measured at HERMES [6].

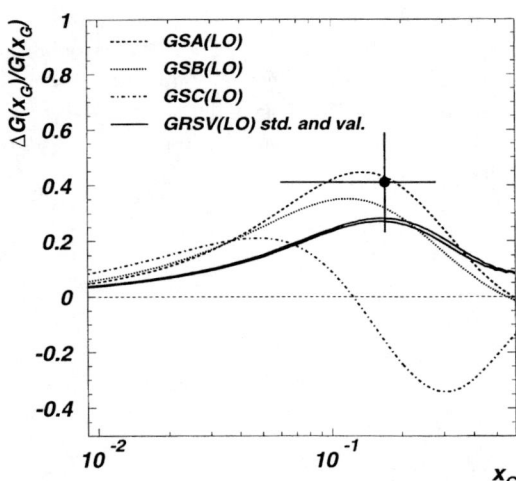

FIGURE 4. The extracted gluon polarization $\Delta G/G$ in comparison to phenomenological fits to $g_1^{p,n}(x, Q^2)$ [6].

does not include the potentially large model uncertainty in this measurement.

AZIMUTHAL SPIN ASYMMETRIES

Azimuthal single-spin asymmetries in semi-inclusive deep inelastic scattering can be due to higher twist effects or due to the presence of a T-odd fragmentation function (Collins mechanism). As shown in Fig. 5, a significant azimuthal asymmetry is observed in π^+ production on a longitudinally polarized hydrogen target at HERMES [7]. No asymmetry is seen in π^- production. The observed effect is a single target spin effect, the corresponding beam-related asymmetries measured on unpolarized targets are found to be consistent with zero. The $x-$ and p_T-dependence of the π^+ asymmetry agrees with the expectation based on the Collins model (assuming the transversity distribution $h_1(x)$ to be equal to the longitudinal spin distribution $g_1(x)$). A much better understanding of these new distribution and fragmentation function will become available when HERMES will measure single-spin asymmetries on a transversely polarized target. These measurements are expected to take place in 2001/2002.

THE UNPOLARIZED STRUCTURE FUNCTION IN NUCLEI AT LOW X

In addition to spin-dependent measurements, HERMES has taken data on unpolarized gas targets of hydrogen, deuterium, ^{3}He, and ^{14}N. The unpolarized semi-inclusive deep inelastic scattering events have been analyzed to extract information on the flavor asymmetry of the light quark sea [9]. The unpolarized inclusive DIS data have been analyzed in terms of ratios to cross sections [8]. The ratio of cross sections from nucleus A and deuterium D is then given by:

$$\frac{\sigma_A}{\sigma_D} = \frac{F_2^A}{F_2^D} \frac{(1+\epsilon R_A)(1+R_D)}{(1+R_A)(1+\epsilon R_D)} \quad (9)$$

where R_A and R_D represent the ratio $\frac{\sigma_L}{\sigma_T}$ for nucleus A and deuterium and ϵ is the virtual photon polarization parameter. For $\epsilon \rightarrow 1$ the cross section equals the ratio of structure functions $\frac{F_2^A}{F_2^D}$. For smaller values of ϵ the cross section ratio is equal to $\frac{F_2^A}{F_2^D}$ only if $R_A = R_D$. A difference between R_A and R_D will thus introduce an ϵ-dependence of $\frac{\sigma_A}{\sigma_D}$.

At HERMES [8], DIS events were extracted from the data by demanding $Q^2 > 0.3$ (GeV/c)^2, $W > 2$ GeV, and $y < 0.85$. Radiative corrections are sizable particularly at low x. The resulting ratio of cross sections is shown in Fig. 6. In the first four bins a striking discrepancy between the HERMES and NMC data is observed. The discrepancy increases with Q^2, but at the same time the average deviation in each x bin decreases with x.

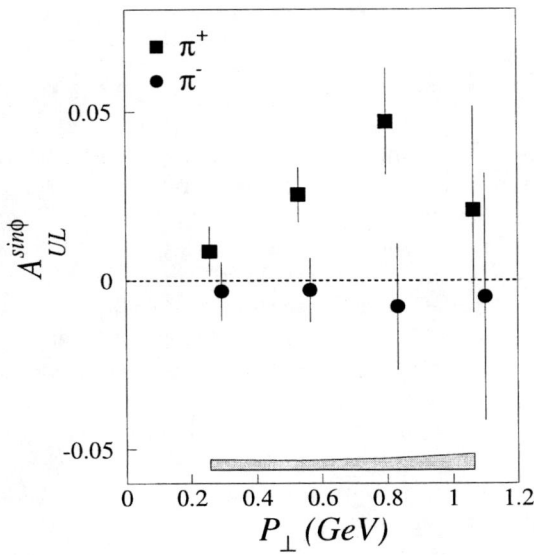

FIGURE 5. Azimuthal single-spin asymmetry in semi-inclusive π^+ and π^- production [7].

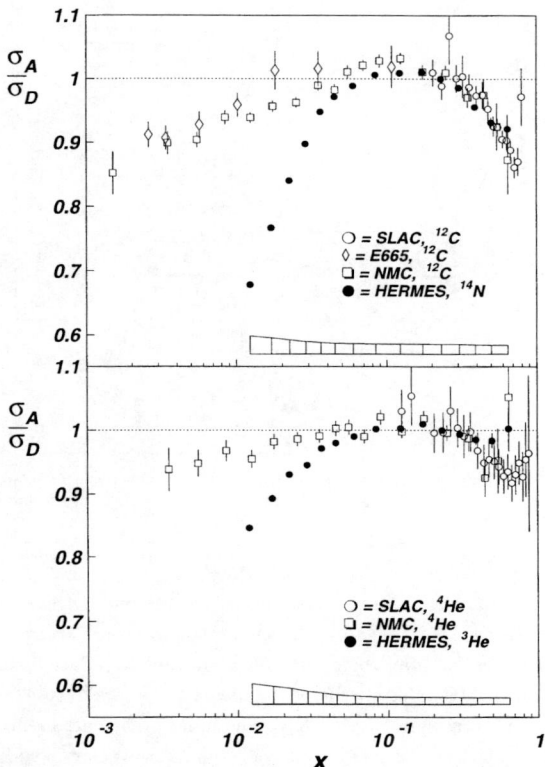

FIGURE 6. Ratio of cross sections of inclusive deep-inelastic lepton scattering from nucleus A and D versus x [8]. The error bars of the HERMES measurement represent the statistical uncertainties, the systematic uncertainty of the HERMES data is given by the error band. The error bars of the NMC, E665, and SLAC data are given by the quadratic sum of the statistical and systematic uncertainties.

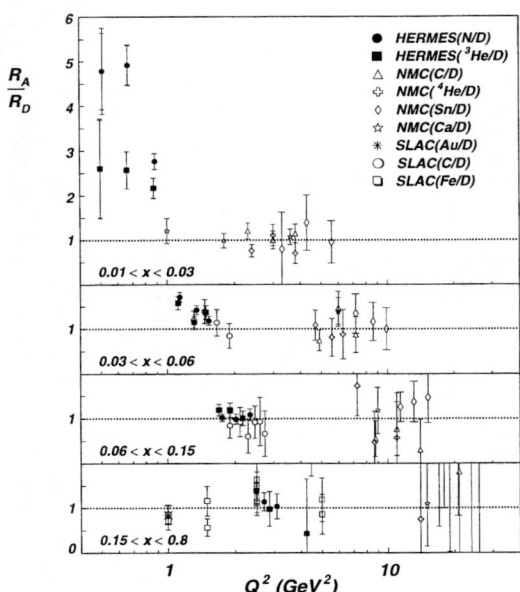

FIGURE 7. The ratio R_A/R_D for nucleus A and deuterium as a function of Q^2 for four different x bins [8]. The HERMES data on ^{14}N(^{3}He) are represented by the solid circles (squares). The open triangles (^{12}C) and crosses (^{4}He) have been derived from the NMC data using the same technique. The other SLAC [28] and NMC [29] data displayed have been derived from measurements of $\Delta R = R_A - R_D$ taking a parameterization [30] for R_D. The inner error bars include both the statistical uncertainty and the correlated error in F_2^A/F_2^D. The outer error bars also include the systematic uncertainties.

As the structure function ratio $\frac{F_2^A}{F_2^D}$ depends only on x and Q^2, the observed difference in the cross section ratios measured at HERMES and NMC/E665 can be explained only by an A dependence of the ratio $R(x, Q^2)$. The resulting values of $\frac{R_A}{R_D}$ are shown in Fig. 7.

SUMMARY

Using semi-inclusive deep-inelastic scattering on polarized and unpolarized targets the HERMES experiment has significantly improved our understanding of nucleon structure. The experiment has confirmed and extended our understanding of the quark contribution to the spin structure of the nucleon. Additionally, HERMES has found direct experimental evidence for a positive and large gluon polarization. The experimental study of the new T-odd distribution and fragmentation functions has just begun and has the promise to provide new insight into the spin structure of the nucleon in the future.

Precise measurements on unpolarized targets by HERMES have yielded a substantial suppression of the ratio of nuclear cross sections for $x < 0.06$ and $Q^2 < 1.5$ (GeV/c)^2. The observed difference is attributed to an A-dependence of the ratio $R = \sigma_L/\sigma_T$ of longitudinal to transverse deep-inelastic scattering cross sections at low X and Q^2.

I wish to thank my HERMES colleagues and, in particular Antje Brüll, for considerable help in preparing this contribution. My research is supported by the United States Department of Energy under Cooperative Agreement DE-FC02-94ER40818.

REFERENCES

1. E-143, K. Abe *et al*, *Phys. Rev.* D **58**, 112003 (1998).
2. E-155, P.L. Anthony *et al*, SLAC-PUB-7983 (1999), hep-ex/9901006.
3. E-143, K. Abe *et al*, *Phys. Rev. Lett.* **74**, 346 (1995).
4. E-154, K. Abe *et al*, *Phys. Rev. Lett.* **79**, 26 (1997).
5. K. Ackerstaff *et al.*, *Phys. Lett.* B **464**, 123 (1999).
6. A. Airapetian *et al.*, hep-ex/9907020.
7. A. Airapetian *et al.*, hep-ex/9910062.
8. K. Ackerstaff *et al.*, hep-ex/9910071.
9. K. Ackerstaff *et al.*, *Phys. Rev. Lett.* **81**, 5519 (1998).
10. SMC, D. Adeva *et al*, *Phys. Lett.* B **420**, 180 (1998).
11. H.L. Lai *et al*, *Phys. Rev.* D **55**, 1280 (1997).
12. L.W. Whitlow *et al*, *Phys. Lett.* B **250**, 193 (1990).
13. D. Diakonov *et al*, *Nucl. Phys.* B **480**, 341 (1996).
14. R. Fries and A. Schäfer, hep-ph/9805509.
15. T. Gehrmann and W.J. Stirling, *Phys. Rev.* D **53**, 6100 (1996).
16. M. Burkardt and R. L. Jaffe, *Phys. Rev. Lett.* **70**, 2537 (1993).
17. R. L. Jaffe, *Phys. Rev.* D **54**, R6581 (1996).

18. P. J. Mulders and R. D. Tangerman, *Nucl. Phys.* B **461**, 197 (1996).
19. ALEPH, D. Buskulic *et al*, *Phys. Lett.* B **374**, 319 (1996).
20. OPAL, K. Ackerstaff *et al*, Eur. Phys. J. **C2** (1997) 49.
21. D. de Florian, M. Stratman and W. Vogelsang, *Phys. Rev.* D **57**, R5811 (1997).
22. NMC, M. Arneodo *et al*, *Phys. Rev.* D **50**, 1 (1994).
23. For a recent review see F.M. Steffens and A. W. Thomas, Phys. Rev. **C55**, 900 (1997) and references therein.
24. M. Glück *et al.*, Z. Phys. **C67**, 433 (1995).
25. A. D. Martin *et al.*, Phys. Rev. **D51**, 4756 (1995).
26. A. D. Martin *et al.*, hep-ph/9803445.
27. E866, E. A. Hawker *et al.*, Phys. Rev. Lett. **80**, 3715 (1998); E866, J.C. Peng *et al.*, hep-ph/9804288.
28. S. Dasu *et al.*, Phys. Rev. D **49**, 5641 (1994).
29. P. Amaudruz *et al.*, Phys. Lett. B **294**, 120 (1992); M. Arneodo *et al.*, Phys. Lett. B **481**, 23 (1996).
30. K. Abe *et al.*, Phys. Lett. B **452**, 194 (1999).

Longitudinally Polarized Electrons in a Storage Ring below 1 GeV

I. Passchier,[1,2] D.J. Boersma,[2] M. Harvey,[5] D. W. Higinbotham,[2,6] H. R. Poolman,[1,2] E. Six,[7] R. Alarcon,[7] P.W. van Amersfoort,[2] Th.S. Bauer,[3] H. Boer Rookhuizen,[2] J.F.J. van den Brand,[1] L.D. van Buuren,[1] H.J. Bulten,[1] R. Ent,[4,5] M. Ferro-Luzzi,[1,2] D. Geurts,[1] P. Heimberg,[1] C.W. de Jager,[4,6] P. Klimin,[8] I. Koop,[8] F. Kroes,[2] J. van der Laan,[2] G. Luijckx,[2] A. Lysenko,[8] B. Militsyn,[2] I. Nesterenko,[8] J. Noomen,[2] B. E. Norum,[6] M.J.J. van den Putte,[2] Yu. Shatunov,[8] J. J. M. Steijger,[2] D. Szczerba,[9] H. de Vries.[2]

(1) Dept. of Phys. and Astr., Vrije Universiteit, NL-1081 HV Amsterdam, The Netherlands
(2) NIKHEF, P.O. Box 41882, NL-1009 DB Amsterdam, The Netherlands
(3) Physics Department, Utrecht University, NL-3508 TA Utrecht, The Netherlands
(4) TJNAF, Newport News, VA 23606, USA
(5) Department of Physics, Hampton University, Hampton, VA 23668, USA
(6) Department of Physics, University of Virginia, Charlottesville, VA 22901, USA
(7) Department of Physics, Arizona State University, Tempe, AZ 85287, USA
(8) Budker Institute for Nuclear Physics, Novosibirsk, 630090 Russian Federation
(9) Institut für Teilchenphysik, Eidg. Technische Hochschule, CH-8093 Zürich, Switzerland

Abstract. We report on the results of studies of the longitudinal electron polarization in the AmPS storage ring at NIKHEF. The ring was operated using a partial Siberian snake at the first magic energy, and a full Siberian snake between 440 MeV and 720 MeV. We have investigated the effect of high beam currents in the ring, and found that some of the electron polarization is lost if the beam current becomes larger than ≈ 120 mA.

In storage rings, electrons will become polarized by the Sokolov-Ternov self-polarization mechanism [1]. However, for the AmPS storage ring at NIKHEF the Sokolov-Ternov polarization build-up time is of the order of 10 h, while the beam lifetime during internal target experiments is of the order of 10 minutes. We obtained a stored polarized beam by injecting pulses from a polarized electron source [2]. It was used in combination with polarized internal targets [3,4] to perform spin-dependent electromagnetic studies of the nucleon and few body systems. The electron source used strained InGaAsP cathodes, which were illuminated with laser

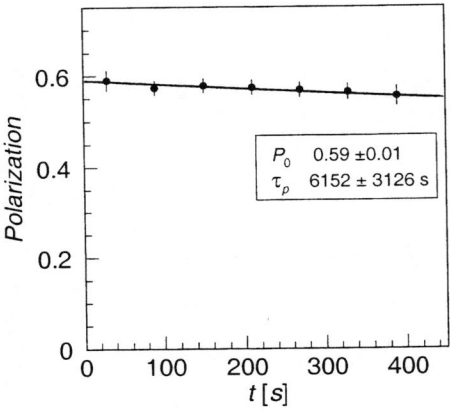

FIGURE 1. Electron polarization vs time for stored electrons with a beam energy of 665 MeV. The line is a fit to the data with the function $P(t) = P_0 e^{-t/\tau_p}$.

light to produce electron with a degree of polarization of up to 80%. The source used a pulsed high voltage power supply, which increased the lifetime of the cathodes from a couple of hours to 180 h. The electron polarization was measured at the source with a Mott polarimeter.

In order to maintain the longitudinal polarization at the internal target, a Siberian snake was implemented at the exact opposite side in the AmPS ring. The longitudinal polarization at the interaction point was determined with a Compton backscattering polarimeter [5]. The electron polarization of the stored beam was optimized by changing the orientation of the spin at the source with a Z-shaped spin manipulator and measuring the polarization in the ring. When the spin orientation in the accelerator was longitudinal, no depolarization was observed. However, up to 10% depolarization was observed when the spin was (partially) transverse to the momentum in the accelerator. This was probably due to the spin precession in the longitudinal magnetic field in the first sections of the accelerator and the spread in the energy of the electrons.

The polarization lifetime was measured with the Compton polarimeter at various beam energies and was always more than 4500 s, much longer than the beam lifetime when an internal target was present, see Fig. 1.

The polarized electron source produces pulses of about 5 mA. To obtain higher currents in the ring, 3-turn injection and pulse stacking is used. Currents in excess of 200 mA have been obtained. The effect of high beam currents on the electron polarization has been investigated using the Compton polarimeter and the internal target via the measurement of a spin correlation parameter. Here, the spin correlation parameter of the reaction $^3\vec{\mathrm{He}}(\vec{e},e'n)pp$ at the quasi-elastic peak was used as a (relative) beam polarization analyzer [6]. The polarization of the ^{3}He target was monitored independently. Figure 2 shows the electron polarization as

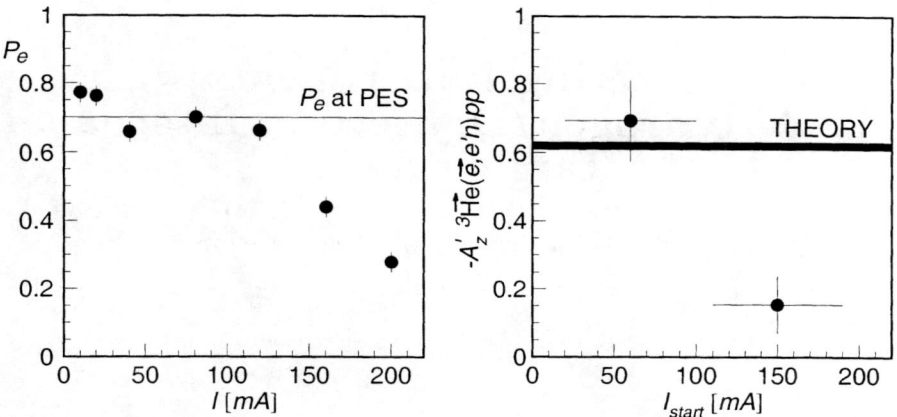

FIGURE 2. Electron polarization vs injection current as measured with the Compton polarimeter (left) and spin correlation parameter vs injection current (right). The spin correlation parameter has been extracted, assuming no loss of electron polarization.

measured with the Compton polarimeter (left) and the extracted value of the spin correlation parameter with the online analysis, assuming no electron polarization is lost (right). The theoretical prediction was calculated using plane wave impulse approximation and the electromagnetic form factors of the neutron and is in good agreement with the extracted value for $I_{start} < 100$ mA. At higher currents, a clear depolarization of the electron beam is observed and a large discrepancy exists between the extracted and predicted spin correlation parameter, indicating a significant loss of electron polarization. This is probably due to a shift of the horizontal and/or vertical betatron tune in the storage ring.

In summary, we have presented studies of the behavior of polarized electrons in a medium energy storage ring. We have shown it is possible to inject and store polarized electrons in a storage ring with a spin lifetime, which makes it possible to use the polarized beam for internal target experiments.

REFERENCES

1. A. A. Sokolov and I. M. Ternov, Sov. Phys.-Dokl. **8**, (1964) 1203.
2. Y. B. Bolkhovityanov *et al.*, in *Proc. of the 12th International Symposium on High Energy Spin Physics*, C. W. de Jager *et al.* (eds.), World Scientific, (1996) 730.
3. H. R. Poolman *et al.*, in *International workshop on polarized beam and polarized gas targets*, H. P. gen. Scheick and L. Sydow (eds.), World Scientific, (1997) 29.
4. M. Ferro-Luzzi *et al.*, in *Polarized gas targets and polarized beams*, R. J. Holt and M. A. Miller (eds.), AIP, (1998) 79.
5. I. Passchier *et al.*, Nucl. Instr. Meth. **A414**, (1998) 446.
6. R. Alarcon *et al.*, NIKHEF proposal 94-05. Spokesman J. F. J. van den Brand.

Electron Scattering Experiments with Polarized Hydrogen/Deuterium Internal Targets

L. D. van Buuren[a,b] for the 97-01 collaboration

[a]*Dept. of Physics and Astronomy, VU, NL-1081 HV, Amsterdam, The Netherlands*
[b]*NIKHEF, NL-1009 DB Amsterdam, The Netherlands*

Abstract. A high-density polarized hydrogen/deuterium internal gas target is presented. The target is based on a setup previously used in electron scattering experiments with tensor-polarized deuterium [1]. To increase the target thickness, new state-of-the-art permanent sextupole magnets and a more powerful pumping system were installed together with a longer (60 cm) and colder ($\sim$70 K) cylindrical storage cell. Electro-nuclear spin observables were measured by scattering longitudinally polarized electrons stored in the AmPS ring (NIKHEF) from the target gas. The product of electron beam and target polarization was determined from the known $e'p$ (quasi) elastic asymmetries. We achieved a target thickness of 1.1×10^{14} atoms/cm^2 which with typical beam currents of 110 mA corresponds to a luminosity of about 7.5×10^{31} cm^{-2}s^{-1}. Target and beam polarizations up to 0.7 and 0.65 respectively were obtained.

INTRODUCTION

Electron scattering experiments exploiting spin degrees of freedom give an enhanced sensitivity to small quantities that enter the electro-magnetic response of nuclei and nucleons. To carry out such experiments, we developed a polarized hydrogen/deuterium target and installed it in the Amsterdam Pulse Stretcher storage ring (AmPS). The polarized gas produced by an Atomic Beam Source (ABS) was injected into a storage cell, through which a 720 MeV longitudinally polarized electron beam was guided. The electron polarization was maintained by a Siberian snake and measured regularly by using a Compton Backscattering Polarimeter (CBP). The scattered electrons were detected in a large acceptance magnetic spectrometer (96 msr) and the ejected hadrons in a TOF detector (250 msr).

TARGET CONFIGURATION

In the ABS, molecular gas (H$_2$ or D$_2$) is dissociated into atoms by a 27.1 MHz RF discharge. A beam is formed by the cooled nozzle, skimmer and collimator (fig. 1).

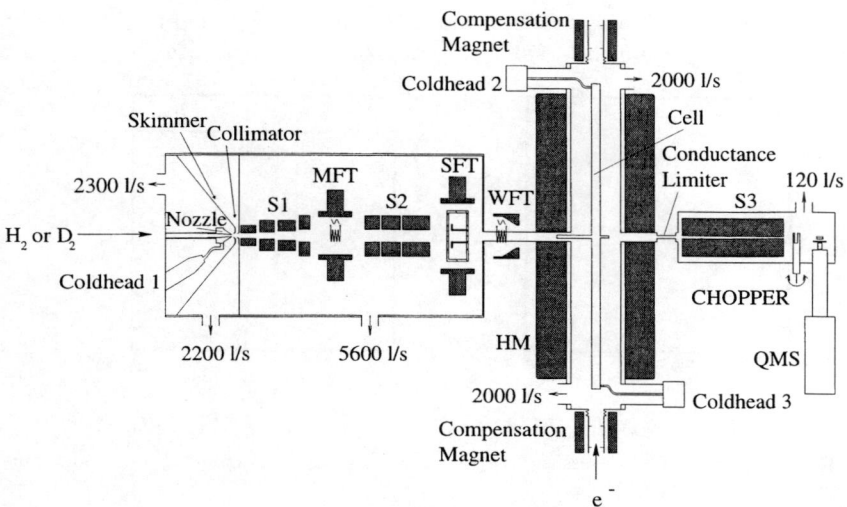

FIGURE 1. Schematic view of the ABS, target region and BRP. Vacuum pumping speeds are for H_2 gas. S1, S2, S3: permanent sextupole magnets; MFT, SFT, WFT: medium, strong, weak field transition units; HM: holding field magnet; QMS: quadrupole mass spectrometer.

Two sets of sextupole magnets (de) focus atoms with electron spin (anti) parallel to the magnetic field. Nuclear polarization is obtained by inducing RF-transitions between selected hyperfine states. In our configuration, vector polarized hydrogen or deuterium as well as tensor polarized deuterium can be produced. Systematic errors in the measured asymmetries were reduced by flipping the polarization every 8 seconds. The polarized atomic beam was injected into a 60 cm long, 15 mm diameter storage cell, fabricated from 30 μm thin Al foil and cooled to $\sim$ 70 K. A magnetic field was produced in the target region to provide a target polarization axis. This axis could be oriented in any direction in the scattering plane. A small fraction of the direct atomic beam flowed into the Breit-Rabi Polarimeter (BRP) via a sample tube. The BRP, which consists of a sextupole magnet followed by a quadrupole mass spectrometer, measured the amount of atoms with electron spin up and was used to tune the RF-units. The main improvements, with respect to the previous internal target experiments at NIKHEF with tensor-polarized deuterium [1], were the implementation of permanent sextupole magnets, a more powerful pumping system in the ABS beam-formation chamber and in the target region, and the use of a longer and colder storage cell.

INTERNAL TARGET PERFORMANCE

An important feature of (polarized) internal gas targets is the low background rate. By comparing rates with and without polarized gas in the storage cell the background contribution was determined. In quasielastic kinematics ($^2\vec{\mathrm{H}}(\vec{e}, e'p)n$) the background contribution was less than 1 % of the total rate.

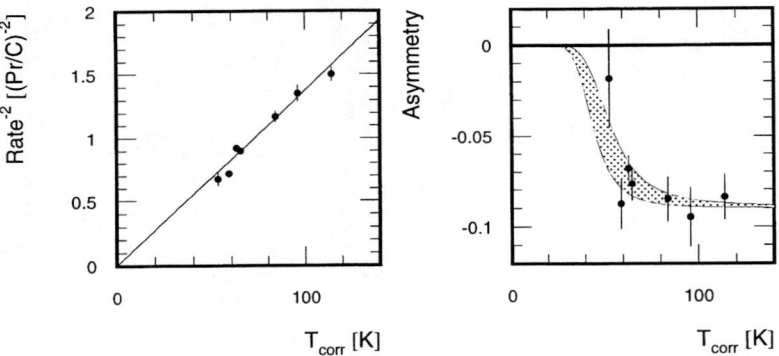

FIGURE 2. Left: (Rate)$^{-2}$ versus T_{cell} for $^1\vec{\mathrm{H}}(\vec{e}, e'p)$. Right: Elastic asymmetry versus T_{cell} for $^1\vec{\mathrm{H}}(\vec{e}, e')p$.

The thickness of the polarized gas target was determined by monitoring the (quasi) elastic coincidence rate, $e'p$. By comparing this rate to the rates obtained with molecular hydrogen (or deuterium) gas of known thickness, an absolute number can be obtained. To increase the target thickness the storage cell was cooled (fig. 2, left). However, depolarization effects become important at temperatures below $\mathrm{T}_{cell} \approx 100$ K. Therefore we performed asymmetry measurements at different temperatures (fig. 2, right) and found an optimal temperature of ≈ 70 K. The measured target thickness at this temperature was $(1.1 \pm 0.1) \times 10^{14}$ atoms/cm^2 for polarized hydrogen and $(1.0 \pm 0.1) \times 10^{14}$ atoms/cm^2 for polarized deuterium. The extracted fluxes for the injected atomic beams are $(6.6 \pm 0.4) \times 10^{16}$ atoms/s for hydrogen and $(4.6 \pm 0.3) \times 10^{16}$ atoms/s for deuterium.

To extract a spin-correlation parameter from an experimental asymmetry, only the product of electron beam and target polarization needs to be known. This quantity was determined on-line by monitoring the asymmetry of (quasi) elastic scattering from $^1\vec{\mathrm{H}}$ (resp. $^2\vec{\mathrm{H}}$) in kinematics for which the spin-correlation parameters are well known. Target polarizations up to 0.7 were extracted from this product and the CBP measurements.

This work was supported in part by the Stichting voor Fundamenteel Onderzoek der Materie (FOM), which is financially supported by the Nederlandse Organisatie voor Wetenschappelijk Onderzoek (NWO) and the Swiss National Foundation.

REFERENCES

1. Z.-L. Zhou et al., Nucl. Instr. and Meth. **A378** (1996) 40.

Multi-Target Operation at the HERA-B Experiment

Yu.Vassiliev[2], V.Aushev[1], K.Ehret[2], M.Funcke[2], S. I. Sever[2],
Yu.Pavlenko[1], V.Pugatch[1], S.Spratte[2], M.Symalla[2], N.Tkatch[1],
D.Wegener[2]

[1] *Institute for Nuclear Research, Kiev, 252028, Ukraine*
[2] *University of Dortmund, D-44221, Dortmund, Germany*

Abstract. The HERA-B internal target consists out of eight target ribbons arranged around the beam. Each target can be moved in radial direction independently in sub-micron steps allowing to compensate relative beam shifts and to steer for the desired interaction rate. The experimental constraints require a stable interaction rate equally distributed over all inserted targets. The actual equalization is based on a measurement of charge originated from the beam-target interaction. The system shows a good linearity with the interaction rate and allows a reasonable distribution of the interaction rate among several wires. To cross check the performance of the multi wire steering the reconstructed tracks and primary vertices in the silicon vertex detector have been used.

INTRODUCTION

HERA-B is a fixed target experiment [1] with the primary goal to study CP violation in the decays of B-mesons [2] into the "gold plated" mode $B^0 \to J/\psi + K^0_s$. The B-mesons are produced in interactions of the 920 GeV protons of the HERA storage ring with an internal target operating in the beam halo close to the beam core at typical distances of a few beam sigma. Due to the small cross section for b-quark production, small branching ratios and limited detector efficiency an interaction rate of 40 MHz is required to achieve a significant CP violation signal within one year of data taking. At the given HERA-bunch frequency of 8 MHz, this leads to 5 simultaneous interactions per bunch crossing. For a good quality event reconstruction one has to distribute them over several spatially separated targets (cp. Fig.1).

TARGET SETUP AND OPERATION

HERA-B operates a set of 8 target ribbons with a typical dimension of 500 μm along the beam axis and 50 μm in the radial direction [3]. The set-up is simple, stable and easy to operate. It provides requested Interaction Rate (IR) of 40 MHz. At 40 MHz IR targets are very close to the beam: up to 4 beam-σ with a typical σ of 400 μm (cp. Fig. 2). At this position the IR is very sensitive to beam movements and target steering has to react fast enough at any critical situation.

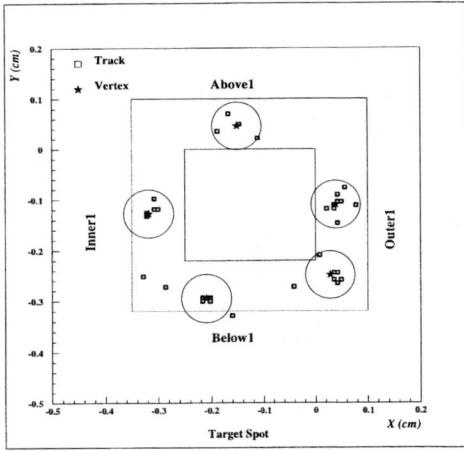 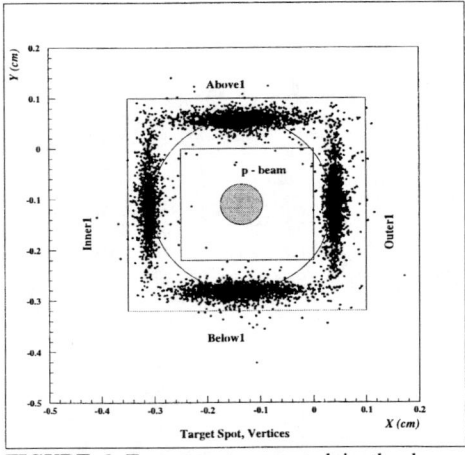

FIGURE 1 5 vertices (indicated by a "star"-marker) distributed over 4 targets. Solid line indicates range of expected target positions. Interaction Rate 40 MHz.

FIGURE 2 Four targets operated in the beam-halo simultaneously. Vertices reconstructed by the SVD at 40 MHz.

The operation of the target has to be extremely reliably as far as the efficient running of HERA for all four experiments is concern. This is the task for the so-called Target Control System (TaCoS) which is supposed to run 10^7 s/year continuously, in a secure and reliable way. TaCoS steers targets every 0.1 s in accordance with the feedback data from Scintillating Hodoscopes (overall IR and Charge Integrators (CI) connected to each single target (partial contribution into the IR) [4]. Fig. 3 illustrates distribution of the IR over the 8 targets provided by CI.

Step-motors drive targets in automatically chosen steps (minimum 0.1 μm) under dynamically varying proton beam features (intensity, position, collimator setting, etc.). Under the nominal operation conditions movement of any target by 10 μm leads to the IR variation by factor of two.

VERTEX DETECTOR RECONSTRUCTION

Nevertheless, the most relevant data about the performance of the target setup emerge from HERA-B sub-detectors. Silicon Vertex Detector (SVD) data (reconstructed tracks and vertices) are used for the target alignment as well as for the evaluation of the beam-halo profile. Fig. 4 shows the distribution of the vertices over 7 targets. Number of vertices at each target is in a good agreement with the IR distribution provided by the CI.

The vertex distribution along the wire gives an important information about the beam and beam-halo shape. It is possible to calculate the absolute beam and target positions using SVD reconstructed tracks information.

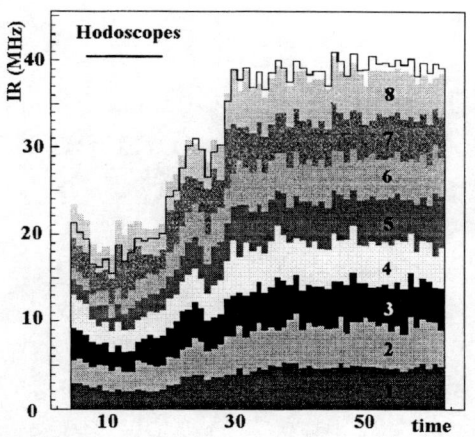
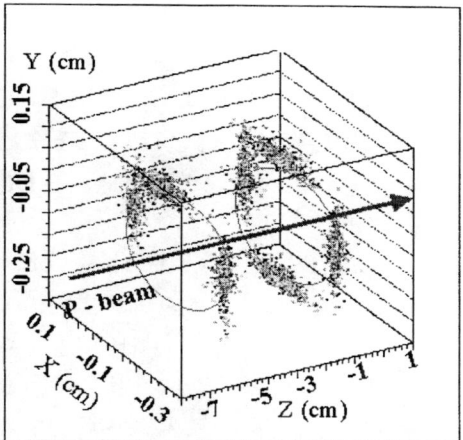

FIGURE 3 Interaction rate distribution over the 8 targets provided by the CI. Individual contribution by every target is shown by different style of hatched histograms.

FIGURE 4 Vertices distributed over all 7 inserted targets seen by the SVD.

In order to keep the interaction rate constant all the inserted targets should moving towards to the beam core with time. This kind of SVD-data will be used on-line at the Second or Third Level Trigger in order to equalize the number of tracks per wire.

CONCLUSIONS

- Since 3 years now HERA-B Multi Target System is in continuous operation.
- The Multi Target System performance is improving continuously.
- The setup including the steering works very reliably.

ACKNOWLEDGMENTS

This work was supported by the BMBF Bonn under contract number 05 7Do55P and 057Bu35I. One of the authors (Yu.V.) is grateful for the support by the Graduiertenkolleg and central public funding organization for academic research in Germany (DFG).

REFERENCES

1. Lohse T. et al., HERA-B Proposal, DESY-PRC 94/02, 1994.
2. Wolfenstein L. Phys. Rev. D54, 1 149-153 (1996).
3. Ehret K. Performance of the HERA-B Target, 1998
4. Hast C et al., NIM A354 (1995), p.224

Recent results from the internal polarized deuterium target experiment at the electron storage ring VEPP-3

I.A.Rachek[1], H.Arenhoevel[2], L.M.Barkov[1], V.F.Dmitriev[1],
M.V.Dyug[1], S.L.Belostotsky[3], R.Gilman[4], R.J.Holt[5], L.G.Isaeva[1],
C.W.de Jager[6], E.R.Kinney[7], R.S.Kowalczyk[8], B.A.Lazarenko[1],
A.Yu.Loginov[9], S.I.Mishnev[1], V.V.Nelyubin[3], D.M.Nikolenko[1],
A.V.Osipov[9], D.H.Potterveld[8], Yu.V.Shestakov[1], A.A.Sidorov[9],
V.N.Stibunov[9], D.K.Toporkov[1], D.K.Vesnovsky[1], V.V.Vikhrov[3],
H.de Vries[6] and S.A.Zevakov[1]

[1] *Budker Institute for Nuclear Physics, 630090 Novosibirsk, Russia*
[2] *Institute für KernPhysik, Johannes Gutenberg-Universität, D-55099 Mainz, Germany*
[3] *St.Petersburg Nuclear Physics Institute, 188350 Gatchina, Russia*
[4] *Rutgers University, Piscataway, NJ 08855, USA*
[5] *University of Illinois at Urbana-Champaign, Department of Physics, Urbana, IL 61801, USA*
[6] *NIKHEF, P.O. Box 41882, 1009 DB Amsterdam, The Netherlands*
[7] *Department of Physics, University of Colorado, Boulder, CO 80309, USA*
[8] *Argonne National Laboratory, Argonne, IL 60439, USA*
[9] *Tomsk Polytechnical University, 645050 Tomsk, Russia*

Introduction

In Novosibirsk the study of the simplest nucleus, the deuteron, is being carried at the 2–GeV electron storage ring VEPP–3 by the international collaboration. The experiments are performed using the superthin internal target technique.

Novosibirsk pioneered in polarized internal target experiments. The first run was conducted back in 1984 at the 400 MeV VEPP–2 electron storage ring [1]. Starting from 1986 the measurements are being performed at the VEPP–3 ring with gradual increase of the luminosity [2]. It is at VEPP-3 that the first successful application of the storage cell filled with polarized nuclei as a target for a storage ring was demonstrated [3].

In spring-summer 1999 a new data taking run has been accomplished. Here we overview the instrumentation used in the experiment and show the preliminary results.

Instrumentation

VEPP-3. The VEPP-3 storage ring was designed for electron-positron colliding experiments, but actually it serves as a pre-accelerator and an injector for the VEPP-4 electron-positron collider. VEPP-3 can also be used for other experiments, including the internal target ones. The main parameters of VEPP-3 are: ring circumference – 74.4 m, operational energy 350 – 2000 MeV, bunch cross section at 2 GeV $\sigma_{rad} \times \sigma_{vert} = 1.4$ mm $\times$ 0.5 mm, maximal beam current 200 mA.

In order to be able to use a small-aperture storage cell we have modified the electron optics in the experimental section of VEPP-3. This was done by installing additional quadrupoles upstream and downstream the experimental section and adjusting the currents in existing lenses. As a result both radial and vertical beta-functions in the experimental section were decreased by a factor 2.5.

Polarized Atomic Beam Source. The new Atomic Beam Source (ABS) has been designed and constructed for this stage of experiments at VEPP-3 [4]. It has a conventional design with sextupole spin-separating magnets and high-frequency transitions units. The main difference of our ABS from other modern sources is the usage of superconducting sextupole lenses instead of permanent magnets. Superconducting coils provide a pole-tip field strength three times larger than that for permanent magnets. In the very first test we got a flux 6.5×10^{16} of polarized deuterium atoms reaching the storage cell per second. We expect to further increase this value in future by proper tuning of the ABS.

Storage cell. The storage cell installed in the experimental section of the VEPP-3 was fabricated from a $30 \mu m$ aluminum foil and its surface was covered by drifilm. It has an elliptical cross section 13×24 mm^2 and a length of 40 cm. The cell was cooled to about $100°K$ by a circulating liquid nitrogen.

Particle detectors. A set of large acceptance non-magnetic particle detectors was used. The detector has two identical systems which are placed in vertical plane, Fig. 1. Each of the systems consists of a segmented CsI + NaI electron calorimeter (total thickness 16 radiation lengths) on one side of the electron beam and a hadron scintillator hodoscope (5 layers of scintillator plastic with a total thickness of 28cm) on the other side. Angular acceptance of each calorimeter is $\theta = 20° \ldots 30°$ and $\Delta\phi = 60°$ and that for each hadron arm is $\theta = 60° \ldots 70°$, $\Delta\phi = 60°$ For the registration of the particle tracks different sets of drift chambers are employed. The detection system includes also the target polarimeter based on the measurement of the asymmetry in elastic electron-deuteron scattering at a small angle (the low-Q polarimeter). Electrons scattered at an angle $\theta = 9°$ are detected in coincidence with recoil deuterons which are registered by the hadron arm.

Spring-summer-1999 data taking run

In March-1999 all the equipment was assembled at the experimental straight section of VEPP-3 and after a two-weeks commissioning period a data-taking run was

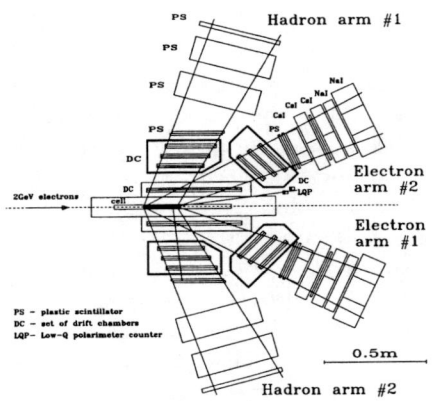
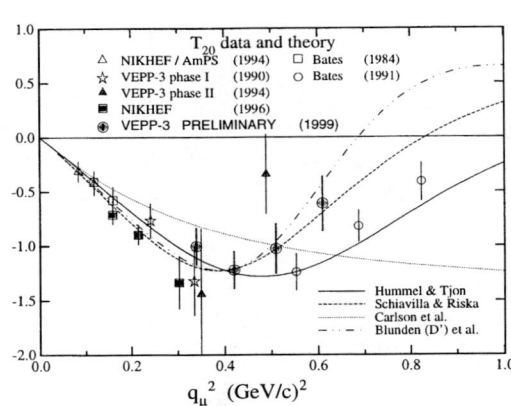

FIGURE 1. Detector package.

FIGURE 2. World data on T20 and preliminary results of the run-1999 at VEPP-3.

carried out. Beam current integral of 108 KCoulomb was collected at a polarized target thickness of about 3×10^{13} at/cm^2 and average beam current 71 mA. The run-averaged tensor polarization of the deuterium atoms inside the storage cell was derived from the measured asymmetry in the low-Q polarimeter and found to be $+0.32$ and -0.64.

Preliminary results on T_{20} in elastic ed scattering for the first region of 4-momentum transfer is shown in Fig. 2. Analyses of the second Q-region as well as of the other reaction channels ($^2\vec{H}(e,e'p)n$, $^2\vec{H}(e,pp)e', \pi^-$) are in progress.

Perspectives

We will have more beam time for the internal target experiment in the fall-1999. We are going at least to double the statistics. Hopefully the target polarization will be higher after we have replaced the storage cell by a new one with better coating.

This work was supported by Russian Foundation for Basic Researches, grants 98-02-17949 and 98-02-17993 and by INTAS, grant 96-0424.

REFERENCES

1. V.F.Dmitriev et al., Phys. Lett. **157B**(1985)143.
2. R.Gilman et al., Phys. Rev. Lett. **65**(1990)1733.
3. R.Gilman et al., NIM **A327** (1993) 277
4. L.G.Isaeva et al., Nucl. Instr. and Meth. **A411** (1998) 201.

Photoneutron Cross Section Measurements on ^{9}Be Using Laser-Induced Compton-Backscattered Photons

Hiroaki Utsunomiya*, Yoriko Yonezawa*, Hidetoshi Akimune*,
Tamio Yamagata*, Masahisa Ohta*, Masatoshi Fujishiro[†],
Hiroyuki Toyokawa[‡], and Hideaki Ohgaki[‡]

*Department of Physics, Konan University, Higashinada, Kobe 658-0072, Japan
[†]Inst. Advan. Sci. & Technol., Osaka Pref. Univ., Sakai, Osaka 593, Japan
[‡]Quantum Radiation Div., Electrotechnical Lab, Tsukuba, Ibaragi 305-8568, Japan

Abstract. Photoneutron cross sections were measured for ^{9}Be in the energy range from 1.77 to 3.75 MeV using quasi-monochromatic γ-rays produced in laser-induced Compton-backscattering. These cross sections are relevant to the reaction rate of the first step of the $\alpha - process$ of Type II & Type Ib supernovae, i.e., $\alpha(\alpha\,n,\,\gamma)^9$Be. Results are compared to the data taken with other photon sources like radioactive isotopes and Bremsstrahlung.

The α - process in a neutrino-powered wind (ejecta) formed at the birth of Type II & Type Ib supernovae begins with the synthesis of ^{9}Be by $\alpha(\alpha\,n,\,\gamma)^9$Be which is followed by ^{9}Be$(\alpha,n)^{12}$C. Under such stellar environment ($T_9 \sim 4$, $\rho \sim 5 \times 10^4$ g/cm^3) this bridges A = 5, 8 mass gap to produce ^{12}C much more efficiently than the triple alpha process [1]. The $\alpha(\alpha\,n,\,\gamma)^9$Be essentially proceeds through the first excited state $(1/2^+)$ lying just above the n - ^{8}Be threshold (1.67 MeV) as well as the second $(5/2^-$ at 2.43 MeV) excited state in ^{9}Be.

Here we report on results of photoneutron measurement on ^{9}Be using a new γ source based on laser-induced Compton backscattering.

Inverse Compton scattering (IC) of laser photons incident on relativistic electrons accumulated in the storage ring TERAS of the Electrotechnical Laboratory (ETL) produces quasi-monochromatic γ-rays in the energy range from 1 to 40 MeV. The IC γ-rays are available as a pencil-like beam with typical flux of 10^4 photons/sec/mm^2 and energy resolution of a few to several percents [2].

A Nd:YLF laser (λ=1053 nm) was used. The laser system was pulse-operated at 1 kHz and produced 100 % linearly polarized light with $\vec{E}$ aligned in a vertical plane.

Electron energy was varied from 315 to 460 MeV to produce γ-rays in the energy range of 1.77 - 3.75 MeV. γ-rays were collimated in 2 mm diameter at 5.5 m from the collision point with a 20 cm thick Pb block. γ-ray energy was measured with a pure-Ge detector. The properties of the quasi—monochromatic photon beams, i.e., the peak energy and energy spread, were investigated with an anti-Compton spectrometer [3]. The minimum γ energy used in the present measurement was 1.77 MeV. Below this energy, a significant fraction of γ-rays always lay below the neutron threshold energy, introducing large ambiguity to the measurement. Photons were directly counted with a BGO detector placed at the end of the beam line. From the measured piled-up spectra, photon flux was determined with use of single photon spectra. Attenuation by the target material was taken into account.

A 4 cm thick ^{9}Be rod (25 mm in diameter) was irradiated. Neutrons were measured with 4 BF_3 counters embedded in a polyethylene cube. Two BF_3 counters were located vertically, while the other two horizontally. Neutron moderation time in the polyethylene was measured in 1 ms time range with a TAC module (OR-TEC 566). Thus, background neutrons that randomly arrived at BF_3 counters were separated.

The efficiency of the neutron detector was measured with a neutron source composed of ^{24}NaOH and D_2O. Irradiation of NaOH by thermal neutrons, its mixture into D_2O and the efficiency measurement were carried out at the Research Reactor Institute of Kyoto University. Details of the efficiency measurement are given in [4]. The resultant detection efficiency was (1.67 ± 0.09) % per BF_3 counter. The Monte

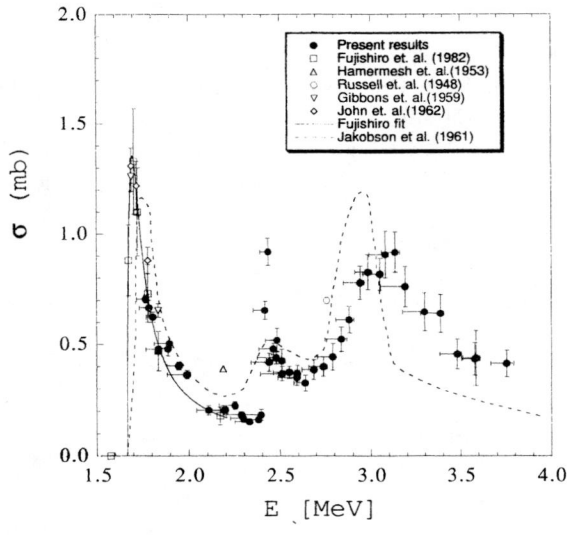

FIGURE 1. Photoneutron cross sections for ^{9}Be.

Carlo code MCNP was used to obtain the dependence of the detection efficiency on neutron energy.

Results of the present measurement are shown by solid circles in Fig. 1 in comparison with other data. Previously photoneutron cross sections were measured using two real photon sources, Bremsstrahlung [5] and radioactive isotopes [6–10]. The Bremsstrahlung measurement lacked great accuracy due to the poor energy resolution, whereas the radioactive isotope measurements were limited to low energies. The present data are in good agreement with those of Fujishiro et. al. [10]. A very sharp peak corresponding to the $5/2^-$ state ($\Gamma = 0.77 \pm 0.15$ keV [11]) was observed at 2.43 MeV. A broad bump consisting of two states, the 2.78 MeV state ($1/2^-$) and the 3.05 MeV state ($5/2^+$) [11], was also observed.

The angular distribution of photoneutrons was previously measured [5,7]. It was isotropic at 1.70 and 1.81 MeV, but not quite isotropic at 2.40, 2.76, and 2.95 MeV. The present data are not corrected for the asymmetric angular distributions. It is noted that the neutron angular distribution was smeared to large extent in the polyethylene moderator. The angular distribution effect will be investigated with help of MCNP.

A least-square fit to the data will be made to deduce resonance parameters. Discussions on the reaction rate of $\alpha(\alpha\, n, \gamma)^9$Be will follow.

We thank Mr. K. Okamoto (the Research Reactor Institute of Kyoto University) for his technical assistance in the irradiation of NaOH by thermal neutrons. We are grateful to Mr. Y. Ogawa (Kinki University) for providing us with the Monte Carlo code MCNP along with kind assistance. This work was supported by the Japan Private School Promotion Foundation.

REFERENCES

1. Woosley, S.E., and Hoffman, R.D., *Astrophys. J.* **395**, 202 (1992).
2. Toyokawa, H., Ohgaki, H., Sugiyama, S., Mikado, T., Yamada,K., Sei, N., Suzuki, R., Ohdaira, T., and Yamazaki, T., *Nucl. Instr. Meth. in Phys. Res.* **A422**, 95 (1999).
3. Ohgaki, H., Sugiyama, S., Yamazaki, T., Mikado, T., Chiaki, M., Yamada, K., Suzuki, R., Noguchi, T., and Tomimasu, T., *IEEE Trans. Nucl. Sci.* **38**, 386 (1991).
4. Yonezawa, Y., and Utsunomiya, H., *Mem. Konan Univ. Sci. Ser.* **46**, 43 (1999).
5. Jakobson, M.J., *Phys. Rev.* **123**, 229 (1961).
6. Russel, B., and Sachs, D., Wattenberg, A., and Fields, R., *Phys. Rev.* **73**, 545 (1948).
7. Hamermesh, B., and Kimball, C., *Phys. Rev.* **90**, 1063 (1953).
8. Gibbons, J.H., Macklin, R.L., Marion, J.B., and Schmitt, H.W.,*Phys. Rev.* **114**, 1319 (1959).
9. John, W., and Prosser, J.M., *Phys. Rev.* **163**, 958 (1967).
10. Fujishiro, M., Tabata, T., Okamoto, K., and Tsujimoto, T., *Can. J. Phys.* **61**, 1579 (1983).
11. Ajzenberg-Selove, F., *Nucl. Phys.* **A413**, 1 (1984).

STORED, COOLED BEAMS

Storage Conundra

R. E. Pollock

Department of Physics and Indiana University Cyclotron Facility (IUCF)

Musings on some of the puzzling inconsistencies brought on by less than careful use of language by storage device practitioners and instances where a resulting conundrum can lead to enlightenment.

INTRODUCTION

Effective use of beams in/from storage rings requires some familiarity with the language and concepts of beam physics. For example, beam temperature or emittance values may be employed as measures of beam quality. The words carry implications which may not be warranted. A beam is certainly not in thermodynamic equilibrium in the laboratory reference frame because one three-momentum component is larger than the other two, a failure to satisfy equipartition. In fact the "temperature" may vary from point to point along the trajectory, being lowest in regions where the beam is large and the envelope slope is small. Using the phase space area is a step forward, (emittance is actually the phase space area divided by the beam momentum) giving a quantity which for beams of low intensity is nearly the same for all points along the trajectory and therefore more robust than the temperature. Normalized emittance extends the utility of this concept to beams undergoing acceleration by deleting the variability of the momentum denominator.

In the experience of the author, there are many pitfalls for the unwary in the use of the language, leading to communication problems between the accelerator community and the user community. On the interface, the difficulty may manifest itself as a puzzling inconsistency involving language, a conundrum; the resolution of the difficulty may come as a revelation or paradigm-shift. What follows is a selection of such "conundra" (since conundrum is not a Latin word, the plural should be conundrums, making conundra itself a conundrum!). The selection is highly subjective, representing personal experience with the struggle to understand the physics underlying the usage. The scope extends beyond beam physics, as particles may be stored in traps as well as rings. The emerging field of nonneutral plasma physics deals with dense clouds of charged particles in traps, with many enlightening similarities to the physics of dense beams. The interface between the physics of nonneutral plasmas and that of beams is rich in conundra (for example, wall impedance, a concept imbedded in discussion of beam stability, diverges as the v/c in the denominator goes to zero, and has to be recast to handle the plasma-wall interaction). A thirst for more comprehensive applicability may lead to new insight in the reformulation of "well-understood" concepts, and may also lower the barriers to cross-discipline information exchange.

The question/answer (or conundrum/revelation) process for achieving enlightenment is well established in physics, used with effect, for example, by Galileo in his "Dialogs Concerning Two New Sciences". A beginning student is less well served by erudite and concise exposition, which is the preferred form among equals, than by puzzles and contradictions whose resolution bestows a qualititive understanding which can be the foundation for more formal development.

SELECTED CONUNDRUMS

The subset makes no pretense to be complete. It does include a few cases where resolving the puzzle has helped the author to a better if still incomplete understanding of the rich field of storage physics.

Terminology Barricades (Where Is The π ?)

We know there are a dwindling few backward countries resisting metric measurements, after more than a century and despite disastrous consequences that make news when rocket scientists can't miss the backside of Mars. The BeV gave way to the GeV after a struggle because a billion means a thousand million in some countries and a million million in others. More culpable is the generation gap that finds SI units still being resisted in the teaching of physics after 40 years. Retire those statvoltmeters.

Within the accelerator community there is a relativity divide, between the electron community that assumes $\beta = 1$, low energy heavy ion groups that set $\gamma = 1$, and those of us that deal with $\beta\gamma$, and have to struggle to insert the missing factors in formulae from the literature of the two extremes.

Matrix optics works with phase space, and the phase space area of a group of particles is a critical measure. For convenience, an ellipse, sized and shaped to contain an agreed fraction of the particles, is often employed as an approximation to the distribution shape. For an upright ellipse the area is the product of two semi-axes and a factor π. Now comes the confusion. Some communities quote the area divided by π and others quote the area itself. To avoid misunderstanding, it has become essential to state the π factor explicitly: "my cooled beam has longitudinal phase space area of 10^9 π h and transverse phase space area of 10^{10} π h" (We all know that phase space area has the same dimensions as Planck's constant h.)

Admittedly, these are all petty quibbles, but they should serve to remind us that we all share the obligation to present information in forms that are readily comprehensible to a wide audience and as free of pitfalls as possible.

Hot Or Cold Waists ?

Consider a group of particles drifting toward a waist (a minimum in the beam size). If interactions among the particles are neglected, the motion is force-free and the momentum of each particle is fixed. We can derive a temperature from the variance of the momentum distribution of the group and this will be constant. So temperature does not change as the waist is reached.

Now consider the fact that the existence of a waist implies an evolving correlation: the distribution in phase space can be enclosed by a tilted ellipse that becomes upright as the waist is reached. We could subdivide the upstream distribution into slices with different positions, each with a reduced temperature relative to the whole group. Each slice would have a different first moment of its momentum distribution giving rise to convergence toward the waist, but this does not contribute in calculating the slice temperature. So we have a number of cool beamlets converging toward a merger at the waist where they are no longer distinguishable. The temperature will be seen to attain a local maximum at the waist where a vertical cut through the center of the upright ellipse shows the greatest momentum spread.

Now include a collective correction. If the particles are similarly charged, each is repelled by the collective electrostatic field of the group as a whole. In the rest frame of the average particle, a radial inward momentum is slowed by this repulsion; the temperature reaching a minimum near the waist.

We have a conundrum. The temperature at the waist increases, decreases, and stays the same.

Part of the possible confusion arises from starting with "low intensity" with negligible collective forces and adding the "intense beam" effects later. The waist region becomes an active optical element, with focussing and coupling between transverse degrees of freedom. The temperature drop is accompanied by a bump in the potential energy of the system. In fact, one way to characterize the beam intensity is to compare the kinetic energy per particle away from the waist in the group frame with the size of the increase in potential energy per particle at the waist. The beam may be deemed intense if the latter is not negligable compared to the former.

The temperature maximum at the waist is more subtle. If the group of particles is treated as indivisible, the temperature is constant. However, because of the momentum-position correlation arising from the tilted ellipse, a septum or septa could be employed to split the beam into two or more cooler parts (except at the waist) without resort to a Maxwell's demon.

Exploitation of such correlations is a developed artform in external beam experiments, where dispersion matching, use of the Rowland circle and other tricks can reduce the deleterious effects of beam energy spread or emittance in selected experiments (1). Bunching or debunching in a beam line or in a ring are related processing operations.

In passing the waist, without interaction, the upper and lower halves of the momentum distribution exchange spatial positions. If collective repulsion is included, at sufficient intensity some subset of the particles (those with small angular momentum) will begin to reflect rather than passing through to the other side of the group. The treatment in two separable transverse degrees of freedom breaks down.

The point is not that well-meaning people can present and defend contrary arguments, but that the student, by pondering a puzzling inconsistency after exposure to the controversy, is pushed to a deeper understanding.

Does One Particle Have An Emittance?

Phase space motion of a single particle is at least as meaningful as the evolution of the phase space distribution of a group of particles. If the phase space area occupied by the group is conserved, the area of the trajectory of each particle must also be constant. With this viewpoint, it is quite meaningful to classify particles as occupying phase space trajectories near the middle of the group, or near the edge, or in a diffuse region well outside the edge.

In this picture, the action of a beam "scraper" is easy to understand. The scraper simply blocks the path of outer region particles, leaving a distribution with a sharper edge, which is advantageous in reducing experimental backgrounds.

Experimentally, it is observed that a scraper is quite effective in high energy rings and much less so at lower energies. A simple explanation is not hard to find. In an imperfect vacuum, Coulomb scatter from background gas particles gives a diffusive mechanism with a long tail which is effective in repopulation of the region cleared by the scraper. In the presence of scattering, the individual particle does not have a fixed phase space trajectory, but rather executes a random walk with a small angle change at each collision. The collision rate decreases with the square of the kinetic energy so redistribution takes much longer at higher energies.

With active cooling, the growth of phase area for the particle group is counteracted, and an equilibrium profile establishes itself, which is gaussian in character near the middle, with long tails maintained by the Coulomb scattering.

In dense beams, scatter of one particle from another gives an additional diffusive mechanism. This intrabeam scattering couples with the ring dispersion pattern to transfer energy from the beam momentum to other degrees of freedom, so it acts as a heating mechanism as well as a driver toward temperature uniformity.

There is a non-stochastic mechanism arising from the interaction of a single particle with variations in the collective field of the particle group that arise from the lumped element focussing. Consider a single particle outside the main distribution and moving toward it. If the beam envelope is varying with time, the varying collective potential can alter the phase space trajectory for that particle. If the particle is a tennis ball, the moving beam edge is a tennis racquet. Beating of the particle's betatron motion with the beam's envelope modulation gives repetitive encounters. After a number of such encounters, the particle which started outside may find itself inside the distribution. Conversely, a particle which was initially within the main distribution can find itself promoted into the outer halo region by the same mechanism.

This mechanism, described by Jameson in 1992 (2), has important consequences in understanding the limits to halo removal in very intense beam accelerators. You can remove the halo again and again, but it keeps coming back as new particles are ejected from the beam core into the outer regions. Unlike the long-tailed scattering mechanisms, the growth is bounded so that a large acceptance may limit detrimental effects such as component activation.

With hindsight, the concept of single particle phase space area conservation is quite an oversimplification. On some timescale, there is a continuing rearrangement of location of each particle relative to the group. Understanding the character and rates of the churning mechanisms is essential to design of both facilities and experiments to exploit their beams.

"You Could Not Step Twice Into The Same Rivers"

As noted by Heraclitus, rivers flow to the sea. An environmental impact at point A impacts all points downstream of A and few points upstream (salmon runs and dams excepted). An experiment mounted in an external beam line can do terrible things to the beam without bothering the upstream accelerator much at all. The effect lies downstream of the cause.

Life is quite different when particles are stored. In a ring configuration, downstream and upstream, cause and effect, become intermingled. The experimenter has to learn to swim in his own effluent. A thick target affects the arriving beam nearly as much as the departing beam. Cooling rings add a new element to the design equation because the heating by a target, if thin enough, can be counterbalanced by the cooling, with a time-independent near-equilibrium state established.

An optical element in the ring anywhere affects the beam everywhere (globally). A new concept emerges. A family of elements may be devised that limits the effect to a portion of the ring local to these elements. The global/local division replaces the cause/effect division in a non-closed path configuration. Orbit bumps and waist-squeezers (localized tune shifters) are dipole and quadrupole family examples of controllable localized optical disturbances.

Beam Equilibrium

Stored particles which are moving along together cannot be in thermodynamic equilibrium because of the failure of equipartition. The longitudinal momentum component remains larger than the transverse pair. As a consequence, beams of particles are intrinsically unstable or metastable, with numerous mechanisms driving the beam toward equilibrium by transfer of energy from the forward motion into other degrees of freedom.

By contrast, stored particles at rest can reach a state very close to a true thermodynamic equilibrium, so trapped nonneutral plasmas, as such beams at rest are called, can be much more stable, despite sharing much of the physics (eg. beam-wall interactions) that governs intense beam behavior.

Consider one particle travelling along the closed orbit within the conducting vacuum wall of a storage ring. The electric field of the particle induces surface charge on the wall which is dragged along with the particle. Wall resistance causes a lag in the position of the wall image leading to a slowing force on the particle. At high velocities, wall shape changes cause radiation fields which also react on the particle. The rate of slowing of a single particle is very small. Now consider N particles travelling as a group. The wall charge is N times larger and acts back on every particle in the group so the slowing is N times faster. However, if the beam particles are continuously distributed as a moving line charge, the wall image charges can be at rest and the slowing disappears. A conundrum.

With more thought, there must be energy available to turn a continuous beam into a bunched beam. Fluctuations in time structure can grow, driven by the departure from equipartition.

If a beam is off-center in the chamber, the induced wall charge lacks cylindrical symmetry, and the beam is attracted to the near side. The result is a "coherent space charge tune shift" which counters the restoring force of the ring lenses and lowers the betatron frequency. The effect grows with beam intensity, and a stability limit can be encountered when the coherent oscillation shifts into resonance with a destructive member of the family of orbit resonance lines.

A trapped nonneutral plasma also induces wall charge and is attracted toward the nearest wall. If confined by a uniform field **B** along the axis of a cylindrical chamber, the radial **E** field of an off-centered plasma leads to a revolution about the trap axis with drift velocity **ExB**/B^2 which grows with particle number. The phenomenon serves as a useful non-destructive measure of particle number, being proportional to line charge density. This plasma rotational motion has neutral stability. Negative electronic feedback can be employed to restore the plasma to the trap axis, in analogy to the "coherent dampers" found in storage rings.

Many other similarities between the behavior of stored beams and trapped plasmas can be found. As one further example, the Schottky spectrum of a well-cooled coasting beam develops a splitting which is the signal of coherent charge density waves moving forward and backward along the beam. The split is proportional to the square root of the beam current. These same waves are easily observed travelling along an elongated trapped plasma and reflecting from the plasma ends to form standing wave modes. Both in beams and plasmas, a measurement of wave velocity gives the space charge potential of the confined population.

Does A Beam Have A Debye Length ?

If a beam is viewed as a moving nonneutral plasma, then we should be able to characterize it by its plasma frequency ω_{pl} and Debye length λ_d, where $(\omega_{pl})^2 = q^2 n/(m\varepsilon_o)$, $(\lambda_d)^2 = \varepsilon_o kT/(q^2 n)$, and the particle density n, charge q, mass m and temperature T appear. The product $\lambda_d \omega_{pl}$ is seen to be the thermal velocity.

One immediate problem is that the density and temperature vary along the trajectory due to envelope modulations, so these plasma parameters are local in nature. With suitable averaging over the ring, one can conclude that the plasma frequency is directly related to the space charge tune shift, and that this frequency must be much smaller than the betatron frequency to avoid resonance crossings. By a similar argument, the Debye length must be larger than the beam size. Otherwise, the transverse density profile becomes flattened as the particles shield the group interior from applied electromagnetic fields, which causes a reduction in the incoherent betatron frequency.

In the quest for a crystallized beam, a necessary first step is to raise the density and lower the temperature so that Debye length is much less than the beam size. This is just the condition that the betatron frequency within the beam becomes much reduced. The usual alternating-gradient ring design is ill-suited to the confinement task as a number of destructive resonances have to be crossed as the temperature is lowered or the density is raised.

Plasma purists insist that a group of particles with Debye length greater than group size is not a plasma, because it lacks too many aspects of the collective nature of the plasma state. A ring designer keeps the space charge tune shifts within reasonable bounds to avoid resonance crossings and thus keeps the beam from becoming a plasma.

At a waist, if the temperature drops and the density peaks, the Debye length has a local minimum, but so does the beam size. In fact, one can convince oneself that the beam is more plasma-like where the envelope is at a maximum. However, the density profile hardly has time to evolve from a flattened distribution in one part of the ring to a gaussian form in another part. There have been interesting recent developments in the theory of equilibrium beam profiles in the presence of periodic focussing that bear on this issue. (3)

The electron beams used for cooling are not stored, and are cold and dense enough to qualify as moving plasmas. Taking the IUCF cooling system as a representative example, the density of $10^{14}/m^3$ and transverse temperature of 0.2 eV lead to a Debye length of 0.33 mm which is much less than the 13 mm beam radius. Note that, up to factors of order unity, the square of the ratio of beam radius to Debye length is the ratio of electrostatic potential energy to thermal kinetic energy, so either ratio will serve as a measure of "space charge domination".

The electron cooling beam is thus dense while the beam it cools is not, because the magnetic confinement of the electron beam is more robust against electrostatic repulsion than is the periodic focussing of the storage ring.

While neutral plasmas have to be hot, nonneutral plasmas of a single species have no recombination channels available, and can be confined over a wide range of temperatures and densities, so that a continuous range of ratios of distribution size to Debye length can be studied. The development of strong correlations and crystal formation in laser cooled nonneutral plasmas is well established (4), while crystal beams remain elusive.

CLOSING REMARKS

Beams from accelerators have been with us about 2/3 of a century. The storage rings and traps used to confine charged particles in motion or at rest came along 1/3 century later. At first, the technology was the dominant concern, and the end use of paramount importance, for example in the drive to higher energy. Only quite recently have we realized that a stored beam or plasma has properties as worthy of study in its own right as is the atom or nucleus or subatomic particle whose study was the original incentive for the technological development.

Those of us involved in the physics of beams and plasmas have an obligation to share our understanding with those doing physics with these objects. If there are readers of this short article who wish to pursue to greater depth issues raised here without rigor, review articles such as that of O' Neil (5), and modern texts such as those of Reiser (6) and Davidson (7) are excellent starting points.

ACKNOWLEDGMENTS

This work supported by the D.O.E. under grant DE FG0297 ER 54433 A001 and by the N.S.F. under grant NSF PHY 96-02872.

REFERENCES

1) Reich, J., Martin, S., Protic, D., Riepe, G.,Proc Seventh Int'l Conf on Cyclotrons and Their Applications, W. Joho, ed., p. 238 (1975).
2) Jameson, R. A., Proc. 1993 Part. Accel. Conf., IEEE 93CH3279-7, p. 3926-3931.
3) Davidson, R. C., Qin, H., *Phys. Rev. Spec. Topics - Accel. Beams* **2** 074401-1-29(1999).
4) Gilbert, S., Bollinger, J. J., Wineland, D. J., *Phys Rev. Lett.* **60** 2022-2025 (1988).
5) O'Neil, T. M., *Nonneutral Plasma Physics*, A.I. P. Conf. Proc. **175** 1-26 (1988).
6) Reiser, M., *Theory and Design of Charged Particle Beams*, Wiley (1994).
7) Davidson, R. C., *Physics of Nonneutral Plasmas,* Addison-Wesley (1990).

News in Electron Cooling: Highlights from ECOOL'99

D. Reistad

The Svedberg Laboratory, Box 533, S-751 21, UPPSALA, Sweden

Abstract. A Workshop on Electron Cooling and Related Topics was organised in Uppsala, Sweden, from 19 to 22 May 1999. The workshop, which incorporated the 5th Workshop on Medium Energy Electron Cooling, included papers on theory, technology, limitations and applications of electron cooling as well as papers on laser cooling and stochastic pre-cooling. The last day of the workshop was devoted to so-called Medium ($2<\gamma<20$, i.e. FNAL Recycler and DESY PETRA) and High (i.e. DESY HERA) Energy Electron Cooling. Reports on measurements and achievements made at a number of electron cooling facilities, including the most recently completed ones, i.e. at SIS (GSI, Darmstadt) and at AD (CERN), were given. There were also reports on electron coolers under construction at the National Institute of Radiological Science (NIRS) at Chiba in Japan and at the Heavy Ion Research Facility (HIRFL) in Lanzhou, China. Work on medium and high-energy electron cooling at FNAL, DESY, and JINR was presented.

INTRODUCTION

The six-dimensional phase-space density of stored beams can not be made to increase because of magnetic fields *or* fields with forces, which don't depend on the velocity. This is a well-known consequence of Liouville's theorem. Synchrotron radiation provides a mechanism to increase the phase-space density of electron and positron beams and very high energy proton beams. Methods to increase the energy of low and medium energy ion and hadron beams have been invented and are referred to as "beam cooling," in analogy with thermodynamics, where molecules of a cooler gas occupy a smaller volume of phase space, than do warmer ones. This use of the word "cooling" here is however confusing. As pointed out by Sessler [1], in beam physics, the concept of "cooling" has a different meaning than in the rest of physics, where "cooling" means the reduction of temperature. As one focuses and defocuses beams, the transverse temperature of the beam particles can easily be changed. This is not what is meant by beam cooling, which implies increase of the six-dimensional phase-space density of the beam.

Three methods to cool stored ion and hadron beams have been achieved experimentally. These are electron, stochastic, and laser cooling. Electron and stochastic cooling have become established tools, used to improve the beam quality and for accumulation, whereas laser cooling is an interesting phenomenon studied for its own sake.

In electron cooling the stored beam is merged, over a fraction of the circumference of the ring, with a "cold" (i.e. mono-energetic and parallel) electron beam of the same velocity. The electron beam is created in an electron gun and accelerated to the appropriate velocity. After the interaction region, the electron beam is decelerated to a potential close to that of the cathode, and collected in a collector. The beam particles experience friction against the electron beam, and are thus "cooled" both longitudinally and transverse at the same time.

The highest electron energy used for electron cooling today is at the CELSIUS ring in Uppsala, Sweden (300 keV), followed by the Indiana Cooler in Bloomington, Indiana (270 keV). These are essentially built according to the same basic principles, which were used to build the world's first electron cooler at the NAP_M ring in Novosibirsk, Russia [2]. Use of electron cooling at much higher energies, which is now considered for the Recycler at FNAL [3], PETRA and HERA at DESY [4], and RHIC [5] will require different technique.

Stochastic cooling makes use of signals from a very high bandwidth pick-up in order to control a kicker placed downstream to the pick-up, in order to kick particles in the stored beam towards the center of the phase-space volume.

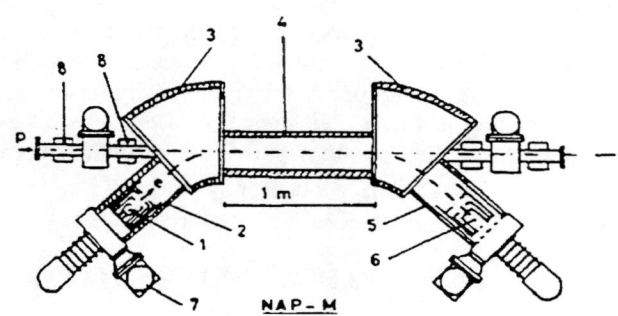

FIGURE 1. Cross section of the world's first electron cooler which was at the NAP-M storage ring in Novosibirsk, Russia. It consisted of an electron gun (1), the interaction region, where the stored beam and the electron beam travel together, and a collector (6), where the electrons are decelerated and absorbed. Solenoids (2, 3, 4, 5) create a longitudinal magnetic field. This is necessary to keep the electron beam together against its space charge. Pumping near the gun and collector was assured with ion pumps (7). Correction dipoles (8) correct the closed orbit of the stored beam.

Laser cooling works only for a few particular ions. In order to be laser cooled, the ions must not be fully stripped, and must have a closed transition between two atomic energy levels (i.e. the population must be confined to these two levels). A laser beam of suitable frequency is placed parallel and/or antiparallel to the stored beam. Ions that are in resonance with the laser beam absorb photons. Each absorbed photon transfers momentum $h\nu/c$ to the ion. This momentum transfer is in the direction of the laser beam propagation. When the ion spontaneously returns to the low energy level it again recoils with momentum $h\nu/c$, but now in a random direction. Thus, there is a net radiation pressure force, directed along the laser beam, on resonant ions. The force depends on the velocity of the ion through the Doppler shift. To get cooling there must be a counter-force, which can be from another laser, from an induction accelerator, or from RF.

THE WORKSHOP

The ECOOL'99 (Workshop on Electron Cooling and Related Topics) took place in Uppsala, Sweden, from 19 to 22 May 1999. The name of the workshop was meant to connect to ECOOL'84, organized by Helmut Poth in Karlsruhe in 1984. It can be seen as one in an informal series of workshops on beam cooling (besides Karlsruhe, Legnaro, 1990 [7], Montreux, 1993 [8], and on board of a ship from Moscow to Nizhny Novgorod 1996 [9]).

The ECOOL'99 workshop incorporated "The 5th Workshop on Medium Energy Electron Cooling," the fifth in a series of small workshops where proposed electron-cooling schemes for the FNAL Recycler and other medium and high energy storage rings have been discussed during the last several years. The workshop attracted some 65 participants from CERN and laboratories in China, Denmark, Germany, Japan, Russia, Sweden, and the United States.

Main topics of the workshop covered the theory, technology and limitations of electron cooling, special problems of electron cooling of heavy-ion beams, cooled beam diagnostics, and beam accumulation by means of electron cooling. Papers were also presented in the related topics of laser and stochastic cooling. A special session dealt with so-called medium (i.e. $2 < \gamma < 20$) and high energy electron cooling; this has not yet been realized but is now seriously considered to be used at FNAL and DESY. Proceedings of ECOOL'99 will be published by NIMA.

PRESENT AND PLANNED USE OF ELECTRON COOLING

The electron coolers presently in operation are listed in table 1. Electron coolers, which are currently under construction, are listed in table 2, and electron coolers, which are under consideration in table 3.

Table 1. Electron coolers in operation.	Electron energy (keV)	Purpose	Ref.
ASTRID (Århus, Denmark)	0.5 – 3	Atomic and molecular physics	11
CRYRING (Stockholm, Sweden)	0.2 – 13	Atomic and molecular physics	12
TSR (Heidelberg, Germany)	0.5 – 16	Atomic and molecular physics	13
AD (CERN)	3 – 30	Deceleration of antiprotons	14
SIS (Darmstadt, Germany)	5 – 35	Nuclear physics, injector to ESR	15
COSY (Jülich, Germany)	20 – 100	Nuclear physics	16
ESR (Darmstadt, Germany)	2 – 240	Atomic, nuclear and accelerator physics	18, 19
"Cooler" (Bloomington, Indiana)	20 – 270	Nuclear and accelerator physics	17
CELSIUS (Uppsala, Sweden)	5 – 300	Nuclear physics	20

Table 2. Electron coolers under construction	Electron energy (keV)	Purpose	Ref.
HIMAC (Chiba, Japan)	3 – 30	Heavy-ion therapy with positron emitters Heavy beams (Fe) for radiobiology Microbeam probe Short-pulsed beams	21
CSRm (Lanzhou, China)	2 – 30	Injector for CSRe Nuclear physics	22
CSRe (Lanzhou, China)	10 – 250	Radioactive ions and highly charged heavy ions for internal target experiments and high-precision spectrometry	22

Table 3. Electron coolers under consideration	Electron energy (keV)	Purpose	Ref.
LEIR (CERN)	2.3	Accumulation of 4.2 MeV/u Pb^{54+} for LHC	23
MUSES-ACR	30 – 250	Accumulation of radioactive ion beams Nuclear, atomic, molecular physics	24, 25
MUSES-DSR	50 – 800 (1,900)	Colliding experiments: ion-ion ion-e^- ion-γ	25
Recycler (FNAL)	4,300	To increase Tevatron luminosity to 10^{33} $cm^{-2}s^{-1}$ by recycling antiprotons and more effective antiproton accumulation	3, 26, 27
PETRA (DESY)	9,800	To increase HERA luminosity	4, 28, 29
RHIC	55,400	To increase RHIC luminosity and lifetime of beams at RHIC	30
HERA (DESY)	450,000	To maintain high HERA luminosity	4, 31

Present

Except for the AD and the electron cooler at SIS the presently operating electron coolers were commissioned around 1990. Atomic and molecular physics is the dominating field of research at CRYRING, TSR, and ASTRID. In these rings, studies of electron cooling and of other electron-ion interactions have been performed using a wide range of ions, both positive and negative, including molecular ions, such as for example OH^+ at TSR, H_2D^+ at CRYRING and C_2^- at ASTRID, and also including highly charged heavy ions, such as Pb^{54+} at CRYRING and Au^{51+} at TSR. In many experiments the electron beam is used not only for phase-space cooling, but also as an electron target. A crucial parameter is the temperature of the electrons, which determines the energy resolution when e.g. recombination cross sections are measured as a function of the relative energy between ions and electrons. Adiabatic expansion of the electron beam (see below) at CRYRING, TSR, TARN-II and ASTRID makes it possible to reduce the electron temperature as compared to the temperature of the cathode.

Activities at COSY, the Indiana Cooler, and CELSIUS are dominated by nuclear physics research. These electron coolers are used for beam accumulation and to improve the conditions for the nuclear physics experiments. At the ESR, a wide range of research includes atomic as well as nuclear physics. CELSIUS has the present world record in electron energy used for electron cooling, 300 keV corresponding to 550 MeV protons.

SIS

The heavy-ion synchrotron SIS at Darmstadt, Germany is filled with 11.4 MeV/u heavy-ions from the Unilac by horizontal multiturn injection. An electron cooler was installed during 1997 and was commissioned during 1997 and 1998 and is now in operation. Its purpose is to increase the intensity of highly charged heavy ions. For highly charged ions the maximum number of particles has been increased from 1×10^8 to

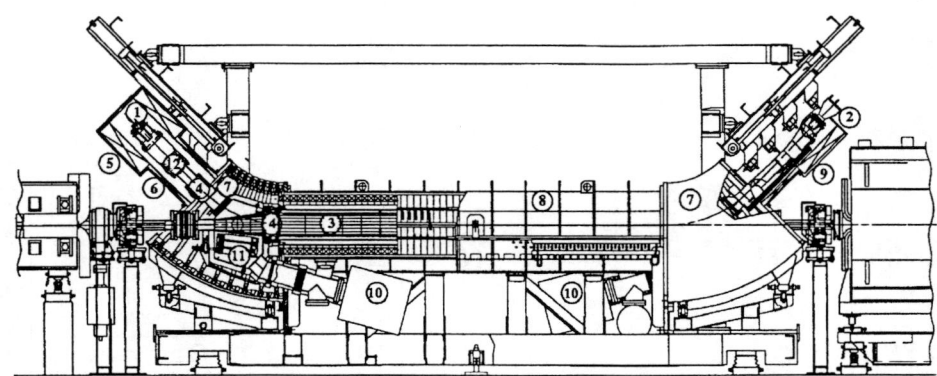

FIGURE 2. Cross section of the SIS electron cooler. 1 electron gun, 2 electron collector, 3 central drift tube, 4 clearing electrodes, 5 gun solenoid, 6 expansion solenoid, 7 toroid, 8 cooling solenoid, 9 collector solenoid, 10 sputter ion pumps, 11 NEG pumps, 12 Ti sublimators.

1×10^9. For lighter ions intensity limitations, which are caused by high phase-space density of the cooled ion beam, are encountered.

The design of this cooler (see fig. 2) is an example of the state-of-the-art of design of low-energy electron coolers. It is remarkable how similar in principle its cross section is to that of the first electron cooler, at NAP-M (fig. 1). Of course, it is made with much larger acceptance, and higher mechanical precision, and modern vacuum components, such as NEG pumps, are used, but the principle is very much the same. One innovation [12], which is used in all new electron coolers, is that of adiabatic expansion of the electron beam, see below.

AD

The most recent electron cooling facility consists in fact of the old LEAR electron cooler, which is used in the new Antiproton Decelerator (AD). This will be a simplified scheme at CERN for the provision of low-energy antiprotons. It uses the existing antiproton production target area and the modified Antiproton Collector (AC) ring in their current locations. Electron cooling will be used to reduce the transverse emittance and the momentum spread at 300 and 100 MeV/c (47 and 5 MeV). This facility is now (November 1999) in a final commissioning stage.

HIMAC

An electron cooler is under construction at HIMAC in Chiba, Japan. It will utilise electron cooling for new fields, outside of physics, namely radiation therapy, biology, and radiochemistry. It will be used to accumulate positron-emitting beams (^{11}C) for heavy-ion therapy with these beams. Benefits include the possibility to do online measurements of the range of the radiation in the body by detecting annihilation γ-rays when a pencil beam irradiates the target. Also, the irradiated volume will be confirmed with high sensitivity with positron emission tomography (PET) after the irradiation. Other applications will be accumulation of heavier beams, such as Fe beams, to be used for radiation biology. Electron-cooled beams will also be used as microbeam probes for cellular radiation response studies and bunched electron-cooled beams will be used to produce short pulses of radiation for time-resolved measurements in radiation chemistry.

HIRFL-CSR

A major new project for heavy-ion and radioactive-beam research has recently been funded in China. It consists of the extension of the HIRFL facility in Lanzhou of two rings, CSRm (161 m circumference, 10.6 Tm rigidity) and CSRe (125 m, 8.4 Tm). The existing cyclotrons SFC $(k = 69\,\text{MeV})$ and SSC $(k = 450\,\text{MeV})$ will be used as injectors.

CSRm will be used for accumulation, cooling, and acceleration of heavy-ion beams from the injector cyclotrons and CSRe will be used for internal target experiments with secondary beams (radioactive ion beams or highly charged heavy-ion beams). Electron cooling will be employed for beam accumulation in CSRm and to provide high-resolution beams for internal target experiments in CSRe.

Collaboration on the design and construction of the electron coolers has been established between the HIRFL and the Budker Institute of Nuclear Physics in Novosibirsk, Russia.

Pb ions at CERN

In addition to proton-proton operation, the LHC will work with lead (Pb) ions. The present lead-ion injection complex, comprising the ECR source, an RFQ and a LINAC with stripper ($Pb^{27+} \rightarrow Pb^{54+}$) is able to produce 25 μA (3×10^6 ions/μs). This is not sufficient to fulfil the desired luminosity goals for lead ions in the LHC. A gain of the order of 150 has to be found. It is proposed to transform LEAR into a low energy accumulator ring (LEIR) to increase the number of ions per bunch. The goal is to accumulate 1.2×10^9 ions per cycle, of 3.6 s duration, at 4.2 MeV/u. After accumulation and cooling, the beam will be bunched, accelerated to 14.8 MeV/u and transferred to the PS.

LEAR has been modified in order to perform tests. Presently, about 6×10^8 lead ions can be accumulated in about 6 s, so a factor of 2 is missing both in the accumulated intensity and the accumulation time. An improvement would come from an improved vacuum in LEAR Other considered improvements would make use of an inclined septum or a "corner" stripper foil, placed at the entrance of LEAR. This would allow stacking in the vertical as well as the horizontal and longitudinal phase planes.

Another idea is to use two electron coolers. Cooler 1 will be used to drag the injected batch from its injection energy to the stack energy. Cooler 2, which will have a smaller electron current and possibly a hollow electron beam, will be used to maintain the stack cooled.

A lead-ion "stripper loop", with which essentially 100 % instead of 16 % of the lead ions will end up in the desired Pb^{54+} charge state, is also discussed.

RIKEN

The construction of a radioactive beam factory was authorised by the Japanese government in 1997, and the construction of a big superconducting cyclotron ($k = 2,500$ MeV) started in 1998 and will go on until 2002. Following this construction period, the construction MUSES will start, and go on until 2008. Electron cooling will be used in the ACR as well as both DSR rings. The ACR electron cooler (250 keV) is already designed in some detail [24].

The Recycler

Fermilab has again set out to upgrade the luminosity of the Tevatron. This will maintain the impressive trend which has been kept up since 1989, see fig. 3. The scarcity of antiprotons is the chief limitation to the luminosity of the Tevatron. The main components of this upgrade are the Main Ring and the Recycler, improved stochastic cooling

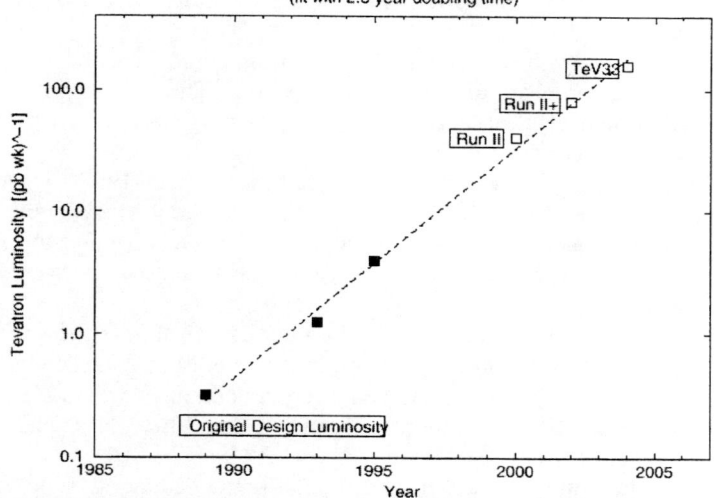

FIGURE 3. Tevatron collider luminosity past and projected on a semi-log scale in the manner of the Livingston plot for accelerator energy. The doubling time is about 2.3 years.

in the Antiproton Source, and increasing the number of bunches in the Tevatron from 6 to 36.

The Recycler is intended to perform two functions. First, it will store the high-intensity antiproton beam prior to transfer to the Tevatron. In previous runs, the stacking rate in the Accumulator decreased after several hours of antiproton stacking. Regular beam transfers to the Recycler will eliminate the decrease in stacking rate in the Accumulator. Second, the Recycler will collect the unspent antiprotons after they have been decelerated in the Tevatron and the Main Injector.

The Recycler uses permanent magnet technology for reduced fabrication, installation and operational costs and for increased reliability. The Recycler will use a fixed-frequency, variable wave-form, radiofrequency system to create barrier buckets to bunch the beam for injection and extraction.

The Recycler will initially use stochastic cooling only, but the performance would benefit strongly from the use of electron cooling also. Electron cooling in the Recycler does however require a much higher energy electron beam, than is used in the present electron coolers. Therefore, in 1995, Fermilab started an R&D programme in electron cooling. This programme has two principal objectives: (1) to determine the feasibility of electron cooling the 8 GeV antiprotons, and (2) to develop and demonstrate the necessary technology.

The ultimate goal is to realise a luminosity of 10^{33} cm^{-2}s^{-1} in the Tevatron by supplying a larger flux of antiprotons.

Fermilab has chosen to accelerate the electrons in a so-called Pelletron accelerator, commercially manufactured by National Electrostatic Corporation (NEC) in Middleton, Wisconsin. This is a van de Graaff type accelerator. A challenge is posed by the small charging current of the Pelletron, which is a few hundred µA. Thus, a very good collection efficiency is required.

A technical goal set for a proof-of-principle demonstration was to maintain a 200 mA MeV-energy electron beam for a period of one hour. The only technically feasible way to attain such high electron currents is through energy recovery of the electron beam, so-called "beam recirculation."

This goal was achieved with a 1.5 MeV electron beam in 1998 by recirculating beam currents of 200 mA for periods up to five hours without breakdowns. Although the recirculation tests used 1.5 MeV and the Recycler needs a 4.3 MeV beam, the demonstration is considered to be relevant because the increased energy does not involve fundamental changes in technology.

The schematic layout of the Recycler electron cooling system is shown in figure 4. The length of the cooling section has been defined to be 20 m and the electron beam current up to 0.5 A. This system is now being reproduced at Fermilab for a full-scale proof-of-principle test. A 5 MeV Pelletron has been ordered from NEC and will be delivered in the summer of 2000.

Traditionally, a longitudinal magnetic field, which may be homogeneous or stronger at the gun than in the interaction region (expanded electron beams). The choice of a

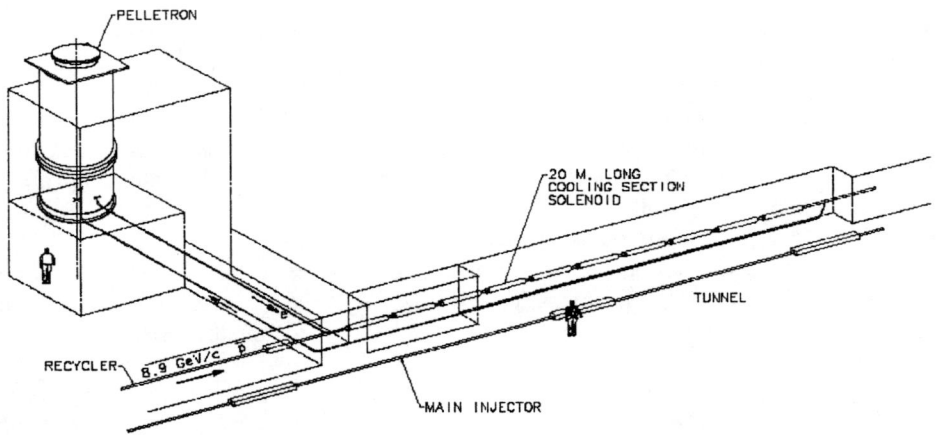

FIGURE 4. Schematic layout of the Recycler electron cooling system.

standard Pelletron accelerator prohibits the use of a continuous magnetic field. The adopted electron transport scheme assumes homogeneous longitudinal magnetic fields in the gun, collector, and cooling sections, but a lumped focusing system in between. To provide zero angular velocity of the electrons inside the cooling section solenoid, the electrons must enter this solenoid with the correct relation between radius and divergence, determined by the Busch theorem.

The project schedule calls for achievement of the R&D goals in the test set-up in the summer 2001. Under the most optimistic scenario, the electron cooling system could be installed in the Recycler in 2002.

DESY

Electron cooling schemes have been proposed also at DESY. In a first step, the proton beam would be cooled at an energy of 15 to 20 GeV in the pre-accelerator PETRA. In a second step the protons would be cooled at the HERA top energy in order to maintain the beam quality.

In the PETRA cooler, the electrons will be accelerated to about 10 MeV in a 208 MHz travelling wave linac. The gun and linac will be surrounded by a solenoid so that emittance growth is avoided. In order to reduce the energy spread of the electron beam and to lengthen it the electrons will be run through a decompressor before they are brought together with the proton beam in a 40 m long cooling section, which will again be surrounded by a solenoid.

The cooler in HERA consists of a small electron storage ring with two long straight sections. Precooling of the protons in PETRA is absolutely necessary because otherwise the cooling times in HERA would be hundreds of hours. This cooler would be more effective on heavy ions; the cooling time for Au^{79+} would be a few hours.

RHIC

Electron cooling of completely stripped 100 GeV/u gold ions is considered for RHIC. The cooler would consist of an electron storage ring. The cooling time for Au^{79+} is calculated to be 15 minutes.

ON THE EXPANSION OF THE ELECTRON BEAM

As was mentioned above, all new electron coolers are equipped with the possibility to adiabatically expand the electron beams. The expansion factors are variable, up to 8 (at SIS), 20 (ASTRID), 25 (TSR), and 100 (CRYRING). The possibility for expansion of the electron beams has been retrofitted in the ASTRID, TSR, and CRYRING electron coolers, and was in the original design of the SIS electron cooler. In fact at CRYRING, the electron cooler was first modified for an expansion factor of 10, then again modified to give an expansion factor of 100. Thus, CRYRING has the coldest electron beam for electron cooling.

Everybody agree that the adiabatic expansion of the electron beam does decrease the electron temperature, and is useful for atomic physics studies where the electron beam is used as an electron target. On the other hand, the usefulness of this adiabatic expansion in order to improve electron cooling seems to be a matter of some confusion. Beutelspacher and Danared agree that varying the expansion factor (keeping the current density constant) does not change the derivative of the force upon the relative velocity, i.e. (for small relative velocities) the cooling times. On the other hand, the maximum of the drag force, which has been measured, improved each time the electron cooler was reconstructed at CRYRING, and also become higher at TSR after the modification than it was before.

These measurements were however made with small electron currents. With high electron currents, the transverse energy of the electrons will be dominated by $\mathbf{E} \times \mathbf{B}$ drift due to the electron beam space-charge; this is not affected by the expansion. The electron temperature will not be important.

Expansion does reduce the ion lifetimes in cases where radiative capture in the electron cooler is significant, and could therefore be of disadvantage. An advantage gained with a variable expansion, besides of any change in electron temperature, is the possibility to optimise the electron beam diameter.

LIMITATIONS IN INTENSITY AND PHASE-SPACE DENSITY

In general, the phase space density of beams can not be increased arbitrarily. Longitudinal and transverse instabilities occur if the phase space density becomes too high. Thresholds such as the "Keil-Schnell" and "Keil-Zotter" are used as benchmarks. This has been the subject of many articles in conferences and schools. An update in this genre, with particular emphasis on electron cooled beams in LEIR and the Recycler was given by Möhl [10].

Burov argued that the observed intensity limits that are encountered during accumulation with electron cooling might be due to the envelope instability [33]. Parkhomchuk on the other hand suggests that the same observed intensity limits can be due to coherent interactions between the cooling electron beam and the stored ion beam [34]. Although these explanations are rather different and their authors don't seem to agree, they both advocate the same cure, i.e. the application of a "hollow" electron beam, previously suggested several times [35, 36].

CONCLUSIONS

Many other important matters were discussed during the workshop. It was apparent that the electron cooling community does not agree on all technical and scientific matters, including matters of electron cooling theory and the best choices of parameters, but on the other hand get along extremely well. It was also clear that electron cooling is a growing field, finding an increasing number of applications, and covering a large energy range. Like the STORI conference, ECOOL'99 attracted a manageable number of participants, and there were excellent discussions and exchange of opinions after every talk. It is the hope of the author that somebody else will pick up the race baton and organise another cool workshop in two or three years or so.

REFERENCES

The proceedings of ECOOL'99 will be published in Nuclear Instruments and Methods in Physics Research A.

[1] A.M. Sessler, "Methods of beam cooling," Proc. 31st Workshop of the INFN Eloisatron Project, Crystalline Beams and Related Issues, Erice, Italy, 12-21 November 1995, World Scientific, D.M. Maletić, A.G. Ruggiero, editors, page 93.
[2] G.I. Budker, N.S. Dikansky, V.I. Kudelainen, I.N. Meshkov, V.V. Parkhomchuk, D.V. Pestrikov, A.N. Skrinsky, B.N. Sukhina, Particle Accelerators, 7 (1976) 197.
[3] A. Burov, J. MacLachlan, J. Marriner, S. Nagaitsev, "Scenario for electron cooling of antiprotons in the recycler," ECOOL'99.
[4] K. Balewski, R. Brinkmann, Ya. Derbenev, K. Floettmann, P. Wesolowski, M. Gentner, D. Husmann, C. Steier, "Studies of electron cooling at DESY," ECOOL'99.
[5] A. Burov, V. Danilov, P. Colestock, Ya. Derbenev, "Electron cooling for RHIC," ECOOL'99.
[6] H. Poth (editor), Proceedings of the Workshop on Electron Cooling and Related Applications, Karlsruhe, Germany, 24-26 September, 1984, KfK 3846.
[7] R. Calabrese, L. Tecchio (editors), Proceedings of the Workshop on Electron Cooling and New Cooling Techniques, Legnaro, Italy, 15-17 May 1990, World Scientific.
[8] J. Bosser (editor), Proceedings of the Workshop on Beam Cooling and Related Topics, Montreux, Switzerland, 4-8 October 1993, CERN 94-03.
[9] I. Meshkov (editor), Proceedings of the Eleventh International Advanced ICFA Beam Dynamic Workshop on Beam Cooling and Instability Damping Dedicated to the 30th Anniversary of Electron Cooling on board a ship from Moscow to Nizhny Novgorod, 18-26 June 1996. Nuclear Instruments and Methods in Physics Research A 391 (1997).
[10] J. Bosser, C. Carli, M. Chanel, N. Madsen, S. Maury, D. Möhl, G. Tranquille, "Stability of cooled beams," ECOOL'99.
[11] J.S. Nielsen, S.P. Møller, L.H. Andersen, P. Balling, M.K. Raarup, "Electron cooling of D⁻ at the ASTRID storage ring," ECOOL'99.
[12] H. Danared, A. Källberg, G. Andler, L. Bagge, F. Österdahl, A. Paál, K.-G. Rensfelt, A. Simonsson, Ö. Skeppstedt, M. af Ugglas, "Studies of electron cooling with a highly expanded electron beam," ECOOL'99.
[13] M. Beutelspacher, M. Grieser, D. Schwalm, A. Wolf, "Longitudinal and transverse electron cooling experiments at the Heidelberg heavy ion storage ring TSR," ECOOL'99.
[14] N. Madsen, S. Maury, D. Möhl, "Equilibrium beam in the Antiproton Decelerator (AD)," ECOOL'99.
[15] M. Steck, L. Groening, K. Blasche, B. Franczak, B. Franzke, T. Winkler, V.V. Parkhomchuk, "Beam accumulation with the SIS Electron Cooler," ECOOL'99.
[16] D. Prasuhn, J. Dietrich, R. Maier, R. Stassen, H.J. Stein, H. Stockhorst, "Electron and Stochastic Cooling at COSY," ECOOL'99.
[17] D.L. Friesel, G. East, T. Sloan, "Status of the IUCF Electron Cooled Storage Ring," ECOOL'99.
[18] F. Nolden, K. Beckert, F. Caspers, B. Franczak, B. Franzke, R. Menges, A. Schwinn, M. Steck, "Stochastic cooling at the ESR," ECOOL'99.

[19] M. Steck, K. Beckert, H. Eickhoff, B. Franzke, F. Nolden, H. Reich, B. Schlitt, T. Winkler, "Lowest temperatures in cooled heavy ion beams at the ESR," Proc. European Particle Accelerator Conference, Stockholm, Sweden, 22-26 June, 1998.
[20] L. Hermansson, D. Reistad, "Electron cooling at CELSIUS," ECOOL'99.
[21] K. Noda, T. Murakami, E. Takada, M. Kanazawa, T. Honma, S. Yamada, T. Nagafuchi, I. Watanabe, H. Kozu, S. Shibuya, K. Ohtomo, S. Sasaki, "Electron cooler for medical and other application at HIMAC," ECOOL'99.
[22] Y.N. Rao, J.W. Xia, Y.J. Yuan, W.Z. Zhang, P. Yuan, W.L. Zhan, B.W. Wei, "The HIRFL-CSR project," ECOOL'99.
[23] J. Bosser, C. Carli, M. Chanel, R. Maccaferri, G. Molinari, S. Maury, D. Möhl, G. Tranquille, "The production of dense lead-ion beams for the CERN LHC," ECOOL'99.
[24] T. Tanabe, T. Rizawa, K. Ohtomo, T. Katayama, A. Yamashita, E. Syresin, I. Meshkov, "Design of an electron cooling device for the Accumulator Cooler Ring in MUSES Project," ECOOL'99.
[25] T. Katayama, S. Watanabe, Y. Batygin, N. Inabe, K. Ohtomo, T. Ohkawa, M. Takanaka, T. Tanabe, M. Wakasugi, I. Watanabe, Y. Yano, K. Yoshida, "MUSES Project of RIKEN RI Beam Factory," Proc. European Particle Accelerator Conference, Stockholm, Sweden, 22-26 June, 1998.
[26] S. Nagaitsev, A. Burov, A.C. Crawford, T. Kroc, J. MacLachlan, C.W. Schmidt, A. Shemyakin, A. Warner, "Status of the Fermilab electron cooling project," Proceedings of the 1999 Particle Accelerator Conference, New York, NY, 29 March - 2 April, 1999.
[27] S. Nagaitsev, A. Burov, A.C. Crawford, T. Kroc, J. MacLachlan, G. Saewert, C.W. Schmidt, A. Shemyakin, A. Warner, "FNAL R&D in medium energy electron cooling," ECOOL'99.
[28] K. Balewski, R. Brinkmann, Ya. Derbenev, K. Floettmann, N. Holtkamp, M. Schmitz, G.-A. Voss, P. Wesolowski, D. Yeremian, "Preliminary study of electron cooling possibility of hadronic beams at PETRA," Proc. European Particle Accelerator Conference, Stockholm, Sweden, 22-26 June, 1998.
[29] P. Wesolowski, K. Balewski, R. Brinkmann, Ya. Derbenev, K. Floettmann, "An injector study for electron cooling at PETRA using a bunched beam," ECOOL'99.
[30] A. Burov, V. Danilov, P. Colestock, Ya. Derbenev, "Electron cooling for RHIC," ECOOL'99.
[31] M. Gentner, R. Brinkmann, Ya. Derbenev, D. Husmann, C. Steier, Nucl. Instr. and Meth. in Phys. Res. A 424 (1999) 277.
[32] H. Danared, G. Andler, L. Bagge, C.J. Herrlander, J. Hilke, J. Jeansson, A. Källberg, A. Nilsson, A. Paál, K.-G. Rensfelt, U. Rosengård, J. Starker, M. af Ugglas, Phys. Rev. Letters 72 (1994) 3775.
[33] A. Burov, S. Nagaitsev, "Envelope instabilities in electron cooling," ECOOL'99.
[34] V.V. Parkhomchuk, "New insights in the theory of electron cooling," ECOOL'99.
[35] D. Anderson, M.S. Ball, V. Derenchuk, G. East, M. Ellison, T. Ellison, D. Friesel, B.J. Hamilton, S.S. Nagaitsev, T. Sloan, P. Schwandt, "Cooled beam intensity

limits in the IUCF Cooler," Proc. Workshop on Beam Cooling and Related Topics, Montreux, Switzerland, 4-8 October 1993, CERN 94-03, p. 377.
[36] A. Sharapa, A. Shemyakin, Nucl. Instr. and Meth. in Phys. Res. A 336 (1993) 6.

STORAGE RINGS: PAST, PRESENT AND FUTURE

A.D. Krisch*

Randall Laboratory of Physics, University of Michigan
Ann Arbor, Michigan 48109-1120

This lecture will attempt to review storage rings: past, present and future. I will spend more time on the past, because the past has produced most of our data, while the present can be rather brief. There is not yet much future data, but there are some plans about what we hope for. Professor Andy Sessler of Berkeley, who recently reviewed the early history of colliders [1], loaned me his slides; for this I have much appreciation.

Storage rings can either collide a stored particle beam with a second stored particle beam or they can store particles for collisions with internal targets. The recent Accelerator Handbook, edited by A. W. Chao and M. Tigner [2], contains a list of all colliders above 1 GeV; this list, which is shown in Fig. 1, mostly contains electron and proton colliders. Since Professor Muenzenberg just told us about some important advantages in having electron on heavy ion colliders, perhaps the next edition of this handbook will include more nuclear facilities.

Location	Name (type)	Max. E_{cm} (GeV)	Start
Stanford/SLAC, USA	CBX[b] (e^-e^- DR)	1.0	1963
	Spear (e^+e^- SR)	5.0	1972
	PEP (e^+e^- SR)	30	1980
	SLC (e^+e^- LC)	100	1989
	PEP-II (e^+e^- DR)	10.6	1999
Frascati, Itali	AdA (e^+e^- SR)	0.5	1962
	Adone (e^+e^- SR)	3.0	1969
	DAΦNE (e^+e^- SR)	1.0	1997
Novosibirsk, Russia	VEP-1 (e^-e^- DR)	0.26	1963
	VEPP-2/2M (e^+e^- SR)	1.4	1974
	VEPP-4 (e^+e^- SR)	14	1979
Cambridge, USA	CEA Bypass (e^+e^- SR)	6	1971
Orsay, France	ACO (e^+e^- SR)	1.0	1966
	DCI ($e^{\pm}e^{\pm}$ DR)	3.6	1976
DESY, Germany	Doris (e^+e^- DR)	6.0	1974
	Petra (e^+e^- SR)	38	1978
	HERA ($e^{\pm}p$ DR)	160	1992
CERN, Europe	ISR (pp DR)	63	1971
	SppS ($p\bar{p}$ SR)	630	1981
	LEP (e^+e^- SR)	190	1989
	LHC (pp DR)	14000	2004
Brookhaven, USA	RHIC (heavy ions DR)	200/u	1999
	RHIC (pp DR)	500	
Cornell, USA	CESR (e^+e^- SR)	12	1979
KEK, Japan	Tristan (e^+e^- SR)	60	1986
	KEK B (e^+e^- DR)	10.6	1999
Beijing, China	BEPC (e^+e^- SR)	3.1	1989
Fermilab, USA	Tevatron ($p\bar{p}$ SR)	1000	1987

Fig. 1   List of Colliders [2]

* Supported by a Research Grant from the U.S. Department of Energy

In his lectures on early colliders, Professor Sessler asked the question, "How did it happen?"; then he pointed out that for many years people knew that it would be much easier, because of relativity, to reach very high center-of-mass energies by having two moving particles collide head-on. He also noted that the famous Norwegian physicist, Wideroe, patented this idea during the War years; however, he apparently never received any money for his patent. Wideroe also apparently invented the LINAC, but that is another story which is not related to STORI99.

Sessler stressed that Donald Kerst, Gersh Budker and Bruno Touschek each played a very important role in storage rings; he loaned me some nice pictures of these three distinguished gentlemen. Donald Kerst is shown in Fig. 2; he was the senior author of the first paper proposing colliding beams and he also invented the betatron. He was a very distinguished scientist who died in 1993; somehow, he did not get a Nobel Prize; I do not understand why. Gersh Budker is shown in Fig. 3; he was the founder of the Novosibirsk Institute and he led the construction of a series of colliders at Novosibirsk, which have been quite important. Bruno Touschek is shown in Fig. 4 along with one of his junior colleagues. He is the only one of the three that I never met; he was very active in the pioneering AdA and Adone facilities in Frascati, Italy.

Fig. 2   Donald Kerst[†]

Fig. 3   Gersh Budker[†]   **Fig. 4**   Bruno Touschek[†]

For colliders to work properly, the strong focusing principle needed to be invented. This is true because, for two beams to collide with a reasonably high collision rate, each beam must have a high brightness. Until about 1960, most accelerators were of the weak focusing type with beams that were rather large and defuse. The weak focusing accelerators worked fairly well for fixed target experiments, but it was clear that they would not work well for colliders because of their low brightness. Fortunately three very bright people, Courant, Livingston and Snyder, proposed in 1952 the strong focusing principle [3]; essentially all modern accelerator rings have been built using this strong focusing principle. Strong focusing is not strictly related to colliders, but colliders probably would not have worked without it. In 1956, some people at the Midwest Universities Research Association (MURA) proposed the somewhat related *Fixed Field Alternating Gradient* particle accelerators in another attempt to increase the brightness of large diffuse beams [4].

The development of colliding beams and storage rings started with a 1956 *Physical Review* paper by a group at MURA which was led by Kerst [5]. Sessler considers this a landmark paper along with another landmark paper [6] by Kjell Johnsen, a second Norwegian, who was the first Director of the world's first high energy proton collider, the ISR. The first page of the Kerst *et al.* paper is shown in Fig. 5; the other authors are: F. T. Cole, H. R. Crane, L. W. Jones, I. J. Laslett, T. Ohkawa, A. M. Sessler, K. R. Symon, K. M. Terwilliger and N. V. Nilsen. Three of these authors, Crane, Jones and Terwilliger, I know well because they were at Michigan; the others authors were mostly at other mid-western universities. This paper, which was published in April 1956, soon led to many other papers.

Figure 6 shows the first page of an April 11, 1956 paper [7] that was written by G. K. O'Neill at Princeton, apparently just a few days after the Kerst *et al.* paper was published. O'Neill's idea was to produce colliding beams not by having two accelerators touching tangentially at one collison point, as Kerst *et al.* proposed, but instead by extracting two beams and then making the two beams collide. This technique was recently used at the SLAC Linear Collider (SLC). Moreover, this technique is now being considered for the huge proposed multi-billion dollar lepton colliders called the TESLA Linear Collider, the Next Linear Collider, the New Linear Collider or the Nippon Linear Collider.

Attainment of Very High Energy by Means of Intersecting Beams of Particles

D. W. KERST,* F. T. COLE,† H. R. CRANE,‡ L. W. JONES,‡ L. J. LASLETT,§ T. OHKAWA,‖ A. M. SESSLER,¶ K. R. SYMON,** K. M. TERWILLIGER,‡ AND NILS VOGT NILSEN††

Midwestern Universities Research Association,‡‡ University of Illinois, Champaign, Illinois

(Received January 23, 1956)

IN planning accelerators of higher and higher energy, it is well appreciated that the energy which will be available for interactions in the center-of-mass coordinate system will increase only as the square root of the energy of the accelerator. The possibility of producing interactions in stationary coordinates by directing beams against each other has often been considered, but the intensities of beams so far available have made the idea impractical. Fixed-field alternating-gradient accelerators[1] offer the possibility of obtaining sufficiently intense beams so that it may now be reasonable to reconsider directing two beams of approximately equal energy at each other. In this circumstance, two 21.6-Bev accelerators are equivalent to one machine of 1000 Bev.

The two fixed-field alternating-gradient accelerators could be arranged so that their high-energy beams circulate in opposite directions over a common path in a straight section which is common to the two accelerators, as shown in Fig. 1. The reaction yield is proportional to the product of the number of particles which can be accumulated in each machine. As an example, suppose we want 10^7 interactions per second from 10-Bev beams passing through a target volume 100 cm long and 1 cm^2 in cross section. Using 5×10^{-26} cm^2 for the nucleon interaction cross section, we find that we need 5×10^{14} particles circulating in machines of radius 10^4 cm.

There is a background from the residual gas proportional to the number of particles accelerated. With 10^{-6} mm nitrogen gas, we would have 15 times as many encounters with nitrogen nucleons in the target volume as we would have with beam protons. Since the products of the collisions with gas nuclei will be in a moving coordinate system, they will be largely confined to the orbital plane. Many of the desired p-p interaction products would come out at large angles to the orbital plane since their center of mass need not have high speed in the beam direction, thus helping to avoid background effects.

Multiple scattering at 10^{-6} mm pressure is not troublesome above one Bev; but beam life is limited by nuclear interaction with residual gas to $\sim$1300 seconds. Consequently, in about 1000 seconds the high-energy beam of 5×10^{14} particles must be established in each accelerator. The fixed-field nature of the accelerator allows it to contain beams of different energy simultaneously. It may be possible to obtain this high beam current in this time by using $\sim10^4$ successive frequency modulation cycles of radio-frequency acceleration, each cycle bringing up 5×10^{11} particles. It is encouraging to learn that Alvarez and Crawford[2] succeeded in building up a ring of protons by successively bringing up several groups of particles to the same final energy by frequency modulation in the 184-in. Berkeley cyclotron.

The number of particle groups which may be successively accelerated without leading to excessive beam spread can be estimated by means of Liouville's theorem.[3] One can readily convince himself that there is adequate phase space at high energy to accommodate

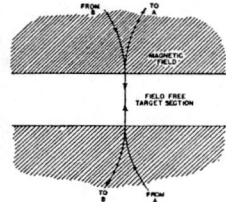

FIG. 1. The target straight section. *B* and *A* can be adjacent or concentric fixed-field alternating-gradient accelerators.

the necessary number, N, of particle groups. Assume for simplicity that synchrotron and betatron phase space are separately conserved, so that for the former

$$(\Delta p)_f(\Delta S)_f = N(\Delta p)_i(\Delta S)_i$$

Fig. 5 First Colliding Beam Paper [5]

THE STORAGE-RING SYNCHROTRON -- A DEVICE FOR HIGH-ENERGY PHYSICS RESEARCH*

Gerard K. O'Neill, Princeton University

April 11, 1956

As accelerators of higher and higher energy are built, their usefulness is limited by the fact that the energy available for creating new particles is that measured in the center-of-mass system of the target nucleon and the bombarding particle. In the relativistic limit, this energy rises only as the square root of the accelerator energy. However, if two particles of equal energy traveling in opposite directions could be made to collide, the available energy would be twice the whole energy of one particle. Kerst, among others, has emphasized the advantages to be gained from such an arrangement, and in particular of building two FFAG accelerators with beams interacting in a common straight section.

It is the purpose of this note to point out that it may be possible to obtain the same advantages with any accelerator having a strong, well focused external beam. Techniques for beam extraction have been developed by Piccioni and Ridgway for the Cosmotron, and by Crewe and LeCouteur for lower-energy cyclotrons.

Fig. 6 O'Neill Paper [7]

MURA -DBL/RGN/MHR-1

INTERSECTING BEAM ACCELERATOR WITH STORAGE RING

D.B. Lichtenberg, R.G. Newton, M.H. Ross

Indiana University and Midwestern Universities Research Association*

April 26, 1956

The idea of an intersecting beam accelerator is attractive for two reasons; first, much higher energies are obtained in the center of mass system than with one beam striking a stationary target, and second, collisions are observed directly in the center of mass system. The principal difficulty is the one of achieving sufficient intensity in the intersecting beams.

The simplest idea to get beams to intersect is to build two machines which accelerate particles in opposite directions. These machines would have a common straight section along which the particles would collide. A perhaps more economical device, however, which achieves the same purpose is a single accelerator and a deflecting mechanism which causes the beam to split into two intersecting orbits. An example of such an accelerator is shown schematically in Fig. 1.

Fig. 7 Lichtenberg, Newton & Ross Paper [8]

The first page of another important colliding beam paper [8], written by Donald Lichtenberg, Roger Newton and Marc Ross, is shown in Fig. 7; all three were then professors here at Indiana and also members of MURA. I am pleased to see Professor Lichtenberg in the audience. They proposed to use a single accelerator, but to somehow use its magnets to steer the particles so that they could collide. This technique is now used in many existing colliders; for example, Fermilab's Tevatron-Collider has only one ring but it contains two counter-rotating beams, one beam of protons and the other beam of antiprotons. For some reason these three gentlemen never published this paper; however, I had heard about it and Professor Lichtenberg recently gave me a copy of it and O'Neill's paper.

Professor Sessler gave me some nice photos of early accelerator and storage rings. Figure 8 shows the MURA Mark I FFAG, which was called the *Michigan Model*. This accelerator looks a bit like a cyclotron, but it did exhibit a type of strong focusing. The FFAG technique has not really been used very much, probably for the same reason that one would not want to build an SSC sized cyclotron: because then one must fill much of the area inside the ring with iron. The MURA *Two Way Model*, which was perhaps the first operating collider, is shown in Fig. 9.

Fig. 8 MURA Mark I FFAG *Michigan Model* †

Fig. 9 MURA *Two Way Model* Collider†

The first Stanford electron-electron collider, which started operating in 1959, is shown in Fig. 10; this was clearly using the idea originally proposed by Kerst *et al.* of having two physically separate accelerator rings with beams which collide where they touch. Figure 11 shows AdA, which Touschek built along with his colleagues at Frascati, in March 1961; AdA was the first electron-positron collider.

Fig. 10 First Stanford e-e Collider[†]   **Fig. 11** AdA at Frascati (1961)[†]

During the 1960s many people were enthusiastically building colliders. The first Novosibirsk electron-positron storage ring VEEP 2 is shown in Fig. 12 in 1967. Perhaps this electron-positron collider was the first use of the technique of Lichtenberg, Newton, and Ross, where one made two particles from a single ring somehow collide. VEEP 4 is the present collider at the Budker Institute of Nuclear Physics (BINP) in Novosibirsk; it has had an active program of experiments.

Fig. 12 VEEP 2 at Novosibirsk[†]   **Fig. 13** ADONE at Frascati[†]

ADONE, which was the first "large" electron-positron storage ring, is shown in Fig. 13; its energy was 1 or 2 GeV. There is a special problem with circular electron storage rings at very high energy, where their synchrotron radiation becomes very large; therefore, one must inject a great deal of RF acceleration energy to compensate for these synchrotron radiation energy losses. Moreover, one needs many RF stations, distributed uniformly around the ring, to inject the RF into the ring uniformly and thus avoid the large energy loss that can occur in a single turn.

The electron model for the ISR, around 1960, is shown in Fig. 14. Kjell Johnsen and his colleagues built this small-scale model of ISR with electrons rather than protons.

Fig. 14 *Electron Model* for ISR[†]   **Fig. 15**   The CERN ISR (1971)[†]

Now I will diverge from Andy Sessler's talk, because I worked on the ISR when it first operated in 1971. Figure 15 shows a photo of ISR, which first extended CERN from Switzerland into France; the ISR's injector was the venerable PS which is also shown. I remember that we had to be very careful not to go into France by accident without showing our passports. Figure 16 shows a diagram of the ISR and the PS; the PS would alternately inject protons into each ISR ring. Notice the eight different crossing regions where the two ISR beams crossed in the horizontal plane.

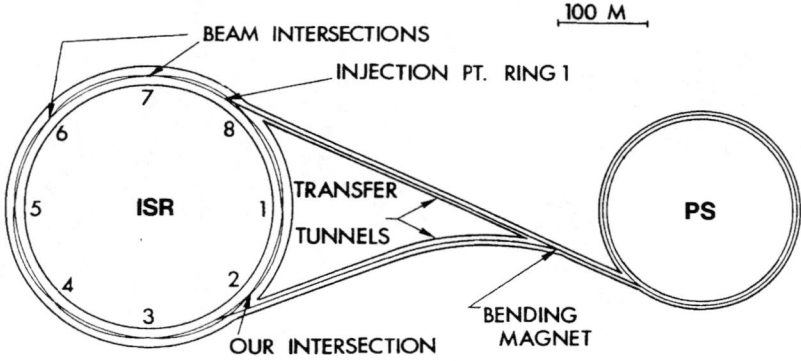

Fig. 16   CERN ISR and PS [6]

Figure 17 shows the I 2 crossing region where we did the ABM experiment, which studied inclusive scattering using a narrow 43-meter-long spectrometer containing three bending magnets and a few scintillation and Cherenkov counters. We measured inclusive cross sections at 1500 GeV laboratory energy equivalent; that was a huge energy jump from the pre-1971 maximum energy of 30 GeV. Moreover, many people expected to find real physical quarks at the ISR. Therefore, this was a very exciting time.

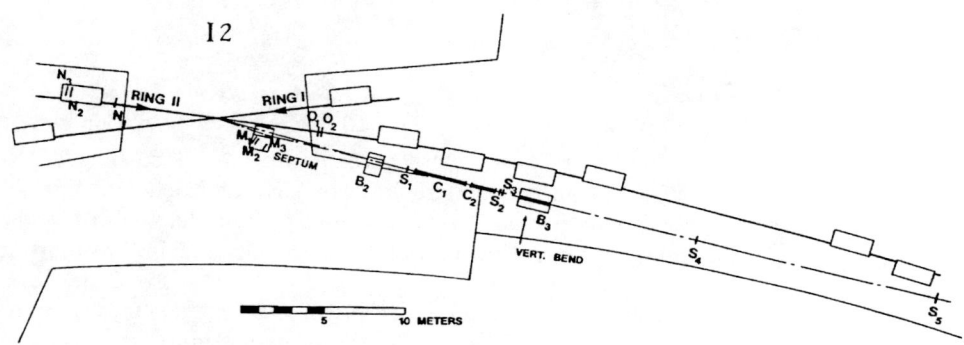

Fig. 17 ISR's I 2 Region with Argonne-Bologna-Michigan Inclusive Experiment

The ISR was an absolutely beautiful machine; it may be the best accelerator facility that will ever be built. It started operating six months ahead of time and apparently reached its design value before it was supposed to first operate. Eventually, the ISR reached a luminosity of about two orders of magnitude above its design value.

We did the first experiment at the ISR, which was slightly strange, because we were mostly an American group at the first really forefront European accelerator. Perhaps the reason was that I have always been an optimist; thus, all our equipment was ready about six months ahead of time and then suddenly the ISR started running. Most other ISR experimenters, such as Carlo Rubbia, were off doing something else; then suddenly the ISR beam was operating and they did not have any working detectors. The year 1971 was great fun and was clearly one of the high points of my career.

The I 2 crossing region is again shown in Fig. 18; notice the ISR ring magnets and the front end of our spectrometer which started with a special septum magnet that we built and brought to CERN. Michigan paid for this septum magnet which was assembled and tested at Argonne by Larry Ratner and Jim Bywater. We had to convince Kjell Johnsen and his colleagues that the septum's fringe field would not destroy the ISR beam, which was only a few centimeters from the septum's corner.

Fig. 18 ABM Septum Magnet **Fig. 19** ABM Cherenkov Counters

The Cherenkov counters in our spectrometer, which were provided by CERN, are shown in Fig. 19. We had our own Cherenkov counters at Michigan, but we used CERN Cherenkov counters because these detectors were under ten or twenty atmospheres of pressure and they could turn into rockets. My friends in Washington were concerned that, if an American-made Cherenkov counter destroyed the ISR, there could be a lot of trouble. Probably the CERN people were also concerned; in any case, they provided the Cherenkov Counters. We brought the other detectors from Michigan; these were standard scintillation counters mounted on transit stands with plumb bobs to make sure that they did not move.

Fig. 20 End of ABM Spectrometer **Fig. 21** ABM Trailer

The curvature of the ISR tunnel can be seen in Fig. 20, along with our spectrometer's final magnet, which was bending vertically up. Also notice our two downstream scintillation detectors on high extended transit stands. Fig. 21 shows our experimental counting house which was then called a trailer. Our ABM Collaboration involved Argonne, Bologna and Michigan; in 1971 there was a fierce debate in the world about anti-ballistic missiles, so we had fun with the name ABM. Our counting house's interior is shown in Fig. 22; notice the state-of-the-art electronics from 28 years ago and a pulse-height analyzer, which was already old in 1971.

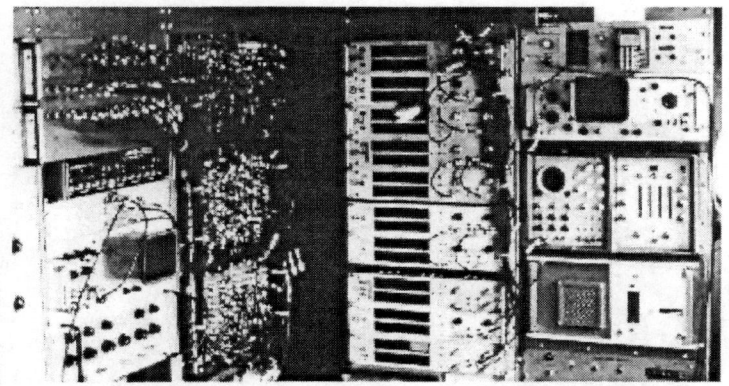

Fig. 22 Inside the ABM Counting House (Trailer)

When the first ISR program was approved, everyone was very concerned about how to measure the luminosity; no one had ever done an experiment with proton storage rings and there was no plan about how to measure luminosity. Our experiment planned to measure inclusive cross sections at very high energy to test the scaling law, which had been proposed independently by Feynman and by Yang. Therefore, we certainly needed an absolute measurement of the luminosity; otherwise, we could not calibrate the 1500 GeV inclusive cross sections to compare them with the 12 and 24 GeV data from the Argonne ZGS [9] and the CERN PS [10]. Every experimenter was quite concerned about this luminosity problem and this concern certainly reached the ISR staff.

To measure the luminosity, a very smart ISR staff member named Simon van der Meer invented the idea of using some small magnets to shift the beam vertically in small steps: the first magnet would bend the beam vertically up; the next magnet would bend it back into the horizontal plane; then, after the interaction region, the third magnet would bend the beam down and the fourth magnet would realign the beam along the original orbit. These four magnets were wired together in series; thus, when one increased the current by perhaps 1 amp, the beam would just move up perhaps 1 millimeter in the interaction region. Figure 23 shows the event rate in our luminosity monitors plotted against the vertical separation in millimeters. This variation of the separation allowed one to measure the effective height h_{eff} of the two ISR beams where they collided. This h_{eff} value allowed an absolute collaboration of the luminosity using the equation:

$$\text{Luminosity} = \frac{I_1 I_2}{h_{eff} \cos \alpha} \quad , \tag{1}$$

where I_1 and I_2 are the intensities of the two beams and α is their crossing angle in the horizontal plane. Thus, Dr. van der Meer allowed us all to measure absolute cross section. I do not recall meeting him in 1971; perhaps Kjell Johnsen and Franco Bonaudi told me of his idea. I recall thinking that van der Meer must be very clever; some years later he did something else clever and got the Nobel Prize.

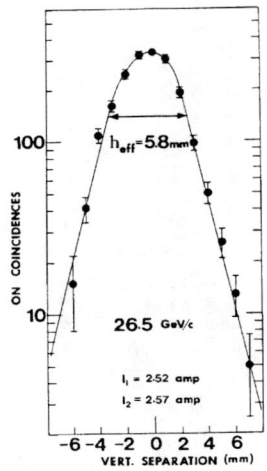

Fig. 23 *van der Meer* Curve with Small Beams

Fig. 24 *van der Meer* Curve with Large Beams

Note that in Fig. 23, the effective height of the beams was 5.8 millimeters, while it was 10.3 millimeters in Fig. 24, when the beams were much larger. Figure 24 also contains some tests that we did together with the ISR staff; in one case, one beam was fixed while the other beam was moved; in the second case the two beams were moved equally. Notice that the two sets of data agree perfectly; this helped to confirm that the *van der Meer Technique* really did work.

These first ISR inclusive results were published in *Physical Review Letters* [11] on 5 July 1971, the paper was submitted 2 June. In 1974 we published a detailed *Physical Review* paper [12]. Figure 25 shows our ISR inclusive cross section data at effective energies of 500, 1100 and 1500 GeV plotted against $P_\perp^2$ and X_F together with the lower energy data. Notice that we measured the inclusive production of π^+ mesons, K^+ mesons and protons, while our Bologna collaborators later published inclusive data on π^-, K^-

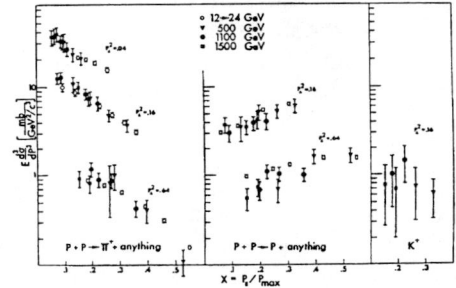

Fig. 25 First Evidence for Feynman-Yang Inclusive Scaling [11,12]

and antiproton production [13]. Our inclusive data points at 500, 1100 and 1500 GeV fall right on top of each other; moreover, they fall right on top of our earlier 12 GeV ZGS and the 24 GeV CERN data. Thus it was clear that Feynman-Yang scaling was correct to first order; this scaling law was proposed independently by Feynman [14] and Yang [15] with totally different models. Yang's model was called Limiting Fragmentation and Feynman's model was just called Feynman's model; our ISR data showed that both models were correct.

In subsequent years, people worked very hard to find and then study small deviations from Feynman-Yang scaling; however, until 1971, there was absolutely no agreement about how inclusive cross-sections would behave in the TeV region. Thus, this first verification of Feynman-Yang scaling seems rather important; however, CERN never seemed to appreciate the importance of this result or of several other significant ISR results.

Now returning to Professor Sessler's lectures, he noted that Ken Robinson and Gus Voss, who were then at CEA, first developed the idea of low beta intersections. This technique involved installing quadrupoles near a collision point to squeeze the beams so that they are very small where they collide; this technique increases the luminosity without changing the intensity of the beams. These low beta intersections are now very important to all colliders.

Fig. 26 ICE Storage Ring at CERN[†]   **Fig. 27** Simon van der Meer[†]

Another important idea that made colliders work was stochastic cooling. This was the second very clever idea of Simon van der Meer; he received the Nobel Prize for it. While the idea itself is rather simple, using it can be rather complex. I will not describe every detail of the technique, but basically one installs a detector on one side of a ring to determine if some part of the beam is properly centered or not; then, while the beam is going around to the ring's other side, one can send a signal across the ring's diameter. Because this signal can arrive before the beam arrives, the signal can tell the ring to energize some magnets, which then recenter the beam. Dr. van der Meer made a complex but very impressive device to test this technique at CERN, called the ICE storage ring, which is shown in

Fig. 26. This stochastic cooling technique is especially important with antiprotons because they normally have a much lower intensity than protons; thus, being able to focus antiprotons to increase their brightness is very important. To improve the antiproton beam's longitudinal, transverse and vertical phase space, ICE had many detectors and many corrector magnets; each magnet was 180° away from its detector. ICE worked beautifully. Stochastic cooling then allowed the SPS antiproton-proton collider to be built at CERN with enough luminosity to discover the intermediate boson. Simon van der Meer is shown in Fig. 27.

The Fermilab Tevatron-collider, which was the next big collider to operate, is shown in Fig. 28; it is just outside of Chicago. At the moment it is the world's highest energy collider; its c.m. energy is almost 900 GeV.

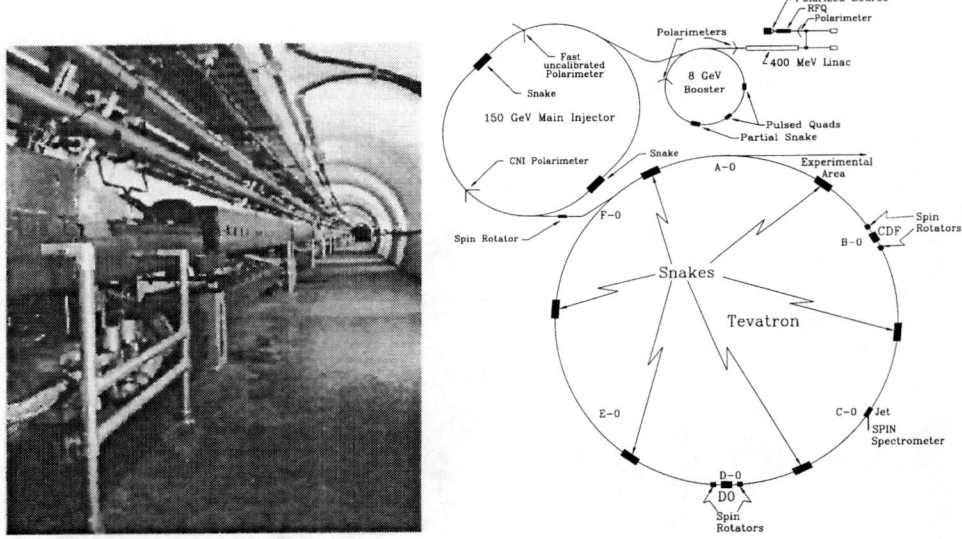

Fig. 28 Fermilab Tevatron Tunnel[†]   **Fig. 29** Fermilab Design for Polarized Protons

Figure 29 is a drawing of Fermilab from our polarized Tevatron study [16]. Starting in 1991 Fermilab commissioned our SPIN Collaboration to design the capability to accelerate polarized protons in the Main Injector and the Tevatron; however, in 1995, Fermilab decided not to do it, at least for now. This was partly due to the $25 Million cost, because there are no appropriate empty spaces for 6 Siberian snakes. Creating these 6 empty spaces would require removing 24 existing Tevatron superconducting ring dipoles and replacing them with shorter higher field dipoles. The Tevatron is about 6.3 kilometers in circumference with an 8 GeV Booster and its new 150 GeV Main Injector which started operating earlier this year.

Another beautiful large collider is the HERA positron-proton collider at DESY which is shown in Fig. 30. HERA is similar in size to Fermilab because its proton energy is similar. HERA recently reached 920 GeV; thus its energy is now slightly higher than Fermilab's. No one has yet reached a TeV, but they are both getting close. HERA collides protons with either electrons or positrons. HERA's center of mass energy is much lower than Fermilab's because HERA's electrons or positrons are at about 30 GeV.

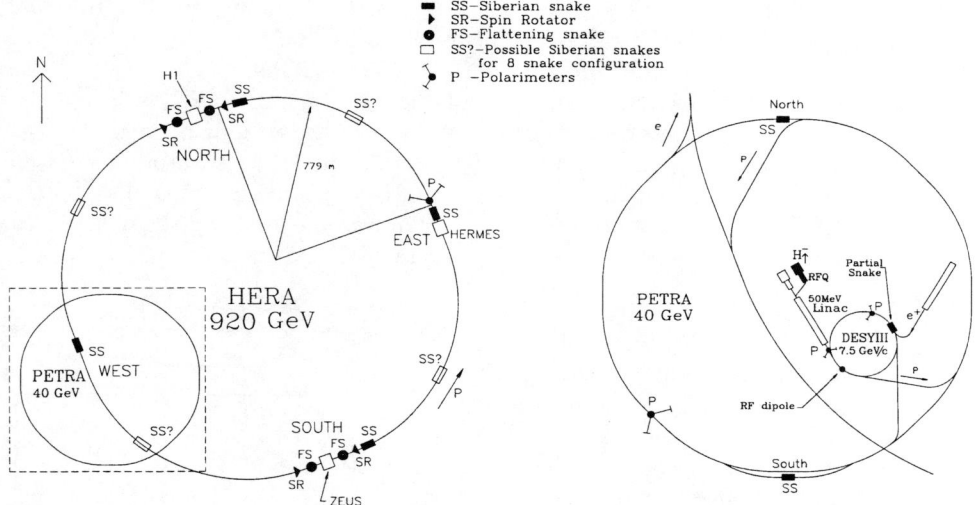

Fig. 30 HERA Design for Polarized Protons

Starting in 1996 we also did a polarized proton study for HERA, which already had polarized positrons. Professor Wilk and the other DESY Directors were interested in possibly polarizing the protons and they commissioned us to study the problem. We concluded that one could easily install four Siberian snakes in HERA; however, four snakes might not be enough to maintain the proton polarization at 920 GeV [17]. Eight Siberian snakes would certainly be able to maintain the polarization but would cost more money. Figure 30 shows the Siberian Snakes and other hardware needed to have polarized protons at 920 GeV; by looking at the names of all HERA's booster rings, one can see a history of DESY. The lowest energy booster is DESY III and then PETRA and then finally HERA itself; thus, we certainly recycle our old accelerators and do not waste them.

The second part of my talk is more directly related to this Conferences main subject, which is storage rings with internal targets. When Prof. Meyer asked me to give this talk, my first concern was what to talk about. Then I realized that I had worked on an early "storage ring" experiments for my thesis at the AGS in 1962-63. We used a rotating CH_2 target about one millimeter thick placed inside the AGS ring to study proton-proton elastic scattering at high $P_\perp^2$. At that time, one did not think of the AGS as a storage ring, but it had a storage time of about one second, during which the beam passed through our target about 500,000 times and was effectively used up. The storage time was somewhat smaller than what is used now, but the AGS was indeed being used as a Storage Ring.

We published a *Physical Review Letter* [18] on the AGS experiment, which is shown in Fig. 31; since it was my thesis, I firmly believe that it was a nice experiment. Notice that it was submitted on 11 November 1963 and published on 1 December 1963. We later published a *Physical Review* paper in 1965 [19]. In those days, one really had to publish a detailed *Physical Review* paper after the *PRL*; otherwise the editor, Professor Goudsmit, would remember and the next time you submitted a *Letter*, he would return it without even reading it.

This AGS data is an important part of Fig. 32, which is a 1967 graph [20] of all high energy proton-proton elastic scattering cross-sections plotted against a scaled transverse momentum variable called $\beta^2 P_\perp^2$ that I derived around 1964 [20] by assuming that protons behave as Lorentz-contracted spheres. This plot of all the world's high energy proton-proton data in 1967, was inspired by the sharp break found in our 1966 Argonne ZGS proton-proton elastic experiment at exactly $90°_{cm}$; that was probably the first evidence for structure inside the proton [21]. Figure 32 also contains the AGS data; note the point near $\beta^2 P_\perp^2 = 13$ (GeV/c)^2; its -t value was about 25 (GeV/c)^2. Since this point was measured 36 years ago, no one has measured an exclusive cross-section at a larger transverse momentum; perhaps, this is because our 1963 AGS experiment [18,19] required a five-month dedicated run with an internal target.

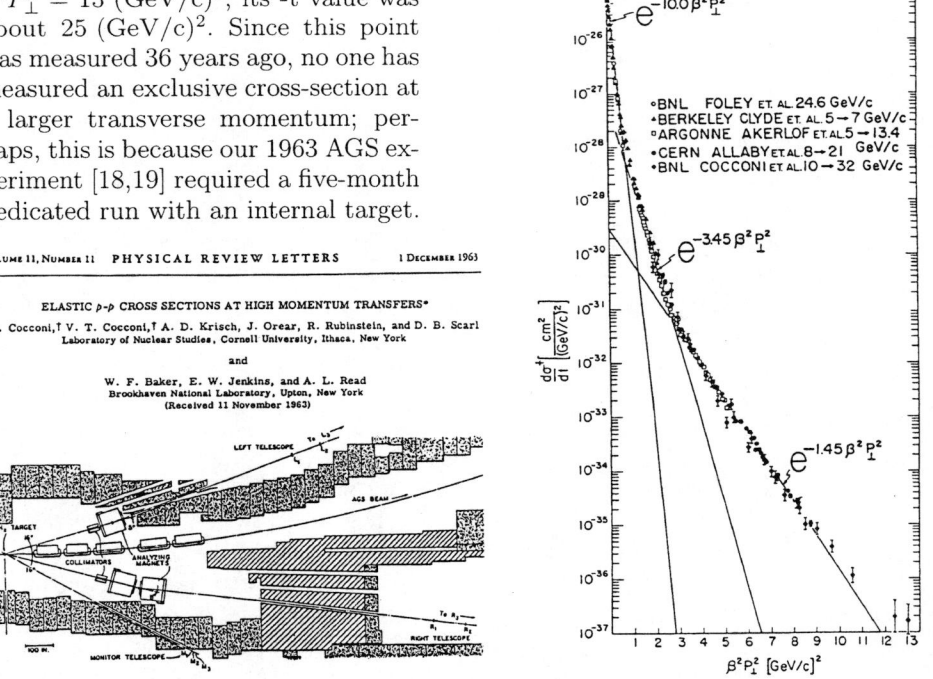

Fig. 31 High-$P_\perp^2$ Elastic p-p Experiment [18,19]   **Fig. 32** 1967 Plot of p-p Elastic Data [20]

Now I will jump from 1963 to 1999 for the HERMES experiment; we heard two talks about HERMES this morning. During the last few years, HERMES has certainly been one of the most impressive fixed target storage ring experiments. The experimental apparatus is shown in Fig. 33; the HERMES group successfully designed and carried out this beautiful experiment with an impressive polarized "storage cell" target. The experiment has produced some important data on the nucleon structure functions. HERMES certainly has a very different scale in geometry and in time than the above experiments in the 1960's and early 1970's.

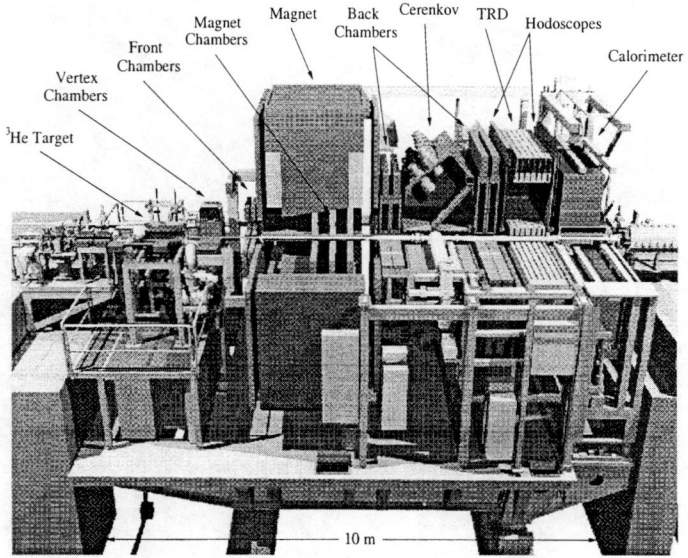

Fig. 33 The HERMES Experiment at HERA

The AmPS facility, which is shown in Fig. 34, has had internal targets and a Siberian Snake which allowed it to carry out some nice experiments. It is very sad that AmPS was recently shut down, like the ZGS, before it was ready to be shut down. I recall that in 1974, when the ZGS was already scheduled to be shut down, Professor Dirac visited Argonne and toured the ZGS; apparently, he was quite impressed and then said "You high energy physicists seem quite foolish; you build beautiful facilities and then shut them down before they have been half-utilized." Of course, Dirac was a theorist and theorists are sometimes not very practical, but perhaps he was wiser than some of our administrators and politicians.

The very nice COSY facility at Jülich is shown in Fig. 35. COSY operates both as a storage ring with some internal targets and as an extracted beam facility. Notice that it uses a cyclotron as its injector. COSY has been running for several years; it has outstanding hardware and it has the potential to become one of the worlds leading intermediate energy facilities.

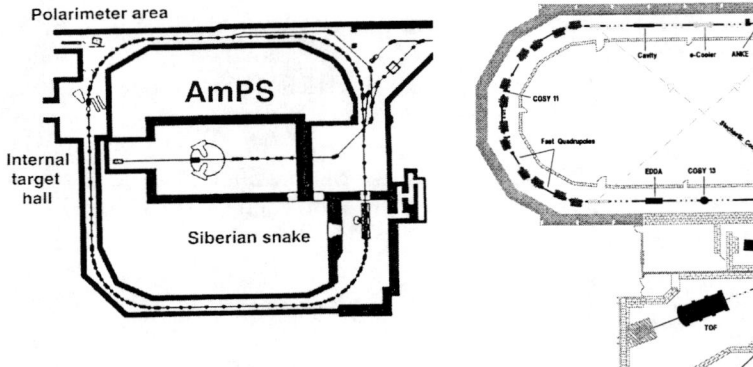

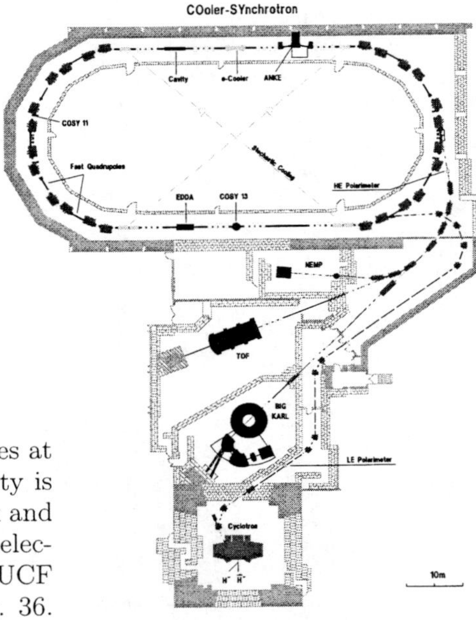

Fig. 34 AmPS at NIKHEF

Next, I will briefly discuss a few activities at the IUCF Cooler Ring. The first activity is the electron cooling work of Bob Pollock and his colleagues; in the 1980s, they used electron cooling to develop the pioneering IUCF Cooler Ring [22], which is shown in Fig. 36. Electron cooling was certainly invented by Budker and his colleagues at Novosibirsk; but the IUCF Cooler Ring demonstrated that electron cooling could provide a stored beam quality that has allowed experiments of unprecedented precision. It seems that the Indiana Cooler Ring has had a large influence over physics in many energy ranges.

Fig. 35 COSY at Jülich

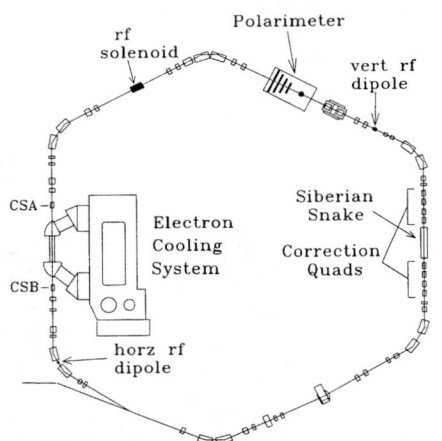
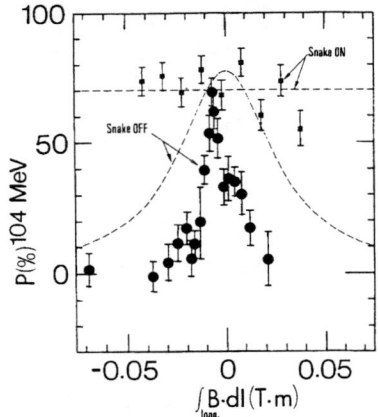

Fig. 36 IUCF Cooler Ring

Fig. 37 First Siberian Snake Test (1989)

Our much-loved Siberian snake was first installed in the IUCF Cooler Ring in 1989. Figure 37 shows the data which first showed that a Siberian Snake really is capable of overcoming depolarized resonances [23]. Notice that with no snake, there is full polarization only if all imperfection fields are exactly corrected; any uncorrected imperfection field destroys the polarization. However, with the Siberian snake turned on, we completely overcame this $G\gamma=2$ imperfection depolarization.

Now I will move to the future. It would be difficult to cover all future storage ring facilities, but I will discuss some that I am familiar with. Starting at the very highest energy, I must mention the SSC. Many people would like to forget it; however, it was very important during 1983 to 1993 when we spent about $2 Billion on it. As shown in Fig. 38, the SSC was huge; it was about 80 kilometers in circumference and contained many rings. The injector system is expanded to show its detail; there were three boosters: the 12 GeV Low Energy Booster; the 200 GeV Medium Energy Booster; and the 2 TeV High Energy Booster. The two 20 TeV Rings would allow many people to do experiments. We spin people feel

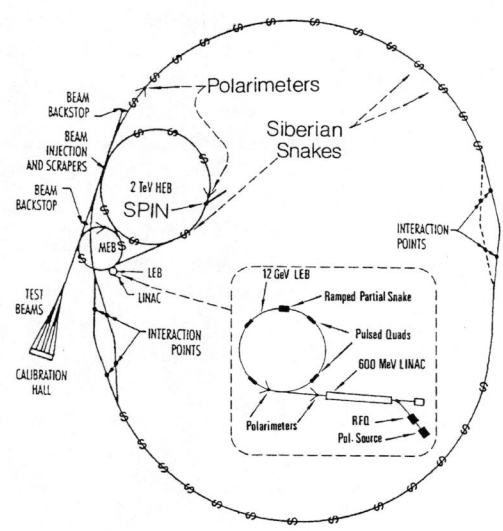

Fig. 38 SSC Design for Polarized Protons

especially badly about the SSC's death, because it was modified to have twenty-six 19-meter-long empty spaces to later install 26 Siberian snakes in each SSC Ring. Of course, the SSC and its empty spaces for snakes are now all gone.

The highest energy facility now being constructed is the LHC at CERN which will be installed in the present LEP tunnel when LEP finishes running. CERN is now building the magnets for LHC, which will be installed in the present 28 km LEP tunnel. Apparently, LHC should first operate for physics runs around 2006. LHC will allow the study of 7 TeV on 7 TeV proton-proton collisions with a luminosity of 10^{33} or 10^{34}. LHC is certainly the largest presently approved collider; because it is quite expensive, many countries are contributing to its cost. Any larger facilities may be even more international because of their high cost.

The LEP electron-positron collider, which is now in this same 28 km circumference tunnel, has been running very effectively. It now operates at about 100 GeV in each ring, which required overcoming a major synchrotron radiation problem. The main cost in CERN's electric bill is for the RF power to overcome LEPs synchrotron radiation; this requires almost 100 MegaWatts. LEP may have reached the upper limit for a circular electron ring, because the synchrotron radiation energy increases as the fourth power of the energy; therefore, doubling LEP's energy to 200 GeV, would increase its synchrotron radiation 16-fold; this might require more than a TeraWatt of electric power, which does not seem practical.

The UNK facility, at the Institute of High Energy in Physics in Protvino, Russia, was to be built in phases, which would eventually produce a 3 TeV on 3 TeV collider. The 21-kilometer-circumference UNK tunnel, which is shown in Fig. 39, is just slightly smaller than the LHC tunnel. However, IHEP was more conservative in its superconducting magnet design; UNK was designed to operate at about 4 Tesla, while CERN is now planning to run LHC at about 7.7 Tesla. UNK contains a huge underground area built for our NEPTUN/NEPTUN-A experiment; we would have used a polarized jet and a 55-meter-long spectrometer to study spin effects in high-$P_\perp^2$ elastic proton-proton scattering, first at 400 GeV and then at 3 TeV. Our Russian NEPTUN collaborators would have several other spectrometers pointing at the Michigan Ultra-cold Polarized Jet which recently started operating. The UNK tunnel is essentially finished and about 80% of the UNK-1 ring magnets are constructed and tested; however, around August 1998, UNK was put on hold by the Russian government due to the financial situation. About 100 people are maintaining the 21 km tunnel in the hope that IHEP may later finish and operate UNK, but the economic situation has already delayed UNK by 5 or 6 years.

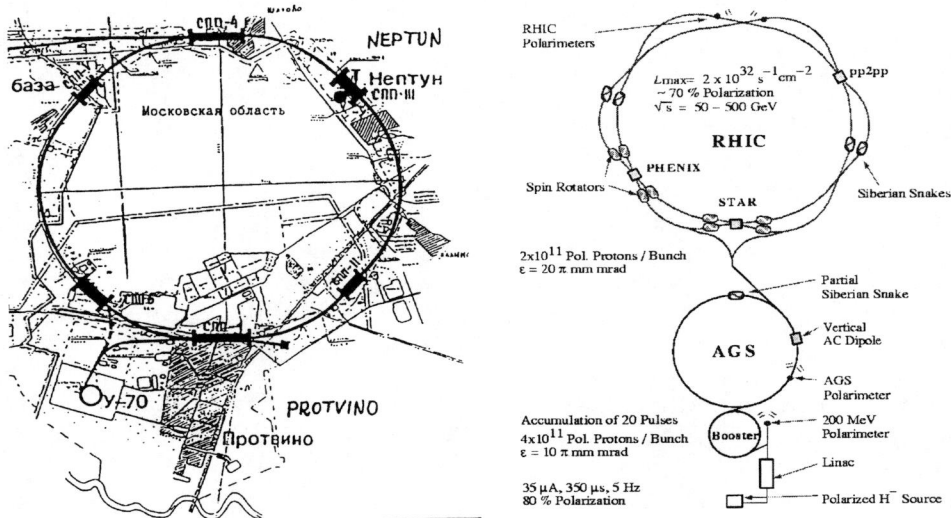

Fig. 39   UNK at IHEP-Protvino   **Fig. 40**   RHIC with Polarized Protons

The RHIC facility is shown in Fig. 40. RHIC seems important for both nuclear and high energy physics, because it will accelerate and collide both Relativistic Heavy Ions and polarized protons. RHIC should have a rather high luminosity and it recently accelerated and stored heavy ions in one ring; the first heavy ion collisions should occur in December 1999. The first polarized proton commissioning run should occur in 2000 and the first polarized proton physics run should occur in 2001. During the next decade, RHIC should produce outstanding hadron scattering data in the 100 GeV per nucleon range for both nuclear and high energy physicists.

The next new storage ring is the MIT-Bates facility, which is shown in Fig. 41. I consider Bates as a future facility because the South Hall Storage Ring, which is a 1 GeV polarized electron storage ring, should start operation early in 2000. In July 1999, the BLAST detector was being installed in the ring. This 1 GeV polarized electron ring seems to have a great deal of capability.

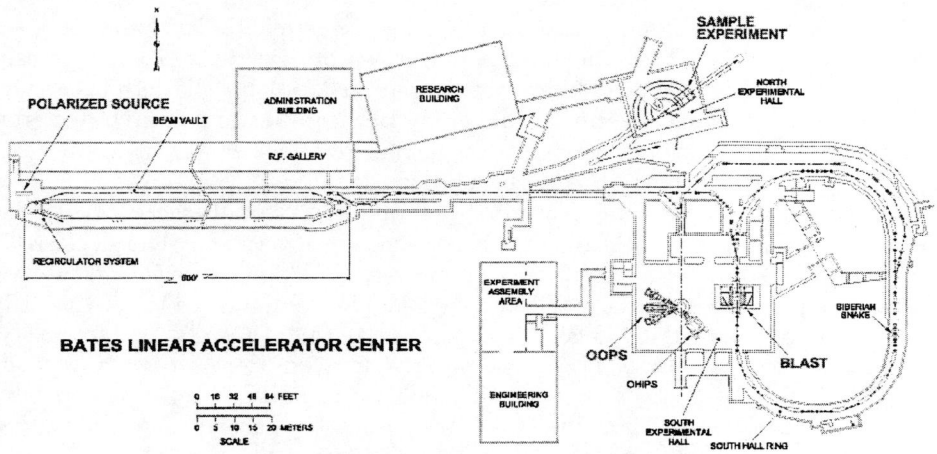

Fig. 41 MIT-Bates Linac and New Storage Ring

I will end with Fig. 42, which shows the Indiana University Cyclotron Facility Cooler Ring. The IUCF name has become a lie because the Cooler Ring no longer has a Cyclotron injector; the Cyclotron is now curing people with eye problems. The new injector for the Cooler Ring consists of two high quality new facilities.

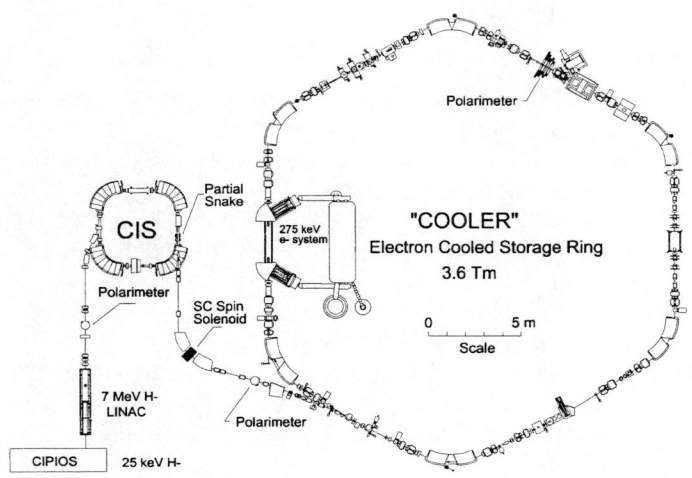

Fig. 42 IUCF's CIPIOS-CIS-Cooler Ring Complex

One is the Cooler Injector Synchrotron (CIS), which is a new small Synchrotron that is a much better injector for the Cooler Ring than the Cyclotron. [A Cyclotron is not a good injector for a Synchrotron because of the bad energy match.] In July 1999 the new CIPIOS polarized ion source started operating; CIPIOS is coupled to CIS through a 7 MeV RFQ and LINAC. The CIPIOS-CIS-Cooler Ring system is now working very well; there were several unpolarized runs in early 1999 and the first polarized run occurred in July 1999. The experimental hardware installed in the Cooler Ring includes: an Illinois laser-driven polarized target, a Wisconsin atomic-beam-source polarized target, our Siberian snake, Bob Pollock's plasma target, the neutron tagging experiment and Andy Bacher's future polarized deuteron experiment. Note that our conference chairman, Professor Meyer, is leading the three-body experiments with the Wisconsin target.

Both MIT-Bates and IUCF recently went through, what Richard Milner described to me as, a near-death experience when our government had decided to shut down Bates on rather short notice. However, by making an effective case for the importance of its physics capabilities, each laboratory managed to convince the government to substantially delay the shut-down date. Certainly, no laboratory, which requires a large budget, can expect to live forever. However, when we consider Bates and IUCF, together with HERA, COSY, RHIC, LHC and hopefully UNK and some other facilities, we can look forward to a broad range of exciting storage ring physics for many years.

REFERENCES

Note: In preparing this manuscript, I did not have time to find every detailed reference. For more detailed references, please consult Prof. Sessler's book [1] and the proceedings of this conference.

† Photo credits: From the private collection of A.M. Sessler.

1. A.M. Sessler, *Lectures at Osaka, Washington and Vancouver*; C. Pelegrini and A.M. Sessler, *The Development of Colliders*, AIP (1995).
2. A.W. Chao and M. Tigner, *Handbook of Accelerator Physics and Engineering*, page 12, World Scientific, Singapore (1999).
3. E.D. Courant, M.S. Livingston, and H.S. Snyder, Phys. Rev. **88**, 1190 (1952).
4. D.W. Kerst *et al.*, *Fixed Field Alternating Gradient Particle Accelerators,* 32, CERN Symposium (1956).
5. D.W. Kerst *et al.*, Phys. Rev. **102**, 590 (1956).
6. K. Johnsen, *The CERN Intersection Storage Rings,* 8th International Conference on High Energy Accelerators, CERN, 79 (1971).
7. G.K. O'Neill, Phys. Rev. **102**, 1418 (1956).
8. D.B. Lichtenberg, R.G. Newton, and M.H. Ross, MURA Report-DBL/RGN/MHR-1 (April 26, 1956), unpublished.
9. C.W. Akerlof *et al.*, Phys. Rev. **D3**, 645 (1971); L.G. Ratner *et al.*, Phys. Rev. Lett. **16**, 855 (1966), and Phys. Rev. **166**, 1353 (1968).

10. J.V. Allaby et al., CERN Report No. 70-12 (1970).
11. L.G. Ratner et al., Phys. Rev. Letters **27**, 68 (1971).
12. L.G. Ratner et al., Phys. Rev. **D9**, 1135 (1974).
13. A. Bertin et al., Phys. Letters **38B**, 260 (1972).
14. R.P. Feynman, Phys. Rev. Letters **23**, 1415 (1969).
15. C.N. Yang et al., Phys. Rev. **188**, 2159 (1969) and Phys. Rev. Lett. **25**, 1072 (1970).
16. *Acceleration of Polarized Protons to 120 GeV and 1 TeV at Fermilab*, SPIN Collaboration, UM HE 95-09 (1995).
17. *Acceleration of Polarized Protons to 920 GeV at HERA*, SPIN Collaboration, UM HE 99-05 (1999).
18. G. Cocconi et al., Phys. Rev. Letters **11**, 499 (1963).
19. G. Cocconi et al., Phys. Rev. **138**, B165 (1965).
20. A.D. Krisch, Phys. Rev. Letters **19**, 1149 (1967); Phys. Rev. **134**, B1456 (1964), Phys. Rev, Letters **11**, 217 (1963).
21. C.W. Akerlof, Phys. Rev. Letters **17**, 1105 (1966); Phys. Rev. **159**, 1138 (1967).
22. R.E. Pollock, IEEE Trans. Nucl. Sci. **30**, 2056 (1983).
23. A.D. Krisch et al., Phys. Rev. Letters **63**, 1137 (1989).
24. J.E. Goodwin et al., Phys. Rev. Letters **64**, 2779 (1990).

APPENDICES

APPENDIX A

PROGRAM

SUNDAY, SEPTEMBER 12, 1999

AFTERNOON: (session chair: J. Cameron, Indiana)

2:00	G.E. Walker, Indiana University: *Welcome*
2:20	R.E. Pollock, IUCF, **Storage Conundra**
3:05	W. Scobel, Hamburg, **pp Elastic Scattering: New Results from EDDA**
4:15	P.D. Eversheim, Bonn, **Preparing the COSY-Ring for a Test of Time-Reversal Invariance**
4:35	P. Moskal, Cracow: **Heavy Meson Production at COSY-11**
5:10	R. Jahn, Bonn, **Near-Threshold Two-Meson Production: pd-->^{3}He$\pi^+\pi^-$ and pd-->^{3}He K$^+$K$^-$**
5:30	H. Machner, COSY, **Meson Production in p + d Reactions**

MONDAY, SEPTEMBER 13, 1999

MORNING: (session chair: W. Haeberli, Wisconsin)

8:30	Ch. Hanhart, Washington, **Meson-Exchange Models for Meson Production**
9:05	J. Balewski, IUCF, **Pion Production with Polarized Beam and Polarized Target at IUCF**
9:50	S.A. Coon, New Mexico, **Off-Mass-Shell Pion-Nucleon Scattering and pp-->ppπ^0**
10:45	N. Kaiser, München, **Phenomenology of Meson Production Near Threshold**
11:30	D. Reistad, CELSIUS: **News in Electron Cooling: Highlights from ECOOL99**

AFTERNOON: (session chair: J. Bisplinghoff, Bonn)

2:00	G. Gabrielse, Harvard, **Observing the Quantum Limits of an Electron Cyclotron Accelerator**
2:45	H. Ströher, COSY, **First Results on Strangeness Production from the ANKE Facility**
4:00	K. Tamura, RCNP, **Pion Production Mechanism in Nucleon-Nucleon Collisions**
4:20	B. Jakobsson, Lund, **Pion Production and Entropy Experiments at Storage Rings**
4:55	H. Ohm, COSY, **Lifetime of Hypernuclei: COSY-13**
5:30	I.A. Rachek, Novosibirsk, **Internal Polarized Deuterium Target Experiments at the Electron Storage Ring VEPP-3**

TUESDAY, SEPTEMBER 14, 1999

MORNING: (session chair: W. Daehnick, Pittsburgh)

8:30	S. Wirth, Erlangen, **Associated Strangeness Production at COSY-TOF**
8:50	M. Wolke, COSY, **Hyperon and Charged Kaon Pair Production Close to Threshold**
9:25	J.M. Laget, Saclay, **Theoretical Issues in Strangeness Production**
10:35	L.D. Knutson, Wisconsin, **Polarization in Meson Production Reactions**
11:20	F. Rathman, Erlangen, **Review of Polarized Internal Targets**
11:55	T. Wise, Wisconsin, **Polarization of Molecular Hydrogen from Recombined Polarized Atoms**

AFTERNOON:

2:00	POSTER SESSION
4:00	*Picnic at Lake Monroe*

WEDNESDAY, SEPTEMBER 15

MORNING: (session chair: S. Vigdor, Indiana)

8:30	H. Lacker, PSI, **Spin Dependence of the Reaction np-->ppπ^-**
8:50	J. Zlomanczuk, CELSIUS, **Pion Production in pN Collisions Near Threshold**
9:10	M. Strikman, Penn State, **QCD Physics with ep Colliders (EPIC99)**
9:45	B. Lorentz, COSY, **Stochastic Cooling and Extraction at COSY**
10:45	P. Golubev, Lund, **CHICSi - A 3π Multi-Detector System for Studying Heavy Ion Interactions at Storage Rings**
11:05	T. Katayama, Tokyo, **Storage of Radioactive Beams**
11:40	I. Mishustin, Niels Bohr Institute, **Studies of Nuclear Reactions and Phase Transitions with Storage Rings**

AFTERNOON: (session chair: M. Miller, Illinois)

2:00	Ch. Scheidenberger, GSI, **Experiments with highly-charged exotic nuclei in the storage and cooler ring ESR**
2:35	T. Radon, GSI, **First Isochronous Time-of-Flight Mass Measurements of Short-Lived Projectile Fragments in the ESR**
2:55	B. Franzke, GSI, **The ESR: Present Status and Prospect for Radioactive Ion Beams**
4:00	H. Calen, CELSIUS, **WASA Detector: Rare Pion and Eta Decays**
4:35	P. Egelhof, Mainz, **Calorimetric Low-Temperature Detectors for High Resolution X-ray Spectroscopy on Stored Highly Stripped Heavy Ions**
4:55	M. Falch, München, **Schottky Mass Spectrometry at the ESR at GSI with a New Data Acquisition System**
6:00	*Banquet*

THURSDAY, SEPTEMBER 16, 1999

MORNING: (session chair: P.V. Pancella, W Michigan)

8:30	T. Roser, BNL, **Acceleration and Storage of Polarized Proton Beam at RHIC**
9:05	G. Westfall, Michigan State, **Experiments at RHIC**
9:40	L.D. van Buuren, NIKHEF, **Electron Scattering Experiments with Polarized Hydrogen or Deuterium Internal Targets**
10:35	R. Milner, MIT, **The HERMES Experiment**
11:20	I. Passchier, NIKHEF, **Longitudinally Polarized Electrons in a Storage Ring at ~ 1 GeV**
11:40	J.L. Matthews, MIT, **The Bates Large Acceptance Spectrometer Toroid**

AFTERNOON: (session chair: C. Ekström, CELSIUS)

2:00	J. Haidenbauer, Jülich, **The reaction $pp \rightarrow p\Lambda K^+$ and $pp \rightarrow p\Sigma^0 K^+$ Near Threshold**
2:15	M. Miller, IU Medical School, **Three-Nucleon Force Effects in 200 MeV p+d Elastic Scattering**
2:35	T. Peterson, IUCF, **Tagged Neutron Production with a Storage Ring**
3:10	G. Muenzenberg, GSI, **Plans for eA Colliders**
4:15	A.D. Krisch, Ann Arbor, **Storage Rings: Past, Present, and Future**
5:00	*Closing of* STORI99

APPENDIX B

LIST OF PARTICIPANTS

Bacher, Andrew
Indiana University Cyclotron Facility
2401 Milo B. Sampson Lane
Bloomington, IN 47405 USA
Telephone: (812) 855-9365
     FAX: (812) 855-6645
bacher@iucf.indiana.edu

Balewski, Jan
Indiana University Cyclotron Facility
2401 Milo B. Sampson
Bloomington, IN 47408 USA
Telephone: (812) 855-0934
     FAX: (812) 855-6645
balewski@iucf.indiana.edu

Bisplinghoff, Jens
Universitaet Bonn, ISKP
Nussallee 14-16
Bonn NNW,   D 53115   Germany
Telephone: 49-228-73-2543
     FAX: 49-228-73-2505
jens@iskp.uni-bonn.de

Bland, Les
Indiana University Cyclotron Facility
2401 Milo B. Sampson
Bloomington, IN 47408 USA
Telephone: (812) 855-6051
     FAX: (812) 855-6645
bland@iucf.indiana.edu

Bredeweg, Todd
Indiana University Cyclotron Facility
2401 Milo B. Sampson Lane
Bloomington, IN 47408 USA
Telephone: (812) 855-5189
     FAX: (812) 855-6645
bredeweg@iucf.indiana.edu

Calen, Hans
Uppsala University
The Svedberg Laboratory, Box 533
Uppsala, S-751 21 Sweden
Telephone: 46-18-471-3846
     FAX: 46-18-471-3513
calen@tsl.uu.se

Cameron, John
Indiana University Cyclotron Facility
2401 Milo B. Sampson Lane
Bloomington, IN 47408 USA
Telephone: (812) 855-5466
     FAX: (812) 855-6645
cameron@iucf.indiana.edu

Chu, Chungming (Paul)
Indiana University Cyclotron Facility
2401 Milo B. Sampson Lane
Bloomington, IN 47408 USA
Telephone: (812) 855-5196
     FAX: (812) 855-6645
cmchu@iucf.indiana.edu

Coon, Sidney
New Mexico State University
MSC 3D, Box 30001
Las Cruces, NM 88003 USA
Telephone: (505) 646-2102
FAX: (505) 646-1934
coon@nmsu.edu

Daehnick, Wilfried
University of Pittsburgh
Dept. of Physics
Pittsburgh, PA 15260 USA
Telephone: (412) 624-9236
FAX: (412) 624-9163
daehnick+@pitt.edu

Egelhof, Peter
GSI Darmstadt
Planckstr. 1
Darmstadt, D-64291 GERMANY
Telephone: 49-6159-712662
FAX: 49-6159-712809
p.egelhof@gsi.de

Ekström, Curt
Uppsala University
The Svedberg Laboratory, Box 533
Uppsala, S-751 21 SWEDEN
Telephone: 46-18-4713112
FAX: 46-18-4713833
ekstrom@tsl.uu.se

Eversheim, Dieter
Universität Bonn
Institut f. Strahlen-und Kernphysik
Nussallee 14-16
Bonn, D-53115 GERMANY
Telephone: 49 228 735299
FAX: 49 228 732505
evershei@iskp.uni-bonn.de

Eyrich, Wolfgang
Universität Erlangen-Nürnberg
Physikalisches Institut
Erwin-Rommel-Str. 1
Erlangen, D- 91058 GERMANY
Telephone: 49-9131-85-27086
FAX: 49-9131-15249
eyrich@physik.uni-erlangen.de

Falch, Markus
Ludwig Maximilians Universität
Sektion Physik der LMU München
Am Colombwall 1
Garching, D-85748 GERMANY
Telephone: 49-89-289-14065
FAX: 49-89-289-14072
mfalch@physik.uni-muenchen.de

Falk, Willie R.
University of Manitoba
Department of Physics
Winnipeg, MB, R3T 2N2 CANADA
Telephone: (204) 474-9856
FAX: (204) 474-7622
falk@physics.umanitoba.ca

Franzke, Bernhard
GSI
Accelerator Physics
Planckstr.1
Darmstadt, D-64291 GERMANY
Telephone: 49-6159-71-2366
FAX: 49-6159-71-2987
b.franzke@gsi.de

Gabrielse, Gerald
Harvard University
Physics Department
235 Lyman Laboratory,
Cambridge, MA 02138 USA
Telephone: (617) 495-4381
FAX: (617) 495-4381
gabrielse@hussle.harvard.edu

Golubev, Pavel
University of Lund
Department of Physics, Box 118
Lund, Skone, S-221 00 SWEDEN
Telephone: 46-46-2220486
    FAX: 46-46-2224015
kosu_golubev@garbo.lucas.lu.se

Haeberli, Willy
University of Wisconsin
Dept. of Physics
1150 University Ave.
Madison, WI 53706 USA
Telephone: (608) 262-0009
    FAX: (608) 262-3398
willy@haeberli.com

Haidenbauer, J.
Forschungszentrum Jülich
Institut f. Kernphysik
Jülich, D-52428 GERMANY
Telephone: 49-2461-614401
    FAX: 49-2461-613930

Hanhart, Christoph
University of Washington
Department of Physics, Box 351560
Seattle, WA 98195 USA
Telephone: (206) 543-3931
    FAX: (206) 685-9829
hanhart@phys.washington.edu

Jacobs, William
Indiana University Cyclotron Facility
2401 Milo B. Sampson Lane
Bloomington, IN 47408 USA
Telephone: (812) 855-8873
    FAX: (812) 855-6645
jacobs@iucf.indiana.edu

Jahn, Rainer
Universität Bonn, ISKP
Nussallee 14-16
Bonn, D-53115 GERMANY
Telephone: 49 228 732236
    FAX: 49 228 732505
jahn@iskp.uni-bonn.de

Jakobsson, Bo
Lund University
Department of Physics, Box 118
Lund, S-221 00 SWEDEN
Telephone: 46-46-222-7708
    FAX: 46-46-222-4015
bo.jakobsson@kosufy.lu.se

Kaiser, Norbert
Technical University Munich
Physics Department T39
James Franckstr
Garching, D-85747 GERMANY
Telephone: 49-89-289-12367
    FAX: 49-89-289-12296
nkaiser@physik.tu-muenchen.de

Katayama, Takeshi
University of Tokyo, CNS
Beam Physics Lab / RIKEN
3-2-1- Midoricho
Tanashi Tokyo, 188-0002 JAPAN
Telephone: 424-69-9509
    FAX: 424-68-5845
katayama@cns.s.u-tokyo.ac.jp

Klepper, Otto
GSI/KPII
Planckstr. 1
Darmstadt, D-64291 GERMANY
Telephone: 49-6159-71-2439
    FAX: 49-6159-71-2902
o.klepper@gsi.de

Knutson, Lynn
University of Wisconsin
Physics Dept.
Madison, WI 53706 USA
Telephone: (608) 262-3096
FAX: (608) 262-3598
knutson@uwnuc0.physics.wisc.edu

Kolster, Hauke
DESY/Hermes
Geb. 1E Notkestrasse 85
Hamburg, D-22603 GERMANY
Telephone: 40-40-89983306
FAX: 40-40-89984034
kolster@hermes.desy.de

Kolybasov, Victor
Lebedev Physical Institute
Theoretical Nuclear Physics
Leninsky Prospect 53
Moscow, 117924 RUSSIA
Telephone: 7-095-1354288
FAX: 7-095-1354288
kolybasv@sci.lebedev.ru

Krisch, A.D.
University of Michigan
Randall Lab.of Physics
Ann Arbor, MI 48105-1120 USA
Telephone: (734) 936-1027
FAX: (734) 936-0794
krisch@umich.edu

Kuhlmann, Eberhard
Forschungszentrum Jülich
Institut f. Kernphysik
Jülich, D-52425 GERMANY
Telephone: 49 2461-61-2381
FAX: 49 2461-61-3930
e.kuhlmann@fz-Jülich.de

Lacker, Heiko
Universität Freiburg
Fakultät f. Physik
Herrmann-Herder-Str. 3
Freiburg, D-79104 GERMANY
Telephone: 49 761 203 5917
FAX: 49 761 203 5705
lacker@hpfr01.physik.uni-freiburg.de

Laget, Jean Marc
CEA / Saclay
DAPNIA / SPhN
GIF-SUR-YVETTE, 91,
F91191 FRANCE
Telephone: 33-1-69-08-75-54
FAX: 33-1-69-08-75-54
laget@phnx7.saclay.cea.fr

Lefort, Thomas
Indiana University Cyclotron Facility
2401 Milo B. Sampson Lane
Bloomington, IN 47408 USA
Telephone: (812) 855-4201
FAX: (812) 855-6645
lefort@iucf.indiana.edu

Lorentz, Bernd
Forschungszentrum Jülich
IKP, Postfach 1913
Jülich, D-52425 GERMANY
Telephone:
FAX:
b.lorentz@fz-Jülich.de

Machner, Hartmut
Forschungszentrum Jülich
Institut f. Kernphysik
Jülich, D- 52425 GERMANY
Telephone: 49-2461-614270
FAX: 49-2461-613930
h.machner@fz-Jülich de

Maier, Rudolf
Forschumgszentrum Jülich
Institut f. Kernphysik
Jülich, D- 52425   GERMANY
Telephone: 49-2461-61-3980
   FAX: 49-2461-61-8259
r.maier@fz-juelich.de

Marcello, Simonetta
INFN TORINO
Via Pietro Giuria n.1
Torino, 10125   ITALY
Telephone: 39 011-6707321
   FAX: 39 011-6707325
marcello@to.infn.it

Matthews, June
Mass. Inst. Technology
77 Mass. Ave., Room 26-433
Cambridge, MA 02139   USA
Telephone: (617) 253-4238
   FAX: (617) 258-5440
matthews@pierre.mit.edu

Meyer, Hans-Otto
Indiana University Cyclotron Facility
2401 Milo B. Sampson Lane
Bloomington, IN   47408 USA
Telephone: (812) 855-2883
   FAX: (812) 855-6645
meyer@iucf.indiana.edu

Miller, Michael
IU Medical School
1111 West 10th Street
Indianapolis, IN 46202-4800   USA
Telephone: (317) 630-7134
   FAX: (317) 630-7449
miller5@iucf.indiana.edu

Milner, Richard
MIT-Bates
Physics Dept, .26-411
Cambridge, MA   02139   USA
Telephone: (617) 253-9200
   FAX: (617) 253-9599
milner@mitlns.mit.edu

Mishustin, Igor
The Kurchatov Inst.
Kurchatov Sq. 1
Moscow,   123182   RUSSIA
Telephone: 7-095-1966281
   FAX: 7-095-1966281
mish@mbslab.kiae.ru

Moskal, Pawel
Jagellonian University
Institute of Physics
Cracow, W. Malopolskie,
30-059 POLAND
Telephone: 48-12-632-48-88
   FAX:
ufmoskal@sigma.if.uj.edu.pl

Muenzenberg, Gottfried
KPII
GSI Darmstadt
Darmstadt, D-64291   GERMANY
Telephone: 49-61-59-71-2733
   FAX: 49-61-59-71-2902
g.muenzenberg@gsi.de

Nann, Hermann
Indiana University
Cyclotron Facility
2401 Milo B. Sampson Lane
Bloomington, IN   47408   USA
Telephone: (812) 855-2884
   FAX: (812) 855-6645
nann@iucf.indiana.edu

Nazaruk, Valery
Inst. For Nucl. Research of RAS
60th October Anniversary Prospect 7a
Moscow,   117312   RUSSIA
Telephone: 095-334-01-88
        FAX: 095-334-01-84
nazaruk@al20.inr.troitsk.ru

Nolden, Fritz
GSI / ESR
Planckstrasse 1
Darmstadt,   D-64291   GERMANY
Telephone: 49-6159-71-2407
        FAX: 49-6159-71-2985
f.nolden@gsi.de

Ohm, Henner
Forschungszentrum Jülich
IKP
Jülich D-52425   GERMANY
Telephone: 49-2461-613723
        FAX: 49-2461-613930
h.ohm@fz-Jülich.de

Opare, Richard
University of Science & Technology
124M Indx Hall, U.S. T-Kumasi
Ghana,   WEST AFRICA
Telephone: 233 81 24106
        FAX: 233 51 60299
ropare@mailcity.com

Pancella, Paul
Western Michigan University
Everett Tower
Kalamazoo, MI 49008-5151   USA
Telephone: (616) 387-4962
        FAX: (616) 387-4939
paul.pancella@umich.edu

Passchier, Igor
NIKHEF
P.O. Box 41882
Amsterdam, 1009 DB NETHERLANDS
Telephone: 31-20-592-2147
        FAX: 31-20-592-5155
igorp@nikhef.nl

Peterson, Todd
Indiana University Cyclotron Facility
2401 Milo B. Sampson Lane
Bloomington, IN 47408   USA
Telephone: (812) 855-5195
        FAX: (812) 855-6645
tpeterson@iucf.indiana.edu

Pollock, Robert
Indiana University Cyclotron Facility
2401 Milo B. Sampson Lane
Bloomington, IN 47408   USA
Telephone: (812) 855-8306
        FAX: (812) 855-6645
pollock@iucf.indiana.edu

Rachek, Igor
Budker Institute of Nuclear Physics
11 Lavrentiev Prospect
Novosibirsk,   630090 RUSSIA
Telephone: 7 3832 394026
        FAX: 7 3832 342163
rachek@inp.nsk.su

Radon, Torsten
Gesellschaft f. Schwerionenforschung
GSI, Abteilung KP II
Darmstadt, D- 64291   GERMANY
Telephone: 49 6159 71 2142
        FAX: 49 6159 71 2902
t.radon@gsi.de

Rathmann, Frank
Forschungszentrum, Jülich
Kernphysik II
Jülich, D-52425   GERMANY
Telephone: 2461-614-558
       FAX: 2461-613-930
f.rathmann@fz-Jülich.de

Reistad, Dag
The Svedberg Laboratory
Box 533
Uppsala, 75121 SWEDEN
Telephone: 46-18-471-3177
       FAX: 46-18-471-3833
reistad@tsl.uu.se

Rinckel, Tom
Indiana University Cyclotron Facility
2401 Milo B. Sampson Lane
Bloomington, IN 47408 USA
Telephone: (812) 855-9365
       FAX: (812) 855-6645
rinckel@iucf.indian.edu

Roser, Thomas
BNL / AGS
Bldg. 911B
Upton, NY 11973 USA
Telephone: (516) 344-7084
       FAX: (516) 344-5954
roser@bnl.gov

Saha, Swapn K.
University of Pittsburgh
Department of Physics
Pittsburgh, PA 15260 USA
Telephone: (412) 624-9240
       FAX:
swapan@vms.cis.pitt.edu

Scheidenberger, Christopher
GSI
Planckstrasse 1
Darmstadt,   D-64291   GERMANY
Telephone: 49-6159-712706
       FAX:
c.scheidenberger@gsi.de

Schwandt, Peter
Indiana University Cyclotron Facility
2401 Milo B. Sampson Lane
Bloomington, IN 47408 USA
Telephone: (812) 855-9365
       FAX: (812) 855-6645
schwandt@iucf.indiana.edu

Scobel, Wolfgang
Universität Hamburg
Institut f. Experimentalphysik
Luruper Chaussee 149
Hamburg,   D-22761   GERMANY
Telephone: 40-89982166
       FAX: 40-89982101
scobel@kaa.desy.de

Snow, William
Indiana University Cyclotron Facility
2401 Milo B. Sampson Lane
Bloomington, IN 47408 USA
Telephone: (812) 855-7914
       FAX: (812) 855-6645
snow@iucf.indiana.edu

Sowinski, James
Indiana University Cyclotron Facility
2401 Milo B. Sampson Lane
Bloomington, IN 47408 USA
Telephone: (812) 855-9365
       FAX: (812) 855-6645
sowinski@iucf.indiana.edu

Stephenson, Edward
Indiana University Cyclotron Facility
2401 Milo B. Sampson Lane
Bloomington, IN 47408 USA
Telephone: (812) 855-5469
FAX: (812) 855-6645
stephenson@iucf.indiana.edu

Strikman, Mark
Pennsylvania State University
Physics Department
104 Davey Lab
University Park, PA 16802 USA
Telephone: (814) 865-7382
FAX: (814) 865-3604
strikman@phys.psu.edu

Stroeher, Hans
Institut f. Kernphysik
Forschungszentrum Jülich
Jülich, D-52425 GERMANY
Telephone: 49-2461--61-4408
FAX: 49-2461-61-3930
h.stroeher@fz-juelich.de

Tamura, Keisuke
Fukui Medical University
Matsuoka -23-3
Fukui, 910-1193 JAPAN
Telephone: 81-776-61-8283
FAX: 81-776-61-8141
tamura@fukui-med.ac.jp

Utsunomiya, Hiroaki
Konan University
Department of Applied Physics
Okamoto 8-9-1, Higashinada
Kobe, Hyogo, 658-0072 JAPAN
Telephone: 81-78-435-2471
FAX: 81-78-435-2539
hiro@konan-u.ac.jp

Van Buuren, Laurens
NIKHEF
P O 41882
Amsterdam,1009 DB NETHERLANDS
Telephone: 31-20-592-2091
FAX: 31-20-592-5155
buuren@nikhef.nl

Vassiliev, Iouri
HERA-B, F15
Notkestrasse 85
Hamburg, D-22607 GERMANY
Telephone: 89982842
FAX:
yuri@mail.desy.de

Vigdor, Steven
Indiana University Cyclotron Facility
2401 Milo B. Samspon Lane
Bloomington, IN 47408 USA
Telephone: (812) 855-9369
FAX: (812) 855-6645
vigdor@iucf.indiana.edu

Viola, Vic
Indiana University Cyclotron Facility
2401 Milo B. Sampson Lane
Bloomington, IN 47408 USA
Telephone: (812) 855-2878
FAX: (812) 855-6645
vicv@iucf.indiana.edu

von Przewoski, Barbara
Indiana University Cyclotron Facility
2401 Milo B. Sampson Lane
Bloomington, IN 47408 USA
Telephone: (812) 855-2913
FAX: (812) 855-6645
przewoski@iucf.indiana.edu

Wagner, Marcus
Universität Erlangen-Nürnberg
Physikalisches Institut
Erwin Rommel Str . 1
Erlangen, D- 91058   GERMANY
Telephone: 49-9131-85-27261
    FAX: 49-9131-15249
mwagner@physik.uni-erlangen.de

Wellinghausen, Arne
Indiana University Cyclotron Facility
2401 Milo B. Sampson Lane
Bloomington, IN   47408   USA
Telephone: (812) 855-2949
    FAX: (812) 855-6645
arne@iucf.indiana.edu

Westfall, Gary
Michigan State University
National Super Conducting Cyclotron Lab.
East Lansing, MI   48824-1321   USA
Telephone: (517) 333-6324
    FAX: (517) 353-5967
westfall@nscl.msu.edu

Whitaker, Thomas
Indiana University
Physics Department
Bloomington, IN 47405 USA
Telephone: (812) 855-0293
    FAX: (812) 855-5533
tjwhitak@indiana.edu

Wirth, Stefan
Universität Erlangen-Nürnberg
Physikalisches Institut
Erwin-Rommel-Str.1
Erlangen, D- 91058   GERMANY
Telephone: 49-9131-85-27261
    FAX: 49-9131-15249
wirth@physik.uni-erlangen.de

Wise, Thomas
University of Wisconsin
Physics Department
Madison, WI   53706   USA
Telephone: (608) 262-6555
    FAX: (608) 262-3598
wise@uwnuc0.physics.wisc.edu

Wissink, Scott
Indiana University Cyclotron Facility
2401 Milo B. Sampson Lane
Bloomington, IN   47408   USA
Telephone: (812) 855-5192
    FAX: (812) 855-6645
wissink@iucf.indiana.edu

Wolke, Magnus
Forschungszentrum Jülich
Institute f. Kernphysik
Jülich, D- 52425   GERMANY
Telephone: 49-2461-61-4280
    FAX: 49-2461-61-3930
m.wolke@fz-Jülich.de

Xia, Jiawen
Institute of Modern Physics
Chinese Academy of Sciences
363 Nanchang Road
Lanzhou, Gansu,   730000 CHINA
Telephone: 86-931-8857647
    FAX: 86-931-8881100
xiajw@csraxp.lzb.ac.cn

Yang, Haichuan
Indiana University Cyclotron Facility
2401 Milo B. Sampson Lane
Bloomington, IN   47408   USA
Telephone: (812) 855-0932
    FAX: (812) 855-6645
haiyang@indiana.edu

Yoshikazu, Maeda
Osaka University
Research Center for Nuclear Physics
10-1 Mihogaoka, Ibaraki
Osaka,   567-0047   JAPAN
Telephone: 81-6-6879-8937
       FAX: 81-6-6879-8899
ymaeda@rcnp.osaka-u.ac.jp

Zhan, Wenlong
Institute of Modern Physics
Chinese Academy of Sciences
363 Nanchang Road
Lanzhou, Gansu,   730000   CHINA
Telephone: 86-931-8854940
       FAX: 86-931-8881100
jwxm@ns.lzb.ac.cn

Zlomanczuk, Jozef
Uppsala University
Dept of Radiation Sciences
Box 535
Uppsala,   S-75121   SWEDEN
Telephone: 46-18-471-1587
       FAX: 46-18-471-3833
jozef@tsl.uu.se

AUTHOR INDEX

A

Adam, H.-H., 65, 143
Adam, Jr., J., 111
Akimune, H., 365
Alarcon, R., 353
Arenhoevel, H., 362
Attallah, F., 263, 305
Aushev, V., 359

B

Balewski, J. T., 37, 65, 143, 210
Barkov, L. M., 362
Barsov, S., 138
Bauer, T. S., 353
Bechstedt, U., 138, 243
Beckert, K., 246, 305
Belostotsky, S. L., 362
Bleile, A., 259
Boer Rookhuizen, H., 353
Boersma, D. J., 353
Bollen, G., 275
Borchert, G., 138
Borgs, W., 138, 157
Bosch, F., 275, 305
Bosh, F., 263
Budzanowski, A., 65, 143
Bulten, H. J., 353
Büscher, M., 138

C

Calén, H., 229
Casares, A., 275
Cassing, W., 157
Coon, S. A., 111

D

Daehnick, W. W., 37, 210
Debowski, M., 138
de Jager, C. W., 353, 362
de Vries, H., 353, 362

Dietrich, J., 243
Dmitriev, V. F., 362
Dolinsky, A., 266
Doskow, J., 210
Drochner, M., 138
Dyug, M. V., 362

E

Egelhof, P., 259
Ehret, K., 359
Ent, R., 353
Erven, W., 138
Eßer, R., 138
Eversheim, P. D., 224

F

Falch, M., 263, 305
Falk, W. R., 114
Fedorets, P., 138
Ferro-Luzzi, M., 353
Flammang, R. W., 37
Franczak, B., 305
Franzke, B., 246, 263, 266, 305
Friesel, D., 210
Fujishiro, M., 365
Funcke, M., 359

G

Gasparian, A., 120
Geissel, H., 263, 275, 305
Geurts, D., 353
Gilman, R., 362
Goodman, C., 65, 143
Gotta, D., 138
Groening, L., 246
Grzonka, D., 65, 143

H

Haeberli, W., 210
Haidenbauer, J., 120
Hanhart, C., 81, 120

Hartmann, M., 138, 157
Harvey, M., 353
Hausmann, M., 263, 305
Heimberg, P., 353
Henn, K., 243
Higinbotham, D. W., 353
Holt, R. J., 362

I

Isaeva, L. G., 362

J

Jahn, R., 168
Jakobsson, B., 49
Jarczyk, L., 65, 143, 157
Jochmann, M., 65, 143
Junghans, H., 138

K

Kacharava, A., 138
Kaiser, N., 96
Kamys, B., 138, 157
Kerscher, T., 263, 305
Kholomeev, A., 275
Khoukaz, A., 65, 143
Kilian, K., 65, 143
Kinney, E. R., 362
Klehr, F., 138
Klepper, O., 263, 269, 305
Klimin, P., 353
Kluge, H.-J., 259, 263, 305
Knutson, L. D., 177
Koch, H. R., 138, 157
Köhler, M., 65, 143
Kolster, H., 210
Komarov, V. I., 138
Kondratyuk, L., 120
Koop, I., 353
Koptev, V., 138
Kowalczyk, R. S., 362
Kowina, P., 65, 143
Kozhuharov, C., 263, 269, 305
Krisch, A. D., 394
Kroes, F., 353

Kulessa, P., 138, 157
Kulikov, A., 138
Kurbatov, V., 138

L

Lacker, H., 34
Laget, J.-M., 125
Lang, N., 143
Lazarenko, B. A., 362
Lehrach, A., 243
Liebisch, U., 259
Lister, T., 65, 143
Löbner, K. E. G., 263, 305
Loginov, A. Y., 362
Lorentz, B., 37, 210, 243
Luijckx, G., 353
Lysenko, A., 353

M

Macharashvili, G., 138
Machner, H., 60
Maeda, Y., 117
Maier, H. J., 157
Maier, R., 138, 157, 243
Matoba, M., 157
Matsuoka, N., 117
Mc Cammon, D., 259
Meier, H. J., 259
Merzliakov, S., 138
Meyer, H. O., 37, 210
Mikirtytiants, S., 138
Militsyn, B., 353
Milner, R. G., 339
Mishnev, S. I., 362
Mishustin, I. N., 308
Moskal, P., 65, 143
Müller, H., 138
Münzenberg, G., 263, 275, 293, 305
Mussgiller, A., 138

N

Nelyubin, V. V., 362
Nesterenko, I., 353
Nikolenko, D. M., 362

Nioradze, M., 138
Nolden, F., 246, 263, 266, 305
Noomen, J., 353
Norum, B. E., 353
Novikov, Y. N., 263, 305

O

Oelert, W., 65, 143
Ohgaki, H., 365
Ohm, H., 138, 157
Ohta, M., 365
Osipov, A. V., 362

P

Pancella, P. V., 37, 210
Passchier, I., 353
Patyk, Z., 263
Pavlenko, Y., 359
Peña, M. T., 111
Peterson, T., 235
Petrus, A., 138
Pollock, R. E., 37, 210, 371
Poolman, H. R., 353
Potterveld, D. H., 362
Prasuhn, D., 138, 157, 243
Pugatch, V., 359
Pysz, K., 138, 157

Q

Quentmeier, C., 65, 143
Quin, P. A., 210

R

Rachek, I. A., 362
Radon, T., 263, 305
Rathmann, F., 37, 138, 193, 210
Reistad, D., 380
Rimarzig, B., 138
Rinckel, T., 37, 210
Roser, T., 213
Rudy, Z., 138, 157

S

Saha, S. K., 37, 210
Santo, R., 65, 143
Schatz, H., 305
Scheidenberger, C., 263, 275, 305
Schepers, G., 65, 143
Schleichert, R., 138
Schnase, A., 243
Schneider, C., 138
Schneider, H., 138, 243
Schrieder, G., 293
Schult, O. W. B., 138, 157
Schwartz, B., 37, 210
Scobel, W., 3
Sebastián, O., 259
Seddik, U., 65, 143
Sefzick, T., 65, 143
Sever, S. I., 359
Sewerin, S., 65, 143
Seyfarth, H., 138
Shatunov, Y., 353
Shestakov, Y. V., 362
Sidorov, A. A., 362
Siemaszko, M., 143
Sistemich, K., 138
Six, E., 353
Smyrski, J., 65, 143
Speth, J., 120
Spratte, S., 359
Stadler, A., 111
Stadlmann,, 305
Stahle, C. K., 259
Stassen, R., 243
Steck, M., 246, 263, 305
Steijger, J. J. M., 353
Stein, H. J., 138
Stibunov, V. N., 362
Stockhorst, H., 243
Stöhlker, T., 259
Ströher, H., 138, 157
Strzałkowski, A., 65, 143, 157
Sümmerer, K., 306
Sun, Z., 305
Symalla, M., 359
Szczerba, D., 353

T

Tamura, K., 117
Thörngren-Engblom, P., 37
Tkatch, N., 359
Tölle, R., 243
Toporkov, D. K., 362
Toyokawa, H., 365

U

Uozumi, Y., 157
Utsunomiya, H., 365

V

v. Przewoski, B., 37, 210
van Amersfoort, P. W., 353
van Buuren, L. D., 353, 356
van den Brand, J. F. J., 353
van den Putte, M. J. J., 353
van der Laan, J., 353
Vassiliev, Y., 359
Vesnovsky, D. K., 362
Vikhrov, V. V., 362

W

Weber, M., 259
Wegener, D., 359
Weick, H., 275, 305
Wellinghausen, A., 210
Westfall, G. D., 324
Winkler, M., 263
Wirth, S., 171
Wise, T., 37, 210
Wolke, M., 65, 143
Wollnik, H., 263, 275, 305
Wüstner, P., 65, 138, 143

Y

Yamagata, T., 365
Yonezawa, Y., 365

Z

Zevakov, S. A., 362
Zipper, W., 143
Złomańczuk, J., 43
Zychor, I., 157